Information Technology and Constructivism in Higher Education:
Progressive Learning Frameworks

Carla R. Payne
Union Institute and University of Vermont College, USA

Information Science REFERENCE

INFORMATION SCIENCE REFERENCE
Hershey · New York

Director of Editorial Content: Kristin Klinger
Senior Managing Editor: Jamie Snavely
Managing Editor: Jeff Ash
Assistant Managing Editor: Carole Coulson
Typesetter: Larissa Vinci
Cover Design: Lisa Tosheff
Printed at: Yurchak Printing Inc.

Published in the United States of America by
Information Science Reference (an imprint of IGI Global)
701 E. Chocolate Avenue
Hershey PA 17033
Tel: 717-533-8845
Fax: 717-533-8661
E-mail: cust@igi-global.com
Web site: http://www.igi-global.com/reference

and in the United Kingdom by
Information Science Reference (an imprint of IGI Global)
3 Henrietta Street
Covent Garden
London WC2E 8LU
Tel: 44 20 7240 0856
Fax: 44 20 7379 0609
Web site: http://www.eurospanbookstore.com

Library of Congress Cataloging-in-Publication Data

Information technology and constructivism in higher education : progressive learning frameworks / Carla R. Payne, editor.
p. cm.

Includes bibliographical references and index.

Summary: "This volume is grounded in the thesis that information technology may offer the only viable avenue to the implementation of constructivist and progressive educational principles in higher education, and that the numerous efforts now under way to realize these principles deserve examination and evaluation"--Provided by publisher.

ISBN 978-1-60566-654-9 (hardcover) -- ISBN 978-1-60566-655-6 (ebook) 1. Education, Higher--Effect of technological innovations on. 2. Constructivism (Education) 3. Web-based instruction. 4. Information technology. I. Payne, Carla R.

LB2395.7.I545 2009
468.3'42102464795--dc22

2008052443

British Cataloguing in Publication Data
A Cataloguing in Publication record for this book is available from the British Library.

All work contributed to this book is new, previously-unpublished material. The views expressed in this book are those of the authors, but not necessarily of the publisher.

Editorial Advisory Board

Table of Contents

Detailed Table of Contents

ways that constructivists should explore, embrace and exploit. This essay explores how Second Life (SL) effectively employs and distills the principles of educational constructivism. SL offers endless opportunities for immersion within user-constructed environments and activities. Educational use of SL may facilitate learner-led activities and yield learning that is prompted by desire and curiosity rather than learning for learning's sake. By exploiting these qualities with constructivist pedagogies, educators create environments that challenge and enable students to engage in the deepest kinds of learning.

In this chapter, a variety of Web 2.0 applications and their affordances are presented and discussed in relation to constructivism in higher education. The aim is to explain how these applications can be used in higher education to promote interactive and engaging learning environments. Recommendations for harnessing the potential of these tools along with practical examples will assist facilitators of higher education with creative means to design their courses and thus promote Learning 2.0.

This chapter examines some ways information technologies (IT) are deployed in higher education courses to help learners create robust mental models. In E-Learning, mental models are created through the following: digitally mediated (1) information visualization work; (2) virtual immersion, and (3) human interactivity. For (1) information visualization work, information technologies afford multi-sensory learning channels: text, visuals, slideshows, screen casts and animations, audio, video, interactivity, and simulations. IT supports the archiving of digital learning artifacts through eportfolios, digital gallery shows, and informational multimedia databases. (2) Virtual immersion has been enabled in 3-D interactive spaces where learners may experience multi-faceted learning. More complex simulations have also been created with animations and long-term continuous learning. To promote (3) human interactivity, IT reifies human intellects and perceptions, and social software uses "swarm intelligence" to support consensus-driven realities. IT creates contexts for co-learning and intercommunications. Collaborative pedagogical strategies—online teaching case studies, mock trials, facilitated games, and various assignments and assessments—highlight the multiple-views, multiple-realities of constructivist learning.

Foreword

Like the Gutenberg revolution of the mid-15th Century, which initiated and made possible a democratization of learning by making mass produced, inexpensive texts widely available, the internet has expanded the amount and kinds of information and the audiences for information beyond the scope or imagination of any previous era. Gutenberg and his successors put printed texts within the reach of a rising middle class, thereby transforming western culture and pedagogy from predominantly oral-aural and visual-oral models to an increasingly literary model. The Internet has closed the circle in this transformation by making images, sounds, and texts widely and immediately available to anyone who can or cares to obtain a public library card or purchase a computer. As the infrastructure for supporting internet communication expands and the costs decline—in other words, as access to the World Wide Web increases—the opportunities for individuals to construct knowledge about the world around them and beyond their immediate physical surroundings also increases. Moreover, as the formats for communication on the internet expand—e-mails, list serves, Wikis, and Weblogs—the opportunities for exchanging knowledge, analysis, and opinion expand as well. These enormous advances in the democratization of knowledge present great opportunities and great challenges for educators and those who seek knowledge.

Traditional models of education have quickly and easily (if at great expense) adapted to the new technologies. The 2005 edition of *Peterson's Guide to Distance Learning* lists more than 4,800 online learning programs at more than 1,000 accredited schools in the United States and Canada. The number is growing, if not exponentially, then surely steadily as colleges and universities everywhere scramble to diversify their programmatic offerings and attract new constituencies of students by adding distance learning opportunities to their more traditional face-to-face classroom schedules. Even traditional residential colleges and colleges that had long ago begun to experiment with brief, intense residencies as alternative models to twice or thrice-weekly class meetings are now also entering the field of online education, where students and teachers meet only on the computer screen or in what are called "blended" formats of virtual classrooms and face-to-face meetings. The fairly straightforward and technologically sophisticated procedures of recording faculty presentations, posting syllabi and assignments, sharing documents, administering examinations, and devising ways to receive, grade, and assess student learning have enabled colleges and universities to emulate or many features of the traditional lecture hall model of education. More complicated are the methods for facilitating interaction between students and instructors or among students; but here, too, information technology has provided opportunities and formats for interaction through threaded discussion, asynchronous online forums, and real-time "chat rooms" that strive to emulate the intimacy of the seminar room.

For progressive educators, especially, the new technologies raise new questions about how students will learn individually and together, and how teachers will participate in student learning.

Among the principles to which educators in the progressive tradition of John Dewey cling most fervently and persistently three are important to consider in relation to the problems and possibilities of distance learning through enhanced technologies. The first is the firm conviction that education must begin with what Dewey himself called "insight into the [student's] capacities, interests, and habits ("My Pedagogic Creed," [1897], 430). This is often now restated as "student-centered learning," where the student is challenged to approach each subject, academic discipline, or assignment through the lens of her or his interests and experiences and where the faculty's role is in part to encourage and guide the student in formulating significant questions and problems that engage and address those interests and experiences. The technologies for communication we now possess can make this process of exploration and discovery possible at a distance as well as face to face, online in one to one communication as well as in more open forum formats. But as anyone who has participated in this form exploratory discussion will attest, doing so at a distance and online is far more time consuming and requires overcoming many more barriers of verbal and nonverbal communication than talking informally across a table. What is required to make this student centered approach to education work in an online or partially online learning platform is a commitment on both sides to the arts of verbal communication.

A second principle of progressive education that deserves some attention—and receives attention in this volume—is what Dewey called the social dimension of education. In "My Pedagogic Creed" Dewey writes: "The school is primarily a social institution. Education being a social process, the school is simply that form of community life in which all those agencies are concentrated that will be most effective in bringing the child to share in the inherited resources of the race, and to use his own powers for social ends" ("My Pedagogic Creed" [1897] Article II—What the School Is). This observation is no less true for post-secondary education than it is for primary education, which was Dewey's primary focus. The Gutenberg revolution helped create a community of scholars not only by making accessible standard texts to which all participants in scholarly dialogue and debate could refer—chapter and verse, page and line—but also by making possible at relatively low cost printed commentaries, responses, and replies that spread rapidly across the continent. The Internet offers promise to revive and revitalize this community of scholarship with enhanced possibilities for communication, comment, editing, and links to resources that are instantly available for reference, use, commentary, and—in some cases—correction.

A third principle of progressive education that needs to be kept in mind as we continue to develop the use and possibilities for technology in education is the insistence that the goal of education is reflective practice. There is, as Dewey reminds us frequently in his writing, an intimate and complex relationship between product and process that is the distinguishing philosophical mark of progressive education. The emphasis on reflective practice depends on this relationship but also on maintaining a distinction between product and process in the construction and application of knowledge and learning. The tools and resources now available via the internet to all students, at all levels and stages of education, make it increasingly easy for an individual student to assemble information into varied and academically rich products. What is harder (but not impossible) to replicate or emulate through information technology is the dynamic of exchange—between student and teacher and among students themselves—that constitutes the matrix for education as a social transaction. And in some sense what is at stake in the future development of online or distance education is the perpetuation of communities of learners. The millions of scholars, bloggers, Wiki contributors and editors who now add their knowledge, information resources, and opinions to the World Wide Web have democratized learning far beyond anything Gutenberg and his successors in the printing trade would have—or could have—imagined. It remains for our generation of post-Gutenberg learners and teachers to devise the ways to meet Dewey's challenge to make education a reflective activity that engages individuals in the social exchange and use of knowledge. This book presents some of the recent reflections and experiments that move us in that direction.

Michael Sherman

Michael Sherman *is a historian, editor, and teacher. He received his B.A. in General Studies in the Humanities, and Ph.D. in history from the University of Chicago. He is the editor of Vermont History, the journal of the Vermont Historical Society; co-author of Freedom and Unity: A History of Vermont (2004); author of articles on sixteenth-century France; editor and contributor to Vermont State Government and Administration since 1965 (1999), and several books of essays on Vermont history and on the humanities and public life. Previous professional positions include Academic Dean of Burlington College; Professor of Liberal Studies in the Adult Degree Program of Vermont College, Union Institute & University; Director of the Vermont Historical Society; Associate Director of the Wisconsin Humanities Committee; and Assistant Professor of History and Humanities at Lawrence University.*

Preface

Most of us are very much aware that the coming of information technology has created major change in higher education. New and underserved populations now have access to a dazzling array of educational opportunities, available almost anywhere and at any time, and catering to every conceivable interest and need. But is this change profound or superficial? Have learning outcomes improved or become more significant to the learner? Or is this merely a new packaging and delivery of old ways of thinking about how we should teach and how we can learn?

This book examines the prospects for progressive higher education in the age of information technology. The topic is especially relevant to the nature of teaching and learning in higher education just because so little of it has actually changed until very recently.

At the beginning of the 20th century, John Dewey and others recognizedthat industrialization and immigration made a revolution necessary in curriculum and in the education of children. Progressive thinkers believed that education was an essential aspect of the growth and preservation of democracy, and that authoritarian practices in the schools were not appropriate to the formation of a citizenry which must look to the future as well as back to a received cultural heritage. Dewey insisted that education must view teaching and learning as "a continuous process of reconstruction of experience." (1938, p. 87)

As Lawrence Cremin has pointed out, the desire to realize Dewey's vision in higher education, by emphasizing the interest and activity of the learner rather than the mere transmission of information, was reflected in the founding of experimenting colleges such as Goddard, Bennington and Black Mountain during the 1920s and 1930s. They worked at reforming the undergraduate curriculum "to take account of changing student needs; . . . to individualize instruction; and . . . to involve students in the making of educational policy" (1964, pp. 308-309). The founding president of Bennington College, Robert D. Leigh, described it as an institution which would "emphasize individuality, direct experience, serious interest, initiative creative and independent work, and self-dependence as educational aims." (Cremin, 1964, p. 310). Some of these progressive colleges, like Antioch and Goddard, persisted to be reinvigorated in the 1960's, and to join in the Union for Experimenting School and Colleges, which in turn gave birth to the Union Institute (now Union Institute & University) in 1964 and the University without Walls (UWW) programs, all eschewing traditional didactic pedagogy and replacing it with learner centered graduate and undergraduate program models. These programs were committed to cultivating active learning, contextual learning, immediate and continual feedback, to the concept of knowledge as adaptability (Spira, 1998), and to individualized study design. Their interpretation of Dewey's educational thinking owed much to William Heard Kilpatrick (1871–1965), a Romantic progressive who believed that teachers should "not suggest any end or plan to the students" for fear of impinging upon their individuality and freedom, and who rejected coercion or prescription of any kind. (Westbrook, p. 503). This point of view had a profound effect on the approach to academic learning taken at institutions such as Goddard College, whose curriculum was to "reflect the world as experienced by student and teachers more than it would reflect academic subject matters." (*The Goddard Bulletin*, 1969, pp. 74-75) Progressivism was implemented in various ways, but common features among the experimenting colleges were the enabling of student control over curriculum, both individual and institutional, and the testing of learning in real-world situations.

Many of the original institutions which rejected conventional educational concepts, organizations and practices, have now disappeared or they have become more educationally traditional. The same is true of programs for adult learners developed in progressivism's second wave in the 1960's and 1970s. And for the most part they have not been at the forefront of the recent changes in higher education. It is outside the scope of this volume to examine the reasons for the failure of progressivism to thrive or to have a significant impact on mainstream institutions, but there are other developments which hold promise for the application and implementation of some of its most essential insights.

The educational research of Piaget (1896-1980), Vygotsky (1896-1934) and Bruner (1915-) gave rise to constructivism as a theory of teaching and learning, first applied in the primary schools, and later to the needs of a growing number of adult learners at the post-secondary level. In defining effective learning and teaching in terms of process, context and the activity of the individual learner, constructivism may be regarded as the descendent of educational progressivism, which also viewed learning as a natural activity that could best be fostered in a rich environment. But even though Ernst von Glasersfeld (1989, p. 3) a principal theorist, calls constructivism "a form of pragmatism" constructivism is not another name for Deweyean or progressive education. Both philosophies reject objectivism, the assumption that there is a reality separate from and "independent of human mental activity" (Jonassen, Mayes, & McAleese, 1993), and share a view of learning as an active process, rather than the passive reception of information, and of knowing as an adaptive function. The underlying subjectivism of cognitive constructivism, however, is not really consistent with the explicitly social program of progressivism, which treated education as preparation for democracy, or with Dewey's explicit view of "all human experiences (as) ultimately social." (1938, p. 38)

The subjectivism and even solipsism of the radical constructivists is not often discussed by educational constructivists, although its implications are at war with their emphasis on the "social negotiation of meaning which is supported by collaborative construction of knowledge" and require the startling postulation of parallel construction by individuals of the "same understanding for any object or event in the external world" (Jonassen, Mayes, & McAleese, 1993). Since his theory also makes social interaction significant for the viability or coping efficacy of the cognizing organism, von Glasersfeld (1989, p. 4) bridges the gap between the "epistemic" or knowing self and those "others" which constitute a social environment by stipulating that the construction of such others takes place as a developmental task,; they are complex "schemes" arrived at by imputing capabilities, perceptual, cognitive, volitional and so on, to "the objects of interaction." (1989, p. 7) The implications of such a model are profound for distributed learning environments where the participants have limited access to each other, and suggest that substantial research is still needed to validate the foundations of constructivist theory,

Social constructivism, which follows Lev Vygotsky in making learning a function of social interaction, is the theory informing much of online education and the design of technologically mediated learning environments. However, in also assuming the ontological and epistemological dualism of "the human mind and the external world" (Liu & Matthews, 2005, p. 386) it favors the implicitly subjectivist orientation in the architecture and use of educational technologies marketed to higher education for their efficiency and cost benefits. There is a primary emphasis on individualized delivery, rather than on inquiry, which undercuts efforts to create collaborative learning situations. A further consequence of the emphasis on reducing costs and reaching wider markets is that Jonassen's vision (1994) of the constructivist use of information technology to empower the learner has not yet been fully realized:

Rather than using technologies by educational communications specialists to constrain the learners' learning processes through prescribed communications and interactions, the technologies are taken away from the specialists and given to the learner to use as media for representing and expressing what they know. Learners function as designers using the technology as tools for analyzing the world, accessing information, interpreting and organizing their personal knowledge, and representing what they know to others.

The coming of information technology has certainly created opportunities and challenges for all education. The fact that its introduction coincided with the popularization of constructivism has opened new possibilities for change. This volume is grounded in the thesis that information technology may offer the only viable avenue to the implementation of constructivist and progressive educational principles in higher education, and that the numerous efforts now under way to realize these principles deserve examination and evaluation. In this connection there are key issues which demand our attention. How widely is constructivist thinking accepted and utilized in today's college classroom? To what extent are IT applications viewed as capable of implementing constructivist education? How successfully do they function when used for these purposes? What are the obstacles to the realization of constructivism at the college and university level?

Perhaps the most exciting impact of information technology on higher education is the attention now being given to teaching itself. What was taught (content) and who was allowed to teach were the traditional preoccupations in academic institutions, and the student was typically not thought of as a learner in the active sense, but as the recipient of information, knowledge and occasionally even wisdom. The outcomes of instruction were measured, but not related to teaching methods, which were generally limited to the didactic. Dewey and the educational progressivists made the poverty of such an approach clear, but this view was not taken very seriously in the academy until the coming of information technology. The richness of possibilities it affords for teaching and learning has highlighted the change in what it means to be a teacher, because we now have a far more complex and nuanced understanding of learning. Of course some of what is now offered in the educational marketplace is merely the translation of conventional classroom and teacher-student relationships into electronic form, but the new sophistication of the learner will make it impossible to maintain the instructor as the sole locus of control and authority. And this is recognized globally; innovation in higher education is happening everywhere in the world, and the scholarship of teaching and learning is taking its place as an important and legitimate aspect of postsecondary concern.

What are the prospects that progressive and constructivist philosophies of education will prevail in transforming higher education by using information technologies as tools? The contributors to this volume offer some answers in their analyses and in the cases they describe of constructivist teaching and learning in the new technological environment. Their experiences and their willingness to innovate for the sake of better learning and more effective teaching should be of interest to educators, administrators, instructional designers and all of those who are committed to furthering progressive thinking.

In Section I, "Constructivism and Progressive Higher Education in the Technological World," a set of chapters focuses on conceptual frameworks. Beginning with the relationship of John Dewey's thinking to the development of constructivism, the authors frame their discussions of constructivist theory in terms of the ways in which information technology can be deployed to support active and collaborative learning. There is also a critical analysis of whether progressive approaches to education are in fact being promoted by IT in educational institutions serving adults, or whether globalization and market forces emphasize efficiency rather than open inquiry.

Perez Cavana, in Chapter I, "Closing the Circle: From Dewey to Web 2.0," considers the ideas of John Dewey as they inspired the development of constructivist and constructionist theory. Practical obstacles of the implementation of constructivism in educational settings are analyzed and weighed against potential benefits.

O'Rourke, Fitzpatrick and Hayes in Chapter II, "Beyond Constriction and Control: Constructivism in Online Theory and Practice," examine assumptions about those benefits resulting from the emphasis on practice, rather than theory, in the teaching profession. In making the creation of knowledge by the active learner its central element, constructivism departs from traditional hierarchical educational and authority structures.

de la Harpe and Peterson in Chapter III, "The Theory and Practice of Teaching with Technology in Today's Colleges and Universities," explore "Mainstream University Teachers' Practice, and Their Use of Educational Technology" to see whether and how they are influenced by constructivist learning theory.

Garrison, Richardson, and Swan in Chapter IV, "A Constructivist Approach to Online Learning: The Community of Inquiry Framework," present their Community of Inquiry model for understanding the online educational experience. The interaction of social presence, cognitive presence, and teaching presence is viewed from a Deweyan perspective as giving rise to online learning.

Lee and Lin, Chapter V, "Applying Constructivism to Online Learning– A New Instructional Design Map," base a new constructivist design map for online learning on five elements: learner, knowledge, learning environment, assessment, and technology, while also discussing its applications in practice.

Rubin in Chapter VI, "Enhancing Authentic Assessment Through Information Technology," points out that authentic assessment is a key aspect of constructivist education, and that its mediation by information technology presents both opportunities for development and limitations. Chronicity and durability are two of the key dimensions which determine the degree of authenticity possible in assessment.

Coulter and Mandell, Chapter VII, "Nontraditional Students and Information Technology: The Siren Call of the Virtual Classroom and its Impact on Progressive Educational Ideals," argue that the use of information technologies has not necessarily promoted progressive educational practices for adult learners. They suggest that progressive and constructivist approaches to learning may be more likely to be found outside formal higher education.

Section II, "Teaching and Learning with Information Technology: Constructivist Approaches," includes analyses of a variety of information technologies as they are being used within constructivist frameworks in higher education. The facilitation of active learning in immersive environments and collaborative settings and the creation of mental models as supported by digital media are described in detail.

Van der Pol in Chapter VIII, "Online Learning Conversations: Potential, Challenges and Facilitation," takes a close look at online conversations as tools for learning. The challenges to their effective use are detailed, and it is demonstrated that anchored discussion can address them by turning opinion-oriented exchanges into meaning-oriented processing of conceptual material.

Nicosia, Chapter IX, "Virtual Constructivism: Avatars in Action," looks at Second Life as providing opportunities for constructivist educational experiences within user-constructed environments and activities. Immersion in SL can facilitate deep learning prompted by desire and curiosity.

Page and Ali in Chapter X, "The Power And Promise Of Web 2.0 Tools," present an overview of Web. 2.0 applications as they can be used to create interactive and engaging learning environments.

Shalin in Chapter XI, "Information Technology and Mental Modeling," examines ways in which information technologies are deployed in higher education courses to help learners create mental models.

E-Learning can facilitate this important constructivist process though visualization work, virtual immersion and enhancing interactivity by the use of collaborative pedagogical strategies.

Section III, "International Approaches to Constructivism: Case Studies," details how higher education institutions around the world are experimenting with the constructivist use of information technology. The case studies in this section illustrate a number of applications used with a variety of academic subject matters, in blended classrooms and in online delivery, as well as the response of an historically progressive college to technology's challenges.

Donnelly in Chapter XII, "Transformative Potential of Constructivist Blended Problem-based Learning in Higher Education," considers a blended approach to problem-based learning, using interactional analysis of the work of academic staff who are engaged in professional development in higher education in Ireland to identify design implications for further development.

Cronin in Chapter XIII, "Critical Survey of Information Technology Use in Higher Education – Blended Classrooms," looks at the blended classroom from the standpoint of lifelong learners in a diploma course in Dublin, Ireland, in European art history, and finds that the constructivist paradigm is a necessary framework for the effective use of information technology in this setting.

Ligorio and Sansone in Chapter XIV, "Structure of a Blended University Course: Applying Constructivist Principles to Blended Teaching," implement constructivist principles in the architecture of a blended university course at an Italian university. They have designed individual, small-group and collective activities to explore and utilize digital identities, E-Tutoring, online role-play, and E-Portfolios.

Lim and Sudweeks in Chapter XV, "Constructivism and Online Collaborative Group Learning in Higher Education: A Case Study," present a hybrid undergraduate course at a university in Australia, which is shaped by sociocultural constructivist principles, providing interactional opportunities and scaffolding for joint construction and appropriation of new knowledge. An exploration of the theoretical roots of constructivism is followed by a course description and the results of a concluding survey on student learning experiences.

Lohr, Eastham, and Kendrick in Chapter XVI, "Constructivist Strategies to Optimize Four Levels of Interaction in a Distributive Learning Environment: A Case Study," describehow a constructivist theory of learning guided the structuring of a graduate level course on instructional design at the University of Northern Colorado. Results were derived from participant surveys, designer observations, and academic performance, and were generally positive with respect to student perceptions of the benefits of the course. Some constructivist strategies appeared to detract from the learning experience, and suggested the need for clarification of expectations and redesign of certain course components.

Gaspar, Langevin, and Boyer in Chapter XVII, "Facilitating Students-Driven Learning of Computer Programming with Technology," discuss the transition from instructivist to constructivist strategies as facilitated by information technology. Their illustrations are drawn from computer programming courses at the University of South Florida Polytechnic, and touch on learning barriers traditionally encountered by novice programmers. They conclude by describing how the integration of technology, and the switch to online environments, has the potential to enable authentic student-driven programming pedagogies as well as facilitate formal computing education research or action research in this field.

Miller in Chapter XVIII, "Designing Asynchronous Discussions to Teach Critical Thinking," finds that the modeling and practicing of critical thinking is a central component of constructivist pedagogy at the college level. The written and asynchronous nature of online discussions enhances its ability to teach critical thinking, but unless they are properly designed they may have the opposite effect. Six basic principles for effective online discussions are set forth, and their implementation is illustrated by examples from two online literature courses at National University in California.

Schulman in Chapter XIX, "'To Be in Occasional Touch': Goddard College's Progressive Principles and Distributed Learning," offers a case study of Goddard College's response to new modes of educational delivery and emerging information technologies. As one of the pioneering experimenting colleges, Goddard, in rural Vermont, continues to espouse the progressive principles which informed its establishment in 1938. How its innovative low residence model for adult college students can make use of IT without compromising its values is the major challenge seen by the author.

Stebick and Rinke in Chapter XX, "Using Blogs to Foster Inquiry, Collaboration, and Feedback in Pre-Service Teacher Education," present a critical case study of the integration of blogs into two constructivist-oriented teacher preparation courses at Gettysburg College in Pennsylvania, with the goal of helping students learn to think like a teacher through enhanced inquiry, collaboration, and feedback. They discuss design principles for combining online and face-to-face environments and offer possibilities for the expanded use of blogs in pre-service teacher education.

Zellermayer, Mor and Heilweil in Chapter XXI, "The Intersection of Theory, Tools and Tasks in a Postgraduate Learning Environment," have created a learning environment for veteran teachers, graduate students at an Israeli institution who are interested in developing professionally as designers and moderators of Information Communication Technology learning environments. The chapter describes the activities and tasks in each of three concentric circles of ICT use, which are based on the assumption that learners construct knowledge and understanding when they are an integral part of the learning environment, and when they are members of local and global communities where learning is conceived as acquiring the necessary skills for participation through their own participation.

REFERENCES

Cremin, L. (1964). *The Transformation of the School: Progressivism in American Education* (pp. 1876-1957). New York, NY: Alfred A Knopf.

Dewey, J. (1938). *Experience and Education*. New York, NY: Macmillan.

Dewey, J. (1958). *Experience and Nature*. New York, NY: Dover.

The Goddard Bulletin (1969). Plainfield, VT: Goddard College.

Jonassen, D. (1994). *Technology as Cognitive Tools: Learners as Designers.* Retrieved August 22, 2008 from http://it.coe.uga.edu/itforum/paper1/paper1.html

Jonassen, D., Mayes, T., & McAleese , R. (1993). *A Manifesto for a Constructivist Approach to Technology in Higher Education.* Retrieved September 26, 2007 from http://apu.gcal.ac.uk/clti/papers/TMPaper11.html 9/26/07

Liu, C. H., & Matthews, R. (2005). Vygotsky's philosophy: Constructivism and its criticisms examined. [Electronic version]. *International Education Journal, 6*(3), 386-399.

Spira, J. L. (1998) Integrating Principles of Progressive Education Into Technology-based Distance Learning [Electronic version]. *The Technology Source.*

von Glasersfeld, E. (1989). *Cognition, Construction of Knowledge, and Teaching.* Retrieved August 22, 2008 from http://www.vonglasersfeld.com/

Westbrook, R. (1991). *John Dewey and American Democracy.* Ithaca, NY: Cornell University Press.

Acknowledgment

It is my pleasure to acknowledge the scholarly debts which have been incurred in the preparation of this volume. Arthur Chickering, Patrizia Ghislandi, Jennifer Holtz, Pankaj Kamthan, Kinshuk, Marina Papastergiou, and David Seelow were kind enough to serve on the Editorial Advisory Board. The informed and incisive critiques provided by the numerous reviewers made the editorial work possible; I thank Radwan Ali, Carol Caldwell-Edmonds, Arthur Chickering, Xenia Coulter, Geoff Danaher, Nicholas Eastham, Patrizia Ghislandi, Rick Hendra, Jennifer Holtz, Pankaj Kamthan, David Kendrick, Gulsun Kurubacak, Elliott Lauderdale, Jennifer Lee, Beatrice Ligorio, Lin Lin, Linda Lohr, Alan Mandell, Kara Monroe, Laura Nicosia, G. Andrew Page, Marina Papastergiou, Andreas Pester, Cornel Reinhart, Beth Rubin, David Seelow, Hai-Jew Shalin, Steven Stahl, Phillipa Sturgess, Violeta Todorovic, and Margaret Wherry. The contributors of the chapters themselves were enthusiastic and cooperative in our joint effort to produce a useful and informative compilation reflecting their pioneering efforts to transform higher education in the 21st Century. Ms. Rebecca Beistline of IGI Global offered timely and knowledgeable editorial and technical support, and constant encouragement.

The support and interest of friends and family were essential to my persevering. I wish to mention in particular Joyce Kadwell, Mary Leahy, and Cornel Reinhart, and my husband, Richard G. Payne, for his patience.

Carla R. Payne

Section I

Constructivism and Progressive Higher Education in the World of Information Technology

Chapter I
Closing the Circle: From Dewey to Web 2.0

Maria Luisa Pérez Cavana
The Open University, UK

ABSTRACT

Taking into account the complexity and multiplicity of constructivist theories, the first part of this chapter focuses on the relationship between epistemology and pedagogy in constructivism, in particular in the radical constructivist position of von Glasersfeld, which is considered a significant referent in constructivism. To overcome some of the shortcomings of radical constructivism, the author have then explored the origins of constructivist theory and practice in the work of John Dewey, whose ideas could be still a source of inspiration for constructivist educational practice. The second part of this chapter analyses the social constructivist development in different internet-based learning platforms and social software and considers at the end some practical difficulties and benefits of online learning for the implementation of constructivist learning theories for learners as teachers.

INTRODUCTION

The definition of constructivist theory or constructivism is an issue in itself. Constructivism has become such a complex topic that just the effort to clarify the different positions and trends is enough to serve as the main subject of some papers where the authors have tried to put some structure into the enormous range of definitions, sections and positions regarding constructivism (Phillips, 2005; Riegler, 2005; Steffe & Gale, 1995). It is not the purpose of this paper to do so, I will however give a brief overview of the complexity and multidimensionality of this field.

The difficulty of defining "constructivism" starts with the question whether constructivism is a theory, an approach, or a perspective. For von Glasersfeld constructivism is a way of thinking (von Glasersfeld (1985, 1992), for Siebert it is a metatheory (Siebert, 2004), Huitt considers the constructivist approach to teaching and learning as based on a combination of cognitive psychology and social psychology (Huitt, 2003), Dougiamas talks about the faces of constructivism (Dougiamas, 1998) as does Philips, considering constructivism as a secular religion within educational theory (Phillips, 1995) or even a magic word (Phillips, 2000) and Duit (1993) regards it as a fashionable and fruitful paradigm. What seems clear from all these studies is that, as Horst Siebert (2005) puts it, constructivism is not a scientific discipline in itself but an interdisciplinary and transdisciplinary "Paradigma," it is a perspective in which reality depends on the observer.

Common to all the approaches and different definitions is the source of the term "constructivism," which is a metaphor of architecture, and is about the building up of structures from pre-existing pieces, possibly specially shaped for the task. This metaphor, as Ernest (1995) points out, describes understanding as the building of mental structures, and it is also contained in the term "restructuring," often used as synonym for accommodation or conceptual change. Ernest also notes a relevant feature of this metaphor: the building blocks are not merely received, they are products of previous acts of construction.

A fundamental component of constructivism is action: knowing is an active process, learners are not passive receivers of learning contents. As Glasersfeld (1989) formulates the first principle of constructivism: "Knowledge is not passively received but actively built up by the cognizing subject" (p.182).

Although constructivism has received significant inputs from specific sciences such as neurobiology with the authors Humberto Maturana and Francisco Varela, and from brain research, two main fields remain the most relevant within the constructivist discussion: epistemology and pedagogy. Epistemology, or theory of knowledge, which considers the genesis and the nature of knowledge, is the main philosophical discipline associated with the term constructivism. The particular vision of theory of knowledge which considers that our knowledge is constructed has been labelled as constructivist epistemology or epistemological constructivism (Heylighen, 1995). In pedagogy the term constructivism applies to different aspects of this discipline such as learning theory, teaching techniques or the general pedagogical approach (Phillips, 2000). Although the theoretical frame of this paper as well as that of this volume is clearly within the educational aspects of constructivism, the first sections of this chapter will look in some detail the most relevant position in constructivist epistemologies to provide an insight into the origins and spirit of constructivism. After this I will look into new developments of online learning tools and to what extent they are within the spirit of constructivism.

CONSTRUCTIVISM AS EPISTEMOLOGY

A first look into the studies on "constructivism" show clearly the two parallel trends within this paradigm: the epistemological and the pedagogical. It is worth researching the extent to which these two positions are parallel – i.e., that they expand and develop in different realms never touching each other, or whether they are intrinsically interrelated.

It seems to make sense that learning theories come after philosophical ones, that is, the first fundamental question is: "what is knowledge?" and then the question arises "how to acquire knowledge?"

Constructivist theory has been associated with epistemology, which is the discipline within philosophy which deals with the question of what we can know, and with the possibility of truth and objectivity. Epistemology addresses the fundamental question in theory of knowledge of the relation between "reality" – the outside world – and our ideas about it. One main epistemological tradition, realism, states the possibility of true knowledge, of objectivity. Realism maintains the existence of a world – independent from the knowing subject - and the correspondence between this world and our ideas about it. Opposed to this tradition, constructivism argues that our beliefs and perceptions of the world are purely human constructs and it denies the possibility of discovering and knowing the world as it is.

It is not clear where the origins of constructivist epistemology can be found, whether its origins can be dated to the 17th or 18th centuries, or even to Socrates. Although the term "constructivist epistemology" seems to have appeared recently (Piaget, 1967), the ideas of constructivist epistemology can be clearly found in the Enlightenment in the 18th Century in Kant's philosophy, even if he does not use this term. This philosopher declared that it is we who construct - and not discover – the known world. In his *Critic of Pure Reason* Kant demonstrated the framework-dependent character of perception and argued that the epistemological claim of knowing an object in "itself", independently of the subject that knows, has to be abandoned. It is impossible – argued Kant – to know the things in themselves, the noumena. This pure knowledge is possible for religion and ethics, but not for science. Science has to do with phenomena, and these are always determined by the categories of our way of knowing objects. Hence even if Kant did not use the word "constructivism," the essence of his epistemology is clearly constructivist as the human cognitive system is responsible for shaping our experience in the categories he described.

Later in the 20th century the rise and development of philosophy of science took as its starting point this critical position in relation to knowledge, and all the different positions that emerged at this time can be considered "nonfoundationalist epistemologies" (Phillips, 2003). Traditional epistemologies on the contrary were foundationalist in the sense that they regarded knowledge as being built upon some solid foundation. In the 20th century the major epistemological trend was to recognize that there is no absolutely secure starting point for knowledge: all knowledge is tentative, as Karl Popper (1968) states. Also the work of Thomas S. Kuhn on scientific revolutions and paradigms has been considered an important influence for constructivist authors, as Kuhn (1962) stressed the active role of scientific communities in science construction and he popularized the notion that inquirers always work within the context of a paradigm.

RADICAL CONSTRUCTIVISM AS A REFERENT

One of the most influential authors within the constructivist context is Ernst von Glasersfeld, whose work is an obligatory reference for every constructivist approach. According to Siegler (2005) Glasersfeld pioneered the philosophical-epistemological approach. The Austrian author explicitly takes Kuhn as a precursor of his position, which he has called "radical constructivism". Glasersfeld, who was also significantly influenced by Piaget's work, based his position on two principles:

Knowledge is not passively received but actively built up by the cognizing subject and the function of cognition is adaptive and serves the organization of the experiential world, not the discovery of ontological reality, Glasersfeld (1989).

He names his position "radical" because he claims that constructivism has to be applied to all levels of description:

The revolutionary aspect of Constructivism lies in the assertion that knowledge cannot and need not be 'true' in the sense that it matches ontological reality, it only has to be 'viable' in the sense that it fits within the experiential constraints that limit the cognizing organism's possibilities of acting and thinking. (Glasersfeld, 1989, p.162)

These two principles of Glasersfeld quoted above have become a reference with respect to other constructivist positions and have, so to speak, a definition building character, that have played a significant role in distinguishing several types of constructivism, depending on the acceptance of the first principle only or of both of them (Ernest, 1995)

According to Glasersfeld (1983) the constructivist approach has had a significant impact on different disciplines, such as psychotherapy, but the main influence has been in educational practice and research. In following the consequences for education that Glasersfeld enumerate will be summarized:

- Educational procedures aim at generating understanding (teaching) and not merely repetition of behaviour (training),
- The interest of the educator will be focused on what can be inferred to be going on inside the student head, rather than on overt responses
- Teachers will realize that knowledge cannot be transferred to the student by language, but language can be a tool to guide the student's construction
- Teachers will be aware that students are attempting to make sense in their experiential world, therefore they will be interested in student's deviation form the expected path.

These features of what Glasersfeld considers to be a constructivist pedagogy seem to be in agreement with the constructivist epistemology: the consequence of the enormous limitations of human knowledge would be to transfer the focus of education in learning how to learn rather than on learning contents. As Zarnik (2001) observes, the educational philosophy of constructivism is usually understood as being motivated by these metatheoretical insights.

The question is whether a constructivist epistemology is the condition for developing a constructivist pedagogy. And there are some reasons to think that it is not a necessary condition. On the one hand – as Rubin (1995) points out – nowadays constructivist epistemology is the only paradigm that survives, and that would mean everyone would belong to the constructivist school, and would turn "constructivism" into a more or less empty concept. On the other hand there are other relevant authors in progressive education – e.g. John Dewey - whose starting point was not radical constructivist epistemology. The shortcomings of Glasersfeld's constructivism have been exposed in respect to his theory of knowledge, as well as to his pedagogical ideas. Vanderstraeten and Biesta (1994) state how radical constructivist theories of knowledge fail to explain the existence of plural worlds: it is not enough to reiterate that the correspondence between reality and knowledge cannot be proved and to acknowledge in general terms the existence of an external world. What is needed is to explain how we know that there is no correspondence, and how we know that there are no external objects out there. Also Howe and Berv (2000) point to the weak, actually shaky bridge between the subjective and intersubjective, when the private worlds of the teacher and her students are merely "taken-as-shared" in radical constructivist theory.

Although Glasersfeld's epistemological position is without doubt a necessary reference within constructivist theory, it is too embedded in epistemological questions to be a reference for a constructivist pedagogy in particular in relation to the more and more significant field of education within CMC (Computer Mediated Communication). Maybe an insight into the work one of the first precursors of constructivist learning theories, John Dewey, could provide a fruitful link of the origins of constructivism and new developments in online learning.

JOHN DEWEY: BACK TO THE BEGINNINGS

There is no doubt that Dewey can be considered one of the developers of educational theories that lead to the evolution of constructivism. As a matter of fact he anticipated most of the relevant issues in constructivist pedagogy. Dewey has been considered a pragmatic social constructivist (Garrison, 1998). His starting point is clearly the primacy of action. In his work *The Quest for Certainty* (1929/1979) he regrets "the depreciation of action, of doing and making" that has been cultivated by philosophers. According to Dewey the reason for this elevation of knowledge above practice is fundamentally connected with the quest for certainty "which shall be absolute and unshakable." (Dewey, 1929/1979, p.6) and the search for certainty is based on the human need for safety, "to escape from peril." Dewey's proposal is, however, not to abandon the idea of certainty but only the one based on pure ideas. He states that many of the long-standing problems of philosophy in general and of theory of knowledge in particular could be addressed by "substituting search for security by practical means for quest of absolute certainty by cognitive means" (Dewey, 1929/ 1979, p.26)

One of the fundamental topics Dewey developed through his works is the intrinsic relation between action and knowledge: "our conceptions and ideas, are designations of operations to be performed or already performed. Consequently their value is determined by the outcome of these operations. They are sound if the operations they direct give us the results which are required." (Dewey, 1929/1979, p.137) For Dewey "knowing is not the act of an outside spectator but of a participator inside the natural and social scene, then the true object of knowledge resides in the consequences of directed action." (Dewey, 1979, p.196)

This last quote also shows Dewey's integration of individual and social dimensions in the construction of knowledge, which is actually a leitmotiv throughout Dewey's work, to reconsider the relationship of organism and environment (Vanderstreten & Biesta, 1998). Dewey's concept of transaction stresses the continuous, intrinsic connection of organism and world on the level of action (Dewey, 1896).

This practical approach to philosophy has been called "pragmatic social constructivism" (Garrison, 1998) and has significant consequences for his educational vision.

It is clear that with his work Dewey built the foundations of constructivist pedagogy which, according to Howe and Berv (2000), has two basic premises: instruction must take as its starting point the knowledge, attitudes and interests students bring to the learning situation, and instruction must be designed so as to provide experiences that effectively interact with these characteristics of students so that they may construct their own understanding. These both premises are extensively developed in the Deweyan educational theory.

In his book *How we think* (1910/1991) Dewey states the need to train the natural capacity of thinking into habits of critical examination and inquiry but also stresses the active role of learners:

In the educational transaction, the initiative lies with the learner (...) If an individual can learn to think only in the sense of learning to employ more economically and effectively powers he already possesses, even more truly one can teach others to think only in the sense of appealing to and fostering powers already active in them. Effective appeal of this kind is impossible unless the teacher has an insight into existing habits and tendencies, the natural resources with which he has to ally himself." (Dewey, 1991, p.29-30)

Dewey stresses the necessity of an active role for learners. When this active principle is neglected and the child "is thrown into a passive, receptive, or absorbing attitude (...) the result is friction and waste." The energy in schools should be "spent in training the child's power of imagery and in seeing to it that he was continual forming definite, vivid, and growing images of the various subjects with he comes in contact with his experience." Dewey (1897), p.5

FROM CONSTRUCTIVISM TO CONSTRUCTIONISM: MORE THAN A WORD GAME

A significant emphasis on the social dimension of constructivism is the main characteristic of constructionism.

This term has been used in many contexts and for different purposes since Peter L. Berger and Thomas Luckmann published their book *The Social Construction of Reality* in 1966. As a learning theory the term has been associated with Seymour Papert. According to Papert (1987), the word constructionism is a mnemonic for two aspects of the theory of science education: From constructivist theories of psychology it takes a view of learning as a reconstruction rather than as a transmission of knowledge; the other aspect is that learning is most effective when part of an activity the learner experiences as constructing a meaningful product.

The simple catchy version of this idea is "learning-by-making", but as Papert (1991) asserts, constructionism is a multifaceted idea: with the play on the words from constructivism to constructionism (the *n* word opposed to the *v* word) constructivism shares with constructionism the connotation of learning as "building knowledge structures," irrespective of the circumstances of the learning. Constructionism then adds that learning occurs especially well when the learner is engaged in constructing something for others. Ackermann (1990) also notes that while most constructivist models of human intelligence remain essentially science-centred and logic-oriented (e.g. Piaget and Vygotsky) for constructionism (e.g. Papert and Turkle) offers a far less canonical view on formal knowledge and revaluates the concrete, more personal knowledge, as well as the art of learning and the importance of tools, media and context.

Gergen (1995) puts more emphasis on the affinities and differences between the two terms. From an epistemological point of view both are sceptical of foundational epistemologies and both challenge the traditional view of the individual mind as a device for reflecting the character and conditions of an independent world; the main difference -, however is that radical constructivism still maintains dualistic formulations traditional to Western epistemologies, while constructionism tries to break with this tradition, as also does Dewey in his practical and social approach to knowledge.

As educational theory constructionism is according to Gergen (1995) "a process in motion" which can be best described with the metaphor of the social dialog. Not only knowledge is considered in constructionism as being in continuous production of dialogues. The idea of the authority of the teacher is also challenged. The constructivist approach in pedagogy since Dewey has questioned in depth the "frontal style" of teaching.

Construcionism goes one step further questioning the myth of the superiority of teachers, regarding them more as a facilitators and as a learners themselves, as they try to know their students and their background.

In spite of these different connotations, the term constructionism does not seem to have been adopted generally by constructivist authors, not even in the context of social constructivism or CMC. For that reason to avoid confusion, I will use the general term "constructivism" or "social constructivism" instead of constructionism.

CONSTRUCTIVISM IN PRACTICE: AN EPIC JOURNEY?

The development of all these constructivist approaches to learning theories involves the rise of new attitudes, practices and techniques within education. It is, however, not a smooth development. The clash between theory and practice in education is impossible to overlook.

On the one hand constructivist approaches are currently broadly accepted within the academic community. Today at educational research conferences, the idea that learners actively construct new knowledge is taken almost as a gospel (Kafai and Resnik, 1996), but in the concrete educational settings and institutions it is very difficult to put into effect. As Dewey (1938) noted long time ago, the road of new education is not an easier one to follow than the old road but a more strenuous and difficult one.

First, the roles of teachers and students have to be redefined and new roles learned, and this causes a significant amount of insecurity, anxiety and resistance. For students it is much easier to continue with the traditional passive role and acquire what the teacher or the book says, than to take an active role in the process by constructing, reflecting on their own experience, collaborating with other students and taking responsibility for their own learning. It is not easy to help students "break away" from their initial "regions of comfort" (Resnick and Ocko, 1991).

Second it is also enormously challenging for teachers to renounce their comfortable, secure and efficient authority position and to develop a very different – much more egalitarian -relationship to the students, more as a facilitator than a teacher. The change is not only difficult from a psychological point of view, as a possible crisis of identity, but is also related to specific didactic complexities within the new learning model. Examples are the necessity of finding out in which types of environments students are most likely to make personal connections with their projects, or exploring how learning activities can be organized to insure that students care about their projects in an deep and personal way (Resnick, 1991). Drawing on my own experience, a constructivist approach implies a bigger workload and time consuming preparation, and sometimes also lack of understanding by colleagues who ask: "Do you really want to do so much work?".

Trying to use a constructivist approach in teaching practice can be like an epic journey where many obstacles have to be overcome: after passing the land of the unwilling students and the land of the identity crisis and hostile colleagues – to use a metaphor from the Lord of the Rings – the last and most difficult obstacle has to be overcome: the land of the rigid institutions.

It is very difficult to introduce changes in the structure of schools, colleges, universities, even if they are small, and it always involves a great deal of effort, energy and time. Educational institutions are very reluctant to challenge their view of learning and to question, for example, ways of assessment or to change the curriculum to allow more flexibility in academic tasks or to facilitate collaborative learning.

Despite all these difficulties it is not only possible to develop constructivist approaches and attitudes in teaching and learning, but it is, I would say, an unstoppable process that has been accelerated through the use of CMC (Computer Mediated Communication) and elearning.

LEARNING MANAGEMENT SYSTEMS (LMS), E-TIVITIES AND WEB 2.0: A PATH WITHOUT RETURN

Constructivist ideas have been further developed and implemented in particular within the field of new technologies and design of learning software.

One of the most relevant authors in this context is Martin Dougiamas with his work on a well known open source virtual learning environment, Moodle.

Other successful Learning Management Systems - for example: Blackboard, eCollege, WebCT, Angel, Desire2learn - have very similar features. They all offer an online environment where students have access to personal functions such as email, calendars, file sharing and basic discussions, and some will offer as well new features encompassed in the realm of Web 2.0.

I will however concentrate on Moodle because its author has explicitly referred to constructivist theories and authors in developing his VLE. The characteristics of LMS described in following can thus be considered general for virtual learning platforms.

According to Dougiamas (1998) the learning process is, as he states about his own developing of ideas, especially fostered when creating texts, web sites or computer software for and with others. This type of social constructivism focuses on the individual development of meaning through construction and sharing of texts and other social artefacts. (Dougiamas, 2003) The pilosophy of Moodle extends the ideas of constructivism into a social group constructing things for its members, collaboratively creating a small culture of shared artefacts with shared meanings.

Dougiamas drew on the work of S. Papert, K.J. Gergen and F. Steier to develop a pedagogy for internet-based tools to support learning. He states as the core of social constructivism five principles (Dougiamas 2008, p.2)

1. All of us are potential teachers as well as learners.
2. We learn particularly well from the act of creating or expressing something for others to see.
3. We learn a lot by just observing the activity of our peers.
4. By understanding the context of others, we can teach in a more transformational way (understanding the background of the students).
5. A learning environment needs to be flexible and adaptable, so that it can quickly respond to the needs of the participants within it.

These five principles focus on the idea of redefining the role of teacher and learner, where the limits become fuzzy or even interchangeable and also stress learning by doing something with and for others. Drawing on these principles, Moodle, the software designed by Dougiamas, aims to be a true collaborative environment to develop the idea that all are potential teachers as well as learners. Many of the activities in Moodle allow students to control common content such as forums, wikis, glossaries, databases, messaging and so on. It also tries to break down further the distinction between teachers and students allowing system administrators and teachers to create new roles with any mix of capabilities they like.

All VLEs also facilitate construction and sharing of learning through all types of collaborative tools and tries also to make possible the understanding of the contexts of others, for example through the use of profiles, individual blogs, discussion starters, etc.

The question is whether the mere use of a LMS would per se imply a constructivist and more interactive way of learning. Tolmie and Boyle (2000) investigate the idea that technologies create environments which shape interpersonal exchange. They look into the notion of *affordances*: "properties of objects or systems which allow certain actions to be readily performed with them and which therefore push behaviour in certain directions" (p.120)

According to their study, high-quality interaction, full participation and reflection do not happen simply by providing the technology. A "little human time and energy to get them to work" in words of Gilly Salmon (2002, p.4) is necessary. This is the starting point for this author in developing the concept of e-tivities and a five-stage framework for e-learning. Computer-mediated technologies offer an enormous potential for active and reflective learning; the key point, however, is activity design. The five steps in Salmon's model of teaching and learning online are: 1. Access and motivation, 2. Online socialization, 3. Information exchange, 4. Knowledge construction and 5. Development. The last two ones are clearly based on the "spirit of constructivism." At stage four, *Knowledge Construction,* participants start interacting with each other in more exposed and participative way, and as conferences unfold many participants engage in some very active learning (Salmon, 2000).

The role of e-moderators at this stage is crucial to pull together participants and to enable development of ideas through discussion and collaboration. E-moderating is, however, according to Salmon, not the same as facilitating a face-to-face group. In CMC at stage four there is much less of a hierarchy, because participants start to become online authors rather than transmitters of information. At stage five, development, e-moderators and participants are essentially using a constructivist approach to learning: they explore their own thinking and knowledge building process.

According to Salmon the role of the e-moderator and activity planning would be essential for these learning platforms to support a constructivist learning approach.

Stephen Dowes goes one step further with his concept of "E-learning 2.0" (Dowes, 2006)

Dowes considers that the learning technology employed in LMSs in almost any higher education institution as a way to deliver online courses and organize learning content is not very different from traditional ways of delivering distance learning, even if the learning platform is supported by quizzes, tests and discussions.

He refers to other technologies that are introducing other ways of more active, learner-centered learning: Web 2.0

This term was coined to describe new developments and collective trends in the use of internet and to distinguish them from the original uses of the Web (1.0). Franklin & Harmelen (2007) explained that while Web 1.0 provided content for an audience of passive readers, in Web 2.0 many more users generate, consume and transform material posted on the web, an innovation referred to as "shared content" or "read/write." The most important feature of Web 2.0 is that the web becomes a platform and enables users to collaborate, work and play together.

The most important applications of Web 2.0 are blogs, wikis, podcasts, social tagging services (del.icio.us, Flicker), social networks and tagging. All these applications have many educational uses (Franklin & Harmelen, 2007) in particular in higher education, because Web 2.0 is a natural platform for open learning materials.

In relation to constructivism Web 2.0 is highly relevant because it represents the web shifting from being a medium in which information was transmitted and consumed, into being a platform, where content is created, shared, repurposed and passed along as (Dowes 2006, p.4) formulates it:

What happens when online learning ceases to be like a medium, and becomes more like a platform? What happens when online learning software ceases to be a type of content-consumption tool, where learning is "delivered," and becomes more like a content-authoring tool, where learning is created? The model of e-learning as being a type of content, produced by publishers, organized and structured into courses, and consumed by students, is turned on its head. Insofar as there is content, it is used rather than read— and is, in any case, more likely to be produced by students than courseware authors. And insofar as there is structure, it is more likely to resemble a language or a conversation rather than a book or a manual.

Although the use of Web 2.0 in higher education is too recent to have relevant and conclusive data about its pedagogical benefits it seems clear that the tools associated with Web 2.0 support social constructivist approaches to teaching and learning. Some studies carried out recently in UK and European universities suggest that students and staff recognized the advantages of Web 2.0 compared to traditional VLEs (Franklin & Harmelen, 2007). Dowes also emphasizes that the use of Web 2.0 takes as a starting point the student's own range of interest and experiences rather than the course topic; he states further,

This approach to learning means that learning content is created and distributed in a very different manner. Rather than being composed, organized and packaged, e-learning content is syndicated, much like a blog post or podcast. It is aggregated by students, using their own personal RSS readrer_or some similar application. From there, it is remixed and repurposed with the student's own individual application in mind, the finished product being fed forward to become fodder for some other student's reading and use. (Dowes, 2006, p.5)

This development in teaching and learning involves significant challenges not only in terms of intellectual property, control and accessibility. From a pedagogical point of view assessment is one of the main issues. The nature of assessment will need to be revised and new forms and new methods will have to be developed, where group work is going to replace individual work.

But the changes introduced by web 2.0 are going to challenge other concepts within the academic community such who is to decide the learning content of a course, or even whether there are objective criteria to compare different learning products? These questions bring us to back to the field of philosophy, specifically to epistemology. Web 2.0 confronts us with a reality that seems to be very near von Glasersfeld's radical constructivism and also in keeping with Dewey's ideal of learning.

CHALLENGES FOR THE FUTURE

Taking into account the issues that I have briefly considered in the previous sections: the difficulties of putting constructivist learning theories into practice and the affordances of new technologies, in particular of Web 2.0, there seems to be a clear shift in direction. New generations of students and course participants are so embedded in new technologies and new social uses of the web that they are using and developing constructivist ideas anyway. The problem of the unwilling students seems to be fading away while the hostile colleagues, and the land of the rigid institutions are likely to become the most important obstacles to deal with. As a recent research report by Becta (2008), on new technologies states, the most basic tension to be experienced in adopting Web 2.0 practices in educational settings concerns a shift in control or management of the educational experience: teachers and institutions may be actively resistant to changes that imply a greater learner autonomy, and a change of the teacher role from 'sage on the stage' to 'guide on the side'. But the change from individual learning to collaborative learning is also going to be a major issue in terms of assessment within the context of academic degrees. It is possible that new approaches to collaborative work within social constructivist approach could also change the traditional idea of assessment.

CONCLUSION

Familiarity with the knowledge about epistemological constructivism in general and radical constructivism in particular, as well as the work of Dewey is important not only to understand the origins and development of constructivist theory and practice, but also as a wider theoretical frame to see and analyze current debates and developments in teaching and learning with new technologies. Kuhn was one important theoretical referent for radical constructivism and now we can observe in the development of uses of new technologies, in particular in Web 2.0, a Kuhnian shift of education paradigms, where the direction is bottom up, where students are being the motor of changes in direction to more constructivist approaches. This trend is also very much in the spirit of John Dewey's practical approach to learning.

In constructivism we are confronted with the old philosophical problem of the relation between theory and practice. This is a tension that can be transformed into a fruitful dialogue between philosophical theories and technical developments, where both can learn and benefit from each other.

REFERENCES

Ackermann, E. (1990). *Piaget's constructivism, Papert's Constructionism: What's the difference?* Retrieved April 15, 2008, from: http://learning.media.mit.edu/

Boudourides, M. A. (1998, July). *Constructivism and Education: A Shopper's guide.* Paper presented at the International Conference on the Teaching on Mathematics, Samos, Greece. Retrieved May 2, 2008, from: http://math.upatras.gr/~mboudour/articles/constr.html

Becta (2008). *Web 2.0 technologies for learning at Key Stages 3 and 4.* Retrieved August 13, 2008 from: http://becta.org.uk

Dewey, J. (1910/1991). *How we think.* Amherst. New York: Prometheus.

Dewey, J. (1929/1979). *The Quest for Certainty.* New York: Paragon.

Dewey, J. (1897) My Pedagogic Creed. *The School Journal, LIV*(3), 77-80. Retrieved April 23, 2008 from http://www.infed.org/archives/e-texts/e-dew-pc.htm

Dougiamas, M. (1998). *A Journey into Constructivism.* Retrieved May 1, 2008, from http://dougiamas.com/writing/constructivim.html

Dougiamas, M. (2008). *Pedagogy.* From MoodleDocs. Retrieved May 1,2008, from http://docs.moodel.org/en/Pedagogy

Dougiamas, M. (2003). *Moodle. Using Learning Communities to Create an Open Source Course Management System.* Refereed paper, presented at EDMEDIA 2003. Retrieved May 2, 2008, from http://dougiamas.com/writing/edmedia2003/

Downes, S. (2006). E-learning 2.0. *ELearn Magazine.* Retrieved from http://elearnmag.org/subpage.cfm?section=articles&article=29-1

Duit, R. (1993). The Constructivist View: A Fashionable and Fruitful Paradigm for Science Education and Practice. In L. Steffe (Ed.), *Epistemological Foundations of Mathematical Experience.* New York: Springer-Verlag.

Eisner, E. (2003). Objectivity in Educational Research. In M. Hammersley (Ed.), *Educational Research. Current issues* (pp. 57-72) London: Paul Chapman Publishing.

Ernest, P. (1995). The One and the Many. In P. Steffe & J. Gale (Eds.), *Constructivism in Education* (pp. 355-366). Hillsdale/New Jersey: Lawrence Erlbaum Associates.

Franklin,T., & van Harmelen, M. (2007). *Web 2.0 for Learning and Teaching in Higher Education.* London/ UK: The Observatory on borderless Higher Education.

Gergen, K. J. (1995). Social Construction and the Educational Process. In P. Steffe & J. Gale (Eds.), *Constructivism in Education* (pp. 355-366). Hillsdale/New Jersey: Lawrence Erlbaum Associates.

Glasersfeld, E. von (1989). Constructivism in Education. In T. Husen, & T. N. Postlethwaite, (Eds.), *The International Encyclopedia of Educa-*

tion, Supplement, 1, 162-163. Oxford/New York: Pergamon Press.

Heylighen, F. (1997). Epistemological Constructivism. *Principia Cybernetica Web.* Retrieved May 5, 2008 from http://pespmcl.vub.ac.be/CONSTRUC.hyml

Hildebrand, D. L. (2008). *John Dewey: A Beginner's Guide.* Oxford, UK: Oneworld Press.

Holec, H. (1981). *Autonomy and Foreign Language Learning.* Oxford: Pergamon.

Howe, K. R., & Berv, J. (2000). Constructing Constructivism, Epistemological and Pedagogical. In D. C. Phillips (Ed.), *Constructivism in Education: Opinions and Second Opinions on Controversial Issues.* Ninety-ninth Yearbook of the National Society for the Study of Education. Part I. Chicago, Illinois, The University of Chicago Press.

Kant, I. (1784). *An Answer to the Question: "What is Enlightenment?"* Konigsberg in Prussia, 30th September, 1784. Retrieved May 27, 2008, from http://philosophy.eserver.org/kant/what-is-enlightenment.txt

Kuhn, T. S. (1962). *The Structure of Scientific Revolutions.* Chicago and London: The University of Chicago Press.

Papert, S. (1987). *Constructionism: A New Opportunity for Elementary Science Education.* Retrieved May 12, 2008 from http://nsf.gov/awardsearch/ showAward.do?AwardNumber=8751190

Papert, S. (1991). Situating Constructionism. In I. Harel & S. Papert (Ed.), *Constructionism* (pp.1-12). Norwood, NJ: Ablex Publisheng Corporation.

Phillips, D. C. (1995). The Good, the Bad and the Ugly: The Many Faces of Constructivism. *Educational Researcher, 24*(7), October, 5-12.

Phillips, D.C. (Ed). (2000). *Constructivism in Education: Opinions and Second Opinions on Controversial Issues.* Ninety-ninth Yearbook of the National Society for the Study of Education. Part I. Chicago, Illinois, The University of Chicago Press.

Phillips, D. C. (2003). Subjectivity and Objectivity: An objective Inquiry. In M. Hammersley (Ed.), *Educational Research. Current issues* (pp. 57-72) London: Paul Chapman Publishing.

Popper, K. (1968). *Conjectures and refutations.* New York: Harper.

Resnick, M. (1991). Xylophones, Hamsters, and Fireworks: The Role of Diversity in Constructionist Activities. In I. Harel & S. Papert (Ed.), *Constructionism* (pp. 1-12). Norwood, NJ: Ablex Publisheng Corporation.

Resnick, M., & Ocko, S. (1991). LEGO/Logo: Learning Through and About Design. In I. Harel & S. Papert (Ed.), *Constructionism* (pp.1-12). Norwood, NJ: Ablex Publisheng Corporation.

Riegler, A. (2005). Editorial. The Constructivist Challenge. *Constructivist Foundations, 1*(1), 1-8.

Rubin, D. (1995). Constructivism, Sexual Harassment, and Presupposition: A (Very) Loose Response to Duit, Saxe and Spivey. In P. Steffe & J. Gale (Eds.), *Constructivism in Education* (pp. 355-366). Hillsdale, NJ: Lawrence Erlbaum Associates.

Salmon, G. (2000). *E-Moderating. The Key to Teaching and Learning Online.* London and Sternling (USA): Kogan

Salmon, G. (2002). *E-tivities. The Key to Active Online Learning.* London and Sternling (USA): Kogan.

Siebert, H. (2004). Sozialkonstruktivismus: Gesellschaft als Konstruktion. *Sowi-online.e.v., 2. R*etrieved May 5, 2008, from http://www.sowi-online.de/journal/2004-2/sozialkonstruktivismus_siebert.htm

Siebert, H. (2005). *Pädagogischer Konstruktivismus. Lernzentrierte Pädagogik in Schule und Erwachsenenbildung.* Weinheim und Basel: Beltz Verlag.

Vanderstraeten, R., & Biesta, G. (1998). Constructivism, Educational Research and John Dewey. *In Paideia Project on-line, 20th World Congress of Philosophy.* Retrieved May 5, 2008, from http://www.bu.edu/wcp/Papers/Amer/AmerVand.htm

Zarnic, B. (2001). Learning to learn: An Epistemological Paradox in Education. *Synthesis Philosophica, 32,* 355-362.

Chapter II
Beyond Constriction and Control:
Constructivism in Online Theory and Practice

Noel Fitzpatrick
Dublin Institute of Technology, Ireland

Nóirín Hayes
Dublin Institute of Technology, Ireland

K.C. O'Rourke
Dublin Institute of Technology, Ireland

ABSTRACT

Constructivism has become the comfortable face of educational theory in recent years, due in no small part to the mainstreaming of learning technologies since the 1990s. Many of these technologies embed constructivism in their aspirations and actual design. But, generally speaking, the uncritical acceptance of constructivism as beneficial to education has been widespread – not surprising, perhaps, given that teaching as a profession has tended to be inspired by practice rather than being theory-led. In this chapter, the authors attempt to explore the uncomfortable tranche between theory and practice which constructivism currently occupies. Education has historically been about hierarchy and the control of knowledge and knowledge-flows, rather than about the construction of knowledge by the "amateur" individual or group. Constructivism conspires to foster active learning and the organic creation of knowledge, a radical departure from the accepted authority of the curriculum, leaning towards learning situated in the context of the learner, which is ultimately non-objective in the traditional manner. Is a bridge between theory and practice possible? In considering this question the chapter draws on the authors' experiences in designing and running an on-line graduate degree program according to constructivist principles. In doing so, it also attempts to describe and evaluate the impact which constructivism as a theory could have on the reality of teaching and learning practice in the early 21st Century.

INTRODUCTION

Since the 1990s, higher-education institutions (HEIs) have embraced learning technologies in a manner that is, at one level, quite astonishing. Despite a reputed aversion to change, the Academy appears to have rushed wholeheartedly into the information age, at least in its adoption of learning management systems. The reasons for such widespread adoption are manifold, but in many cases, the initial impetus at institutional level was an urge towards financial reward. However, the high-profile failures of many early ventures, coupled with unrealistic expectations surrounding the transformative natures of the technologies themselves, quickly shattered any dreams of a quick route to riches in the online education market.[a] What did become clear fairly quickly was that, despite the optimism of many, technology of itself lacked the power to change what happens in the lecture hall: the culture of learning, teaching and academic practice in essence continues to remain largely unchanged, and the main impact of technology has been to provide an efficient way to distribute content (Zemsky & Massey, 2004).

But despite some early failures it appears that technology is beginning to demonstrate some beneficial effects on student learning (JISC, 2008; Harvard University, 2008). New approaches to the educational process itself might help to change the situation. Traditional theories of learning treat it as a concealed and inferred process, something that "takes place inside the learner and only inside the learner" (Simon, 2001, p. 210). The theory behind learning technologies is broadly labeled constructivism which, according to Thorpe (2002), is probably the most widely recognized social position within eLearning research, having come to dominate the field over the last decades. Precisely what is understood by constructivism, however, can vary widely across disciplines, across campuses and across borders.[b] The unpredicted emergence of blended learning as the dominant model on campus can be read, in many ways, as a result of the underlying confusion between theory and practice. Nonetheless, it may well be that we are now witnessing the beginning of a transition to a newer, more expanded educational space as a result of the impact of learning technologies, and with it the opportunity to revisit educational theories and pedagogical practices. Attempts continue at various levels within the Academy to realign educational practice to constructivist theory, highlighting the desire to change higher education from mere knowledge transfer to a place where academic staff and students work together to create knowledge. And while there is general agreement among educational theorists that a shift to a constructivist model of practice will ultimately be good for students, there appears to have been – at least in the European context – a lack of informed critical debate around the reasons for adopting the constructivist model, as well as the issues concerning the wider impact which technology may have on the educational process itself. Moving effectively to a constructivist environment will necessitate a major shift in thinking about the purpose and practice of higher education, not in itself an unwelcome aspiration.

In many ways, constructivism could have a radical impact on learning and teaching practices, but its impact is not necessarily consistent with answering the perceived needs of the knowledge economy, which at its most basic level favors knowledge transfer and a culture of compliance (Jamieson & Naidoo, 2004). Additionally, while support among theorists for constructivism as promoting active learning continues, those of a philosophic bent often stand back in horror at the ultimate relativism that constructivism embodies (Boghossian, 2006). One way or another, enthusiasm for constructivism is widespread among educators, and an opportunity to put theory into practice was presented to the present authors in 2006 in a European-funded project to develop *ab initio* an international, inter-institutional Master's degree program in early childhood education and care.[c] While our experience was in the main positive, it instilled a realization that the challenges facing higher education adopting constructivist models of education will be large ones, and in many cases the Academy will have to relinquish the tacit control which it continues to exert inside and outside the lecture theatre.

THE EUROPEAN CONTEXT

The context within which we in Europe operate is one of change following the influence of the 1999 Bologna Accord impacting on programme development and design in higher education (Conference of European Ministers of Education, 1999). Europe has traditionally boasted a multiplicity of education systems and degrees, and with it policies of mutual non-recognition of qualifications. The aim of the Accord, which 46 countries have now adopted, is to create a European common market in higher education, with mobility of staff, researchers and students, and a single coherent system with "European dimensions" – borderless education projects and common systems of categorization and nomenclature. A credit system is part of the move to make European higher education more transferable. In Ireland the response is well advanced, primarily through the work of the National Qualifications Authority of Ireland (NQAI) and the influence of the National Qualifications Framework (NQAI, 2003). The advantages identified for students include

- transparency of course structures
- flexibility of curriculum
- equality of workload
- recognition of achievement rather than failure
- academic recognition for study abroad
- enhanced benefits of mobility

In response to Bologna, HEIs are slowly moving away from a system of teacher-driven provision, and towards a student-centered concept of higher education which embraces features of constructivism. The reforms are laying the foundations for a credit system relating two key elements – student workload and learning outcomes. Despite increased use of credits, however, a majority of institutions continue to rely on traditional end-of-year examinations to assess student knowledge. As the assessment of learning outcomes is required for credits to be awarded, this continuing reliance on summative assessment raises questions about how profoundly programs have been restructured when introducing the system. In many instances, the reaction has been mechanical rather than philosophical in import, and former programs have simply been repackaged to fit the new mould.

The use of a European framework of qualifications and its translation into national frameworks in some countries in Europe has generated a debate around the use of competencies, skills and know-how in curriculum design. In some cases this had led to a purely reductionist approach, and the creation of a stringent set of categories and lists of active verbs, usually based around Bloom's taxonomy of learning, to be used to describe the learners' expected outcome. Quality assurance procedures and processes often appear to be dictating the description of learning, and resistance by certain elements of the Academy to a learning-outcomes approach stems from a misconception about the purpose and intent of the Accord. Understanding and integrating the use of a learning-outcomes based approach remains a key medium-term challenge (Crosier, Purser & Smidt, 2007).

A parallel action taken by many institutes of higher education in response to Bologna was to embrace learning technologies. In English-speaking countries the learning management system became known as the virtual learning environment (VLE), and hopes were high that the technology would drive change. The fact that here, as elsewhere, learning management systems have been primarily successful as an efficient mechanism for the distribution of content may be tentatively explained by program design at undergraduate level. Such programs tend to be content-driven, responding to a student desire to learn the techniques required for passing assessments. In our experience, undergraduate students generally *expect* to be taught rather than to engage in a form of do-it-yourself education which constructivism at its crudest appears to be. Meeting student expectations is high on the list of priorities of student recruitment and retention policies, and so it is probable that a radical change away from the didactic culture widely embodied in our primary and secondary schools will not prove the most successful, and could lead to reduced revenues from tuition fees, never welcome in the economic discourse of educational administration. Academic development programs in learning and teaching have a tendency to present theory in a reductionist manner as a set of behavioral rules: learning outcomes are defined, assessments are mapped to these expected outcomes and students complete the loop. Since they supply students with the necessary content that they must learn in order to be successful in the inevitable assessment, the widespread use of learning management systems as distribution mechanisms for course materials is understandable. Learning management systems have been very successful in enabling content to be delivered to students more effectively, and student reaction to the accessibility of such information continues to be positive. The underlying concept of education as knowledge transfer has not changed. Such shallow understanding of reconsidering curriculum and practice on the basis of learning outcomes has led to an underlying epistemological confusion between theory, research and practice, and certainly has not capitalized on the vision of the Bologna Accord (Crosier, Purser & Smidt, 2007). It seems fair to say that HEIs and their staff are still at the early stages of realizing the potential of EU reforms (Witte, 2008).

A CONSTRUCTIVIST APPROACH TO PROGRAM DESIGN

The online graduate program that provides the context for this paper is in the field of early childhood education. This field, from its roots in the late eighteenth century, has been one of cross-national development. Within the European Union (EU) ensuring broad access to early education across member states is high on the political agenda, providing a growing requirement for a strong European research perspective (Woodhead 2006; OECD 2001; OECD 2006). Since the 1990s there has been an emerging discourse on equal opportunities for all EU-citizens, which has linked the debate to issues of the economy and the labor market as well as to issues of children's rights and citizenship. From the perspective of research and professional development, such commitments imply a need to explore ways of enabling young children's active participation in the process of early education itself.

Concurrent with substantial investment in the early childhood education and care (ECEC) systems throughout the EU has been a renewed focus on professional development and a review of educational approaches. Within this global attention to developing early childhood services, there is a danger that the rich European heritage in early education may get lost in local-policy contexts due to the dominance of economic and labor-market approaches underpinning provision and the absence of a forum to further develop the European dimension in the practice field. In 2004 a team of six academics from Germany, Ireland, Malta, Norway, Scotland and Sweden identified gaps between the political agenda and the actions that are being taken towards research, shared learning and development in the sector. With particular interest in professional development they came together for the purpose of designing a Masters program with a large underlying online component under the curriculum innovation fund of the EU Socrates scheme.[d]

Although there is evidence that the quality of the early-years workforce is the crucial point for the quality of provision, the efforts to achieve a highly educated workforce have varied across Europe. From team discussions and research the need for an integrated European approach to ECEC was identified, especially with regard to its practices, theories, critical themes and research at a post-graduate level. The project was designed to contribute to achieving the goal of a unified teaching, learning and research environment and as such both challenge and strengthen the Bologna process.

When submitting the original proposal for consideration by the EU, the team noted that curriculum development and implementation would be committed to the inseparable nature of learning and teaching. From a pedagogical perspective, it was proposed that this would be achieved by involving program participants (students) as researchers in their own right. They would systematically become involved in a process of co-creation and be encouraged to take responsibility for their individual and joint process of professional and personal development as well as for the development of the program itself. In addition the team recognized that they too would be learners in the process and participants in the extension of knowledge. Key features of the approach included

- applying the principle of "sustained shared thinking" on all levels of the program
- creating a learning environment committed to research and professional reflection
- making extended use of information technology and providing an eLearning environment
- agreeing upon a professional code of ethics in ECEC

The course design was influenced by a number of mutually agreed, informing and integrating principles. These include recognition of children's rights; issues of diversity, gender, equality; historical and comparative dimensions; practice understandings and transition issues. The course was designed with eight core modules of equal credit value and a dissertation. Within each subject there was scope for presentations, critical reflection and academic writing as outlined in Figure 1.

Shared contribution to, and ownership of the program across national groups was expected to bring difference and disagreement as well as shared directions. A hallmark of the program was the way in which students were expected to critically position themselves in relation to the discourses presented.

Working within the current EU framework, and cognizant of the constraints and limitations as well as the potential of the structures, the team devised some headline program learning outcomes to be achieved. These learning outcomes and the associated competencies to be acquired by the Masters students were very broad and included

- professional reflection
- autonomy regarding academic decisions and discursive positionings
- an enriched perspective on children and childhoods
- a higher level of academic understanding of the professions of ECEC across Europe across time and place
- a proficiency in critical analysis and research
- increased personal articulation, in both spoken and written language, in home languages and in English
- abilities regarding the initiations of changing professional practices, policies and research

Curriculum was structured so that each of the subject areas was presented to the students

Figure 1. Matrix of the overall program structure

SUBJECT AREAS & UNDERPINNING THEMES	Themes	Practical aspects	Children's rights	Diversity, equality, and inclusion	Comparative Issues	Historical issues	Transitions
Core Modules (80 ECTS)							
Aesthetics of children's culture			√	√	√	√	√
Addressing diversity & equality		√	√	√√	√	√	√
Perspectives on childhoods & children in contemporary Europe		√	√√	√	√	√	√
Play & learning		√	√	√	√	√	√
Professional Reflections		√	√	√	√	√	√
Questioning curriculum & critical issues		√	√	√	√	√	√
Research Methodologies		√	√	√	√	√	√
The evolutions of early childhood research, practices & policies		√	√	√	√	√√	√√
Thesis (40 ECTS)							

as an independent module, with short written summaries of the indicative content, learning outcomes, assignments, assessment criteria and a list of recommended texts (articles, chapters and books). The integrity of the program was maintained by the active involvement of all the team – to differing degrees, across all modules. The program structure included core modules only and two intensive study visits. The democratic decision-making process allowed the team to consider the specific learning outcomes and competencies needed, and match leadership of each part of the course to academic specialization and availability of staff from each of the participating countries. A general policy was that theories would relate to practices, and that the research practices would take account of contemporary policy developments and political emphases within early childhood education.

Online collaborative learning is often cited as an example of constructivist pedagogy in action, whereby the learners through computer-mediated communication engage in discussions, examine case studies, share experiences and ultimately negotiate and construct meaning/knowledge together. Central to the project was the asynchronous on-line forum, set up for students and staff to work through the modules, present ideas, experiment with issues and technology, write early drafts and get feedback on other people's thinking.[e] Emphasis was placed on the use of the communication tools as a means of exchange and dialogue between peers and students with their tutor, on the understanding that online collaborative learning is underpinned by an essentially constructivist approach to the creation of knowledge through social collaboration and interaction. The online cooperation/collaboration opens up a new space to negotiate learning between peers and teacher, and the power imbalance inherent to any teaching situation can be offset to some degree by a community of learners supporting one another. Cowie (1992, p.158) describes the benefits of cooperative learning as opening up "the possibility of a different balance of power in the classroom where pupils can explore their own meanings and may challenge those of others in a supportive community of peers." Yates (1997) goes further to suggest that the benefits of such cooperation go far beyond what is supposedly measured by learning outcomes and can be related to a "democratic theory" of social and democratic development of the individual: Learning is seen therefore as a fundamental social activity. It was clear that it would be necessary to move away from course models where the VLE is used primarily to distribute course notes and handouts, and a conscious effort was made to mirror the dialogic structure of the learning process in the exchanges and tasks. The course design was therefore developed incorporating online communication tools as well as access to reading materials and presentations. eLearning theorists and practitioners alike seek to incorporate into their research and practice the new means of communication so favored by the digital natives. Social networking systems are beginning to come to the fore in learning communities, whether they be part of institutionally based learning management systems or other tools provided by companies such as Facebook or Second Life (Batson, 2008). There has been among theorists of technology-enhanced learning an opportunity to observe the process of exchange and dialogue online, in which the process of new meaning making and learning can be observed directly. Research into online learning has the advantage of studying learning in settings in which learning is observable and accountably embedded in collaborative activity. One concern, therefore, is with the unfolding process of meaning-making within these settings rather than with learning outcomes (Koschmann, 2001, p. 19).

There are a number of frameworks or models available to research into technology-enhanced learning, one of the most popular being the stepped model of online interaction described by Gilly Salmon (2000; 2002), the different phases of online interaction leading to the goal of independent learning and critical thinking. In this model the eModerator disengages little by little from the process, forcing the learner to actively engage or actively struggle. The dialogue with the eModerator is surpassed by a dialogue with the self: Self-supervision and critical thinking are the highest level of this framework. However, for such dialogue to be of benefit to the participants it has to be an informed one – for learning to take place the student needs to come to the process with their homework done. The construction, therefore, can be seen to take place at different levels of dialogue: Individual internal dialogue of meaning-making, the dialogue through exchange with the other, and the guided or instructed dialogue with the teacher. Whether this informed dialogue can be successful at undergraduate level raises issues about the students' ability to engage at a high level with the learning process. It is hence questionable whether the models of eModeration widely advocated at the moment are suitable for students who do not expect a high level engagement in the learning process. In practice, when the undergraduate student expectation of attending lectures and receiving copies of course materials presented in the lecture hall is not met, confusion and dissatisfaction can abound, along with the perception that the lecturer involved is shirking responsibilities. Complaints to the lecturer and to program and department heads have been the result of such "experimental" practices. At postgraduate level the student is expected to engage at a high level of meta-cognition to develop the necessary critical thinking skills to sustain an informed dialogue. Perhaps it is not surprising that our experiences have led us to adopt graduate-level work as a starting point in which we could pilot and implement what we believed to be a constructivist model of education. Unconstrained by a narrowly prescribed curriculum, the construction of knowledge at graduate level becomes more visible and the educational experience itself becomes much more creative.

For online collaborative learning to be successful, establishing social cohesion in the learning group is highly desirable; in the case of blended-learning modules this can be done by scheduling face to face sessions at the beginning and end of the module. Online presence can also be established by posting photographs and a short biography of the participants. The project team decided that students and tutors should meet face to face initially to ensure that the social cohesion of the group was fostered before entering into online activities as a group. A two-week induction course was carried out in Malta at the beginning of the program. Once the induction of the group had taken place successfully, it was argued, a community of learners with a distinct identity would exist and the online activities which would follow would have a better chance of success. It was essential, therefore, that the online element of the course was designed to sustain the community of learners and maintain an integrated learning experience across all modules. While the participants had differing linguistic backgrounds it was agreed that the principal language of communication for the online activities would be English. The very nature of online communication, lacking pragmatic linguistic elements, can be a cause for misunderstandings and unintended implied meanings. It was therefore seen as necessary to allow space for the negotiation of meaning during the different online interactions. It has also been demonstrated (Thorne, 2003) that in supported online collaborative learning, the speed of interaction and computer skills also need to be taken into account when designing the synchronous discussion sessions with online participants. The scrolling and speed of typing can be a problem for participants particularly if the group of students is a large one, and the result can be an online cacophony where are number of "voices" intermingle. The natural dialogue turn-taking sequence can be interrupted by screen renewals and participants entering the chat room space; the means of online communication can hinder the flow of communication or indeed lead to missed communication and miscommunication. Learning in this manner of dialogue, it would appear, needs to be rule bound. Structuring the dialogue was therefore a central element to the design of the online element of the modules.

The module designed and managed by the present authors was the first fully online module of the program following the face-to-face induction. Entitled "Perspectives on childhoods and children in contemporary Europe," it took place over a six-week period in the final quarter of 2007, with 29 students/participants from six universities throughout Europe, all professionals working in the field of early childhood education. Each participant was expected to take part in asynchronous discussion activities related to four unassessed tasks, and a final graded assignment was issued in December. In order to introduce students to this pedagogical approach and to structure the overall learning process for the module a study flow-chart was developed which captured visually the iterative nature of knowledge construction, as outlined in Figure 2.

Participants were also expected to reflect in their online journals on their role as professionals and how they contributed to the lived childhoods of the children they worked with. Each of the four tasks was designed to lead and prepare the participant for the final assignment. The structure of the discussion groups was designed to allow for discussion among peers, and the result of the discussions was a joint posting to the forum. The

Figure 2. Perspectives on Childhood and Children in Contemporary Europe. A suggested model for study on this module to help students weave together understandings from each input and task.

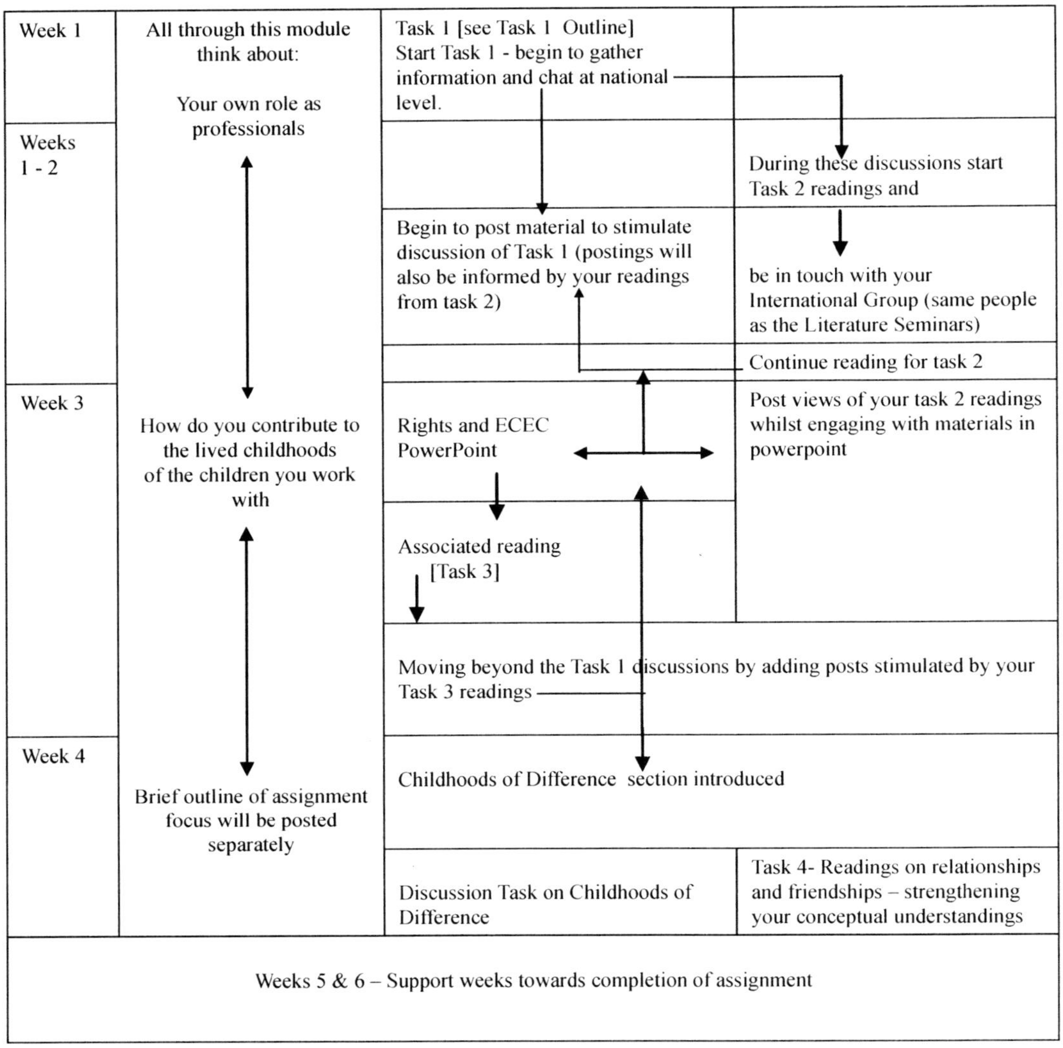

tasks were therefore not individually assessed by the course tutor, but because they were group postings were peer assessed. In addition to asynchronous online discussions, the participants were also given the opportunity to have a synchronous chat with the module leader once a week for two hours. The live interaction on a weekly basis with the tutor, acting variously as informant, critic and prompter, was also designed to provide a glimpse at how the social interaction was construed (Roxanne-Hiltz & Goldman, 2005, p.21).

CONSTRUCTING IDENTITY

The logs generated by these interactions, both synchronous and asynchronous, indicate the nature of the interaction on the discussion boards and in the chat room. As is typical with online interactions, the absence of face-to-face interaction leads to an emphasis on establishing other forms of presence.[f] References were made in the discussion forums to the context of the interaction, and the physical space of the speaker was a point of interest in numerous conversation openings. The geographic dispersion of the participants was also frequently commented upon. The visible unfolding of meaning can be quite hard to grasp. By structuring the interaction in a task-based fashion the participants were quickly able to come to contact with and to engage with the relevant principles, theories and research in the field of early childhood education. However, one must ask the question which is a central preoccupation of the constructivist approach: "What are the means by which knowledge is constructed, legitimated and circulated?" (Conole & Oliver, 2007, p.22). The discussions logs would appear, on first reflection, to be the places where traces of this construction should be found. However, interestingly, as has happened to a large extent on other online courses, the participants choose other unobserved means of communication to engage in dialogue with one another, outside the formal learning environment. In fact it can be seen clearly that the participants on the module used the discussion threads to demonstrate evidence of their learning. Although these postings were not formally assessed, the dialogue between peers was conscious of the learning management system as the panopticon, with its third eye/I looking in from the outside for evidence of learning.

There is, therefore, an implicit triad of communication between the interlocutors, the peer-to-peer implicitly acknowledging the presence of the observer. This could be explained by the very nature of the context of the discussion in the formal educational space of the module. Students circumvented the official channel of educational discursive activity by other preferred means: Reference is made to discussions that took place elsewhere, on instant messaging or Skype or text messaging. The preference for other means of communication could, of course, have a number of different explanations, not least of which might be a preference for a more immediate means of communication. However, the construction has taken place outside as well as within the view of the virtual learning environment, and therefore seems also to be happening outside the control of the formal online learning space. For educators the natural inclination is towards control of student learning in order to quantify and grade it. But to be successful in the constructivist world which the information age offers we must be willing to relinquish a large amount of the power which traditionally lies with the instructor, enabling students to reach beyond mere course content into a more creative way of exploring, exchanging and constructing knowledge. We can provide frameworks within which they can work and structures from which they can hang their ideas, and ultimately we must assess their performance. But the balance of power must shift, and this shift must be acknowledged by both student and lecturer. Just where that balance should lie is a moot point, which perhaps is where tensions among versions of constructivism reside.

CONCLUSION

Constructivism is now regularly cited as a means to foster active learning and to encourage the organic creation of knowledge, both of which are lauded by official policy at institutional and government levels. Information technologies and eLearning are the conduit by which constructivism is becoming accepted as the norm in our universities and schools. Our experience has convinced us that constructivism offers many opportunities for higher education, but that there is much to do in order to embrace and cultivate an atmosphere of constructivist learning. In the first instance, educators and students alike must make themselves aware of what it is we are dealing with. The largely uncritical acceptance of constructivism in academic circles belies the radical departure from the authority of the curriculum and control of learning traditionally exerted by the Academy. If, as has been suggested, *Wikipedia* – perhaps the most visible face of constructivism in contemporary practice – is as good a source of information as established primary sources, the fallout for education is profound (Giles, 2005). Philosophers such as Paul Boghossian (2006) have argued that acceptance of constructivism is ultimately an acceptance of the non-objective nature of knowledge, and therefore it would not necessarily seem to be the best theory to underpin accepted educational paradigms and establishments. But such views continue to be in the minority.

Control is the legacy of educational practice, from manipulation of the masses and the preservation of an intellectual hierarchy, to quality assurance and the commercialism of education in today's online universities. The locus of control will certainly shift in the future, but it appears unlikely that it will be lost to the gatekeepers of higher education. In an era when information on almost any subject imaginable is available within a few keystrokes, and leading universities make the content of their programs freely available on the web, it is obviously a mistake to believe that education is content-driven. Technology has clearly demonstrated that education is much more than content. In Europe, we need to move beyond the idea of learning outcomes as being about the imparting of knowledge and towards the intellectual development of the person who will achieve those outcomes. Rather than prescribing content in this era of content overload, academics will need to define parameters and to foster information literacy among students to ensure that the construction of knowledge occurs within those confines. However, such a sea change will not occur overnight.

The realization of widely adopted constructivist models of education, if and when it unfolds, will probably happen very slowly. Historically, the advent of mass education in the nineteenth century had as much to do with controlling and readying people for work in the factories of the industrialists as it did with enlightening and liberating citizens. The pressure on academia to produce workers for the knowledge economy is having a profound impact on how universities organize and conduct themselves. The debate concerning the nature of higher education itself in the information age is beginning to become more visible in the haze of everyday academic discourse. Ultimately, we believe that our role as academics and intellectuals is to cause people to think and to question, as well as to find truth: the process is as important as the end product achieved. We should start by thinking more about the wider implications of constructivism and the long-term effects of its adoption by the Academy. Constructivism will in all likelihood benefit our students but may in the process change higher education forever.

REFERENCES

Ansell, B. W. (2008). Traders, Teachers and Tyrants; Democracy, Globalization and Public Investment in Education. *International Organization, 62*(2), 289–322.

Batson, T. (2008). Is Web 2.0 "Designed for Education?" *Campus Technology*, March 5, 2008. Retrieved May 28, 2008, from http://www.campustechnology.com/article.aspx?aid=59341

Boghossian, P. (2006). *Fear of Knowledge: Against Relativism and Constructivism*. Oxford: Clarendon Press.

Cannella, G. (1998). Early Childhood Education: a call for the construction of revolutionary images. In W. Pinar (Ed.), *Curriculum: Towards new identities* (pp. 157-184). New York: Garland.

Conference of European Ministers of Education (1999). *Bologna Declaration. 19 June 1999.* Retrieved May 28, 2008, from http://www.eua.be/eua/jsp/en/upload/OFFDOC_BP_bologna_declaration.1068714825768.pdf

Conole, G., & Oliver, M. (Eds.) (2007). *Contemporary Perspectives in e-Learning Research:*

themes, methods, and impact on practice. Abingdon: Routledge.

Cowie, H. (1992). Peer commentaries on Keith Topping's overview of cooperative learning and peer tutoring. *The Psychologist: Bulletin of British Psychology Society, 5*, 158-9.

Crosier, D., Purser, L., & Smidt, H. (2007). *Trends V: Universities Shaping the European Higher Education Area*. Brussels: European Universities Association.

Giles, J. (2005). Special Report: Internet encyclopaedias go head to head. *Nature, 438*, 900-901 (15 December 2005).

Harvard University (2008). *Instructional Technology Survey, January–February 2008*. Department of Romance Languages and Literatures. Retrieved October 15, 2008 from http://blogs.law.harvard.edu/digitalnatives/files/2008/03/instructionaltechnologysurvey.pdf

Jamieson, I., & Naidoo, R. (2004). How the market economy is undermining HE performance. *Management in Education, 18*(2004), 13–16.

JISC (Joint Information Services Committee, UK), (2008). *Exploring Tangible Benefits of eLearning: Does investment yield interest?* Northumbria: JISC infoNet publication. Retrieved October 15, 2008, from http://www.jiscinfonet.ac.uk/publications/publications/info/tangible-benefits-publication

Koschmann, T. (2001). Revisiting the paradigms of instructional technology. In G. M. Kennedy, C. McNaught & T. Petrovic (Eds.), *Meeting at the Crossroads. Proceedings of the 18th Annual Conference of the Australian Society for Computers in Learning in Tertiary Education* (pp. 15-22). Melbourne: Biomedical Unit, The University of Melbourne. Retrieved October 15, 2008, from http://www.ascilite.org.au/images/papers/conferences/melbourne01/pdf/papers/REVISITING%20THE%20PARADIGMS%20OF%20INSTRUCTIONAL%20TECHNOLOGY.pdf.

Livingston, D. (2003, November). *Reclaiming Early Childhood Teacher Education: A Critical Constructivist Approach*. Paper presented at the Annual Meeting of the Southeastern Regional Association of Teacher Educators, Savannah, GA.

NQAI (National Qualifications Authority of Ireland) (2003). *The National Framework of Qualifications – An Overview*. Dublin: NQAI. Retrieved October 15, 2008, from http://www.nqai.ie/docs/publications/13.pdf

OECD (Organization for Economic Co-Operation and Development) (2001). *Starting Strong: Early childhood education and care*. Paris, OECD.

OECD (Organization for Economic Co-Operation and Development) (2006). *Starting Strong II: Early childhood education and care*. Paris, OECD.

Phillips, D. (1995). The good, the bad and the ugly: The many faces of constructivism. *Educational Researcher, 24*(7), 5-12.

Roxanne-Hiltz, S., & Goldman, R. (Eds.) (2005). *Learning Together Online: Research on Asynchronous Learning Networks*. Mahwah, NJ: Lawrence Erlbaum Associates.

Salmon, G. (2000). *E-moderating: The Key to Teaching and Learning Online*. London: Kogan-Page.

Salmon, G. (2002). *E-tivities: The Key Active Online Learning*. London: Kogan-Page.

Simon, H. (2001). Learning to research about learning. In S. Carver & D. Klahr (Eds.), *Cognition and Instruction: Twenty-fice years of Progress* (pp. 205-26). Mahwah, NJ: Lawernce Erlbaum Associates.

Thorne, S. L. (2003). Artifacts and cultures-of-use in intercultural communication. *Language Learning and Technology, 7*(2), 38-67.

Thorpe, M. (2002). From independant learning to collaborative learning: new communities of practice in open, distance and distributed learning. In M. Lea and K. Nicoll (Eds.), *Distributed Learning: Social and Cultural Approached to Practice* (pp. 131-52). London: RoutledgeFalmer.

Witte, J. (2008). Aspired convergence, cherished diversity: Dealing with the contradictions of Bologna. *Tertiary Education and Management, 14*(2), 81–93.

Woodhead, M. (2006). Changing perspectives on early childhood: theory, research and policy, background paper for UNESCO EFA Global Monitoring Report 2007. *International Journal of Equity and Innovation in Early Childhood, 4*(2), 5–48.

Yates, S. (1997). Gender, identity and CMC. *Journal of Computer Assisted Learning, 13*(4), 281-90.

Zemsky, R., & Massey, W. F. (2004). *Thwarted Innovation: What Happened to eLearning and Why.* Philadelphia, PA: The Learning Alliance, University of Pennsylvania.

ENDNOTES

[a] For example, Columbia University's Fathom.com folded in 2003, and the UK's eUniversity followed suit a year later, both with reputedly large financial losses.

[b] Just what Piaget meant by construction is itself a matter of debate, with possible semantic slippage in the translation: *constuire* in French can be translated by "construct" or "construe", both containing the notion of fabrication.

[c] Coincidently, constructivism is much debated in the field of early childhood education, where it has been likened to a secular religion with many sects (Phillips, 1995). The argument has been made that focusing solely on the psychological interpretation of constructivism in early childhood teacher education has reduced pedagogy to a series of simple teaching techniques thus compromising the richness of the early-years curriculum (Livingstone, 2003) and it has been suggested that constructivism should be reclaimed as a theory for radically reforming democratic change in early education (Cannella, 1998).

[d] See http://www.em-ec.eu and http://ec.europa.eu/education/programmes/socrates/minerva/index_en.html (both accessed October 15, 2008).

[e] The program utilized an established learning management system/virtual learning environment for the delivery of the modules.

[f] An interesting comparison could be made here with the notion of absence and presence in Jacques Derrida's work on speech as absence, as opposed to presumed presence and absence of the text.

Chapter III
The Theory and Practice of Teaching with Technology in Today's Colleges and Universities

Barbara de la Harpe
RMIT University, Australia

Fiona Peterson
RMIT University, Australia

ABSTRACT

There is a strong move worldwide for a constructivist theory to underpin the way teaching and learning are viewed in today's colleges and universities. In this chapter, the authors explore the interconnections (or not) between constructivist theory and mainstream university teachers' practice, and their use of educational technology. They suggest that at the heart of ongoing transformation of teaching with technology is a supportive context and inspirational leadership that (re)engages the academic heartland with constructivist learning theory, so that teaching with technology is constructivist based. Technology can thus serve as a powerful catalyst for reinvigorating the inter-relationships between theory and practice.

INTRODUCTION

There have been a number of shifts in colleges and universities over the years. The most obvious shift has seen these institutions move from a provider-driven, industrial-age model to a learner-centric, competency-based, information-age one (Dolence & Norris, 1995). The traditional model of higher education, which was supply led and typical of the industrial age, is giving way to a model that is demand driven and more suited to a knowledge-based economy.

In this model, Derek Bok, L. Dee Fink and John Tagg suggest that teaching franchises become learning franchises. Institutions, rather than being provider driven with a set time for learning, become customer focused with learning individualized. Taking set times out for education is replaced with just-in-time and just-for-me, anywhere-anytime learning. Traditional courses, degrees and academic calendars become modularized learning experiences based on learner needs. A technology push becomes a learning pull. Productivity is measured in terms of meeting the needs of different types of learners, rather than in terms of the numbers of students enrolled, funding gained, student completions, or grades achieved. (Bok, 2006; Fink, 2003; Tagg, 2003).

According to Newman (in Fink, 2003), the four driving forces behind these changes include the advent of new technologies and new modes of delivery, notably online; increased competition by diverse and new providers in the market place; the globalization of higher education; and new kinds of students who want their education to be convenient, inexpensive, and when and where they want it. As a result of these forces in a competitive market, Newman predicts that the traditional instructivist or teacher-centered approach to teaching will continue to give way to a constructivist or learner-centered one.

In this chapter we reflect on a constructivist theory of learning and its role in shaping the practice of teaching with technology in higher education institutions worldwide. We explore whether the introduction of new technologies has been a catalyst for reflecting on "old" pedagogy, and for prompting a transformation of teaching practice. Given all the attention paid to constructivism in teaching and learning, we wonder if constructivist theory and its implications for practice are understood on the ground.

BACKGROUND

The term "learner-centered teaching" now pervades institutional strategic documents, as well as extensive literature and research on learning and teaching. In addition, much effort is devoted to trying to get faculty, who are at the coalface in higher education institutions around the world, to view teaching as constructivist or learner-centered.

Constructivism vs. Instructivism

In a constructivist or learner-centered model, faculty are seen as facilitators and co-constructors of knowledge; and students are active, self-regulated and self-directed learners, with capacity to search, select and synthesize information and to construct their own knowledge and understanding (Tagg, 2003).

In constructivism, learning is an active process with learners seen as architects building their own knowledge structures, rather than absorbing knowledge transmitted by the instructor. In the process learners restructure prior knowledge and make connections with what is already known and what is learned, depending on their view of the world. Knowledge is seen as changing, and what is learned may not always be what is intended (Biggs & Tang, 2007; Merriam & Brockett, 2007; Tagg, 2003). In social constructivism, interaction and collaboration between learners are seen as fundamental to learning, with knowledge being co-constructed through opportunities for peer interaction and discussion (Light & Cox, 2006; Woolfolk, 2008).

Teaching that is learner centered places the attention fairly and squarely on student learning and on teaching that is aimed at facilitating learning (Tagg, 2003; Weimer, 2002). When learners have learned something, they have a clear mental picture of it, are able to talk about it or explain it to another person, and can use it to solve new problems. Learning for understanding is better than simply memorizing facts, because facts may become obsolete, they tend to be rapidly forgotten, they do not help in problem solving, and they do not provide a sound basis for further learning.

In a learner-centered approach the instructor focuses on what the learner is learning, how they are learning it, the conditions under which learning is occurring, and whether learners are retaining and applying what they have learnt. It also includes the instructor focusing on their teaching and whether the approach and strategies they are using are facilitating learning, and equipping learners for future life and work. As Weimer (2002) points out, in learner-centered teaching the spotlight is on the students, that is, on what the students are doing, with the instructor focusing on selecting and adapting teaching strategies in support of this.

There are a number of key constructivist teaching approaches, including active learning, problem-based learning, enquiry learning, cooperative learning, and collaborative learning – all of which are underpinned by student-centered, self-directed and self-regulated learning (Prince, 2004). Based on a synthesis of Jonassen's (1999) "student-centered learning environments" and Fink's (2003) "significant learning experiences," the key characteristics of constructivist teaching approaches include:

- teachers focus on learners and their learning;
- teachers mentor, facilitate and support learners, helping them to (co)construct knowledge;
- teachers include a value add component that helps prepare students to contribute to society, life and work;
- students focus on their learning, being engaged and energized
- students are active, interactive and reflective knowledge (co)constructors; and
- students change in significant and lasting ways.

There is extensive empirical evidence that supports constructivist teaching methods. Constructivist teaching is believed to lead to increased understanding of knowledge, concepts and principles; mastery of skills; and the ability of learners to adapt and apply their learning in different settings. A number of authors (e.g. Bonwell & Eison; Felder; McKeachie) have shown that the use of constructivist methods is associated with improved attitudes, motivation, retention of material, thinking and writing skills (Prince, 2004).

In contrast, in an instructivist or teacher-centered model, faculty are presumed to be the fount of knowledge and are expected to transfer content to students who have limited or no knowledge in the area. In this model, learning focuses on the acquisition or accumulation of facts, skills, concepts, and strategies. There is a fixed body of knowledge to acquire and learning occurs when content is transmitted from the teacher to the learner. Instructors holding this view favor lecturing/direct instruction. They believe that teaching should be directed and controlled by them; content should be organized, linear and structured systematically; and teaching should focus on the content and on students memorizing and recalling it, above all else.

In Table 1, the characteristics of a constructivist learner-centered view of learning and teaching are compared and contrasted with an instructivist teacher-centered view, including assumptions about the learning process; the roles of faculty, students and peers; what the course and learning outcomes focus on; and the teaching and evaluation methods used.

Proponents of teacher-centered teaching see learner-centered teaching as "too unstructured and unsystematic!," while proponents of learner-centered teaching see teacher-centered teaching as "too rigid and teacher centered!" (Roblyer,

Table 1. Key characteristics of constructivist and instructivist views of learning

	View of learning	
	Instructivist (teacher-centered)	**Constructivist (learner-centered)**
Knowledge	Fixed body of knowledge to acquire, exists outside the learner	Changing body of knowledge, individually constructed, built on what learner brings. In social constructivism knowledge is constructed through interaction with others
Learning process	Acquisition of facts, skills, concepts, and strategies, occurs through the effective application of cognitive strategies	Active construction of personal knowledge and restructuring of prior knowledge through active learning. In social constructivism knowledge and values are collaboratively co-constructed
	View of teaching	
Role of faculty/ teacher	Mainly telling about and covering content knowledge, directing learners toward more "accurate" and complete knowledge; telling learners when they have it "wrong" and correcting their misconceptions; faculty knows "all"	Mainly guiding and facilitating learning; scaffolding; challenging thinking; listening for learner ideas, thinking, misconceptions, and helping them construct different interpretations; acting as a resource, model or helper; constructing knowledge with learners
Role of student/ learner	Passive recipient, organizes and reorganizes information on own, memorizes and recalls content, depends on instructions and being given information from others	Active participant in (co)constructing knowledge; active thinker, explainer, interpreter, questioner, and reflector; independent, self-directed lifelong learner; in social constructivism, a collaborator
Role of peers	Not necessary	In social constructivism, learning from and with others a fundamental part of process of knowledge construction
Focus of course	Knowledge presented in sequence, hierarchical curriculum	Knowledge, skills and attitudes developed flexibly, actively and interactively; spiral curriculum
Focus of learning outcomes	Knowledge accumulation, how much content can be recalled, what does the learner know	Knowledge, skills and attitudes, how has learner developed, grown, changed as a person, as well as how much/what content has been learned
Teaching methods	Lecturing, teacher demonstrations, individual practice, set reading	Punctuated lectures, interactive demonstrations/ discussions/presentations, group work/collaborative learning, set readings with follow-up discussion
Evaluation methods	Tests and examinations, multiple choice, short answer, essay, true/false; not negotiable, mainly summative	Projects, oral presentations, logs and journals, essays, writing samples, portfolios, simulations, performances, interviews; may be negotiated, mainly formative

Source: Adapted by permission from H. H. Marshall (1992, April), Reconceptualising learning for restructured schools. Paper presented at the Annual Meeting of the American Educational Research Association, April 1992 in Woolfolk, A. E. (1995), Educational Psychology (6th ed.). Boston: Allyn & Bacon. Table 7.5, p.278, with additions from Jonassen (2000), pp.89-93; Newby, Stepich, Lehman & Russell (2000), Table 1-1, pp.7, 225-231; and Roblyer (2003), Figure 3.2, p.54.

2003, p.53). While the two views are often presented as polar extremes, it has been shown that they can vary across a five-point continuum from a teacher-centered view that sees teaching as (1) imparting information or (2) transmitting structured knowledge, to a learner-centered one that sees teaching as (3) an interaction between the teacher and the student, (4) facilitating understanding, or (5) bringing about conceptual change/intellectual development (Kember, 1997; Kember & Kwan, 2000).

It is also now widely accepted that learner-centered views are more sophisticated than teacher-centered ones, with instructors who hold a learner-centered view more likely to have a deeper understanding and awareness of their own teaching and their students' learning (Entwistle, Skinner, Entwistle, & Orr, 2000; Prosser, Ramsden, Trigwell, & Martin, 2003).

Instructors who hold a learner-centered (constructivist) view of teaching are able to adapt their teaching approaches to different methods and modes in support of student learning, depending on the learner's situation and stage of learning. Thus, at times they may use instructivist approaches such as lectures or direct instruction when topics are new to learners and for teaching foundational skills; for example, in the beginning stages of reading, writing and mathematics (Westwood, 2008). This means that while their dominant view of learning and teaching is a constructivist one, they may also use instructivist strategies such as lectures and tests, as well as traditional face-to-face modes at appropriate times. Merriam and Brockett (2007) support faculty adopting such "...an eclectic position, choosing compatible aspects of different theories to explain and guide practice" (p.47), which epitomizes a highly sophisticated approach to teaching practice and exemplifies scholarly teaching.

On the other hand, those who hold a teacher-centered (instructivist) view of teaching are unable to move beyond seeing teaching as the delivery and presentation of content by the instructor and they are also more likely to assess frequently using tests and quizzes, whereas teachers adopting a learner-centered approach are likely to assess much more flexibly (Kember & Kwan, 2000).

In summary, a learner-centered view is underpinned by a constructivist epistemology that sees knowledge as being subjective and constructed by individual learners through participation in active and interactive learning activities, with construction influenced by prior knowledge, beliefs about learning, background, experiences and abilities. On the other hand, an instructivist or teacher-centered view is underpinned by an objectivist epistemology that sees knowledge as existing independently and learning as transferring knowledge to students' minds. While constructivism in its pure form privileges subjectivism, for many there are undisputed sui generis facts that are in a class of their own, existing independently of how others perceive or think of them at any time as well as existing independently of a learner's ability to comprehend them (Shafer-Landau, 2003). Thus, in constructivism it is the role of the learner to make personal sense of these facts, for the facts to exist for the learner. Unlike in instructivism, such facts cannot simply be transferred to the learner.

Above we have explored the shift to a learner-centered approach to teaching, and how constructivism underpins such a shift. Next, we explore how the introduction of new technologies has been seen as one of the drivers for the shift and has been a catalyst for re-visiting pedagogy in colleges and universities worldwide.

TECHNOLOGY IN PARALLEL

It is now widely accepted that technology has had a profound effect on the course of education. The impact and influence of technology has been driven by a number of forces, including the dramatic increase in the number and type of technological tools and resources available, as well as the societal shift to a knowledge economy (Roblyer, 2003). There are many voices in higher education circles loudly championing the role of technology in learning, highlighting the positive influence it has had on initiating conversations about learning and teaching and rethinking teaching practice.

Many Say Yes to Technology

There are many faculty who are optimistic about the use of technology and its potential for supporting learning, enhancing teaching, providing greater access and flexibility to learning, facilitating faculty communication and collaboration, delivering on student rights to education, and even revisiting teaching altogether.

A number suggest that technology provides new opportunities for creating exciting learning and teaching environments, offering as yet undiscovered and unimagined ways of creating, integrating, and disseminating knowledge (Gregorian, 2005). Technology is seen as "...offering the opportunity to do completely new things – and to move from modes of organization to modes of discovery and integration." (Rankin, 2007, p.1).

Technology is also seen as offering affordable access anytime from anywhere and thus is seen as profoundly increasing access for many who in previous times may have been unable to participate (Ivins, 2008; Moore, Sener & Fetzner, 2006). For example, the US eArmyU initiative, in collaboration with 29 colleges and universities, offers certificates through to masters degrees online asynchronously to over 57,000 mobile soldier students (Stotskopf & Moorash, 2005). Furthermore, technology can help those who require the flexibility of asynchronous learning to complete their studies, when time rather than distance is a barrier (Otte, 2007), with many students opting for this mode because it reduces their travel time (Cavanaugh, 2005).

Many recognize the ability of technology to enable greater and easier communication between faculty and students (Roblyer, 2003). The power and potential of networked technologies are seen to hold out especially great promise for connecting learners and enhancing learning, as "...the software becomes easier to use and more sophisticated, and the entry skills of the users better honed" (Salmon, 2001, p.97). Technology is also seen as enabling ongoing communication and participation no matter what the circumstances. For example, after Hurricane Katrina's devastating effects on New Orleans, technology enabled many students to stay connected and continue their learning (Schweber, 2008).

Some suggest technology goes even further by helping deliver on every student's right to education and achieving their full potential. In this way access to and use of educational technology is seen to support independent inquiry, collaborative inquiry and formative inquiry methods as learners develop expert knowledge independently (Lin, 2008).

There are also those who suggest that technology is a catalyst for discussions around learning and teaching, including web-based and other instructional approaches (Huber & Hutchings, 2005). Many believe that technology has been a magnet for transforming the way faculty think about learning and teaching, attracting and drawing together faculty in large numbers "...to trade insights from innovations in their own classrooms" (Huber & Hutchings, 2005, p.80). For example, the online professional teacher education and professional development program, Seminars on Science (SoS), a collaboration between the City University of New York and the American Museum of Natural History, connects teachers across the United States and internationally to research and resources on teaching science (Picciano & Steiner, 2008).

Technology is also heralded as having the power to alter fundamentally the way faculty work, allowing them to "work smarter" which will result in more opportunities and time to teach innovatively, as well as to undertake research (Moss, Fearnley-Sander & Hiller, 2001).

Others, like George Gilder (2000) in his book *Telecosm: How Infinite Bandwidth will Revolutionize Our World,* have even gone way beyond the conventional, envisaging technology as having the power to transform us into celestial beings or electronic angels capable of transcending the physical realm!

On the other hand, there are those who voice a word of caution.

Many are Cautious about Technology

A number of faculty do not accept the promise of technology, suggesting that technology is not necessarily enhancing learning and teaching practices in our colleges and universities.

Some question whether technology is indeed providing greater freedom and personal control or just "dazzling" and mindlessly consuming us. They suggest that it is reducing the need for personal as well as spontaneous interaction, resulting in isolation and dislocation from the real world (David Denby in Roblyer, 2003, p.ix). They caution that unless real attention is paid to enhancing "communication, connection, and collaboration" the introduction and use of technology is inadequate (Rankin, 2007, p.1). This raises concerns that universities may not have the motivation or capacity to ensure "social networking" as part of the learning experience (De Rossi, 2007, p.1).

Many are also questioning the financial viability and sustainability of online learning and whether universities have sufficient information around business strategies and models (Meyer, 2008; Vignare, Geith & Schiffman, 2006). Others fear that in the rush to adopt technologies in response to market-place pressures, universities will be distorted, becoming "a home depot of courseware…an academic superstore, a collection of courses marketed much like sinks and lumber" (Gregorian, 2005, p.94).

Similarly, there are those who raise issues around whether using technology is actually resulting in better learning and achievement for students, with mixed results reported (Steimberg, Ram, Nachmia & Eshel, 2006). There is a concern that further marginalization and inequalities are being created for the African American students by the "digital divide" (Booth, 2006; Rovai & Ponton, 2005). In addition, research is also showing that students themselves are not convinced that fully online is meeting their learning needs (Meyer, 2007; Moloney, Hickey, Bergin, Boccia, Polley & Riley, 2007).

The use of technology is seen by many faculty to be associated with extra work, and concerns are often raised about the suitability and functionality of the medium (notably LMSs) for certain disciplines and for some learners. Technology is seen as being unable to support the kinds of activities that many faculty want to implement, given the perceived limitations of some enterprise systems as well as cost imperatives. What is more, it may be felt that there is a lack of support and well developed policies for teaching with these systems (Hiltz, Shea & Kim, 2007; Puzziferro-Schnitzer & Kissinger, 2005).

Technology is perceived by many as disruptive, challenging faculty's traditional notions of teaching and learning and lacking in theoretical frameworks to guide informed and sound practice (Arbaugh, 2007; Swan, 2007). Questions are also raised about the roles and expectations of faculty; for example, "…how skilled should academic staff be in web page design and development…is it reasonable to expect them to become web publishers?" (Bruce & Goodall, 2001, p.112) and, if not, are they "…to be reduced to the role of…'content providers'?" (Gregorian, 2005, p.90). Many even question whether it is a faculty role to integrate technology into their teaching, perceiving it as an unrealistic expectation, and pointing out that some may not be sufficiently technologically savvy to do so.

Others point to a lack of a "…deep understanding of this expressive medium" on the part of most educators in order to use the medium to its full advantage in support of learning (Lever-Duffy, McDonald & Mizell, 2003, p.xiii). In fact, it is suggested that many faculty may need to rethink and transform their roles "…pedagogically, socially and technologically if they are to establish a more engaging and fruitful online environment for learning" (Liu, Bonk, Lee & Su, 2005, p.1).

These comments, in contrast to the "many say yes to technology" section above, paint a less than optimistic picture for technology being accepted and integral to learning and teaching in colleges and universities, with many raising issues and expressing caution about the use, viability and impact of technology.

So we may ask, if it is indeed the case that constructivist theory is informing the use of technology on the ground in our colleges and universities.

Transforming Teaching Practice with Technology

In our experience, faculty find it challenging to design learning environments regardless of mode or medium. Technology adds another layer of complexity to teaching, which is exacerbated if faculty do not have a fundamental understanding of learning theory, and instead all they have to rely on is their own experience. We have found that it is very difficult to find faculty who can point to the literature on teaching approaches that underpin effective learning. It is also very difficult to implement innovative strategies and to convince some faculty of their value, if they see that there is an increased workload involved (Bruce & Goodall, 2001).

When support is provided by the institution for advancing theoretical understanding and for technical innovation, criticism is often leveled at those in these support roles. Faculty may comment that while "...good educational theory is all very well..." it is not realistic in their teaching situation, and that they are being required to implement technology for institutional imperatives rather than for educational purposes (Webster & Murphy, 2001, p.120). Thus, even though those in supporting roles are simply implementing and not determining university directions, they may be perceived as interfering and ill informed, with a lack of real-world teaching experience, and may be "...wide open to (often misinformed) criticism concerning a perceived lack of awareness of the constraints to teaching in the 'real world'." (Webster & Murphy, 2001, pp.119-120).

Most faculty do not turn to the educational literature to inform their teaching, or to find answers to issues they are experiencing in their teaching practice (Prince, 2004), despite all encouragement to do so. A review of 406 articles across 17 higher education journals found that in the majority of cases there was no explicit mention of theoretical underpinning; if any engagement with theory was present, it was at best implicit. In the few cases where theoretical perspectives were explicit, the researchers were most likely to be in the areas of social sciences or academic development (Tight, 2004). When faculty are urged or prompted to turn to theory to inform their teaching practice, that is, to engage in scholarly teaching, it is met by some with cynicism and derision. It is not uncommon for "[v]eteran teachers [to] often smile wryly, when told to do this or that, whispering asides about another faddish pendulum swing, closing their classroom doors, quietly going about their business." (Wilson & Peterson, 2006, p.1).

Regarding this theory/practice divide, Bok (2006) states that "[t]hroughout undergraduate education, a great wall separates the world of research from the world of practice – even though the practitioners involved are professors, trained in research, who would seem ideally prepared to take full advantage of whatever findings empirical investigators have to offer." (p.9). Despite all the recent gains made in the research on learning, "...the truth is that most professors have gained relatively little from [this research]...[t]here is a tremendous gap between empirically validated theories and pedagogical practice." (Halpern, 2002, p.42).

The discussion so far has highlighted the fact that technology raises fundamental and difficult questions about the role of faculty and of the university itself. As shown, impediments to progress may include a lack of theoretical knowledge of learning and teaching, and resistance to engage with theory and support provided. Despite this, there are many examples of extraordinary practice in the literature, including countless individual and often whole university success stories of implementing and embracing exciting and innovative reforms to practice. When used appropriately, technology can support effective teaching and learning. However, technology must continue to be seen as a lever to help support the shift from a teacher-centered paradigm to a learning-centered one, in order to enhance teaching and learning. In this process, the goal must be to ensure that faculty view teaching as student centered and learning oriented, given that how faculty view teaching is a powerful determinant of how they will teach using technology and thus how their students learn (Entwistle et al., 2000; Prosser et al., 2003).

In much of the research to-date, technology-based approaches, including blended learning, are seen as holding great promise for facilitating this shift. Blended learning is the theoretically sound integration of face-to-face and online learning modes (Garrison & Kanuka, 2004; Garrison & Vaughan, 2008; Littlejohn & Pegler, 2007), the key being "sound", which requires knowledge of learning theory and technology, and how to use them both appropriately. This makes blended learning demanding of rigor, yet provides enormous potential for creating innovative designs of learning environments.

Blended learning is, therefore, emerging as the way of the future, with its promise to act as a catalyst for transforming learning and teaching. For example, Oliver and Herrington (2003) believe that technology can support constructivist learning approaches through the ability to enable the design and development of resource-based student-centered learning environments, and by enabling learning to be related to both context and practice. Blended learning is also shown to be associated with positive learning outcomes; in fact, it can contribute to "...a quantum shift...in terms of the nature and quality of the educational experience" (Oliver & Herrington, 2003, p.97). Garrison and Kanuka (2004, p.99) point out that "...blended learning does not represent more of the same...[and] is not just finding the right mix of technologies or increasing access to learning....[b]lended learning inherently is about rethinking and redesigning the teaching and learning relationship.".

So we ask, what is needed in mainstream academia to continue to encourage even greater uptake of technology, underpinned by sound constructivist theory?

FUTURE TRENDS

Faculty are at the heart of any attempts at teaching innovation and change in our colleges and universities, since faculty are the ones who "live" the curriculum with their students.

It is futile to expect thoughtful and sophisticated use of technology if faculty do not have a sound theoretical base for learning and teaching, or a willingness to expend effort and energy in developing their knowledge and skills in contemporary theories of learning, notably constructivism. An integral part of transformation in teaching practice involves faculty developing a deep understanding of underpinning theories of learning, and how they are applied in context to bring about effective student learning. This includes interrogating tacit assumptions and engaging with recent and emerging research and theory to "...make progress on that lifelong journey of becoming accomplished practitioners" (Wilson & Peterson, 2006, p.15). It also requires a supportive context, given the complexities of university life, pressures on time, multiple responsibilities and competing demands, which are not necessarily conducive to helping faculty change their beliefs. Thus, in order to bring about meaningful innovation in teaching and learning processes, extensive changes are needed to the ways in which learning is managed and organized, as well as a rethinking of the ways in which change is brought about in academic contexts and the role of leadership in this process (Ramsden, Prosser, Trigwell, & Martin, 2007; Tagg, 2003).

It is equally futile to expect thoughtful and sophisticated use of technology, if those in leadership positions and those charged with bringing about change in higher education do not have the same understanding of theories of learning, for leadership credibility. Academic leaders are pivotal in creating an environment which encourages and supports effective teaching. Garrison and Vaughan (2008) suggest that, only with sustained, decisive, collaborative leadership across the institution, will blended learning provide an opportunity to return to the fundamental values and practices of higher education. Dispelling myths about teaching in higher education, encouraging teamwork, modeling good practice, and providing recognition and reward for effective teaching are important strategies. Those in leadership positions should also be informed and guided by an understanding of learning and how people learn from research in the cognitive sciences (Halpern, 2002).

Of course – in addition to good leadership – highlights from selected technology case studies (Murphy et al., 2001) reveal a number of factors or aspects that are critical when new technologies are involved. They include planning and teamwork, innovating (including taking risks and pushing boundaries), developing skills in moderating, being open to criticism (listening to students), undertaking appropriate staff development and accessing institutional support, and coping with technological deficiency and failure.

However, at the center of successful transformations with technology is the ability for faculty to work cooperatively and collaboratively, and to be brave and bold (Murphy et al., 2001). In addition, the importance and centrality of faculty beliefs about teaching cannot be overlooked, since it is the beliefs about learning and teaching that faculty and those in leadership positions hold, their willingness to change these beliefs where required, and their attitudes towards engaging in transformational activities, which will determine whether deep, meaningful, lasting and impactful reform is possible (de la Harpe & Radloff, 2008).

Being confronted with a shift to a learning-centered paradigm can be very challenging. It is especially so for those who have grown up with and thus are operating in a content-centered paradigm, with its focus on an ever increasing amount of content to be covered. This may leave many feeling that time-honored traditional disciplinary ways of teaching are under attack, and that they are being pressured to abandon the aspect of their teaching that they believe matters most – the content.

A shift cannot be achieved overnight or by individual faculty "going it" alone (Bottino & Robotti, 2007). Somekh (2001) suggests that "[i]f we want higher education to make effective use of new technologies we need a much more integrated approach. Focusing on the delivery of content through discipline specific courseware is insufficient. There needs to be a whole-institution development plan…[d]epartmental structures also probably need to be redesigned to include interdisciplinary teams rather than academic cultural cabals." (Somekh, 2001, p.128).

Underpinning all transformational and strategic change management endeavors to encourage faculty to use technology in their teaching, is the changing of hearts and minds. To this end, Gardner (2006) provides salutary advice for those involved with this kind of transformation and change. He points out that "[n]ew evidence alters beliefs" (p.190). He adds that when engaging in wholesale reform "[t]he trick…is to accept the changes that will happen anyway, acknowledge that certain other changes may be impossible, and concentrate one's efforts on those changes of mind that are important, won't occur naturally, but can be achieved with sufficient effort and motivation." (p.209). He puts forward seven steps for leaders to follow when charged with

Figure 1. Key characteristics of instructivist and constructivist learning and teaching practices

Step 1, Research:	investigate, gather information about and "learn from and emulate other people's examples" (p.98)
Step 2, Resistances:	challenge "...directly the prevailing resistances – the ideas that are stale, erroneous or harmful" (p.99)
Step 3, Resources and Rewards:	use available resources "...such as an appropriate reward system, to initiate new policies and practices" (p.100)
Step 4, Representational Redescription:	describe the "...vision in a variety of ways and [provide] non-threatening opportunities for people to 'try on' [the] new vision..." and to reach as many people as possible (pp.100-101)
Step 5, Reason:	present a well-reasoned argument, rationale and analysis of the facts for the change, "...weighing the pros and cons..." (p.102)
Step 6, Resonance:	make sure that what is required to be done resonates with those involved, so that the "gut" feeling is right, and with leaders' "...actions [reflecting their] convictions..." (p.103)
Step 7, Real World Events:	link the desired change to real world or significant events, or other external drivers, given that "...[f]actors outside the control of the leader can have [a] huge impact..." (p.77)

bringing about institutional change (such as the integration of technology into teaching), as is shown in Figure 1.

Gardner (2006) notes that reform succeeds where there is weak resistance and is only possible when the resistance is overcome or minimized. Changes of mind are unlikely to occur where resistance is strong. In cases of strong resistance, it is better to direct efforts at understanding the resistance, and turn to the vast literature on organizational change to inform efforts at overcoming resistance.

At a personal level, when being confronted with the directive to use technology in teaching, Gardner (2006) points out that we can change our minds and embrace technology use "...either because we ourselves want to change them or because something happens in our mental life that warrants a change..." (p.173). All of the steps above can play a role in change of individual minds. He also wisely advises that the process of mind-changing can take considerable time, may not be long lasting, or even recognized as occurring; however, once a mind change is deeply embedded, it is likely to become integral to all thoughts, feelings and actions.

When leading change aimed at the integration of technology into teaching practice, we need to heed Roblyer's advice that the use of technology can make good teaching even better, with the caveat that,

...technology-using teachers never can be a force for improved education unless they are first and foremost informed, knowledgeable shapers of their craft. Before integrating technology into their teaching, educators must know a great deal, for example, why there are different views on appropriate teaching strategies, how societal factors and learning theories have shaped these views, and how each strategy can address differing needs (Roblyer, 2003, p.ix).

CONCLUSION

In this chapter, we have contrasted constructivist (learner-centered) and instructivist (teacher-centered) paradigms. We contend that, in order for technology use to go further than replicating and perpetuating "old pedagogy," faculty need to embrace constructivist theoretical approaches. The current repositioning of the "*e*" in e-l"*e*"arning, as exemplified in blended learning, holds great promise for renewed attempts at integrating constructivist theory into teaching practice today. To achieve this we steadfastly maintain that – using Gardner's seven steps for transformational change – all efforts must be directed at (re)engaging and (re)invigorating the academic heartland. Faculty should feel inspired to embrace theory and practice equally, and with similar rigor. And leadership and contexts should support them to do so. Only then will technology use become a shining example of true learning in our colleges and universities, rather than intuitive, coincidental practice.

REFERENCES

Arbaugh, J. B. (2007). An empirical verification of the community of inquiry framework. *Journal of Asynchronous Learning Networks, 11*(1). Retrieved July 27, 2008, from Sloan-C-Publications database http://www.sloan-c.org/publications/jaln/index.asp

Biggs, J., & Tang, C. (2007). *Teaching for quality learning at university* (3rd ed.). Maidenhead, Berkshire: Society for Research into Higher Education & Open University Press.

Bok, D. (2006). *Our underachieving colleges: A candid look at how much students learn and why they should be learning more*. Princeton, NJ: Princeton University Press.

Booth, W. (2006). Insuring the nation's destiny: reducing the digital divide. *Journal of Asynchronous Learning Networks, 10*(3). Retrieved July 27, 2008, from Sloan-C-Publications database http://www.sloan-c.org/publications/jaln/index.asp

Bottino, R. M., & Robotti, E. (2007). Transforming classroom teaching & learning through technology: Analysis of a case study. *Educational Technology & Society, 10*(4), 174-186.

Bruce & Goodall (2001). I have some pages up! In D. Murphy, R. Walker, & G. Webb (Eds.), *Online learning and teaching with technology: Case studies, experience and practice* (pp. 107-113). London: Kogan Page.

Cavanaugh, J. K. (2005). Are online courses cannibalizing students from existing courses? *Journal of Asynchronous Learning Networks, 9*(3). Retrieved July 27, 2008, from Sloan-C-Publications database http://www.sloan-c.org/publications/jaln/index.asp

de la Harpe, B., & Radloff, A. (2008). Developing graduate attributes for lifelong learning – how far have we got? In D. Orr, P.A. Danaher, G. Danaher, & R. E. Harreveld (Eds.), *Lifelong Learning: Reflecting on successes & Framing Futures* (pp. 136-142). Refereed proceedings of the Lifelong Learning Conference, Yeppoon, Queensland: Central Queensland University.

De Rossi, L. C. (Ed.) (2007, November 9). Online Social Networking and Education: Study Reports on New Generations Social and Creative Interconnected Lifestyles. *Robin Good's MasterNewMedia*, (pp. 1-11). Retrieved May 30, 2008 from http://www.masternewmedia.org/.

Dolence, M., & Norris, D. (1995). *Transforming higher education: A vision for learning in the 21st Century*. Ann Arbor, Michigan: Society for College and University Planning.

Entwistle, N., Skinner, D., Entwistle, D., & Orr, S. (2000). Conceptions and beliefs about "good teaching": An integration of contrasting research areas. *Higher Education Research & Development, 19*(1), 5-26.

Fink, L. D. (2003). *Creating significant learning experiences*. San Francisco, CA: Jossey-Bass.

Gardner, H. (2006). *Changing minds: The art and science of changing our own and other people's minds*. Boston, MA: Harvard Business School Press.

Garrison, D. R. (2007). Online community of inquiry review: social, cognitive, and teaching presence issues. *Journal of Asynchronous Learning Networks 11*(1). Retrieved July 27, 2008, from Sloan-C-Publications database http://www.sloan-c.org/publications/jaln/index.asp

Garrison, D. R., & Kanuka, H. (2004). Blended learning: uncovering its transformative potential in higher education. *The Internet and Higher Education, 7*(2), 95-105.

Garrison, D. R., & Vaughan, N. D. (2008). *Blended learning in higher education: Framework, principles, and guidelines*. San Francisco, CA: Jossey-Bass.

Gilder, G. (2000). *TELECOSM: How infinite bandwidth will revolutionize our world*. New York: The Free Press, Simon and Schuster.

Gregorian, V. (2005). Six challenges to the American university. In R. H. Hersh & J. Merrow (Eds.), *Declining by degrees: Higher education at risk* (pp. 77-96). New York: Palgrave Macmillan.

Halpern, D. F. (2002). Cognitive science and the work of reform. *New Directions for Higher Education, 119*(Fall 2002), 41-43.

Hiltz, S. R., Shea, P., & Kim, E. (2007). Using focus groups to study ALN faculty motivation. *Journal of Asynchronous Learning Networks, 11*(1). Retrieved July 27, 2008, from Sloan-C-Publications database http://www.sloan-c.org/publications/jaln/index.asp

Huber, M. T., & Hutchings, P. (2005). *The advancement of learning: Building the teaching commons*. San Francisco, CA: Jossey-Bass.

Ivins, T. Z. (2008). Microfranchising microlearning centers: a sustainable model for expanding the right to education in developing countries? *Journal of Asynchronous Learning Networks, 12*(1). Retrieved July 27, 2008, from Sloan-C-Publications database http://www.sloan-c.org/publications/jaln/index.asp

Jonassen, D. H. (2000). Revisiting activity learning theory for designing student centered learning environments. In D. H Jonassen & S. M. Land (Eds.), *Theoretical foundations of learning environments* (pp. 89-122). Mahwah, NJ: Lawrence Erlbaum Associates.

Jonassen, D. (1999). Designing constructivist learning environments. In C. Reigeluth (Ed.), *Instructional design theories and models: A new paradigm of instructional theory, II*, 215-239. Mahwah, NJ: Lawrence Erlbaum Associates.

Kember, D. (1997). A reconceptualisation of the research into university academics' conceptions of teaching. *Learning and Instruction, 7*(3), 255-275.

Kember, D., & Kwan, K. P. (2000). Lecturers' approaches to teaching and their relationship to conceptions of good teaching. *Instructional Science, 28*(5), 469-490.

Lever-Duffy, J., McDonald, J. B., & Mizell, A. P. (2003). *Teaching and Learning with Technology*. Boston, MA: Allyn and Bacon.

Light, G., & Cox, R. (2006). *Learning & teaching in higher education: the reflective professional*. London: Sage Publications.

Lin, L. (2008). An online learning model to facilitate learners' rights to education. *Journal of Asynchronous Learning Networks, 12*(1). Retrieved July 27, 2008, from Sloan-C-Publications database http://www.sloan-c.org/publications/jaln/index.asp

Littlejohn, A., & Pegler, C. (2007). *Preparing for blended e-learning*. London: Routledge.

Liu, X., Bonk, C. J., Lee, S., & Su, B. (2005). Exploring four dimensions of online instructor roles: a program level case study. *Journal of Asynchronous Learning Networks, 9*(4). Retrieved July 27, 2008, from Sloan-C-Publications database http://www.sloan-c.org/publications/jaln/index.asp

Merriam, S. B., & Brockett, R. G. (2007). *The profession and practice of adult education: an introduction*. San Francisco, CA: Jossey-Bass.

Meyer, A. K. (2008). If higher education is a right, and distance education is the answer, then who will pay? *Journal of Asynchronous Learning Networks, 12*(1). Retrieved July 27, 2008, from Sloan-C-Publications database http://www.sloan-c.org/publications/jaln/index.asp

Meyer, A. K. (2007). Student perceptions of face-to-face and online discussions: the advantage goes to... *Journal of Asynchronous Learning Networks, 11*(4). Retrieved July 27, 2008, from Sloan-C-Publications database http://www.sloan-c.org/publications/jaln/index.asp

Moloney, J. F., Hickey, C. P., Bergin, A. L., Boccia, J., Polley, K., & Riley, J. E. (2007). Characteristics of successful local blended programs in the context of the Sloan-C pillars. *Journal of Asynchronous Learning Networks, 11*(1). Retrieved July 27, 2008, from Sloan-C-Publications database http://www.sloan-c.org/publications/jaln/index.asp

Moore, J. C., Sener, J, & Fetzner, M. (2006). Getting better: ALN and student success. *Journal of Asynchronous Learning Networks, 10*(3). Retrieved July 27, 2008, from Sloan-C-Publications database http://www.sloan-c.org/publications/jaln/index.asp

Moss, J., Fearnley-Sander, M., & Hiller, C. (2001). Techno hero fiasco. In D. Murphy, R. Walker & G. Webb (Eds.). *Online learning and teaching with technology: Case studies, experience and practice* (pp. 155-161). London: Kogan Page.

Murphy, D., Walker, R., & Webb, G. (Eds.). (2001). *Online learning and teaching with technology: Case studies, experience and practice*. London: Kogan Page.

Newby, T. J. N., Stepich, D. A., Lehman, J. D., & Russell, J. D. (2000). *Instructional technology for teaching and learning: Designing instruction, integrating computers, and using media* (2nd ed.). Upper Saddle River, NJ: Merrill Prentice Hall.

Oliver, R., & Herrington, J. (2003). Exploring the technology-mediated learning from a pedagogical perspective. *Journal of Interactive Learning Environments, 11*(2), 111-126.

Otte, G. (2007). Online instruction as *local* education: CUNY's online baccalaureate. *Journal of Asynchronous Learning Networks, 11*(1). Retrieved July 27, 2008, from Sloan-C-Publications database http://www.sloan-c.org/publications/jaln/index.asp

Picciano, A. G., & Steiner, R. V. (2008). Bringing the real world of science to children: a partnership of the American Museum of Natural History and the City University of New York. *Journal of Asynchronous Learning Networks, 12*(1). Retrieved July 27, 2008, from Sloan-C-Publications database http://www.sloan-c.org/publications/jaln/index.asp

Prince, M. (2004). Does active learning work? A review of the research. *Journal of Engineering Education, 93*(3), 223-231.

Prosser, M., Ramsden, P., Trigwell, K., & Martin, E. (2003). Dissonance in Experience of Teaching and its Relation to the Quality of Student Learning. *Studies in Higher Education, 28*(1), 37-48.

Puzziferro-Schnitzer, M. & Kissinger, J. (2005). Supporting online adjunct faculty: a virtual mentoring program. *Journal of Asynchronous Learning Networks, 9*(2). Retrieved July 27, 2008, from Sloan-C-Publications database http://www.sloan-c.org/publications/jaln/index.asp

Ramsden, P., Prosser, M., Trigwell, K., & Martin, E. (2007). University teachers' experiences of academic leadership and their approaches to teaching. *Learning and Instruction, 17*(2), 140-155.

Rankin, W. (2007). Transformations in 21st-Century education. *ITHINKED*. Retrieved May 30, 2008, from http://ithinked.com/archives/2007/10/transformations-in-21st-century-education-william-rankin/

Roblyer, M. D. (2003). *Integrating educational technology into teaching* (3rd ed.). Upper Saddle River, NJ: Merrill Prentice Hall.

Rovai, A. P., & Ponton, M. K. (2005). An examination of sense of classroom community and learning among African American and Caucasian graduate students. *Journal of Asynchronous Learning Networks, 9*(3). Retrieved July 27, 2008, from Sloan-C-Publications database http://www.sloan-c.org/publications/jaln/index.asp

Salmon, G. (2001). *E-moderating: The key to teaching and learning online*. London: Kogan-Page.

Schweber, C. (2008). Determined to learn: accessing education despite life-threatening disasters. *Journal of Asynchronous Learning Networks, 12*(1). Retrieved July 27, 2008, from Sloan-C-Publications database http://www.sloan-c.org/publications/jaln/index.asp

Shafer-Landau, R. (2003). *Moral Realism*. Oxford: Clarendon Press.

Somekh, B. (2001). The great courseware gamble. In D. Murphy, R. Walker & G. Webb (Eds.). *Online learning and teaching with technology: Case studies, experience and practice* (pp. 121-128). London: Kogan Page.

Steimberg, Y., Ram, J., Nachmia, R., & Eshel, A. (2006). An online discussion for supporting students in preparation for a test. *Journal of Asynchronous Learning Networks, 10*(4). Retrieved July 27, 2008, from Sloan-C-Publications database http://www.sloan-c.org/publications/jaln/index.asp

Stotskopf, L. D., & Moorash, A. (2005). EArmyU: expanding education access and excellence to highly mobile online learners. *Journal of Asynchronous Learning Networks, 9*(2). Retrieved July 27, 2008, from Sloan-C-Publications database http://www.sloan-c.org/publications/jaln/index.asp

Swan, K. (2007). Research on online learning. *Journal of Asynchronous Learning Networks, 11*(1). Retrieved July 27, 2008, from Sloan-C-Publications database http://www.sloan-c.org/publications /jaln/index.asp

Tagg, J. (2003). *The learning paradigm college*. Bolton, MA: Anker Publishing Company Inc.

Tight, M. (2004). Research into higher education: an a-theoretical community of practice? *Higher Education Research & Development, 23*(4), 395-411.

Vignare, K., Geith, C., & Schiffman, S. (2006). Business models for online learning: an exploratory survey. *Journal of Asynchronous Learning Networks, 10*(2). Retrieved July 27, 2008, from Sloan-C-Publications database http://www.sloan-c.org/publications/jaln/index.asp

Webster, L., & Murphy, D. (2001). The reluctant software developers. In D. Murphy, R. Walker & G. Webb (Eds.), *Online learning and teaching with technology: Case studies, experience and practice* (pp. 114-120). London: Kogan Page.

Weimer, M. (2002). *Learner-centered teaching: Five key changes to practice*. San Francisco, CA: Jossey-Bass.

Westwood, P. (2008). *What teachers need to know about teaching methods*. Melbourne: ACER Press.

Wilson, S. M., & Peterson, P. L. (2006). *Theories of learning and teaching: What do they mean for educators?* Best practices working paper, Washington: National Education Association.

Woolfolk, A. E. (1995). *Educational psychology* (6th ed.). Boston, MA: Allyn & Bacon.

Woolfolk, A. E. (2008). *Educational psychology: Active learning edition* (10th ed.). Boston, MA: Allyn & Bacon.

Chapter IV
A Constructivist Approach to Online Learning:
The Community of Inquiry Framework

Karen Swan
Kent State University, USA

D. R. Garrison
University of Calgary, Canada

Jennifer C. Richardson
Purdue University, USA

ABSTRACT

This chapter presents a theoretical model of online learning, the Community of Inquiry (CoI) framework, which is grounded in John Dewey's progressive understanding of education. The CoI framework is a process model of online learning which views the online educational experience as arising from the interaction of three presences – social presence, cognitive presence, and teaching presence. Each of these three elements in the CoI framework are described and related to Dewey's work, and research findings and issues concerning them reviewed. The development of a common CoI survey measure that promises to address some of these issues is described and discussed. The chapter concludes with emerging findings from new studies which use the CoI survey, directions for future research, and practical uses of the CoI framework.

INTRODUCTION

Inquiry and community were at the core of John Dewey's educational philosophy and practice. Dewey (1959) believed that an educational experience must fuse the interests of the individual and society, that individual development was dependent upon community. He believed the essence of community was the organic fusion of the public and our private worlds. He also believed that the process of inquiry went to the heart of the educative experience. For Dewey, inquiry involved the generalization of the scientific method to practical problem solving and worthwhile learning. It defined the relationship between thought and action. For Dewey, inquiry was also an essentially social activity. Dewey believed that through collaboration that respected the individual, students would assume responsibility to actively construct and confirm meaning. It is this collaborative constructivist approach that is worthy of further exploration in online learning.

Online learning has evolved from early computer conferencing experiences where the big question was whether we could create and sustain a learning community in a largely asynchronous text based environment. Naturally, much attention in these early days was directed toward the question of whether we could establish the social presence essential to a viable online community of learners. As we shall see subsequently, considerable research established that it was possible to develop both social presence and online learning communities. Once this was established, the focus began to shift to the purposeful nature of a community of learners collaboratively constructing meaning.

Higher education has traditionally emphasized constructivist approaches to learning in the sense of individual students taking responsibility for making sense of their educational experiences. What is less common is the collaborative construction of knowledge in a community of learners. This social construction of knowledge must be reasserted considering the fact that the traditional ideal in higher education has been discourse and reflection in a collaborative community of scholars. It is argued here that constructivist approaches and community are necessary for creating and confirming meaning and are essential for achieving effective critical thinking. Therefore, constructivist approaches and community must be necessary parts of higher education. In online higher education, building community is particularly important because it cannot be taken for granted, nor, for that matter, can inquiry. As Garrison and Archer (2000) note "construction of meaning may result from individual critical reflection but ideas are generated and knowledge constructed through the collaborative and confirmatory process of sustained dialogue within a critical community of learners" (p. 91).

This chapter will present and discuss the Community of Inquiry (CoI) framework (Garrison, Anderson & Archer, 2000), a process model of online learning, the core of which is a collaborative constructivist approach. Each of the three elements of the CoI framework – cognitive presence, social presence and teaching presence -- will be described and related to Dewey's work, and research findings concerning them reviewed. Issues emerging from CoI research and the development of a common CoI survey measure that promises to address at least some of these issues will then be discussed. Emerging findings from new studies which use the CoI survey will also be reported. The chapter will conclude with a summary and several directions for future research.

BACKGROUND: THE COMMUNITY OF INQUIRY FRAMEWORK

The CoI framework (Garrison, Anderson & Archer, 2000) is a process model of online learning. It assumes that effective online learning, especially higher order learning, requires the development

Figure 1. CoI Framework (Garrison, Anderson & Archer, 2000)

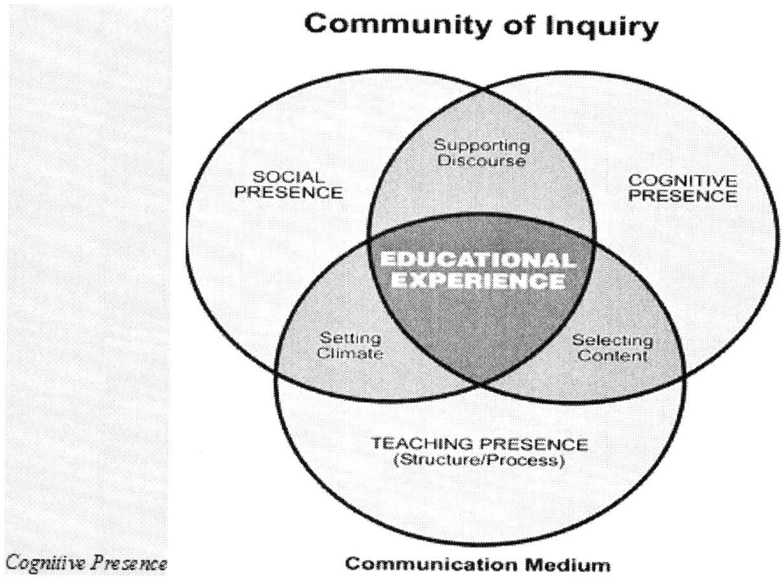

of community, and that such development is not a trivial challenge in the online environment. The CoI framework is a dynamic model of the necessary core elements for both the development of community and the pursuit of inquiry, in any educational environment. Its three core elements -- cognitive, social and teaching presence – described in the sections which follow, are viewed as multidimensional, interdependent, and dynamic. The overlapping nature of these elements is depicted in Figure 1. At their core is the unity of a collaborative constructivist learning experience consistent with the legacy of John Dewey. Together, the two constituting notions of community and inquiry form a pragmatic organizing framework of sustainable principles and processes for the purpose of guiding online educational practice. The CoI framework has provided perspective and guidance to a good deal of important research on purposeful online learning over the last decade.

Cognitive Presence

Dewey's generalization of the scientific method in the form of reflective thinking provided the foundation for the critical thinking movement that is the hallmark of higher education. Dewey (1933) described the complete cycle of reflective activity in terms of a pre-reflective state which starts with a problem, followed by five phases of reflective thought (suggestion, intellectualization, guiding idea, reasoning, and testing), and ends with a satisfactory resolution. Dewey believed that reflective inquiry has practical value in providing meaning to experience, and so described a practical method of inquiry, in addition to the full explanation of reflective inquiry, on which he believed an educational experience should be based. It is this concept that was the genesis for the Practical Inquiry model described below which operationally defines cognitive presence in the CoI framework.

Figure 2. Practical Inquiry Model (Garrison, Anderson & Archer, 2001)

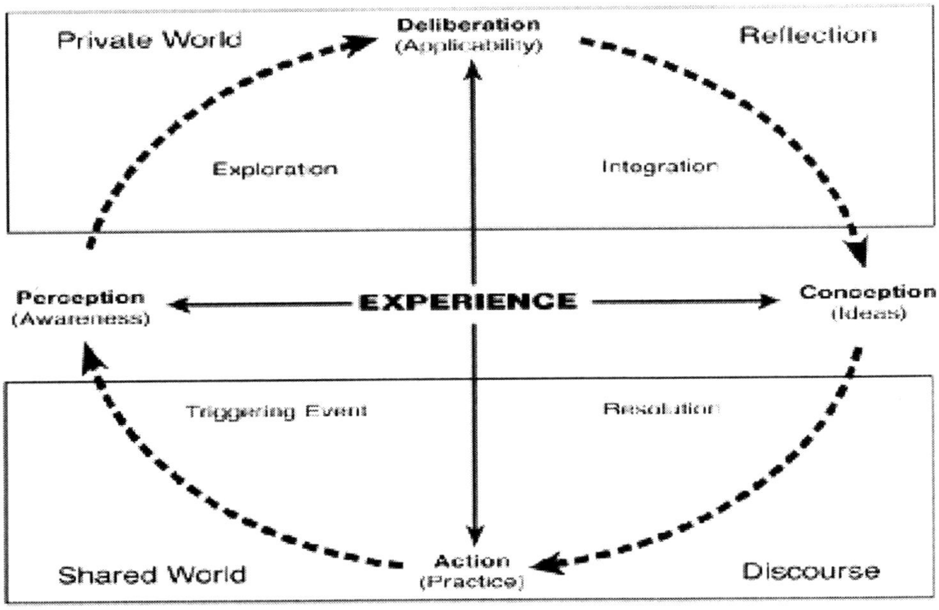

The Practical Inquiry model (Garrison, Anderson & Archer, 2001) is framed along two dimensions (see Figure 2). The vertical axis represents the psychological and sociological sides of the educational process identified by Dewey. This reflects the individual's private and reflective world juxtaposed with the community's shared world of discourse. Practical inquiry iterates imperceptibly between these two worlds. It is a process that includes both deliberation and action. That is, practical inquiry is shaped by the rigorous and purposeful process of reflection and discourse to construct meaning and confirm knowledge. The second dimension of the model defines the divergent process of perception and analysis contrasted with the convergent process of conception and synthesis. The points of perception and conception are points of insight and understanding. At each of these points we see the true fusion of the psychological and sociological and the unity of the educational experience that Dewey advocated.

The dimensions of the Practical Inquiry model provide its conceptual structure; the phases focus on the pragmatic dynamics of the inquiry process. Practical inquiry begins with a *triggering event* in the form of an issue, problem or dilemma that needs resolution. As a result of this event, there is a natural shift to *exploration,* the search for relevant information that can provide insight into the challenge at hand. As ideas crystallize, there is a move into the third phase – *integration* -- in which connections are made and there is a search for a viable explanation. Finally, there is a selection and testing (through vicarious or direct application) of the most viable solution and *resolution.* At each of the stages there may be a need to return to a previous stage for new direction or information. The four phases described in the model are a telescoping of Dewey's phases of reflective thinking for the purposes of parsimony and understanding. Consistent with Dewey's rejection of dualism, the phases should not be seen as discrete or linear. In an actual educational experience, they would be very difficult to label, as those that have used this model to code transcripts will attest (Garrison, Anderson & Archer, 2001).

In the CoI framework, cognitive presence is defined as the extent to which learners are able to construct and confirm meaning through sustained reflection and discourse (Garrison, Anderson & Archer, 2001). In the CoI framework, cognitive presence is seen as consisting of the four phases of practical inquiry. Researchers have been able to find evidence of practical inquiry in online discussion, but initial studies of cognitive presence revealed that most postings in an online discussion forum concentrated at the exploration phase where participants shared information and brainstormed ideas (Garrison & Arbaugh, 2007). Indeed, several studies in this area have found that inquiry revealed in online discussion rarely moves beyond the exploration phase (Kanuka & Anderson, 1998; Luebeck & Bice, 2005; Meyer, 2003; 2004; Murphy, 2004).

While various explanations have been explored, it is most likely that much of this has to do with the nature of the assignments and instructional direction provided (Garrison & Arbaugh, 2007; Akyol & Garrison, 2008). In studies in which students were challenged to resolve a problem and explicit facilitation and direction provided, students did progress to resolution (Meyer, 2003; Murphy, 2004; Shea & Bidjermo, 2008). Further research exploring the link between teaching presence and cognitive presence is thus clearly indicated. This and other issues surrounding cognitive presence will be discussed later in this chapter.

Social Presence

Dewey believed that learning results from experience that is contextually based and socially situated. Moreover, Lipman (1991) argued that "the reflective model is thoroughly social and communal" (p. 19). Social presence is therefore viewed as directly impacting the development of community and collaboration in online courses, and so an integral part of the CoI framework.

Social presence, the degree to which participants in computer-mediated communication feel affectively connected one to another, is the longest researched of the three presences in the CoI framework, predating the creation of the CoI model. It arose from concerns among some communications scholars that computer-mediated communication might prevent students from developing the sense of belonging with other students, instructors, and programs of study which social learning theories suggest support learning. Short, Williams and Christie (1976) originally coined the term "social presence" to refer to the differing capacities various media have for transmitting non-verbal and vocal cues, and so, in their opinion, for communicating the affective and emotional (the social) aspects of learning interactions.

As educators began experimenting with online discussion in the late 1980s and early 1990s, however, they quickly noted what Walther (1994) refers to as the "hyperpersonalness" of online discussion. Participants in online discussion, they maintained, project their personalities into online discussion using text alone (Gunawardena, 1995). They thus argued that social presence was more a matter of individual perceptions than an objective quality of the medium, and so the concept of "social presence" evolved to the notion more common among online educators today.

Gunawardena and Zittle (1997), for example, defined social presence as "the degree to which a person is perceived as 'real' in mediated communication" (p 8). They developed survey items to measure students' perceptions of the social presence of others in an online computer conference and found that perceived social presence predicted 60% of the variance in students' satisfaction with the conference. A number of studies followed which identify the perception of interpersonal connections with virtual others as an important factor in the success of online learning (Picciano,

2002; Richardson & Swan, 2003; Swan, 2002; Tu, 2000; Wegerif, 1998).

It is this sense of social presence that Rourke, Anderson, Garrison, & Archer (1999) incorporated into the CoI framework. Their research team (Rourke, Anderson, Garrison & Archer, 2001) identified three categories of social presence indicators, based on research on immediacy (Weiner & Mehrabian, 1968) and their own observations. Although the elements of social presence have been variously defined, in this chapter (and in the CoI survey discussed), we identify them as *affective expression*, where learners share personal expressions of emotion, feelings, beliefs, and values; *open communication*, where learners build and sustain a sense of group commitment; and *group cohesion,* where learners interact around common intellectual activities and tasks.

Swan (2003) used Rourke et. al.'s categories and similar indicators of social presence to examine the ways in which social presence developed among online students enrolled in an online graduate course in education. Swan's coding noted all uses of paralanguage, emotion, value, humor, self-disclosure (affective indicators), greetings and salutations, vocatives, group reference, social sharing, reflection on course itself (cohesive indicators), acknowledgement, agreement and disagreement, approval, invitation, personal advice (interactive indicators) in discussion postings. She documented the use of an average of almost six social presence indicators per message in a sample of 235 discussion postings and found changes in the relative frequencies of text-based social presence indicators employed over time.

Danchak, Walther and Swan (2001) hypothesized that people communicating online use such textual immediacy indicators to maintain a sense of affective equilibrium in their interactions. Danchak, et. al. argued that participants in environments with less affective communication channels available will evoke more verbal immediacy behaviors to affect a kind of equilibrium of social presence with which they are comfortable.

Noting the relationship between perceived presence and success in online courses, Tu (2000) linked the development of social presence in online courses to course design. Based on elements of social learning theory, he distinguished three dimensions of course design which influenced the development of social presence – social context, online communication, and interactivity, which includes reciprocal communication patterns and timely responses. Tu and McIsaac (2002) found support for these dimensions of social presence in a factor analysis of student responses to an online survey concerned with computer-mediated communication tools.

Swan and Shih (2005) also found support for the impact of course design on perceptions of social presence in a study of its development of four classes (two different courses taught by two different instructors). Their findings further revealed an overlap in perceptions of instructor and peer presence and indicated that the perceived presence of instructors was a more influential predictor of student satisfaction than the perceived presence of peers. Additionally, Swan and Shih found that students perceiving the highest social presence also employed the greatest number of social presence indicators to project themselves into online discussions.

Of course, meaningful research instigates further research. Perhaps the most significant is whether social presence is really a necessary precursor of cognitive presence. Most researchers in this area agree that it is, with the caveat that social presence must be directed toward learning outcomes (Garrison, 2007). This has led to a revision of the original social presence categories and indicators to reflect academic purposes (Garrison, Cleveland-Innes & Fung, 2004). Currently, the elements of social presence are conceptualized as consisting of affective/personal communication, open communication (interaction), and group cohesion and collaboration (Vaughn & Garrison, 2006), but the interaction of social and cognitive presence is still largely unknown and clearly under-researched. Many other questions remain and these will be discussed more fully later this chapter.

Teaching Presence

Dewey (1959) stated "that the educational process has two sides – one psychological and one sociological; and that neither can be subordinated to the other or neglected without evil results following" (p. 20). This clearly reflects the cognitive and social presence elements of the CoI framework. He also explicitly addressed the need for purpose, structure and leadership; that is, *teaching presence.* Dewey (1938) argued that it is the responsibility of the educator to establish aims and activities, but not to be straight-jacketed by them. To establish and sustain a community of inquiry, he maintained, educators must be knowledgeable, flexible but focused, and comfortable with uncertainty. In this regard, he stated, "thought needs careful and attentive educational direction" (Dewey, 1933, p. 22). Dewey (1938) was also aware of the need to facilitate appropriate social relationships by giving as much attention to the organization of the social environment of the classroom as to its physical environment. Teaching presence is established by attending to cognitive and social presence challenges in a collaborative CoI.

In the CoI framework, teaching presence is defined as "the design, facilitation and direction of cognitive and social processes for the purpose of realizing personally meaningful and educationally worthwhile learning outcomes" (Anderson, Rourke, Garrison & Archer, 2001). The collaborative construction of knowledge in any educational context is particularly challenging. However, it is extraordinarily difficult to manage the multifaceted responsibilities of teaching presence in a largely text-based online environment. For this reason, we need to understand the dimensions of teaching presence and their roles in the dynamics of a collaborative constructive educational experience.

The three dimensions or categories of teaching presence alluded to in the previous definition emerged from a review of the literature related to the classification of online teaching responsibilities (Anderson et al., 2001). The first category, *design and organization*, cannot be neglected in an online learning environment. This is particularly true in terms of the selection of worthwhile collaborative learning activities. In an asynchronous nonverbal context, expectations with regard to discourse must be clear. For example, message length and focus of comments must be stated and modeled. The second category of teaching presence is *facilitating discourse*. Regardless of how clearly stated the expectations for online discussion, there will be a need to guide that discussion in a meaningful manner, ensuring that students stay focused. This is essential for students to stay engaged and to build a collaborative community of inquiry. Moreover, in a formal educational context, there will be times when it is necessary to intervene directly to correct misconceptions, provide relevant information, summarize the discussion and/ or provide some metacognitive awareness. This is the third category of teaching presence – *direct instruction* (Anderson et al., 2001).

Researchers have documented strong correlations between learner's perceived and actual interactions with instructors and their perceived learning (Jiang & Ting, 2000; Richardson & Swan, 2003; Swan, Shea, Fredericksen, Pickett, Pelz & Maher, 2000), and between all three elements of teaching presence and student satisfaction and perceived learning in online courses (Shea, Pickett & Pelz, 2004). Teaching presence has also been shown to be linked to the development of a sense of community in online courses (Shea, Li, Swan & Pickett, 2005). Indeed, the body of evidence attesting to the critical importance of teaching presence for successful online learning continues to grow (Garrison & Cleveland-Innes, 2005; Meyer, 2003; Murphy, 2004; Swan & Shih, 2005; Vaughn & Garrison, 2006; Wu & Hiltz, 2004).

There is, however, some confusion about the elements of teaching presence. In his study of teaching presence and sense of community, Shea (2006) found that items developed to measure the three dimensions of teaching presence yielded a two factor solution he interpreted as *instructional design and organization* and *directed facilitation* (merging the instructor behavior dimensions of the construct). Conversely, in their study of MBA students' perceptions of teaching presence, Arbaugh and Hwang (2006) found support for the three-dimensional teaching presence construct. This and other issues surrounding the teaching presence construct will be discussed in greater detail later this chapter.

SUMMARY

Table 1 below summarizes the three presences in the CoI framework and their respective categories. It also gives examples of indicators of the presences and categories that have been used over the years to test the structure of the framework (Arbaugh & Hwang, 2006; Garrison, Cleveland-Innes and Fung, 2004; Shea, Li and Pickett, 2006), to explore various aspects and issues associated with online learning (Garrison & Arbaugh, 2007) as well as providing the inspiration for the development of a quantitative CoI questionnaire (Swan, Richardson, Ice, Garrison, Cleveland-Innes & Arbaugh, 2008).

As noted above, research on each of the presences that combine to form the CoI framework has revealed issues unique to them. Research on cognitive presence, for example, has found that inquiry revealed in online discussion rarely moves beyond the exploration phase (Garrison, Anderson & Archer, 2001; Kanuka & Anderson, 1998; Luebeck & Bice, 2005; Meyer, 2003; 2004; Murphy, 2004). Several explanations might account for such results; it might simply be that integration and resolution take place in other parts of online courses (e.g., essay assignments) or internally in learners' minds (Arnold & Ducate, 2006). Meyer (2003) suggests that sufficient time to reach these phases is not often available in online discussions. Recent research, however, suggests that moving inquiry through to integration and resolution, or the lack thereof, is related to aspects of teaching presence (Garrison & Arbaugh, 2007). There is growing evidence that discussions organized around problem solutions, the production of artifacts, or the completion of tasks are more likely

Figure 3. CoI categories and indicators; (Garrison & Anderson, 2003) Issues Related to the Community of Inquiry Framework

ELEMENTS	CATEGORIES	INDICATORS (examples only)
Cognitive Presence	Triggering Event Exploration Integration Resolution	sense of puzzlement information exchange connecting ideas application
Social Presence	Affective Expression Open Communication Group Cohesion	self projection/expressing emotion trust/risk free climate collaboration/interactivity
Teaching Presence	Design & Organization Facilitating Discourse Direct Instruction	setting curriculum & activities shaping constructive exchange focusing & resolving issues

to move to resolution (Arnold & Ducate, 2006; Murphy, 2004), and that progression through inquiry requires direction from instructors or tutors (Vaughn & Garrison, 2005).

Most studies of social presence have noted the highly democratic nature of online discussion (Harasim, 1990; Kanuka & Anderson, 1998) and accordingly conceptualized social presence as a single construct with an emphasis on perceptions of the presence of peers. As noted above, there is some indication that instructor presence may be equally important (Swan & Shih, 2005). While the social presence of instructors has been considered in explorations of "teaching presence" (Anderson, Rourke, Garrison & Archer, 2000; Shea, Pickett & Pelz, 2003), it has not been isolated therein. In addition, while most studies of social presence implicitly locate its development in online discussion, survey questions have not explicitly addressed it in that context. Indeed, Richardson and Swan (2003) found strong correlations between perceived social presence and student satisfaction and perceived learning in a variety of course activities outside course discussions including such seemingly asocial activities as quizzes and tests, individual assignments, and lecture notes. Their findings may be yet another example of the overlap and interchange that exists between the presences. Finally, recent research on emotion in online learning suggests a principal role for acknowledgement and support of student emotion. Cleveland-Innes and Campbell (2006) define emotional presence as the extent to which learners and teachers transform their behavior to accommodate the overt and covert presence of emotion. While Garrison, Anderson and Archer (2000) suggest that emotional expression is a subcomponent of social presence, Cleveland-Innes and Campbell's research findings suggest that emotion plays a role in all three elements of the CoI framework.

As noted previously, another issue that has arisen in the literature involves the categories associated with teaching presence. While some studies have confirmed the presence of three categories (Arbaugh & Hwang, 2006; Akyol & Garrison, 2008), Shea, Li and Pickett (2006) conducted a large scale factor analysis that suggested a two factor solution was most interpretable. The two factors (i.e., categories) were labeled "design" and "directed facilitation." The second factor was clearly an amalgamation of facilitation and direct instruction. The question is whether this was simply contextual (e.g., these students could not distinguish between facilitation and direct instruction) or whether the structure of teaching presence needs revision. Although it is becoming clear as to the importance of teaching presence, much more empirical research in a variety of contexts is required to conclusively define the categories of teaching presence.

While the CoI framework holds great promise for bringing order and a theoretical base to research in online learning, there is obviously much work to be done before it can meet that promise. Perhaps most importantly, it needs to be kept in mind that the theoretical foundation of the CoI framework is that of a collaborative constructivist educational experience. As such, it is a dynamic model that is in constant search for balance among the presences. That is, the influence of each of the presences and their categories interact and shift over time, and across courses. Most research to date has concentrated on single presences in the CoI framework, while its theoretical strength lies in the dynamics of the whole community. In particular, better understanding of evolving interactions among the CoI presences and their respective categories is needed. In addition, much of the CoI research has employed different measures and sometimes different terminology, especially as regards the elements of social presence. Finally, with some important exceptions, the research has mostly involved single institutions and often single courses. Inter- and intra-institutional research is needed, both to validate the model as a whole and to make use of the model in a myriad of studies that could move online learning research significantly forward.

NEW DEVELOPMENTS: THE COMMUNITY OF INQUIRY INSTRUMENT

To address some of the issues noted above and to move CoI research forward in many other ways, a group of CoI researchers from a variety of institutions and with expertise in the various components of the model collaborated in the development of a CoI survey instrument. The resulting instrument was adapted from measures group members had successfully used to measure individual presences in the CoI framework. It includes twelve items designed to measure cognitive presence (3 for triggering events, 3 for exploration, 3 for integration, and 3 for resolution), nine items designed to measure social presence (3 for affective expression, 3 for open communication and 3 for group cohesion), and thirteen items designed to measure teaching presence (4 for design and organization, 6 for facilitation of discourse, and 3 for direct instruction). In the summer of 2007, the survey was tested in graduate courses at four institutions located in the United States and Canada using principal component factor analysis and the three factor (presences) construct predicted by the CoI framework was supported (Swan, et al., 2008).

Results of the factor analysis provide evidence that, as currently defined and operationalized, an online community of inquiry emerges out of social, cognitive and teaching presence. Student responses to the survey's statements about their online experience clustered around items as defined by the theory. The results also validate a measurement tool of agreed upon and statistically confirmed items that operationalizes the concepts in the CoI framework (Swan, et al., 2008). This measurement tool may be used for continued explication of concepts in the model, and can serve as a ground for more qualitative investigations in mixed methods studies.

For example, a recent study by Akyol and Garrison (2008) attempted to understand how a community of inquiry changes over time. The authors employed content analysis to explore the dynamics of social presence as students' progressed through an online course. As expected, social presence was initially the most frequent type of response, but was over taken by cognitive presence by the end of the course. There were also changes in the relative frequencies of the social presence categories as the course progressed, with group cohesion responses gradually increasing and open communication and affective expression decreasing over time. Similarly, within teaching presence facilitation responses declined and direct instruction increased as the course progressed. Akyol and Garrison suggest that as open communication was established and group cohesion grew, there was less need or perhaps time for affective expression. The focus appeared to be on the task and so direct instruction and cognitive presence grew. It would also appear that as students adjusted to their roles they needed less encouragement and support.

This study was perhaps the first to employ the recently developed CoI instrument to measure the relationships among the three presences, students' satisfaction and perceived learning. The results showed significant relationships between teaching presence and cognitive presence ($p=.001$); between teaching presence and perceived learning ($p=.03$); between teaching presence and satisfaction ($p=.011$), between cognitive presence and perceived learning ($p=.007$); between cognitive presence and satisfaction ($p=.009$), and between social presence and satisfaction ($p=.038$) (Akyol & Garrison, 2008). However, the analysis did not find a significant relationship between social presence and perceived learning. It is also worth noting the concurrent rise of particular categories in each of the three presences. That is, group cohesion, integration and direct instruction responses all increased over the time of the course of study. The question is whether there is a reinforcing effect: Is it that group cohesion (i.e., social presence) and direct instruction (teaching presence) sets the stage for integration (cognitive presence)?

Evidence to this effect, and more insight into the lack of finding a relationship between social presence and perceived learning, has been provided by another new study using the CoI survey by Shea and Bidjermo (2008). Shea and Bidjermo (2008) used Structural Equation Modeling (SEM) to study the impacts of teaching and social presence on cognitive presence, all as measured by the CoI survey. The results of their analysis are shown graphically in Figure 3. They reveal that teaching and social presence together account for 70% of the variation in students' reported level of cognitive presence. However, the authors also found that the development of social presence was contingent on the establishment of teaching presence; that is, social presence did not in itself directly affect cognitive presence but rather served as a mediating variable between teaching presence and cognitive presence. This finding helps explain those of Akyol and Garrison (2008). Shea and Bidjermo concluded that the "teaching and social presence represent the processes needed to create paths to epistemic engagement and cognitive presence for online learners." (p. 14)

FUTURE AND EMERGING TRENDS

As one might expect, several institutions have begun to look to the CoI framework as a development tool for their online courses (Shea, P., & Bidjeramo, T., 2008; Vaughan & Garrison, 2006). In general the CoI model, while considered valuable by researchers and practitioners alike, has not generally taken root at an institutional level for course development. With the validation of the framework and the development and validation of the CoI instrument (Swan, et al., 2008; Shea and Bidjermo, 2008) the framework can be more easily operationalized by developers and practitioners, especially those moving beyond the traditional online course format to a more constructivist one.

One of the fastest growing areas of non-traditional online learning is blended learning. As higher education institutions are looking at asynchronous online discussions as a means to increase students' higher order thinking skills and outcomes they often utilize blended learning environments. Traditionally blended learning environments have mirrored face-to-face teaching

Figure 4. SEM analysis of effects of social & teaching presence on cognitive presence; Detail adapted from Shea & Bidjeramo, 2008; "© 2008, Peter Shea. Used with permission."

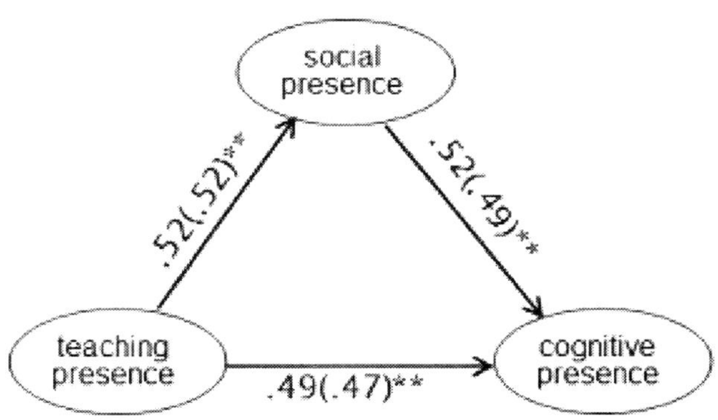

and learning, with the exception of having online discussions and using the online environment as a repository. However, several institutions have begun to look to the CoI framework as a model for these blended learning environments (Garrison & Vaughan, 2008; Vaughan & Garrison, 2006). For example, in a work by Vaughan and Garrison (2006) the authors concluded that, "The key to creating a cohesive, purposeful and worthwhile community of inquiry is the integration of social, teaching and cognitive presence... Each of the three presences manifest themselves and evolve in different ways in a face-to-face or online context." (p. 150)

As research continues on the CoI framework and how it can improve online courses, CoI research is also branching out into other environments. For example, a research group at Purdue University (Radcliffe, D., Strobel, J., Brophy, S & Richardson, J.C., personal communication, April 21, 2008) is looking to the CoI framework as a means for examining and categorizing virtual communities of practice, specifically communities using HUB-Zero technology. The hope is that the CoI framework will be able to serve as a tool for developing successful virtual communities of practice for discovery and learning.

Finally, the CoI framework is also being tied to other emerging online technologies that can help develop the presences within courses. For example, a recent study by Ice, Curtis, Philips & Wells (2007) looked at the effects of audio feedback from instructors embedded in student assignments to enhance teaching presence and learning. They found that audio feedback was associated with the perception that the instructor cared more about the student and that students were three times more likely to apply content for which audio commenting was provided in class projects than was the case for content for which text based commenting was provided. In two subsequent multi-institutional studies (n= 287, n=1138) employing the CoI survey, Ice (2008) found significant differences between students in courses in which embedded feedback and students in courses where it wasn't used that favored embedded feedback on three teaching presence items. Interesting, significant differences favoring embedded feedback were also found on one social presence and two cognitive presence items as well, again providing evidence for the integration of the three presences conceptualized in the model.

CONCLUSION

The CoI framework, stemming from Dewey's emphasis on collaborative constructivism and practical inquiry holds promise as a theoretical and practical model for online learning. At the heart of the CoI framework is the idea that community, critical reflection, and knowledge construction are integral to learning, especially learning online (Garrison & Archer, 2000, p. 91). Moreover, the CoI framework, based on the constructs of cognitive, social, and teaching presence, takes into account the various stakeholder groups involved in an online course (student, instructor, designer) and what each can do to make their course a successful learning experience.

To date, research related to the CoI framework has focused mostly on single presences instead of the framework as a whole. It has also focused on individual courses and institutions with few exceptions, but the research tells us that the influence of each of the presences and their categories interact and shift over time, and possibly across courses (e.g. disciplines, course level). As such, research needs to be conducted not only at an inter-and intra- institutional level but also across disciplines and course levels. Researchers have developed and validated a "common" CoI survey, which has in turn validated the CoI framework. The instrument not only allows for continued investigation of the constructs in the model (cognitive, social and teaching presence) but can also serve as an evaluation tool for online courses utilizing the model for development.

We are seeing the CoI framework evolve into a tool for environments outside of its original purpose, namely virtual communities of practice, blended learning environments, and as a theoretical basis for tool development to strengthen the presences in online learning. Finally, current and future research in this area abounds, and with each new research study more is learned about the elements and components of the model, the shifts in CoI over time, and how the model can be used by and for practitioners developing online courses.

REFERENCES

Akyol, Z., & Garrison, D. R. (2008). The development of a community of inquiry over time in an online course: Understanding the progression and integration of social, cognitive and teaching presence. *Journal of Asynchronous Learning Networks, 12*(3), 3-22.

Anderson, T., Rourke, L., Garrison, D. R., & Archer, W. (2001). Assessing teaching presence in a computer conferencing context. *Journal of Asynchronous Learning Networks, 5*(2), 1-17.

Arbaugh, J. B., & Hwang, A. (2006). Does "teaching presence" exist in online MBA courses? *The Internet and Higher Education, 9*(1), 9-21.

Arnold, N., & Ducate, L. (2006). Future foreign language teachers' social and cognitive collaboration in an online environment. *Language, Learning & Technology, 10*(1), 42–66. Retrieved January 14, 2006 from http://llt.msu.edu/vol10num1/pdf/arnoldducate.pdf

Cleveland-Innes, M., & Campbell, P. (2006). *Understanding emotional presence in an online community of inquiry.* Paper presented at the 12th Annual SLOAN-C ALN Conference, Orlando, Florida.

Danchak, M. M., Walther, J. B., & Swan, K. (2001). *Presence in mediated instruction: bandwidth, behavior, and expectancy violations.* Orlando, FL: Paper presented at the Seventh Annual Sloan-C International Conference on Online Learning.

Dewey, J. (1933). *How we think* (rev ed.). Boston: D. C. Heath.

Dewey, J. (1938). *Experience and education* (7th printing, 1967). New York: Collier.

Dewey, J. (1959). My pedagogic creed. In J. Dewey (Ed.), *Dewey on education* (pp. 19-32). New York: Teachers College, Columbia University. (Original work published 1897).

Garrison, D. R. (2007). Online Community of Inquiry review: Social, cognitive and teaching presence issues. *Journal of Asynchronous Learning Networks, 11*(1), 61-72.

Garrison, D. R., & Anderson, T. (2003). *E-Learning in the 21st century: A framework for research and practice.* London: Routledge/Falmer.

Garrison, D. R., Anderson, T., & Archer, W. (2000). Critical inquiry in a text-based environment: Computer conferencing in higher education. *The Internet and Higher Education,* 2, 87–105.

Garrison, D. R., Anderson, T., & Archer, W. (2001). Critical thinking, cognitive presence and computer conferencing in distance education. *American Journal of Distance Education, 15*(1), 7-23.

Garrison, D. R., & Arbaugh, J. B. (2007). Researching the community of inquiry framework: Review, issues, and future directions. *Internet and Higher Education, 10*(3), 157-172.

Garrison, D. R., & Archer, W. (2000). *A transactional perspective on teaching-learning: A framework for adult and higher education.* Oxford, UK: Pergamon.

Garrison, D. R., & Cleveland-Innes, M. (2005). Facilitating cognitive presence in online learning:

Interaction is not enough. *American Journal of Distance Education, 19*(3), 133–148.

Garrison, D. R., Cleveland-Innes, M., & Fung, T. (2004). Student role adjustment in online communities of inquiry: Model and instrument validation. *Journal of Asynchronous Learning Networks, 8*(2), 61-74. Retrieved August 13, 2005, from http://www.sloan-c.org/publications/jaln/v8n2/pdf/v8n2_garrison.pdf

Garrison, D. R., & Vaughan, N. (2008). *Blended learning in higher education.* San Francisco: Jossey-Bass.

Gunawardena, C. (1995). Social presence theory and implications for interaction and collaborative learning in computer conferencing. *International Journal of Educational Telecommunications, 1*(2-3), 147 – 166.

Gunawardena, C., & Zittle, F. (1997). Social presence as a predictor of satisfaction within a computer mediated conferencing environment. *American Journal of Distance Education, 11*(3), 8-26.

Harasim, L. (1990). *On-line Education: Perspectives on a New Environment.* New York: Praeger.

Ice, P. (2008). *The impact of asynchronous audio feedback on teaching, social and cognitive presence.* Paper presented at the 2008 Conference of the Canadian Network for Innovation in Education, Banff, Alberta, Canada.

Ice, P., Curtis, R., Philips, P., & Wells, J. (2007). Using asynchronous audio feedback to enhance teaching presence and students' sense of community. *Journal of Asynchronous Learning Networks*, 11(2). Retrieved May 6, 2008, from http://www.sloan-c.org/publications/JALN/v11n2/v11n2_ice.asp

Jiang, M., & Ting, E. (2000). A study of factors influencing students' perceived learning in a web-based course environment. *International Journal of Educational Telecommunications*, 6(4), 317-338.

Kanuka, H., & T. Anderson. (1998). Online social interchange, discord, and knowledge construction. *Journal of Distance Education 13*(1), 57–75.

Lipman, M. (1991). *Thinking in education.* Cambridge: Cambridge University Press.

Luebeck, J. L., & Bice, L. R. (2005). Online discussion as a mechanism of conceptual change among mathematics and science teachers. *Journal of Distance Education, 20*(2), 21–39.

Meyer, K. A. (2003). Face-to-face versus threaded discussions: The role of time and higher-order thinking. *Journal of Asynchronous Learning Networks*, 7(3), 55-65.

Meyer, K. A. (2004). Evaluating online discussions: Four different frames of analysis. *Journal of Asynchronous Learning Networks*, 8(2), 101-114.

Murphy, E. (2004). Identifying and measuring ill-structured problem formulation and resolution in online asynchronous discussions. *Canadian Journal of Learning and Technology*, 30(1), 5-20.

Richardson, J. C. & Swan, K. (2003). Examining social presence in online courses in relation to students' perceived learning and satisfaction. *Journal of Asynchronous Learning Networks, 7*(1), 68-88.

Picciano, A. G. (2002). Beyond student perceptions: Issues of interaction, presence and performance in an online course. *Journal of Asynchronous Learning Networks, 6*(1), 21-40.

Rourke, L., Anderson, T. Garrison, D. R., & Archer, W. (1999). Assessing Social Presence in Asynchronous, Text-Based Computer Conferencing. *Journal of Distance Education, 14*(3), 51-70.

Rourke, L., Anderson, T., Garrison, D. R., & Archer, W. (2001). Assessing social presence in asynchronous text-based computer conferencing. *Journal of Distance Education, 14*(2), 50-71.

Shea, P. (2006). A study of students' sense of learning community in online environments. *Journal of Asynchronous Learning Networks, 10*(1), 35-44.

Shea, P., Li, C., Swan K., Pickett, A. (2005) Developing Learning Community in Online Asynchronous College Courses: The Role of Teaching Presence. *Journal of Asynchronous Learning Networks, 9(*4). Retrieved May 1, 2008 from: http://www.sloan-c.org/publications/jaln/v9n4/v9n4_shea.asp

Shea, P. J., Pickett, A. M., & Pelz, W. E. (2003). A follow-up investigation of "teaching presence" in the SUNY Learning Network. *Journal of Asynchronous Learning Networks, 7*(2), 61-80.

Shea, P., & Bidjeramo, T. (2008). *Community of inquiry as a theoretical framework to foster "epistemic engagement" and "cognitive presence" in online education.* Paper presented at the Annual Meeting of the American Educational Research Association, New York, March.

Shea, P., Li, C. S., & Pickett, A. (2006). A study of teaching presence and student sense of learning community in fully online and web-enhanced college courses. The *Internet and Higher Education*, 9(3), 175-190.

Short, J., Williams, E., & Christie, B. (1976). *The Social Psychology of Telecommunications.* Toronto: Wiley.

Swan, K., Shea, P., Fredericksen, E., Pickett, A., Pelz, W., & G. Maher (2000). Building knowledge building communities: consistency, contact and communication in the virtual classroom. *Journal of Educational Computing Research, 23*(4), 389-413.

Swan, K. (2002). Building communities in online courses: the importance of interaction. *Education, Communication and Information, 2*(1), 23-49.

Swan, K. (2003). Developing social presence in online discussions. In S. Naidu (Ed.), *Learning and Teaching with Technology: Principles and* Practices (pp. 147-164). London: Kogan Page.

Swan, K., & Shih, L-F. (2005). On the nature and development of social presence in online course discussions. Journal of Asynchronous Learning Networks, 9(3), 115-136.

Swan, K., Richardson, J.C, Ice, P., Garrison, D. R., Cleveland-Innes, M., & Arbaugh, J. B. (2008). Validating a measurement tool of presence in online communities of inquiry. *E-Mentor, 2*(24), 1-12. http://www.e-mentor.edu.pl/e_index.php?numer=24&all=1

Tu, C. H. (2000). On-line learning migration: From social learning theory to social presence theory in CMC environment. *Journal of Network and Computer* Applications, *23*(1), 27–37.

Tu, C-H. & McIsaac, M. (2002). The relationship of social presence and interaction in online classes. *The American Journal of Distance Education, 16*(3), 131–150.

Vaughan, N., & Garrison, D. R. (2006). How blended learning can support a faculty development community of inquiry. *Journal of Asynchronous Learning Networks, 10*(4), 139-152.

Walther, J. (1994). Interpersonal effects in computer mediated interaction. *Communication Research, 21*(4), 460-487.

Wegerif, R. (1998). The social dimension of asynchronous learning networks. *Journal of Asynchronous Learning Networks, 2*(1), 34-49.

Weiner, M. & Mehrabian, A. (1968). *Language Within Language: Immediacy, a Channel in Verbal Communication.* New York: Appleton-Century-Crofts.

Wu, D., & Hiltz, S. R. (2004). Predicting learning from asynchronous online discussions. *Journal of Asynchronous Learning Networks, 8*(2), 139–152.

Chapter V
Applying Constructivism to Online Learning: A New Instructional Design Map

Jennifer Lee
University of North Texas, USA

Lin Lin
University of North Texas, USA

ABSTRACT

Based on constructivist principles, this chapter provides a new instructional design map for online learning environments. This instructional design map includes considerations of five elements, namely, learner, knowledge, learning environment, assessment, and technology. Considerations of these elements are based on analyses of the past and existing instructional design models, online learning models, and constructive principles. Applications of the instructional design map are also discussed in the chapter.

INTRODUCTION

Dijkstra (2005) defines instructional-design theory as a set of statements that interpret why an instructional program leads to the acquisition of knowledge, skills, and attitudes. Although practitioners and researchers readily agree with this definition, there is much disagreement over which instructional theory or model can best accomplish the objective. This is not surprising as there has been considerable debate in the teaching and learning communities, over many decades about how students think and best learn.

In the past 50 years or so, three major influences have largely accounted for the many changes and debates surrounding instructional design theories and models. Dijkstra (2005) cites "the rediscovery of epistemology, the renewed interest in nature and knowledge acquisition, and the invention of the computer and the development of information and communication technology" (p. 187) as the dominant influences in the field of cognition and learning since the 1950s.

In this chapter, we start by briefly discussing the three major influences in the field of instructional design to underscore the importance of understanding the lineage of contemporary design theories in our quest to prepare a new generation of learners. We will then provide an overview of the ontological, epistemological, and methodological principles that guide the design of our proposed instructional design map theory for online teaching and learning. Although instructional design theories and online learning models target technology-enhanced teaching and learning environments, the developments of the two have evolved separately from one another. Therefore, we aim to bring the two together by examining each through the constructivist lens in the new media and technology environments. In the final section, we will explain how the new design map is expected to result in effective learning and why this is an improvement over existing theories. Discussions on how to apply the theory in practical settings and assessment strategies will also be outlined.

THREE MAJOR INFLUENCES IN THE INSTRUCTIONAL DESIGN FIELD

Research on Cognition and Learning which Started in the 1950s

"There are two main theoretical approaches in the field of instructional design: the systematic approach and the constructivist approach" (Fardanesh, 2006, p.3). According to Tennyson and Schott (1997), instructional design was driven by behaviorism in its early stages. Behaviorism places an emphasis on producing observable and measurable outcomes in students (Ertmer & Newby, 1993). Learning occurs when learners demonstrate the desired behavior in response to a stimulus (Smith & Ragan, 1999). Much of the research and work in 1950s and even 1960s focused on how to design systematic instructions to help learners achieve learning objectives through reinforcement, reward, and punishment

Cognitivism gradually succeeded behaviorism in the 1970s and 1980s as researchers studied the influence of mental processes on learning. Cognitivists believe that learners participate actively in the process of knowledge acquisition and construction. Tennyson and Schott (1997) noted that the definition of instructional design at this point "shifted to considerations of learning theory and the development of models linking these theories to the design of instruction" (p.7). Some of the models and theories introduced during this period are still influential today.

In the last decade, Otting and Zwaal (2006) observed that the "traditional views of education and the focus on absolute truth have been gradually replaced by a diversity of more dynamic, pluralist, and relativists conceptions" (p. 347). Constructivists, identifying themselves with Piaget and Vygotsky and representing an array of perspectives, view learning as a process of constructing meaning from experience. Unlike the traditional instructional systems approaches of designing instruction, constructivists hold that "all knowledge is socially constructed" (Molenda, 1997, p.46). According to Gold (2001, p.37), "knowledge is not separate from but rather embedded within experiences and interpreted by the learner." Karagiorgi and Symeou (2005, p.17) further stated that "constructivism makes a different set of assumptions about learning and suggests new instructional principles." Learning is seen to be an internal construction of reality by the individual. The cognitive process of meaning making is emphasized both as an individual mental activity and a socially interactive interchange.

Rediscovery of Epistemology: The Renewed Interest in Nature and Knowledge Acquisition

With the proliferation of multiple competing theories and viewpoints, the landscape of instructional design is populated by various behaviorists, cognitivists, and constructivists. Even though the merits of each competing theory, models, and approaches have been avidly discussed in multiple learning communities, little attention has been paid to the the underlying philosophical assumptions until the recent decade (Molenda, 1997; Dinter, 1998; Spector, 2000; Schuh & Barab, 2008). Some instructional designers dismiss the nature of such discussions as purely an academic exercise. However, it is our belief that instructional designers should examine their own epistemological and ontological beliefs about the process of learning in order to better understand their frame of reference in instructional design for the particular subject matter being investigated, while utilizing the available recursivity that can be brought to bear for each individual learning task.

Why is this important? Many learning theories and models are often vague in terms of philosophical roots (Schuh & Barab, 2008). The lack of transparency in our philosophy of learning creates inconsistency when we translate our theoretical beliefs into practices (Dinter, 1998; Barab & Duffy, 2000; Schuh and Barab, 2008). Hannifin et al. (1997) remind us why instructional design should be based on a defensible theoretical framework. The authors believe that frameworks help us identify assumptions that are consistent with our beliefs and methods of design. We run the risk of presenting conflicting and incoherent views when we attempt to merge theories with little regard for the philosophical underpinnings supporting such approaches.

Audi (2002) notes that epistemology is the study of knowledge and justification. For Schuh and Barab (2008), epistemology deals with the question of "how come we know about what exists" (p. 71). Our theories about what, if anything, can we know, or which beliefs are justified and which are not, play a significant role in how we view teaching and learning. Guarino (1998) defines ontology as "particular system of categories accounting for a certain vision of the world" (p.2). Epistemology and ontology provide important linkages "between how learning occurs and how the learning process is facilitated" (Schuh & Barab, 2008, p.76).

Over half a century ago, instructional designers viewed learning as a "process of knowledge transmission" (Jonassen and Land, 2000, p.iii). As such, "educators believe that improving learning is a matter of more effectively communicating ideas to learners by improving the clarity of the message" (Jonassen and Land, 2000, p.iii). The emphasis of knowledge transmission meant that "instructional content can be preplanned, organized, and programmed with specific outcomes defined" (Schuh & Barab, 2008, p.77). Jonassen (2005) noted that instructional design models based on an objectivist epistemology tend to be limited in their ability to support a broad range of human activities.

In contrast, constructivism moves away from prescriptive instructions driven by collective learning outcomes. Boghossian (2006) explains that "for the constructivist, each person's subjective experience is just as valid as anyone else's, and no one has an epistemically privileged viewpoint" (p.714). "Constructivists choose to consider knowledge as an internally coherent internally coherent system that we actively build up from within for our own purposes, coping with the world of our individual experience and participating in building a collective, coherent knowledge base" (Staver, 1998, pp.505-506). Constructivists subscribe to the view that "our knowledge and beliefs are influenced by that community and their beliefs and values" (Jonassen and Land, 2000, p.vi).

Hew (2004) notes that the "constructivist paradigm is most aligned with the interpretive ontological perspective" (p.4). It was founded on the principle that our individual experience is subjective in nature and that the source of knowledge is constructed internally. Therefore the notion of truth differs from one individual to another. As such, knowledge construction happens differently for all learners. Wilson and Lowry (2000) point out that constructivist learning happens all the time online and offline. It is not reserved for a specific type of learning mode because "learners are constantly trying to make sense of their environments and the information presented to them" (p.81). It is clear that epistemology – the theory of knowledge – plays a key role in how we think of instructional design, and consequently, affects how we teach and learn.

Invention of the Computer and the Development of Information and Communication Technology

Seel (2005) noted that "several paradigm shifts of psychology as well as new societal and technological demands have challenged instructional design as both a discipline and a technology in the past two decades" (p.65). The field of instructional design is heavily influenced by the rapid development of new media and technology. Technology has changed people's assumptions and expectations on where and how learning can take place (Du Mont, 2002). The rapid development of new media and technologies, particularly, web-based technologies, web 2.0 and web 3.0 technologies, has made it necessary for designers to rethink their approaches to teaching and learning.

In the early days of the Internet, we celebrated the very existence of course materials online as cutting edge. It did not take long for designers to realize that converting our traditional courses or supplementary resources into electronic versions was nothing more than the digitization of our face to face lessons. The first phase of the internet mirrored behaviorist roots in education. The digitization of our curricula represented the *knowledge acquisition* model as we put up information on the web and expect our students to absorb the knowledge.

As we gradually move towards what many experts would call Web 2.0 technologies, we see a paradigm shift in how knowledge is represented. Stated by Natriello (2005), more and more public are involved in knowledge creation and transmission; the division of knowledge creators and knowledge receivers is eroding quickly; and the role of transmitting knowledge is eroding rapidly because basic transmission tasks are handled by knowledge carrying objects. Similarly, learners expect and are expected to take an active role in *knowledge creation*. We see users taking on new roles that challenge the traditional view. Craig (2007) noted that we are witnessing the shift from the "anyway, anytime" online learning approach to "more radical perspectives of community-based appropriation, editing, and self-publication" (p.5). In many ways, modern technologies that are social by nature enrich the learning experience by encouraging students to build knowledge from multiple perspectives and sources whether it is through blogging, modeling, or simulation (Craig, 2007).

What is Web 2.0 and how does it change the way we design our online courses? Web 2.0 has become "a collective term for a mass movement in society: a movement toward new forms of user engagement supported by Web-based tools, resources, services and environment" (Collis and Moonen, 2008, p.94). According to the Wikipedia entry on Web 2.0, it "is a term describing changing trends in the use of World Wide Web technology and web design that aims to enhance creativity, information sharing, and collaboration among users" (http://en.wikipedia.org/wiki/Web_2.0, last accessed on August 30, 2008).

There are differing opinions on the definition of Web 2.0. On the one hand, skeptics led by Tim Berners-Lee see it as "a piece of a piece of jargon, nobody even knows what it means. If Web 2.0 for you is blogs and wikis, then that is people to people. But that was what the Web was supposed to be all along" (quoted from Wikipedia at http://en.wikipedia.org/wiki/Web_2.0, last accessed on August 30, 2008). On the other hand, supporters like Tim O' Reilly (2005) view Web 2.0 as an agent of change. According to O'Reilly (2004), "Web 2.0 is the era when people have come to realize that it's not the software that enables the web that matters so much as the services that are delivered over the web" (quoted from http://www.readwriteweb.com/archives/web_20_is_not_a.php, last accessed on August 30, 2008). As much as we love debates on the real definition of Web 2.0, Alexander (2006) reminds us of what is most important: "The label 'Web 2.0' is far less important than the concepts, projects, and practices included in its scope" (Alexander, 2006, p. 32). For institutions of higher learning, the Web "provides near-ubiquitous access to quantities of content that are many orders of magnitude larger than that provided in any other medium" (Anderson, 2004, p.66). We need a new instructional map to deal with is being described as a "radically fluid environment characterized by media convergence, expectations for participation, and a shift towards collective intelligence" (Craig, 2007, p.7).

APPLYING INSTRUCTIONAL DESIGN IN ONLINE LEARNING ENVIRONMENTS USING CONSTRUCTIVIST PRINCIPLES

It is no coincidence that the rise of constructivism parallels new innovations in technology. Reigeluth (2005) noted that the shift from manufacturing to a knowledge-based economy requires higher levels of learning and thinking. He added that in the past, we have focused on lower level cognitive skills which are truly antiquated and inadequate to handle today's complex economy. We need to generate new instructional design theories to support the construction and application of knowledge in real-life contexts.

The number of students taking online courses tells the same story from a different perspective. In a 2007 report titled "Online Nation: Five Years of Growth in Online Learning," Ellen and Seaman (2007) found that 19.8% of students took at least one online course in Fall of 2006. This is an increase from 2003 where 9.7% of students were reported as being enrolled in at least one online course (Ellen and Seaman, 2007). The report also indicated that the online enrollment grew at a rate of 9.7 percent as opposed to the 1.5 percent growth rate for the overall student population in 2006.

It is time for us to examine our approach to instructional design in view of the rapid expansion of the knowledge-based economy along with the current and projected growth of online education. In 2008 the U.S Department of Labor - Bureau of Labor Statistics expected service-oriented employment to increase in the next decade as goods-producing employment continues to decrease. It anticipated that 15.6 million jobs would be added by 2016 and they would largely come from technology and service-oriented sectors.

We have witnessed the mismatch between technology affordances and pedagogical opportunities in higher education from the early promises of microcomputers in the 1970s to the Internet in the late 1990s (Collis and Moonen, 2008). Despite the prevalent use of Web 2.0 tools among post secondary students, "the potential for pedagogical innovation through the affordances of technology is not (much) reflected in institutional practice" (Collis and Moonen, 2008, p.96). Our institutions of higher learning have failed to harness the full pedagogical potential of the social web.

Where do we start? Merrill (1999) contends that instructional theory is concerned with two primary goals: what to teach and how to teach. The first part deals with the knowledge components needed for a specific type of instruction. Deciding this is a reasonably easy task. The second part, however, specifies the ways in which we present the selected knowledge components to engage our students and promote learning. Gold (2001) points out that "the difficulty of online instruction is not in the transfer of knowledge but in creation of the most apt learning environments for students to acquire this knowledge" (p.42).

In traditional instructional design, learners were taught to master skills and knowledge in a sequential manner. They progress from one chapter to another, completing assignments, and reading exercises in a structured environment where instructors rarely stray from a prescribed order of teaching. With web-based instruction, the rules of engagement can be different. Learners traverse through a series of non-sequential links to view information displayed on dynamic pages.

Among constructivists, there are diverse views and opinions on how to design effective instructions. As a result, we use labels like social constructivist, radical constructivist, and cognitive constructivist to describe our own beliefs about how knowledge is constructed. On one end of the spectrum, there are researchers who feel that knowledge construction is driven by socially-negotiated processes. Others argue that knowledge is driven by the cognitive process of the human brain.

Despite the philosophical differences, Applefield, Huber, and Moallem (2001) reckoned Bruning et. al (1995) and indicated that constructivists tend to agree on four characteristics as central to all learning: learners construct their own learning; new learning depends on students' existing knowledge; social interaction plays a critical role; and authentic tasks are needed to ensure meaningful learning" (p. 8). Using the four characteristics as the framework for constructivism, we will discuss how to apply them to online learning.

In many ways, the Internet is an ideal vehicle for the constructivist designers. Swan (2005) notes that it "provides a knowledge-centered environment where there is less emphasis on memorization of unconnected facts and procedures" (p.7). It offers what constructivists prize most in any learning environment: authentic learning context coupled with real world tasks and unique individual experiences. In order to create such an environment, Jonassen (1994) suggests that learners are presented with real-world problems that allow multiple representations of content and perspectives. In addition, instructional goals should be socially-negotiated.

Experienced online instructors and researchers agree that the main indicator of success in online teaching and learning is a well-designed course that fosters interactivity through the creation of a virtual learning community (Bender, 2003; Collison et al., 2000; Palloff & Pratt, 2001; Preece, 2000; Salmon, 2001; Swan, 2003). That is, the active communication, interaction, online presence, moderated discussions, and formation of an online learning community are the key elements for high quality online education.

Rourke et. al (1999; 2002) and Garrison, Anderson, Archer (2001; 2003) developed a Community of Inquiry model that highlights the importance of the development of social presence in the process. Their model of critical thinking and practical inquiry describes learning as occurring through the interaction of three overlapping core components: cognitive presence, teaching presence, and social presence. Swan (2003) adapted Rourke et al.'s (1999; 2002) model and highlighted interactivity as a feature of online environments that matters or is made to matter in learning. Swan provided suggestions for organizing interactivity in five areas: the learners' interactions with course content, the learners' interactions with instructors, the interactions among the classmates, the learners' interactions with computer and course interfaces, and the learners' virtual interactions (Swan, 2003). Lin (2008) proposed an online learning model – independent inquiry, collaborative inquiry, and formative inquiry towards expert knowledge – that would help learners to achieve their full rights to education and learning with new media and technologies.

These online teaching and learning models are in alignment with the constructivist principles of teaching and learning, although they did not originate from the instructional design models.

THE MERGING OF CONSTRUCTIVIST PRINCIPLES, INSTRUCTIONAL DESIGN MODELS, AND ONLINE TEACHING MODELS: INSTRUCTIONAL DESIGN MAP THEORY

We believe that there is a need to connect constructivist principles, instructional design theories, online learning models, and new media and technologies for web-based teaching and learning, rather than treating them as separate entities and disciplines. Such connections help create a synergistic effort for improving instructional design enhanced with new media and technologies. In order to tie these entities together, we will use Bransford et al. (2000)'s model as the framework because we believe Bransford et al.'s model provides a comprehensive context to start our discussions.

In "How People Learn," Bransford et al (2000) discussed the four components of an effective learning environment: learner-centered, knowledge-centered, assessment-centered, and community-centered. For our instructional design map model, we add the fifth element, that is, the ubiquitous new media and technology environment. Therefore, our map is made up of the following five components:

1. Learning-centered environments
2. Knowledge-centered environment
3. Assessment-centered environment
4. Community-centered environment
5. Ubiquitous media and technology environment

The word "map" is chosen to describe our proposed instructional design theory because this will allow the instructional designers to choose multiple paths to achieve their goals without sacrificing the quality of effective instructions. The instructional map provides a set of guidelines that each designer can use to achieve one or more units of instructional goals.

Instructional goals are often derived from the curriculum of a course. A single unit instructional goal is made up of authentic tasks, tools, and interactions to help learners construct new knowledge. Each unit is designed to stimulate thinking in learners that results in meaningful learning, deep understanding, and transfer to real-world context (Applefield, Huber, and Moallem, 2001). We will explain each of the components below.

Learner-Centered Environment

The first but often overlooked component of an instructional map is the learner profile, specifically in terms of learning styles and strategies. Designers often dive into the process of creating instructions without any clear indicators of who their learners are. The one-size-fits-all approach fails to take into consideration that people learn best when designers take the time to understand their educational and individual needs. These one-size programs tend to be more teacher-centered and instructive in nature. Since wholesale instructional design is clearly ineffective, designers must ask questions like "who will use the online course?" and "is the audience comfortable with technology?" in order to match instructional goals to learners.

One area of considerable interest is the issue of how learning styles impact instructional designs. Chen and Macredie (2002) broadly categorize learners into two groups. When it comes to the organization of learning activities, learners who are field-dependent tend to prefer fixed pathways to learning as opposed to the

Figure 1. A model of the instructional design map

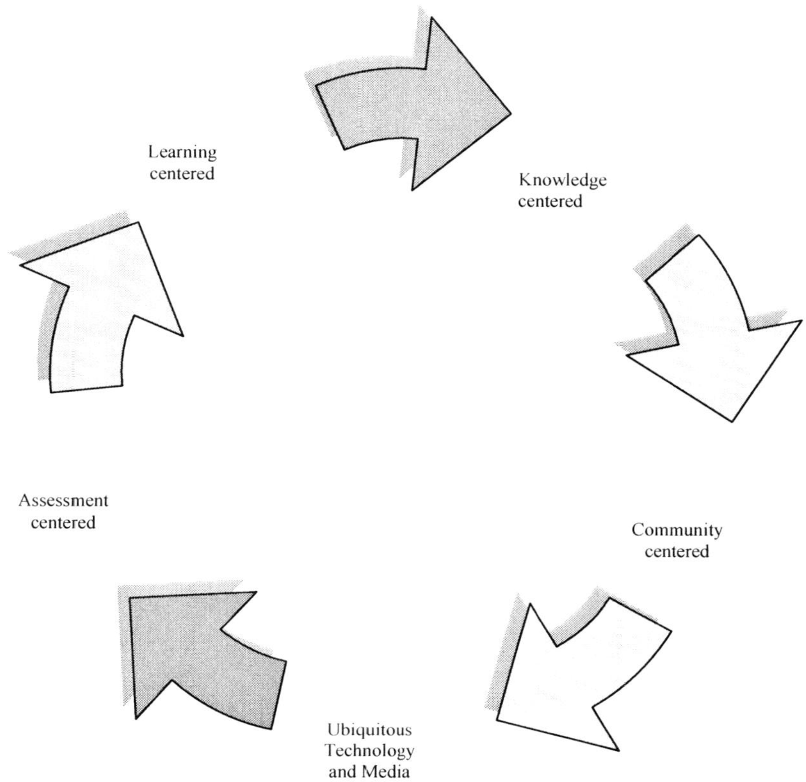

field independent learners. Along the same line, field-independent students tend to welcome the availability of multiple paths in online learning where they can access information in a nonlinear fashion. The field-independent learners tend to thrive in self-directed learning environments as compared to field-dependent learners. In order to accommodate both groups successfully, we must create an inclusive design that blends the desire for guidance (field-dependent) and the preference for a self-discovery approach (field-independent) by presenting multiple learning affordances to the learners.

Lin, Cranton, and Bridgall (2005) found connections between learners' personality and learning preferences while learning through online communications. For instance, more introverted learners find it most beneficial to have the time to think and reflect before sharing their thoughts, whereas more extroverted learners find it most satisfying to view multiple perspectives otherwise unavailable in their learning processes. The study pointed out that asynchronous written communication creates a bridge for some learners between conceptual and visual understandings. Web 2.0 technologies such as blog and wiki can be great tools to facilitate learner-centered environment because they take learners' initiatives into consideration while providing an open and flexible environment for self direction, collaboration, and guidance when necessary.

Knowledge-Centered Environment

In the early days of online course development, most courses were simply text-driven. Instructors would put up their notes or slides online with the hopes that learners would absorb the knowledge automatically. There were little to no discussions on how to create a knowledge-centered environment to promote learning. The issue of inert knowledge is not a modern invention. It has been around for more than 50 years. The online learning environment can extrapolate the risk of traditional education where learners face the problem of knowing something but failing to use it (Vosniadou & Ortony, 1989). Bransford et. al (1989) found that while fact-oriented acquisition may permit learners to retain information, it fails to help learners apply their knowledge to solve new problems. Von Glasersfeld (2005, p. 5) reminds us what is often wrong with traditional teaching:

Too often teaching strategies and procedures seem to spring from the naïve assumption that what we ourselves perceive and infer from our perception is there, ready-made for students to pick up, if only they had the will to do so.

In the constructivist learning environment, learning is seen as contextual (Ally, 2004). As such, situated learning is emphasized over structured learning. Situated learning involves learners constructing "their own knowledge rather than accepting that given by instructor" (Ally, 2004, p.19) through authentic tasks in relevant contexts. In addition, students are encouraged to create their own knowledge by exploring multiple representations of reality through blogging or podcasting.

Instructors play an important role in facilitating knowledge construction in an online learning environment. The constructivist facilitator can create instructional goals, offer diverse knowledge construction paths, promote diversity in ideas, and encourage creative and inclusive social learning processes. In contrast, the traditional transmission approach tends to reduce teaching to a single and less meaningful role. In order that learning be "effective, efficient, or engaging," Merrill (2002) argued that we need to have the following learning principles: 1) observation is a powerful way to learn; 2) learning happens when we apply new knowledge; 3) task-centered instructions engages learners; 4) prior knowledge should be activated; and 5) we learn better when we integrate new knowledge in our daily activities. While an interactive web offers the potential to create these kinds of engagements, we go a step further by noting that it also offers immediate use of the recursive nature of the subject matter and the ability to reinforce the fidelity of the learning point through social networking and simulation. In this respect, the open knowledge environment such as MIT Open Courseware, Wikipedia as well as Web 2.0 technologies such as Delicious and Classroom 2.0 are important knowledge-centered environments to be introduced to students. Students will then be able to use them as references and contribute their own knowledge when they aspire to do so.

Community-Centered Environment

Interactions with classmates are also critical in promoting community-building activities (Swan, 2004). Community-building can come in many forms. Swan (2004) suggests small group collaboration, online discussions, and building trust among learners. In any cases of community-building, authentic activities must be used to support instructional goals. Brophy and Alleman (1991) define activities as "anything students are expected to do, beyond getting input through reading or listening, in order to learn, practice, apply, evaluate, or in any other way respond to curricular content (p. 9)."

In the online environment, authentic tasks must provide "opportunities for collaboration and social interaction during learning and reflection after learning" (Chen, 2007, p.77). Young (1993) recommends the following test of authenticity. Learning situations should include some of the characteristics of real-life problem solving. There should also be an opportunity to distinguish between relevant and irrelevant information and "active/generative engagement in finding and defining problems as well as in solving them" (p.3). For Reeves, Herrington, and Oliver (2002), characteristics of authentic activities include real-world relevance, ill-defined activities, complex tasks that can be completed over a period of time, opportunity for learners to consider multiple perspectives and outcome diversity, collaboration, reflection, integration of multidisciplinary areas, and realistic assessment.

As much as we argue that the Internet is an ideal platform for constructivist learning, creating a rich online learning environment and populating it with authentic collaborative tasks are not easy by any means. Moellem (2001) concurs by suggesting that the task of designing web-based courses is inherently more difficult than it is to design traditional instructions. The virtual environment places more demand on the "instructor's ability to mentor, interpret, coach, scaffold, manage, and help" (p. 12). A community-centered environment is not just about creating a virtual space for a group of students to create new knowledge. The space must be easy for students of varying technical abilities to direct their own learning and participate as a member of the community.

Learners are constantly trying to make sense of their environments and the information presented to them (Wilson and Lowry, 2000). To facilitate the learning tasks, designers must create an effective web interface to manage the contents of their online courses. Learners will certainly benefit from a carefully planned open learning channel with multiple paths through cases and tools for them to determine their own objectives and destinations (Moallem, 2001).

Since we expect learners to spend a substantial amount of time and energy to directing their own learning, our web interface must succeed in helping the learners develop their curiosity, sustain learning interest, and make meaningful connections between content, media formats, and other course participants. Navigation of learning content should be easy and intuitive for the learners. Researchers have often cautioned instructors against using complicated menus, vague metaphors, unfamiliar navigations buttons, or confusing options to prevent users from being frustrated with the interface. Insufficiencies of concepts, consistency, screen layout, interface design, navigation strategies, low-level of interactions can result in inadequate learning environments (Klett, 2002).

In addition, designers are challenged to solve the issue of disorientation among users. Learners (especially beginners) often complain about being "lost" in the navigation system. Unlike a book where one progresses steadily from one page to the next, the ability to "roam" freely in the virtual environment is a double–edged sword. Inexperienced linear learners often find themselves disoriented in a spherical or three-dimensional learning space. Oliver, Herrington, and Omari (1997) suggest three strategies to aid orientation: placement cues to provide visual clues, hierarchies and indices for ease of access, and semantic tools to indicate relationships as well as functions. The strategies will help reduce confusion among users who are not familiar with multiple learning pathways. Hence, online learners can take advantage of open and free web technologies such as Slideshare, Google Docs, Zoho, Second Life, and Ning to create their learning communities and focus on their learning goals without worrying too much about the technical aspects of technologies.

Assessment-Centered Environment

Learners must be assessed in a meaningful context. Swan (2005) reiterates that "good assessment practices emphasize learning with understanding and the application of knowledge, and not the memorization of isolated facts and procedures" (p.8). In online learning, designers must integrate assessments into real-world tasks to gauge whether students understand how the desired knowledge may be applied in everyday life.

Traditional assessment has been paper-based, classroom-based, formalized, synchronized, and controlled. With online learning, there is a great deal of emphasis on self-directed learning, collaborative tasks, multiple paths to knowledge construction and application help students develop their problem solving and metacognitive skills. Unlike the traditional models of assessment, constructivism examines the thinking process (Karagiorgi and Symeou, 2005). We are interested in how students construct and apply knowledge in real-world situations. Designers are encouraged to integrate assessment tools like portfolios, team reports, web pages, and blogs into authentic activities to promote what Swan and others call "meaning making."

Gaytan and McEwen (2007) suggest a variety of online assessments to ensure that "there is a high level of interaction between students and faculty and among students" (p.130). Some of the suggestions include exams, quizzes, projects, self-tests, and peer assessments. They believe that timely and meaningful feedback is the most important component of assessment. It is not uncommon for instructors to take weeks to provide feedback to students on a project or an assignment. By then, students would have moved on to the next learning unit or project and the impact of the feedback is at best minimal. The advantages of using web 2.0 tools are that the services are typically inexpensive or free, easy to maintain, and very scalable.

Like its traditional counterpart, online assessment has its own share of criticism. Researchers have noted that online assessment often mimic classroom assessment. All assessment tools create both intended and unintended consequences and the resultant behaviors in organizations and in learners. Often the learning metrics flow from the organizational metrics. Online learning assessments need to be focused more on the learning objectives and less on the organizational agenda so as to serve the end user in authentic ways intended by the curricula. Web technologies such as digital portfolio, blogs, wikis, and websites have documented as more inclusive, comprehensive, and authentic assessment tools.

A Learning Environment Equipped with Ubiquitous New Media and Technologies

In our Instructional Design Map, new media and technologies are viewed as a means to an end. In the last few years, we have seen just how diverse these technologies are. Unlike the traditional model of online learning that was built around a closed system, i.e., a course or a course management system, the current trend suggests a push towards an open learning system. The open system has tapped into popular social networking technologies or web 2.0 technologies such as blogs, wikis, podcasts, and 3-D learning environments to promote the constructivist agenda. The most important criterion for any technology selection is that the technology must add value to the educational environment. The purpose of using such technologies must be also obvious to the learners. Experienced instructional designers routinely surpass novice instructional designers in this area more than in any other. By their creative use and selection of the most innovative and appropriately matched technologies and simulations, they demonstrate the ontological and recursive aspects of the subject matter. These experts bridge the knowledge gaps of the learners by positioning the learners where they can discover the truths for themselves.

New media and technologies, in particular, Web 2.0 technologies, enable learners to be creators, co-creators, and owners of their learning and communication environments (Lin, 2008).

The learners' efforts are enabled by the ease of use, ease of creation, and ease of participation afforded by the new media and technologies. Weblogs or blogs, perhaps the most familiar example in the Web 2.0 realm, allow users to create a web presence (e.g., to post reflections, images, links without having to code or learn to use web authoring programs), receive comments, and connect to other like-minded people. Wiki technologies, as another example, allow users to write collaboratively on topics, publish them immediately, and update them frequently. Wikipedia, for instance, has not only become a web-based free content encyclopedia, but also a community with more than 75,000 active volunteer contributors working on some 9,000,000 articles in more than 250 languages (http://en.wikipedia.org/wiki/Wikipedia:About, last accessed on August 30, 2008). These new but rapid developments and affordances of new media and technologies are in alignment with the constructive principles, and should be actively incorporated in constructive teaching and learning environments.

FUTURE TRENDS AND CONCLUSION

In this chapter, we have argued that constructivism is an ideal approach for online learning. As in many any other changing fields, we are constantly reassessing and revising our own dynamic theories to determine which engender best practices in designing instructions for a learner-centered environment. In predicting the future trends for instructional technologies, Reiser and Dempsey (2002) envision learning environments that are "immersive, interactive, (and) adaptive to individual learners." Others see the inevitable shift towards an open-ended learning environment where students make autonomous decisions about their learning process (Alessi and Trollip, 2001). We happen to think that we are witnessing the convergence of both the constructivist and the open visions. From a practical aspect, we believe that instructional designers and online teachers are afforded with multiple and flexible channels and technologies to craft their designs for an interactive online learning environment. Undoubtedly, more research needs to be done to determine ways to evolve constructivism for advancing the cause of online education.

REFERENCES

Alexander, B. (March/April 2006)., Web 2.0: A new wave of innovation for teaching and learning? http://net.educause.edu/ir/library/pdf/ERM0621.pdf. *EDUCAUSE Review, 41*(2), 32–44.

Allesi, S., & Trollip, S. (2001). *Multimedia for Learning: Methods and Development.* Massachusetts: Ally & Bacon.

Ally, M. (2004). The Theory and Practice of Online Learning. In T. Anderson & F. Elloumi (Eds.), *Foundations of Educational Theory.* Athabasca, Canada: Athabasca University.

Anderson, T. (2004). Toward a theory of online learning. In T. Anderson & F. Elloumi (Eds.), *Theory and practice of online learning* (pp. 33-60). Athabasca, AB: Athabasca University.

Applefield, J., Huber, R., & Moallem, M. (2001). Constructivism in Theory and Practice: Towards a Better Understanding. *High School Journal, 84*, 35-53.

Audi, R. (2002). *Epistemology: A Contemporary Introduction to the Theory of Knowledge.* Routledge.

Barab, S. A., & Duffy, T. (2000). From practice fields to communities of practice. In D. Jonassen & S. M. Land (Eds.), *Theoretical Foundations of Learning Environments* (pp. 25-56). Berkley, CA: McCutchan Publishing.

Bruning, R. H., Schraw, G. J., & Ronning, R. R. (1995). *Cognitive psychology and instruction*, 2nd ed. Englewood Cliffs, NJ: Prentice Hall.

Bureau of Labor Statistics, U.S. Department of Labor, *Occupational Outlook Handbook, 2008-09 Edition, Tomorrow's Jobs*. Retrieved August 29, 2008 from http://www.bls.gov/oco/ocos114.htm

Bransford, J. D., Franks, J. J., Vye, N. J., & Sherwood, R. D. (1989). New approaches to instruction: Because wisdom can't be told. In S. Vosniadou & A. Ortony (Eds.), *Similarity and analogical reasoning* (pp. 470-497). NY: Cambridge University Press.

Bransford, J. et. al. (2000). *How people learn*. Washington, DC: National Academy Press.

Bender, T. (2003). *Discussion-based online teaching to enhance student learning: Theory, practice and assessment*. Sterling, VA: Stylus.

Boghossian, P. (2006). *Behaviorism, Constructivism, and Socratic Pedagogy*, *38*, 713-722.

Brophy, J. (Ed). (2002). *Social Constructivist Teaching: Affordance and Constraints*. Elsevier: London.

Brophy, J., & Alleman, J. (1991). Activities as instructional tools: A framework for analysis and evaluation. *Educational Researcher, 20*(4), 9-23.

Chen, S. (2007). Instructional Design Strategies for Intensive Online Courses: An Objectivist-Constructivist Blended Approach. *Journal of Interactive Online Learning, 6*(1).

Chen, S., & Macredie, R. (2002). Cognitive Style and Hypermedia Navigation: Development of Learning Model. *Journal of the American Society for Information Science and Technology*. Retrieved April 16, 2008 from http://www.brunel.ac.uk/~csstsyc/sherry_paper.pdf

Collis, B., & Moonen, J. (2008). Web 2.0 tools and processes in higher education: quality perspectives. *Educational Media International, 45*(2), 93. doi: 10.1080/09523980802107179.

Collison, G., Haavind, S., & Tinker, R. (2000). *Facilitating online learning: Effective strategies for moderators*. Madison, WI: Atwood.

Craig, E. M. (2007). Changing paradigms: managed learning environments and Web 2.0. *Campus-Wide Information Systems, 24*(3), 152 - 161. doi: 10.1108/10650740710762185.

Dijkstra, S. (2005). Cognition and Instructional Design for Problem-Based Learning. In Spector, M., Ohrazda, C., Van Schaack, A., Wiley, D. (Eds.), *Innovations in Instructional Technology*, (pp. 187-206). New Jersey: Lawrence Erlbaum Associates.

Du Mont, R. (2002). *Distance Learning – A Systems View: An Assessment and Review of Literature*. Retrieved February 12, 2008 from http://www.rcet.org/research/ATT-OLN/Dumont-finalreport.pdf

Dinter, F. (1998). Constructivism in Instructional Design Theory. *Journal of Structural Learning & Intelligent Systems, 13*(2). Retrieved January 19, 2007 from Academic Search Premier.

Ellen, E., & Seaman, J. (2007). *Online Nation: Five Years of Growth in Online Learning*. Sloan Foundation. Retrieved May 22, 2008 from http://www.sloan-c.org/publications/survey/pdf/online_nation.pdf

Ertmer, P., & Newby, T. (1993). Behaviorism, cognitivism, constructivism: Comparing critical features from an instructional design perspective. *Performance Improvement Quarterly, 6*(4), 50-72.

Fardanesh, H. (2006). *A Classification of Constructivist of Instructional Design Models based on Learning and Teaching Approaches*. Retrieved March 27, 2008 from ERIC database.

Garrison, D. R., Anderson, T., & Archer, W. (2001). Critical thinking, cognitive presence and computer conferencing in distance education. *American Journal of Distance Education, 15*(1), 7-23.

Garrison, D. R., Anderson, T., & Archer, W. (2003). A theory of critical inquiry on online distance education. In M. G. Moore & W. G. Anderson (Eds.), *Handbook of Distance Education* (pp. 113-127). New Jersey, London: Lawrence Erlbaum Associates, Publishers.

Gaytan, J., & McEwen, B. C. (1997). Effective online instructional and assessment strategies. *The American Journal of Distance Education, 21*(3).

Gold, S. (2001). A constructivist approach to online teaching for online teachers. *Journal of Asynchronous Networks, 5*(1). Retrieved April 8, 2007 from http://www.sloan-c.org/publications/jaln/

Guarino, N. (1998). Some ontological principles for designing upper level lexical resources. In *Proceedings of the First International Conference on Lexical Resources and Evaluation*. pp. 1-8. Granada, Spain, May 28-30.

Hannifin, M. J., Hannifin, K. M., Land, S. M., & Oliver, K. (1997). Grounded practice and the design of constructivist learning environments. *Educational Technology Research & Development, 45*(3), 101-117.

Hew, K. (2004). Past Technologies, Practice and Applications: A Discussion on How the Major Developments in Instructional Technology in the 20th Century. *Association for Educational Communications and Technology.* Retrieved March 31, 2008 from ERIC database.

Jonassen, D. (1994). Thinking technology. *Educational Technology,* (pp. 34-37).

Jonassen, D. H., & Land, S. M. (2000). Preface. In D. H. Jonassen & S. M. Land (Eds.), *Theoretical foundations of learning environments* (pp. iii-ix). Mahwah, NJ: Lawrence Erlbaum Associates

Jonassen, D. (2005). Problem Solving: The Enterprise. In Spector, M., Ohrazda, C., Van Schaack, A., Wiley, D (Eds), *Innovations in Instructional Technology,* (pp. 91-110). New Jersey: Lawrence Erlbaum Associates.

Karagiorgi, Y., & Symeou, L. (2005). Translating Constructivism into Instructional Design: Potential and Limitations. *Educational Technology & Society, 8*(1), 17-27.

Klett, F. (2002). Visual Communication in Web-based environment. *Educational Technology & Society, 5*(4). Retrieved March 8, 2008 from http://www.ifets.info/journals/5_4/klett.pdf

Lin, L. (2008). An online learning model to facilitate learners' rights to education. *Journal of Asynchronous Learning Networks (JALN), 12*(1), 127-143.

Lin, L., Cranton, P., & Bridgall, B. (2005). Psychological type and asynchronous written dialogue in adult learning. *Teachers College Record. 107*(8), 1788-1813.

Moallem, M. (2001). Applying Constructivist and Objectivist Learning Theories in the Design of A Web-Based Course: Implications for Practice. *Educational Technology & Society,* 4(3), 113-125.

Merrill, M.D. (1999). Instructional Transaction Theory (ITT): Instructional Design Based on Knowledge Objects. In Charles M. Reigeluth (Ed.), *Instructional Design Theories and Models: A New Paradigm of Instructional Technology.* Lawrence Erlbaum Associates.

Merrill, M. D. (2002). First principles of instructions. *Educational Technology Research and Development, 50*(3), 43-59.

Molenda, M. (1997). Historical and Philosophical Foundations of Instructional Design. In Tennyson, R., Schott, F., Seel, N., & Dijkstra, S. (Eds), *In-*

structional Design Volume 1, 41-54. New Jersey: Lawrence Erlbaum Associates.

Natriello, G. (2005). Modest changes, revolutionary possibilities; distance learning and the future of education. *Teachers College Record, 107*(8), 1885-1904.

Oliver, R, Herrington, J., & Omari, A. (1997). Creating effective instructional materials for the World Wide Web. In *Proceedings of AusWeb 97 Conference*.

Otting, H., & Zwaal, R. (2006). Critical task characteristics in problem-based learning. *Industry and Higher Education, 20*, 347.

Palloff, R. M., & Pratt, K. (2001). *Lessons from the cyberspace classroom: The realities of online teaching.* San Francisco: Jossey-Bass.

Preece, J. (2000). *Online communities: Designing usability and supporting sociability.* Hoboken, NJ: Wiley.

Reeves, T., Herrington, J., & Omari, A. (1996). *Creating effective instructional materials for the World Wide Web.* Retrieved May 31, 2008 from Reeves, T., http://elrond.scam.ecu.edu.au/oliver/docs/96/AUSWEB1d.pdf

Reeves, T., Herrington, J., & Oliver, R. (2002). Authentic Activities and Online Learning. *Australian Journal of Educational Technology, 19*(1), 59-71

Reigeluth, C. (2005). New Instructional Theories and Strategies for a Knowledge-Based Learning. In Spector, M., Ohrazda, C., Van Schaack, A., & Wiley, D (Eds.), *Innovations in Instructional Technology*, (pp. 207-218). New Jersey: Lawrence Erlbaum Associates.

Reiser, R., & Dempsey, J. (2002). *Trends and Issues in Instructional Design and Technology.* New Jersey: Pearson.

Rourke, L., Anderson, T., Archer, W., & Garrison, D. R. (1999). Assessing social presence in asynchronous, text-based computer conferences. *Journal of Distance Education, 14*(3), 51-70.

Rourke, L., & Anderson, T. (2002). Exploring social presence in computer conferencing. *Journal of Interactive Learning Research, 13*(3), 259-275.

Salmon, G. (2001). E-*moderating: The key to teaching and learning online*. London: Kogan Page.

Schuh, K., & Barab, S. (2008). Handbook of Research on Educational Communications and Technology. In J. M. Spector, M. D. Merrill, J. van Merrienboer, & M. Driscoll (Eds.), *Philosophical Perspective* (pp. 67-81). New York, NY: Taylor & Francis.

Seel, N. (2005). Cognition Designing a Model-Centered Learning Environments: Hocus Pocus or the Focus Must Strictly be on Locus. In M. Spector, C. Ohrazda, A. Van Schaack, & D. Wiley, (Eds.), *Innovations in Instructional Technology*, (pp. 95-90). New Jersey: Lawrence Erlbaum Associates.

Smith, P., & Ragan, T. (1999). *Instructional design* (2nd ed.). New York: John Wiley & Sons, Inc.

Spector, M. (2000). *A Philosophy of Instructional Design for the 21st Century.* Invited paper presentation at the American Educational Research Association Annual Meeting, 27 April 2000, New Orleans, Louisiana.

Staver, J. R. (1998). Constructivism: Sound theory for explicating the practice of science and science teaching. *Journal of Research in Science Teaching, 35*(5), 501-520.

Swan, K. (2003). Learning effectiveness: What the research tells us. In J. Bourne & J. C. Moore (Eds.), *Elements of quality online education: Practice and direction (*pp. 13–45). Needham, MA: Sloan Center for Online Education.

Swan, K. (2005). A constructivist model for thinking about leaning online. In J. Bourne &

J. C. Moore (Eds.), *Elements of Quality Education Online: Engaging Communities*. Needham: Sloan C.

Tennyson, R., & Schott, F. (1997). Instructional Design Theory, Research, and Models. In R. Tennyson, F. Schott, N., Seel, & S. Dijkstra, (Eds.), *Instructional Design Volume 1*, (pp. 1-18). New Jersey: Lawrence Erlbaum Associates.

Von Glasersfeld, E. (2005). Introduction: Aspects of Constructivism. In C. Fostenot, (Ed.), *Constructivism: Theory, Perspective, and Practice.* New York: Columbia University.

Vosniadou, S., & Ortony, A. (1989). *Similarity and Analogical Reasoning.* Cambridge University Press.

Wilson, B., & Lowry, M. (2000). *Constructivist Learning on the Web. New Directions for Adult and Continuing Education.* Retrieved February 19, 2007 from http://www3.interscience.wiley.com/cgi-bin/jhome/86011352

Young, M. (1993). Instructional design for situated learning. *Educational Technology Research & Development, 41*(1), 43-58.

Chapter VI
Enhancing Authentic Assessment Through Information Technology

Beth Rubin
DePaul University, USA

ABSTRACT

Constructivist education usually involves authentic assessment, which is affected by the media used to teach. Information technology can enhance or hinder the authenticity of assessment in key ways. This chapter provides a framework to analyze the opportunities to enhance authenticity when assessment is mediated by information technology, as well as the limitations of IT mediation on authenticity. The degree of authenticity possible is determined by several aspects of the competence being assessed: the chronicity of access to and use of information, the durability of the display, the use of written, oral and non-verbal communication, and computer use. This framework is used to identify IT tools that enable more authentic assessment, as well as sample approaches. It identifies aspects of competences that may not be authentically assessed via information technologies, limitations on the use of technology, and future trends.

INTRODUCTION

For more than twenty years, the preeminent theoretical approach to understanding and improving education has been constructivism. Many researchers have built upon the work of Piaget, Dewey, Vygotsky and others to develop different, occasionally competing paradigms of how people think, learn and know within this umbrella concept, such as radical constructivism, social constructivism, critical constructivism and information-processing constructionism (Dougiamas, 1998; Downing & Holtz, 2008; Gergen, 1995; Steffe & Gale, 1995; von Glasersfeld, 1995).

Although there are many differences among these approaches, most share several key principles about how people know and learn:

- Knowledge is adaptive; that is, it helps people to deal with and respond to the world around them.
- Learners actively construct knowledge, rather than passively receiving it from outside experts. Learners need to actively engage with concepts, processes and facts.
- Learning comes from the interaction of the person and his/her environment, particularly with other people; it often socially constructed.
- Learners are not "blank slates" but rather complex thinkers. New knowledge must fit into, and enhance the complexity of, the knowledge and schemas that they already have. Learners perceive all new information through a lens of their history, beliefs, values and perceptions, which are continually evolving.

There are many pedagogical implications of constructivism. These include the importance of practical and applied work, as well as collaboration and social networking. Learning happens best with scaffolded educational materials that support developing expertise over time; as well as flexible and student-centered course designs that provide multiple paths to learning. Learners should have the autonomy to regulate their own learning; and assessment should be authentic (Anderson, 2008; Bonk & Cunningham, 1998; Dougiamas, 1998; Downing & Holtz, 2008; Hirumi, 2002). This chapter will focus on authentic assessment as a constructivist pedagogical approach, and how it is enhanced – or hindered – when mediated by information technology.

Authentic assessment has been defined as "engaging and worthy problems or questions of importance, in which students must use knowledge to fashion performances effectively and creatively. The tasks are either replicas of or analogous to the kinds of problems faced by adult citizens and consumers or professionals in the field" (Wiggins, 1993, p. 229). Authentic assessment is critical if students are situated in the center of the learning, rather than having education revolve around the faculty, and when learning is socially constructed (Albon, 2003; Bransford, Brown & Cocking, 1999) because it is meaningful, realistic, complex, and demonstrates competence through the performance of "real world tasks" (Mueller, 2006). It measures not just what students know, or can do, but what they actually do.

Key principles of authentic assessment include: 1) students are required to "to be effective performers with acquired knowledge"; 2) students address "the full array of tasks that mirror the priorities and challenges" of real world use; 3) there are high standards and mastery of complex skills, because students must "craft polished, thorough and justifiable answers, performances or products"; and 4) assessments "involve 'ill-structured' challenges and roles," reflecting the requirements of real world use (Wiggins, 1990, What Is Authentic Assessment section, para 4-9).

Constructivism and Authentic Assessment

In most respects, authentic assessment reflects a constructivist approach to instruction. It places the student in the center of the learning activity, and involves meaningful and important applications of knowledge. It requires active engagement with information, sometimes socially, and builds on existing knowledge of the learner. It provides for situated learning, and has many opportunities for scaffolding. It gives students a degree of autonomy over their learning process and product.

However, in most work places, work products are produced with limited autonomy. Supervisors and work places have specific requirements for the processes and tools that must be used, and standards for the products. For example, legal briefs must follow a specific format, and medical tests must be conducted and reported in specific ways. When a supervisor requests a PowerPoint presentation with Visio process flow diagrams using specific data, employees who provide expository essays with Venn diagrams using different data soon find out how limited their autonomy truly is; they will be producing essays for their next employer.

In this way, truly authentic assessment may conflict with some tenets of constructivism: it is not entirely student-directed. In some cases, there are single processes that must be closely followed, and products may need to be structured in a specific, limited way. Templates may be required, and specific standards, whether qualitative or quantitative, may be essential. Assessment that is authentic to these requirements, in fields ranging from nursing to criminal justice, may limit learner autonomy in violation of constructivist principles. This is one of the practical challenges of implementing both authentic assessment and constructivist pedagogy. Reflection on the essential nature of each approach can be one route to integrating both pedagogical approaches.

BACKGROUND

When instruction and assessment are mediated by information technologies, it changes the nature of the relationships between the instructor and students, and those among the students. This can increase or decrease the ability to teach in a constructivist manner, particularly in terms of the authenticity of assessment. Understanding the different information technologies, and the ways in which they can be used to increase authenticity of assessment, can increase the ability to use constructivist pedagogy in online courses.

Information Technologies Used to Mediate Instruction and Assessment

Increasingly, many post-secondary classes are presented using information technologies. According to the Sloan Foundation (Allen & Seamen, 2007), twenty percent of post-secondary students took at least one online course in the fall of 2006, which represented a nearly 10% increase from 2005. In addition to distance education courses, many universities are also offering web-enhanced or hybrid courses (Allen & Seamen, 2007). Assessment of student learning conducted through information technology in such courses may involve a range of approaches, from online quizzes to electronic transmission of project requirements and submission of completed work. Students may use asynchronous discussions to explore topics, collaborate on group projects, and conduct and analyze research, all of which can be used for assessment.

Using information technology increases geographical access to teaching materials and student work through Learning Management Systems (LMSes) such as Blackboard (http://www.blackboard.com/), Angel (http://www.angellearning.com/) and eCollege (http://www.ecollege.com/). These tools present content and enable asynchronous discussion and posting of information, as well as announcements, online grade books, group sharing spaces, and some tools for synchronous (same time) collaboration such as written chat. Both LMSes and electronic portfolios (Barrett, 2006) provide more flexibility and control over the timing of doing work because students do not have to wait until class time to view materials or turn in assignments, which increases student autonomy.

Since 2004, many innovations in information technology have been developed, often referred to as Web 2.0 technologies (O'Reilly, 2005). These tools support distributed communication, community development, synchronous and asynchronous collaboration, and distributed generation of knowledge by a community of users (Barrett, 2006; Barrett, 2008; O'Reilly, 2005), and many are well suited to constructivist post-secondary education (Anderson, 2008c; Bonk & Cunningham, 1998; Dougiamas, 1998; Hirumi, 2002). With these tools, the development of everything from web pages to bibliographies is decentralized, using automation to coordinate among many people.

For example, social bookmarking uses web-based tools such as del.icio.us, which allow people to store internet bookmarks and share them with others; each person contributes to and taps into a network of relationships (http://del.icio.us/). Another example is the use of wikis, which are shared websites jointly developed by a group (see http://www.commoncraft.com/video-wikis-plain-english); information is generated by anyone given access, and the site is constantly modified with all changes automatically tracked. Blogs, short for "Web Logs," are asynchronous discussion boards outside of time-limited classes, generated by community members. Another such innovation is the use of RSS ("Really Simple Syndication") feeds, which involve automated transmission of emails or links from one internet location to others. These are used to link people and automate the distribution of information, which improves the ease of connection and potential for social learning (Anderson, 2008c).

Educational Social Software, such as ELGG (http://classic.elgg.org/), includes a variety of tools to support learners in developing a learning community, sharing resources and jointly developing and applying knowledge (Anderson, 2008c). There are a number of free or inexpensive tools to support video and audio recordings by faculty and students, such as Camtasia (http://www.techsmith.com/camtasia.asp) and Audacity (http://audacity.sourceforge.net/). Tools such as Skype (http://www.skype.com) transmit telephone calls and images over the internet for very low cost, using Voice Over Internet Protocol (VOIP).

Webinar software such as Webex (http://www.webex.com/smb/elearning.html), Elluminate (http://www.elluminate.com) and Wimba (http://www.wimba.com/) allow synchronous meetings, video conferencing, shared virtual whiteboards and archiving capabilities. Other new tools that can be used to support post-secondary education include Multi-User Virtual Environments, known as MUVEs, such as Second Life (http://secondlife.com/) and ActiveWorlds (http://www.activeworlds.com/).

Other innovations include downloadable audio and video podcasting; and use of handheld devices such as Blackberries and iPhones to read, watch videos, text message, and communicate with a range of media. Many computers have built-in video cameras, speakers and microphones, and web cameras are inexpensive and easily plug into computers. Digital video cameras produce files that can be uploaded into websites such as YouTube and widely shared. The use of such tools is growing exponentially, and they can support both learning and assessment processes.

With these tools, the locus of expertise is spread among many people because they enable distributed production of knowledge. All class members can contribute to the developing knowledge base, post using a variety of media beyond the written word, insert links to multiple sites, and use RSS feeds to aggregate information from a wide range of internet sources (Anderson, 2008c; Farmer, 2004). Users can capture a wide range of expressions and types of performance, and compile them in online portfolios (Barrettt, 2006; 2008; 2008b).

Many of these tools have multi-sensory outputs, time controls, built-in portfolio frameworks to capture tasks in which students demonstrate competence, and a great deal of flexibility. These new tools can enable authentic assessment of some skills and competences. However, the existence of these tools does not mean that they will be effectively applied to support authentic assessment. As Anderson (2008b) notes:

These innovations have greatly improved the ability to use information technology in support of authentic assessment. However, there are still many limits. There is a challenge to developing awareness and comfort with technologies, selecting technology that will be stable and easy to use, and matching appropriate technology to assessment needs. We are now entering an era, however, where streaming video, video, and audio-conferencing, pod and videocasts, and immersive worlds are readily available for educational use. Thus, online learning theory needs to help educators decide which of the numerous technological options is best suited for their application. (p. 54)

In addition, when assessment of student learning is mediated by information technology, it can create barriers of computer links and written words between teacher and student, increasing artificiality. It can make frequent developmental feedback difficult by requiring many clicks and transaction time, which hinders faculty input into the construction of knowledge. However, information technology can also enhance the authenticity of assessment.

Authentic assessment that is mediated by information technology must reflect key aspects of the competences taught. The next section develops a framework that can be used to identify competences that can be authentically assessed through technological mediation at a distance, and to select appropriate tools.

USING INFORMATION TECHNOLOGY TO SUPPORT AUTHENTIC ASSESSMENT

There are several aspects of every competence that affect how it can be authentically assessed via information technology, and which tools and media to use. In some cases, information technology can increase the authenticity over that available in standard in-class assessments; in others, information technology reduces the authenticity.

These aspects of competences include: 1) the chronicity, or timing, of obtaining or responding to information; 2) the durability of the demonstration or use of the competence; 3) the role of oral and visual non-verbal communication, haptics, proxemics and kinesics; and 4) the role of computers intrinsic to the skill. These are summarized in Figure 1 below, and described in detail in subsequent sections. This figure also presents examples of authentic assessments that can be used for representative tasks, and information technologies that can be used to increase the authenticity of assessment.

Chronicity

Some skills developed in higher education, such as textual analysis, involve dealing with information or stimuli that are present at one point in time. Other skills require that learners deal with incomplete information or procedures that unfold over time. For example, diagnosing medical conditions and managing employees both involve interpreting information that changes over time. Authentic assessment would involve including the same chronal characteristics in assessment that exist in actual use of the skills.

Two aspects, in particular, are critical:

1. The time frame over which information is made available, ranging from simultaneous access to all information, as with a research paper, to information unfolding over weeks with iterative decisions to be made with different levels of knowledge, as when public policy decisions are made during crises.
2. The time available to deal with information, ranging from open-ended time for analysis, as with a research paper, to very short time limits as with medical decisions in life-threatening situations.

Figure 1. Nature of competence and information technology support for authenticity of assessment

Aspect of Competences Assessed	Nature of Competences	Examples of Competences	Examples of Authentic Assessments	IT Support for Authentic Assessment	Examples of IT Tools
Chronicity & control of access to information	Information simultaneously available	- Strategic planning - Giving effective summative feedback to students	- Develop a business plan, - Integrate multiple assessments into a single grade with feedback	- Information posted online or emailed for geographically distributed access - Record and track student access of information	- Learning Management System (LMS) - Wiki
	Information distributed over time	- Medical diagnosis - Conflict management	- Problem-Based Learning (PBL) case of making a medical diagnosis - Exercise in mediating a workplace conflict	- Information released with controlled timing - Information released by task/ module completion - Record and track student access of information	- LMS post, announcement or assignment materials - Wiki with RSS - email
	Information needs determined by learner	- Organizational diagnosis - Financial investment modeling	- PBL case of diagnosing organizational problems - Case analysis of making a financial decision	- Information generated or posted by learners - Record and track student access of information	- LMS - Wiki with RSS - Educational Social Software
Chronicity of processing or acting on information	Limited time to process	- Applying medical treatment - Conflict management	- Simulation of treating a patient in crisis - Role play of addressing disruptive student behavior	- Tasks & resources released with controlled timing - Product (paper, analysis, report) required within controlled time	- LMS - Online testing program - Video or webconference - Voice conference - RSS feed - Educational Social Software
	Extended time to process	- Analysis of historical data	- Compare religious practices of two civilizations	- Information posted online or emailed - Tasks & resources always available	- LMS - Wiki - Blog
Durability of demonstration of competence	Ephemeral process	- Artistic Performance - Public speaking - Negotiation - Critical thinking	- Perform a dance, song or performance art - Class presentation - Role play a negotiation - Debate a current topic	- Record and synchronously transmit recording - Record and asynchronously transmit recording - Synchronous verbal or visual interaction	- Desktop-based recording software - Web-based recording tools - Voice or video recording hardware - Webinar or teleconference
	Durable process	- Creative Writing - Legal analysis - Textual or historical analysis - Computer programming	- Write a screenplay - Write a legal brief - Develop a glossary - Write and test a computer program	- Record and track access of information - Transmit work and feedback to and from individual or group - Check for academic integrity	- LMS - Wiki with RSS - Blog - Automated system to identify plagiarism

continued on following page

Figure 1. continued

Aspect of Competences Assessed	**Nature of Competences**	**Examples of Competences**	**Examples of Authentic Assessments**	**IT Support for Authentic Assessment**	**Examples of IT Tools**
Oral Communication	Intrinsic to competence	- Teaching - Public speaking	- Lead a class activity - Give a lecture - Cross-examine a witness	As with "Ephemeral process" above	As with "Ephemeral process" above
	Not required	As with "Durable Process" above	As with "Durable Process" above	As with "Durable Process" above	As with "Durable Process" above
Non-verbal (visual) Communication and Kinesics	Intrinsic to competence	- Counseling - Bedside manner - Negotiation	- Hold a counseling session - Role play of giving news to a patient - Cross-cultural negotiation simulation	- Record and synchronously transmit recording - Record and asynchronously transmit recording - Synchronous visual interaction	- Desktop-based visual recording tools - Web-based video recording tools - Video recording hardware - Webinar software with web camera - Traditional videoconference
	Not required	As with "Durable Process" above	As with "Durable Process" above	As with "Durable Process" above	As with "Durable Process" above
Proxemics and Haptics	Intrinsic to competence	- Child care - Acting - Nursing skills - Electrical engineering	- Calm an upset child - Give a theatrical performance - Draw blood - Wire an electrical panel	- Record and synchronously transmit recording of face-to-face interaction - Record and asynchronously transmit recording of face-to-face interaction - Virtual simulation	- Web-based recording tools - Video recording hardware - MUVEs (Second Life, Active Worlds, Croquet)
	Not required	As with "Durable Process" above	As with "Durable Process" above	As with "Durable Process" above	As with "Durable Process" above
Computer use	Intrinsic to competence	- Computer programming - Computer animation	- Write and execute computer program - Design short computer animated video	- Use software and hardware	- Desktop software - Remote servers (e.g., Citrix) with software applications
	Not required	N/A	N/A	N/A	N/A

Chronicity of Access to Information. Many professional tasks involve the ability to analyze situations and make decisions with limited information. This competence is needed in many fields, from medicine to business, social work to law. In many situations, professionals obtain critical information in small pieces over time; for example, when counseling patients, new information unfolds as they share it; when making managerial decisions, employees and consumers iteratively react to changes in operations and the environment; when making public policy decisions such as how to deal with a major hurricane, critical information is often delivered over time.

Often professionals must identify which information is critical in making decisions and requisition it, such as ordering medical tests or performing due diligence investigations of business operations before deciding to acquire a firm or make an investment. This information typically filters in over time, as the different analyses or tests are conducted. Jumping the gun on medical diagnoses without full information, for example, can be life-threatening.

In other kinds of tasks, such as historical or textual analysis, all information can be gathered and analyzed together. This is the case for many other kinds of workplace tasks; for example, teachers set final grades after a class has finished, and many business decisions, such as acquisition of new technology, are made after comprehensive analysis and preparation. Obtaining all information at one point in time allows for all connections to be explored, and for a fully integrated analysis.

Authentic assessment should correspond to the nature of the competences being taught. In areas where decisions must be made based on partial information, or people must identify critical information, then assessment should reflect this: all information should not be available simultaneously. If appropriate, students should determine what information is critical, and students should indicate what interpretations and decisions can be made with what kinds of information. All of these elements are facilitated with IT-mediated assessment.

Case analysis is one tool that can be used for authentic assessment of these skills, either via case-based exams (as typical in law schools) or through ongoing cases analyzed as part of class assignments (as in many business school classes). In these circumstances, case materials are developed and distributed to students. While they may include questions about what additional information would be useful, it is very difficult to deliver information over time, and to track what information has been requisitioned or accessed.

In contrast, most learning management systems make this easy. Resources and documents such as financial records, annual reports, results of medical tests, descriptions of patient behavior, and many other kinds of resources can be made available. Most LMSes track and report on which aspects of online materials students have accessed.

Some LMSes allow for time-based release of information; for example, students cannot access materials on the third unit until that unit begins. Therefore, cases can be constructed iteratively, so that students are asked to diagnose problems and recommend solutions at different points in time, with different kinds of information available at that point; each unit or week could build on the information available and the complexity of analysis.

Learning resources that provide key information can be pre-loaded to be released over time or at certain checkpoints, through an LMS or emails set for delivery at specific times and dates; these commonly available tools makes implementation easy. Alternatively, faculty can simply post different materials on an LMS or Wiki sequentially, with RSS feeds delivering the information to students immediately. With information technology, materials can be released hourly, daily, weekly, or over a longer period depending on the nature of the assessment. For example, exams could include timed cases, with a series of questions or analyses. Over time, different materials would be released or sent to students to incorporate into their analysis.

These practices reduce student control and autonomy over their work, which runs counter to some constructivist principles. However, they increase the authenticity of assessment, and therefore the meaningfulness, realism and ultimate skill to do the work. Because of the structure of LMSes, it is easy to hold reflective discussions about these restrictions. (For example, "Why do you think you were asked to make X decision in Week 3 without having the test results? Do you think this is realistic? How did it affect you? How might you deal with this situation in your work?") This constructivist practice of inviting reflection about the design of learning and assessment activities can increase the salience of chronal control, and therefore student awareness of the skills that they are developing.

In Problem-Based Learning, as opposed to faculty-controlled cases, students explore real-world problems and apply information and analytical processes to resolve them. Students can identify critical missing information, post their informational needs, conduct research and share information in LMSes, Wikis and Educational Social Software. They are not limited to one or two class periods per week, but can coordinate, communicate, and evolve the necessary information throughout the week. This provides for far more authentic assessment than traditional classroom settings.

All well-designed authentic assessments would include reflection, so that at the end of an assignment or course students would be invited to compare their initial analysis with their final analysis. This kind of reflection highlights learning by the contrast between the accuracy and understanding at different points in time. LMSs, Wikis and Blogs make this easy by capturing early responses and enabling later comparisons. Because all communication is made durable, comparison between "what I thought then" and "what I think now" is more accurate and learning can be crystallized. This is both authentic and constructivist, putting the student at the center of the learning and encouraging self-directed and reflective learning.

Chronicity of Analyzing Information. Many real-world tasks differ in the amount of time available to process information. Teachers and police officers must make snap decisions about how to handle developing conflicts, and medical personnel must make immediate decisions about how to handle life-threatening emergencies. The same is true for a host of fields and professions; the ability to analyze and integrate information quickly, and to make quick decisions, is a critical competence and an element of many different kinds of post-secondary education.

Therefore, authentic assessment should vary along with the competences. The types of decisions that need to be made quickly should be assessed under similar circumstances. Traditional post-secondary assessment occurs either in an open-book, extended fashion through projects or homework, or in a time-limited fashion within a class or test period. Smaller time frames are very difficult to control in a face-to-face post-secondary setting, although simulations, role plays and presentations can be used.

Information technology makes it easy to expand or contract the time frame in which students must demonstrate their competence. There are several routes to limiting the time frame available to process information:

1. Technology-mediated synchronous sessions such as webinars or traditional videoconferences can be used as a medium for presentations, role plays, and other simulations. Telephone conferences, either Voice Over Information Protocol (VOIP) such as Skype, or institutional systems such as Intercall can be used for voice-only live meetings. These can be time-limited to the same extent that traditional in-class assessments and activities can be.
2. Asynchronous work can be managed by a Wiki with RSS feed, or Educational Social Software, to distribute new information to all class members and invite collaboration on responses. These tools typically do not have time-release mechanisms, so the time frame available would be manually controlled.
3. Asynchronous work can be managed by an LMS within a controlled time frame of anywhere from minutes to months. Information and learning resources can be released at a specific point in time, and analyses, applications or reports are required within a specific time frame.
4. Asynchronous online testing programs often have the ability to control the time available on a test or exam down to the minute. This includes testing systems built into LMSes, as well as free-standing systems such as

Software Secure (http://www.software-secure.com) and Kryterion (http://www.kryteriononline.com).

These tools can be used to support authentic assessment if they are used for responding to cases or other real-world applications within a limited time frame. Again, they may reduce student control over their learning process for the sake of realism.

Durability

Some competences developed in post-secondary education are demonstrated through producing outcomes that last, such as papers or projects. Others are ephemeral or not durable; they are expressed or demonstrated during one point in time which then ends. These include competence in performing music or dance, performance art, and interpersonal skills such as giving developmental feedback or negotiating a contract, dispute resolution, teaching style or "bedside manner."

Competences that are demonstrated or enacted without durability are generally assessed in person in traditional classrooms, and sometimes videotaped or audiotaped for later review and assessment. This requires additional equipment and personnel not typically available in the classroom. Demonstrations in small groups or informal settings are not captured, and therefore cannot be used consistently for reflection or assessment.

On the other hand, when assessment is transmitted via information technology, it can generally be recorded. For example, webconferencing software allows class presentations, role plays or discussions to be synchronously conducted. Students can demonstrate a range of ephemeral skills such as teaching, counseling, presenting, singing, coaching, and others. These can be recorded for later review and assessment by faculty and students. Many telephone conference call systems allow for recordings as well.

There are many other information technology tools to enable capture and sharing of ephemeral tasks. These include downloadable software that allows students to make voice recordings, which can be used to record and share speeches, poetry, foreign language practice, and other authentic tasks. Other software enables voice annotations of computer files or images (e.g., Camtasia, Profcast, Adobe Acrobat or PowerPoint). Hardware such as mp3 recorders, digital cameras or cell phones can also record demonstrations of competence.

The resulting video or audio recordings can be transmitted to other students and faculty via an LMS, email or Wiki with RSS; or uploaded onto platforms for storage and sharing. These range from systems with access control, such as Ooyala (http://www.ooyala.com/) and iTunes University (http://www.apple.com/support/itunes_u/)., to open systems such as YouTube.

Oral Communication

There are many competences developed in post-secondary settings that involve the spoken word, such as foreign language instruction, public speaking, or learning to write or read poetry with appropriate meter and rhyme. To develop mastery of such competences, it is essential to speak rather than write. Paralinguistic characteristics such as intonation, timing, pitch, volume, enunciation, vocalizations such as "um," and other factors (Manninen & Kujannpaa, 2003) are essential components of competence mastery. They must be performed by a learner for authentic assessment, and must be observed by the assessor or instructor for effective formative or summative feedback.

In contrast, many competences commonly taught in post-secondary education can be demonstrated with only the written word. Critical analysis, academic or college-level writing, research, and other such skills can be demonstrated entirely in writing. In addition, knowledge of many subjects, from physics to human resource management and from history to art, can be assessed via written communication.

Technology-mediated learning is ideally suited for assessment of competences that can be demonstrated in writing. The most ubiquitous information technologies used to support remote learning are LMSes such as Blackboard or eCollege, which focus on the written word through asynchronous written discussions and posting of announcements and assignments.

If courses are designed to require regular use of the discussion, there is far more opportunity to perform authentic assessment of writing and critical thinking in a technology-mediated class than in traditional classrooms. Students can demonstrate mastery through discussion-based tasks including analysis of concepts, applications to real-world events, and integrations of theories or concepts through a variety of assignments. These can include analyzing case studies; summarizing research, developing work products, and many other approaches to authentic assessment (Barrett, 2008; Mueller, 2006b).

Many Web 2.0 technologies can be used for authentic assessment of competences that involve speaking. The tools and websites that allow the capture of ephemeral demonstration of competence generally capture voice. These include the voice and video recorders, the still or video annotators, the desktop screen video capture programs, and direct recording hardware. The recordings can easily be transmitted to other students and faculty, as described above.

Some competences require not only one person's spoken words, but interactions among people; for examples, debating, leading groups, counseling and teaching require synchronous oral interaction among different people. Although in practice these activities may be conducted over the internet, most often they are conducted in person, and authentic assessment requires that they be demonstrated. This requires not just recording voices, but synchronous communication by voice. Many tools support this, including VOIP telephony, traditional telephone conferencing, or webconferencing.

Non-Verbal Communication and Kinesics

Another channel of communication that is critical to many competences are those expressed without words. This includes visual non-verbal communication and physical movement. Such competences include interpersonal communication skills such as counseling, mediating, negotiating, teaching, or arguing points of law.

Manninen and Kujannpaa (2003) cite Thalmann's (2001) finding that up to 65% of information communicated in face-to-face interaction is conveyed through non-verbal means. Key elements include: kinesics, or "all body movement except touching" (p. 2) including head nods, posture, body movement and gestures of the hands, arms and feet; facial expressions; and occulesics, or eye movement and eye contact (Manninen & Kujannpaa, 2003). Together, these are often called "body language" (Guye-VuiliEme, Capin, Pandzic, Magnenat Thalmann & Thalmann, 1999). To master intricate and subtle kinds of interpersonal communication, learners must demonstrate, get feedback, and learn to control these communicative elements. Authentic assessment of these competences requires that students and faculty be able to observe face and body movements. Other competences taught in post-secondary settings involve gross body movement for purposes of expression or sport rather than communication, such as with dance, martial arts, or performance art.

These aspects of communication can be captured and communicated through web cameras and digital video cameras. Simultaneous exchange and interaction can be captured by webconferencing software; with web cameras, students and faculty can observe and assess both oral and non-verbal communication. They are also recordable for later review, allowing assessment of growth and reflection on their improvement over time. This is extremely uncommon in traditional classroom settings.

Proxemics and Haptics

Some aspects of non-verbal communication cannot be easily captured by information technology, and therefore assessed with the same degree of authenticity as other aspects. Two elements are proxemics, or closeness, and haptics, or touch. Proxemics refers to "the way people handle the space around their body and situate themselves next to other people in space.... [including] the distance and angle chosen by the individuals..." (Guye-VuiliEme, Capin, Pandzic, Magnenat Thalmann & Thalmann, 1999, p. 51) .These key aspects of non-verbal communication include behaviors such as looming over someone, or casual touch of an arm or a shoulder. At this point, they cannot be assessed with technology-mediated learning with a high degree of fidelity or authenticity. Videos and webconferencing can be used, but typically each geographically isolated student can only position him/herself in relation to the video camera. While it is possible to record interaction with others, as when student teaching, it is awkward and limited.

In addition, competences that involve physical manipulation of objects cannot be directly assessed through information technology. For example, nurses need to learn to draw blood, laboratory technicians must manipulate equipment, and police officers must learn to fire weapons and disarm an opponent. The actual object and physical interaction must be observed and assessed. This is not possible with most currently available information technology systems.

Increasingly, some post-secondary instruction involves use of Multi-User Virtual Environments, or MUVEs, such as Second Life and Active Worlds. In these "virtual worlds," students learn to control a virtual representations of themselves, called an "Avatar." They can control the Avatar's movements, including proxemics, kinesics and touch. However, these representations do not replace real behavior because they must be consciously manipulated and do not involve control over one's own body. Manipulating an avatar to touch someone or to take blood is not the same as actually doing the task.

On the other hand, in virtual environments, students can disguise or manipulate their visual image. This can eliminate the effects of sex, race, ethnicity and other factors that can affect interpersonal perceptions, and allow students to assume different social roles. Authenticity is reduced for the sake of educational equity and understanding.

Use of Computers

Many competences taught in post-secondary settings require students to use computers. This includes everything from writing articles with word processing programs; creating and manipulating data sets; and creating images with drawing programs. Other competences require students to access and manipulate large quantities of complex information, such as statistical analysis of data sets. Competences in math or computer science require students to develop computer programs. These competences virtually require information technology for authentic assessment.

Some software can be purchased by students and run from their personal computers, while other software is prohibitively expensive or requires extensive computing power to run. Online universities such DeVry operate a Citrix server onto which they load professional programs such as costly graphic simulators and electronic engineering software. Students can remotely access these programs using the Internet, for both skill development and authentic assessment.

TECHNOLOGY LIMITATIONS ON AUTHENTICITY OF ASSESSMENT

There are many limitations of the use of information technology which must be addressed in order to use them in support of authentic assessment. These include hardware and software requirements, limitations of different systems, faculty and student skill with technology, time to develop assessments, expertise in development, and cost of investing in technology.

Storing and accessing audio and video files requires high speed connections. Some audio and video systems, such as iTunes University, require students to download files in order to view them; this demands significant computer memory and storage space. Other web-based systems do not require downloads; however, streaming videos requires high speed internet connections for the videos to play. Most webconferencing systems have similar connectivity requirements.

LMSes have different capabilities and limitations. For example, eCollege and Angel allow materials to be associated with specific modules that are not accessible until prior tasks have been completed, or until certain points in time. Other LMS systems do not include these capabilities. Similarly some LMS systems facilitate open-ended, developmental feedback, while others require extensive manipulation and many "clicks" to reach a screen. This puts barriers in the way of formative feedback, which is critical to constructivist education and skill development.

Information technology systems vary in their ease of use. Some have been designed specifically for non-technical users. Others require more training in order to design and implement courses and authentic assessments. Use of white space, san serif fonts, "WYSIWYG" (What You See Is What You Get) editors that allow the user to use toolbars and view visual effects such as spacing, bold and italics, simple interfaces with clear and obvious directions, along with immediate feedback about actions (Neilson, 2000), all support easy use.

Server problems, ISP problems, hardware and software problems can all compound limitations of user knowledge. It is critical that the technical support be provided consistently, and that students be able to make up work if they experience technical difficulties. When technology is mediating interaction, the focus should be on the competence being assessed rather than the technological interface. For active student engagement in learning, systems must be easy to use.

Some systems are more difficult to support than others, due to complexity and reliability. MUVEs such as Second Life require very fast computers with a great deal of memory and sophisticated graphics cards, as well as high speed internet access, in order to work effectively. Even then, they are often unreliable. They also require many hours of training in order to master the commands and move freely around the virtual world. Unless students and faculty will use the virtual world extensively, it can be unrealistic to expect them to upgrade their computers and master the interface for a single class or assessment.

FUTURE TRENDS

Information technology support for learning continues to grow, which will have many effects on learning and assessment in the future. As simulations become more common, both students and faculty will be more comfortable with the hardware and software, enabling educational use. Already, the simulated reality computer game series "The Sims" has sold over 100 million copies (Boyes, 2008), and there were 1.6 million active uses of Second Life in 2007 (Winters, 2007). The barriers of experience and comfort will fade as entertainment uses become more common.

Second, the limits of haptic and proxemic factors are likely to be reduced. One area of explosive growth in information technology is in haptic interfaces: joysticks, physical manipulation of tools and dials. For example, the "Wii" (http://us.wii.

com/) is a child's toy that allows simulated physical experience of playing sports such as tennis, bowling, driving, swimming, and other physical interactions. While mechanical simulators have been used by the military for pilot training since before World War II, and by civilian airlines since 1948, only recently have these kinds of haptic interfaces become common for other uses. The simple joystick of Nintendo has developed into a range of interface tools that simulate physical sensations. Virtual worlds can both feel and look more real, as three-dimensional simulations and computer games are supported with Virtual Reality goggles and sensors; such goggles are now available for less than $100.

Tools to capture oral and visual communication are also becoming more common. Many cell phones have utilities to support voice and video recording, as well as connection to the internet. As computer memory becomes ever cheaper and more areas are covered by wireless networks, it will be ever easier to record interactions and send them to classmates and faculty.

Perhaps the greatest limitation in information technology for education is in one of the areas of greatest promise: MUVEs. Currently, it is expensive and difficult to develop a virtual world. Universities such as Ohio University, along with private companies, are developing more sophisticated and complex virtual worlds. Over time they will likely create virtual buildings and civilizations that can be "rented out" much as LMS space is currently rented. This will allow more authentic exploration and assessment in the study of history, archeology, and other areas. Any competence in which interaction with physical structures or devices is impossible or dangerous, from exploring ancient Rome to designing Occupational Safety and Health rules to protect against biohazards, can be taught – and mastery assessed – in a virtual world. Over time, this development will likely be centralized and sold to educational institutions for instruction and assessment, much as computer games are packaged for entertainment today.

CONCLUSION

Technological innovations have sharply improved the ability to conduct authentic assessment of many competences in post-secondary settings. The ability to control events over time is a key element of some types of information technology, which enables more authentic assessment of competences where the chronal elements do not match the on-ground, two-hour-class-session model. Information technology can also increase the ability to capture ephemeral behaviors, supporting reflection and assessment.

Skills that involve speaking, such as foreign language instruction, visual non-verbal communication, such as counseling, or movement, such as dance, can be shared via information technology; this allows for authentic assessment, reflection and social construction of learning. However, competences that require touch or physical proximity cannot be fully assessed, although they can be simulated using virtual worlds or webcameras. The authenticity is sharply reduced over traditional, in-person settings. Competences involving computers can be authentically assessed through information technology.

Optimal assessment will match key aspects of task performance. In many cases, information technology allows a much more authentic performance than the traditional classroom, and this will only grow as such technology becomes more ubiquitous. This enables more constructivist instructional pedagogies, where students actively engage in meaningful realistic tasks, build on their learning, collaborate, reflect on their learning, and develop true mastery.

REFERENCES

Albon, R. (2003). *Assessment drives the learning: Raising the bar.* Paper presented at the Evaluation and Assessment Conference, Adelaide, Australia, Nov. 2003.

Allen, I. E., & Seaman, J. (2007). *Online Nation: Five Years of Growth in Online Learning.* The Sloan Consortium. Retrieved June 8, 2008 from http://www.sloan-c.org/publications/survey/index.asp

Anderson, T. (Ed) (2008a). *The Theory and Practice of Online Learning, Second Edition.* Edmonton, Alberta: AU Press. Retrieved June 9, 2008 from http://www.aupress.ca/books/Terry_Anderson/TerryAndersonEntireBook.pdf

Anderson, T. (2008b). Towards a Theory of Online Learning. In T. Anderson (Ed.), *The Theory and Practice of Online Learning* (2nd Ed., ch. 2). Edmonton, Alberta: AU Press. Retrieved June 9, 2008 from http://www.aupress.ca/books/Terry_Anderson/TerryAndersonEntireBook.pdf

Anderson, T. (2008c). Social Software to Support Distance Education Learners. In T. Anderson (Ed.), *The Theory and Practice of Online Learning* (2nd Ed., ch. 9). Edmonton, Alberta: AU Press. Retrieved June 9, 2008 from http://www.aupress.ca/books/Terry_Anderson/TerryAndersonEntireBook.pdf

Barrett, H.C. (2008). *Using Technology to Support Alternative Assessment and Electronic Portfolios.* Retrieved June 9, 2008 from http://electronicportfolios.org/portfolios.html

Barrett, H. C. (2008b). *The REFLECT Initiative: Researching Electronic portFolios: Learning, Engagement and Collaboration through Technology with a focus on high school students in the Arizona Education Professions program.* Paper presented at the American Educational Research Association Annual Meeting, New York, NY.

Barrett, H. C. (2006). *Authentic Assessment with Electronic Portfolios using Common Software and Web 2.0 Tools.* Retrieved January 31, 2008 from http://electronicportfolios.org/web20.html

Bonk, C. J., & Cunningham, D. J. (1998). Searching for learner-centered, constructivist, and sociocultural components of collaborative educational learning tools. In C. J. Bonk & K. S. King (Eds.), *Electronic Collaborators: Learner-Centered Technologies for Literacy, Apprenticeship, and Discourse* (pp. 25-50). Mahwah, NJ: Lawrence Erlbaum. Retrieved on August 25, 2008 from http://www.publicationshare.com/docs/Bon02.pdf

Boyes, E. (2008). The Sims Passes 100-Million Mark. *Gamespot News*, April 16, 2008. Retrieved June 9, 2008 from http://www.gamespot.com/news/6189330.html

Bransford, J., Brown, A., & Cocking, R. (1999). *How people learn: Brain, mind experience and school.* Washington, DC: National Research Council. Retrieved June 9, 2008 from http://www.nap.edu/html/howpeople1/

Downing, K. F. & Holtz, J. K. (2008). *Online science learning: best practices and technologies.* Hershey, PA: Information Science Publishing.

Dougiamas, M. (1998). *A Journey into Constructivism.* Retrieved August 25, 2008, from http://dougiamas.com/writing/constructivism.html.

Farmer, J. (2004). *Communication dynamics: Discussion boards, weblogs and the development of communities of inquiry in online learning environments.* Retrieved June 9, 2008 from http://incsub.org/blog/2004/communication-dynamics-discussion-boards-weblogs-and-the-development-of-communities-of-inquiry-in-online-learning-environments

Gergen, K. J. (1995). Social Construction and the Educational Process. In L. P. Steffe & J.E. Gale (Eds.), *Constructivism in Education.* Hillsdale, NJ: Lawrence Erlbaum.

Hirumi, A. (2002). Student-Centered, Technology-Rich Learning Environments (SCenTRLE): Operationalizing Constructivist Approaches to Teaching and Learning. *Journal of Technology and Teacher Education, 10*(4), 497-537. Retrieved on August 25, 2008 from http://www.ryerson.ca/activelearning/files/student-centered.pdf

Manninen T., & Kujanpää T. (2002) Non-Verbal Communication Forms in Multi-player Game Session. In X. Faulkner, J. Finlay, & F. Détienne, (Eds.), *Proceedings of HCI 2002 Conference* (pp. 383-401). Springer-Verlag,. Retrieved June 9, 2008 from http://www.tol.oulu.fi/~tmannine/publications/HCI2002_Tuppi3D_analysis_final.pdf

Mueller, J. (2006). *What Is Authentic Assessment?* Retrieved June 9, 2008 from http://jonathan.mueller.faculty.noctrl.edu/toolbox/whatisit.htm#definitions

Mueller, J. (2006b). "Authentic Assessment Toolbox." Retrieved June 9, 2008 from http://jonathan.mueller.faculty.noctrl.edu/toolbox/

Nielson, J. (2000). *Designing web usability: The practice of simplicity.* Indianapolis: New Riders Publishing.

O'Reilly, T. (2005). What Is Web 2.0: Design Patterns and Business Models for the Next Generation of Software. Downloaded June 8, 2008 from http://www.oreillynet.com/pub/a/oreilly/tim/news/2005/09/30/what-is-web-20.html

Pellegrino, J. W., Chudowsky, N., & Glaser, R. (Eds.) (2001). *Knowing what students know: The science and design of educational assessment.* Washington, D.C.: National Academies Press.

Ridgway, J., McCusker, S., & Pead, D. (2004). *Literature Review of E-assessment.* Futurelab Retrieved June 8, 2008 from http://dro.dur.ac.uk/1929/

Show, Don't Tell: Video and Accountability (2002). Coalition of Essential Schools National Web. Retrieved June 9, 2008 from http://www.essentialschools.org/cs/resources/view/ces_res/229

Steffe, L. P., & Gale, J. E. (Eds.) (1995). *Constructivism in Education.* Hillsdale, NJ: Lawrence Erlbaum.

Guye-VuiliEme, A., Capin, T. K., Pandzic, S., Magnenat Thalmann, N., & Thalmann, D. (1999). Nonverbal Communication Interface for Collaborative Virtual Environments. *Virtual Reality*, 4, 49-59. Retrieved June 9, 2008 from http://www.springerlink.com.ezproxy2.lib.depaul.edu/content/q2767j5xw7216555/fulltext.pdf

Von Glasersfeld, E. (1995). A Constructivist Approach to Teaching. In L. P. Steffe, & J. E. Gale, (Eds.), *Constructivism in Education.* Hillsdale, NJ: Lawrence Erlbaum.

Wiggins, G. P. (1993). *Assessing student performance.* San Francisco: Jossey-Bass.

Wiggins, G. P (1990). The case for authentic assessment. *Practical Assessment, Research & Evaluation*, *2*(2). Retrieved June 8, 2008 from http://PAREonline.net/getvn.asp?v=2&n=2

Winters, C. (2007). Second Life User Statistics. *Social Signal*, March 13, 2007. Retrieved June 9, 2008 from http://www.socialsignal.com/blog/catherine-winters/second-life-user-statistics

Chapter VII
Nontraditional Students and Information Technology:
The Siren Call of the Virtual Classroom and its Impact on Progressive Educational Ideals

Xenia Coulter
SUNY Empire State College, USA

Alan Mandell
SUNY Empire State College, USA

ABSTRACT

The adult college student, caught between the competing demands of work and home, has recently become a valuable commodity in today's fast-changing American universities. The authors argue that the response of the university to the personal circumstances and credentialing needs of adult learners, accentuated by the forces of globalization and the availability of new information technologies, particularly the Internet, has been to focus upon the efficient delivery of information deemed important in our post-industrial society. This response, particularly well exemplified by the virtual classroom, is not conducive to the fluid and open-ended inquiry associated with progressive education. In the end, the authors speculate, adult students may taste the true progressive and constructivist approaches to learning better outside the confines of formal higher education.

INTRODUCTION

For many years we have taught at a college established to meet the educational needs of adults. This institution[a] sought to adopt for adult learners a teaching philosophy that incorporated Dewey's innovative ideas about individualization, democratic ideals, and experiential learning. When the college first opened its doors, these students were essentially outcasts in the world of higher education. Today, more than 35 years later, they have moved into the spotlight. Over this same period of time, the role of universities has become central in a society increasingly dependent upon ever-expanding technologies to support the delivery of an always growing pool of information. It is our judgment that questions about the impact of information technologies upon progressive educational practices cannot be fully addressed without considering the changing role of the university and the growing importance of the adult student.[b] Thus, in this chapter, we will discuss the ways these students, technology, issues of access, the commercialization of education, and the information delivery teaching model, have together contributed to a new climate of institutionalized learning. In this context, we will argue, the realization of progressive ideals in higher education is gradually drifting out of reach.

WHO IS THE ADULT STUDENT?

Much of the literature about American higher education today focuses on the characteristics and needs of what Arnett (2000) has referred to as the "emergent adult." He proposes that this developmental stage, occurring roughly between the ages of 18 and 25, is a unique period in life that should be considered separately from "true" adulthood. It is a time when, ideally, young people can prepare themselves to take on adult responsibilities, such as marriage, serious employment, and community involvement, often through formal educational experiences that include guidance from adult teachers and mentors. Isolated as much as possible from everyday pressures, they are encouraged to explore and shape their place in the world,[c] particularly in universities and colleges, which see themselves as providing the premiere environment in which such preparation for adulthood should take place (e.g., Bok, 2006; Hersh & Merrow, 2005; or see Magolda, 2004).

Adults who have passed beyond this developmental stage are typically referred to in higher education as "nontraditional" students.[d] While some universities have sought to accommodate them with add-ons such as special "night" or "weekend" classes, most conventional institutions, particularly those that primarily serve full-time resident students, have been far less welcoming. Their focus on "campus life" and personal development is not particularly appropriate for these adults who have homes of their own and have already been shaped by their families, communities, and, often, regular employment. While they deserve and will benefit from advisement that helps them navigate the complex university bureaucracy and the complicated language of college curricula, they do not need an environment, or a curriculum, designed to prepare them for adult life.

Why adults seek additional education varies considerably (e.g., Cross, 1992; Merriam, Caffarella, & Baumgartner, 2006, Pusser et al, 2007). Some see the return to college as an opportunity to make intentional changes in the direction of their lives. A small minority put their adult lives on hold in order to be full-time students, playing the role, so to speak, of emergent adults. However, most come to college because they have been forced to change directions due to job loss, escalating credential expectations, economic fragility, disruption in their family situation (particularly for women), and other life-shaking events. What they all have in common is that they are mature adults, with complicated lives and numerous responsibilities. In fairly sharp contrast to emergent adult students for whom college is expected to be their primary commitment, nontraditional students regard college study as only one more activity they must add to their already busy lives.

Within such a lifeworld, it is easy to understand why adult learners are not willing or able to linger with us too long and why they seem sometimes excessively focused upon how efficiently they can obtain their degree. Even taking one or two courses at a time requires them to stretch their lives to create new spaces for a commitment they can justify only if it is a temporary one. Women cannot expect their husbands to substitute for them as cooks or babysitters forever; men cannot spend every evening away from their families reading textbooks; a single parent cannot ask her children to always fend for themselves while she sits in front of a computer. And everyone wants the increased paycheck (or perhaps just a regular paycheck) that comes with a completed degree, as soon as possible. Thus, it's not unreasonable that a significant number of adult students worry, not so much about what they're going to learn, but that whatever learning they must complete takes as little time as possible.

DOES PROGRESSIVE EDUCATION APPLY TO THE EXPERIENCE OF ADULTS?

The progressive educational ideal was originally conceptualized to meet the particular needs of immature, or not yet fully formed, individuals.[e] The language of "discovery" often associated with constructivism makes good sense when educating children or, for that matter, emergent adults. In both cases, the possibilities are unlimited, and with a progressive orientation attuned to the ideals of a democratic society, students are encouraged to explore a wide range of educational options. At the same time, students' individual experiences become the essential building blocks from which they can develop, on their own and in their own way, a path toward adulthood that integrates the knowledge that society, and educators, expect them to master. Learning arranged in this spirit is unhurried, lightly but thoughtfully guided, and open to student-initiated inquiries.

Already a long way down the path of discovery before they come, or return, to college, adult learners, in contrast, often already have a particular educational goal in mind and are keenly aware of what the world expects of them. Unlike emergent adults, they are not in an obvious state of "role confusion" looking to experiences in college to help them establish their identities (i.e., Erikson, 1959). Instead they tend to suffer from a sense of "feeling stupid" (Bloom, 1996, p.44), embarrassment that they earlier failed to earn a college degree, worry about the amount of knowledge they lack, and the realization that without a higher learning credential, they are truly stuck. Note how these concerns focus strongly upon the need for new knowledge. Is it therefore surprising that these adult learners see college not as an institution devoted to progressive growth, but as a repository, or as Freire put it, a "bank" (1972, chap. 2), filled with the information they understand they must have?[f]

Importantly, it is not that adult learners are uninterested in expanding their perspectives or that a progressive educational approach cannot be successfully adapted to their needs. Many welcome the opportunity to reflect upon their current situations or past experiences (e.g., Brookfield, 1986, 1987a; Daloz, 1991). They also are stimulated by exposure to new ideas that excite their imaginations and encourage new ways of thinking (e.g., Brookfield, 1991; Herman & Mandell, 2004). But it requires special teaching skills (e.g., Fiddler & Knoll, 1995, as cited in Hung, Bailey, & Jonassen, 2003, p.18), extra assurances, and much effort (e.g., Hung et al, 2003) to convince busy students to participate in ill-defined, open-ended, educational inquiry. Overwhelmingly when asked, they report that lack of time is a major impediment to their ability to succeed in college (e.g., Lore & Tait, 2004; Valentine & Darkenwald, 1990).[g] The idea of taking college courses in order to make meaning of their lives or to strive for transformative experiences are judged by many over-extended adults, already "in over their heads" (i.e., Kegan, 1994), to be unrealistic, unsustainable, and not really necessary.[h] No matter that it can be an intellectually rich process, even just figuring out on their own what they want to learn (a first step in any progressive educational system) is often judged as too inefficient and time-consuming than simply doing whatever they are told.

Adult students are also sufficiently experienced with contemporary mores to appreciate the importance of an expert when entering new territory. If you're sick, you go to a doctor to fix the ailment; if you're under-educated, you go to a teacher to find out what you need to know. Thus, most adult students come to college wanting answers (not new questions), clear directions (not more options), and access to an expert's knowledge for which they are investing cherished time and hard-earned money. Since teachers too are not immune to social pressure or a wish for recognition, it is not easy for them to completely eschew the role of expert. Indeed, the more a teacher is a content specialist, the more difficult the transition from professor to guide turns out to be (Herman & Mandell, 2004, Chapter 9; Kaufman & Holmes, 1999, as cited in Hung et al, 2003. p.19). Moreover, even college teachers with constructivist inclinations hold assumptions about the educational direction students *ought* to pursue and in subtle ways move them in that direction, unaware that they are responding to questions they would ordinarily agree should be more deeply investigated by the students themselves.

Note too that the educational expectations of adult students are not out of line with contemporary understandings. The developed world has graduated from the Age of Industrialization into a new Age of Information. Within this new "world of transformations" (Giddens, 2000, p. 24), it is easy to see why the university is now regarded, not as an ivory tower in which to contemplate, but as the major source and distributor of the incessant accumulation of new information that continually reshapes our world. In this regard, it is interesting to observe how the meaning of "life-long learning," a concept that seemed to peak in popularity a number of years ago, has also changed (Finger & Asun, 2001). The original idea promoted by educators such as Cyril Houle (e.g., 1961), Alan Tough (1968), or Malcolm Knowles (e.g., 1976), saw adults seeking further education as a form of self-enrichment. The current view is that adults must undergo frequent reskilling in order keep up with job changes and the constant flow of new information (e.g., Pusser et al, 2007; Sennett, 2006). Today, in a society absorbed in discussions about information technology, information systems, information delivery, and knowledge management, the message is clear: College is where you learn what information is important, where to obtain it, and how it can be used. Thus, the somewhat "unprogressive" educational expectations of most adults are wholly consonant with what the contemporary world tells us is important.

HOW DOES THE ADULT STUDENT FIT IN THE UNIVERSITY TODAY?

The university has not remained immune to these changes (e.g., Newfield, 2008). Only a few years ago a degree was not considered necessary for the majority of working adults, but today it is an essential prerequisite for achieving a full time professional job let alone a successful life. Thus, the completed degree has come to represent considerable economic value, and the university has not hesitated to market it as such. No longer ivory towers that claim to stand outside the workaday world, universities, as "knowledge factories" (Aronowitz, 2000), openly identify themselves as major corporate players (Giroux, 2007). Important economic assets in the communities in which they reside, universities now espouse many of the same goals as any commercial enterprise (e.g., Stein, 2004). With the degree as its ultimate high-demand product, a university's success is now measured by its enrollments, graduation rates, robust development offices, physical growth, and proprietary research (Burgan, 2006). That students come to a university to learn is certainly not forgotten, but this goal is regarded as a "service" the university "delivers," and, increasingly, this service is treated as much a marketing tool as a social good.

Within this framework, the adult – the previously nontraditional student – has undergone a major change in status (e.g., Blumenstyk, 2008). According to demographics studies, the emergent adult population is declining (e.g., Ashburn, 2008); in contrast, the number of mature adults who still do not possess a degree is tantalizingly large.[i] Often at least partially subsidized by their employers, readily available for recruitment, with little need for the extra-curricular activities and facilities important to full-time traditional-aged students, and often desperate for credentials, the older student in this light is clearly a plum recruit[j]. Although well recognized colleges that attract more traditional-aged students than they can accept continue to market themselves to young people, the considerably larger number of public and community colleges, and second and third tier universities, have changed their tactics. Suddenly, no longer a second-class citizen, the adult student has become an important "market segment" within the newly commercialized version of higher education.

To attract these students, colleges and universities offer not only professional programs of particular interest to working adults, but various conveniences that address their practical needs and goals. Ease of access, relevance, and speed are attributes that dominate recruitment materials, billboards, and e-ads. Even "individualized" and flexible degree program are marketed less as an opportunity for creatively constructing a customized curriculum than as a way of avoiding required courses for traditional-aged students that adult learners often consider a waste of their time. Some colleges also offer "fast tracks" – accelerated, but concentrated, learning programs (Wlodkowski & Kasworm, 2003). Another form of acceleration increasingly directed toward adults is the assessment of knowledge gained from experience (Michelson & Mandell, 2004). While the process of reflecting upon what one has learned outside the academy contains rich opportunities for deep reflection and new learning (e.g., Andersson & Harris, 2006), a significant number of returning students see the possibility of "life credits" merely as another shortcut to earning a degree.

We should note that the discovery of new or critical perspectives, the possibility of transformative experiences, the opportunity to address open-ended problems, or the excitement of co-constructing new knowledge, are not typically advertised in recruitment materials. That a truly meaningful education demands more than merely following a set of simple and transparent steps is not part of the message. We cannot but wonder whether these various marketing responses to the perceived pragmatic needs of adults may soon spill over to affect the expectations of traditional students.[k] Reinforced by often trumpeted statistics showing the positive effect of a college degree upon income and status, these messages might well be contributing to a significant shift in society's understanding of what *any* college education is supposed to achieve.

HOW DOES TECHNOLOGY AFFECT HIGHER EDUCATION? THE INTERNET

By all accounts (e.g., Burbules, 2000; Mihyoshi, 1998; Welton, 2005), the changes occurring in higher education reflect the forces of globalization, and globalization itself has clearly depended upon, if indeed it is not the expression of, the rapid development of the information technologies (e.g.,

Friedman, 2000; Negroponte, 1996). Although computers were originally designed to perform operations on data, the importance of being able to transfer the data from one computer or location to another was immediately recognized. Over a relatively brief period of time, the amount of data, the speed at which it could be transformed and transmitted, and the miniaturization of the machinery involved were greatly enhanced. Today, for almost anyone in any country, the Internet and the world-wide web can be accessed[l] in their homes, at work, in libraries, in public offices, in schools, and, with the advent of wireless connections, in many commercial locations and even in properly wired outside parks.

One important outcome of this revolution in technology, aside from a vastly improved system of communication, has been broad and immediate access to an immense store of knowledge. Increasingly, the Internet allows us to enjoy, almost instantly, the visual arts, music, literature, scholarly articles, maps, sports events, newspapers from across the world, scientific tools of study, government documents, medical advice, library holdings, movies, educational videos, and information from literally across the globe. Indeed, most writers on technology see the quantity of information available from the Internet, and the speed with which it can be accessed, as representing a qualitative step forward in the cultural history of humankind, much as the printing press was seen as a similar advance 400 years ago (e.g., Postman, 1985, p. 29).

And it is the Internet, and what it represents, that has most significantly impacted education today (e.g., Bauerlein, 2008; Bowers, 2000).[m] On the face of it, the Internet would appear to be the perfect tool for any educational philosophy that emphasizes active student involvement in the process of discovery. To the extent that such pedagogies as inquiry-based (Manning, Manning & Long, 1994) or problem-based learning (Knowlton & Sharp, 2003) can be seen as outgrowths of Dewey's ideas, easy access to unlimited information clearly enhances the utility of these pedagogies. The Internet is also useful as a place where students can share their developing thoughts (e.g., post their papers or research) and, through wikis or blogs, receive instant comment or correction. In turn, these tools make it possible for whole communities of scholars (or students) to construct new knowledge together (Susstein, 2006; Tapscott & Williams, 2008). Given Dewey's (1916) emphasis upon the communal nature of knowledge and core features of social constructivist theory, a subsequent addition to the progressive agenda (Vygotsky 1934, 1978), the Internet has singlehandedly broadened the meaning of community in ways that would have been unimaginable during the time of either Dewey or Vygotsky. If the Internet makes it easier for students working together to locate the raw materials necessary for reconstruction and meaning-making, it would then seem to be an ideal tool for progressive education.

However, there is a serious downside. Increasingly easy access to ubiquitous information makes it increasingly difficult for students to understand that information alone is only a small part of a much more ambitious endeavor (e.g., Roszak, 1986). The scope of the Internet – the sheer enormity of information it provides us – may simply reinforce the view that education consists primarily of searching for, sifting through, and organizing what is already known. Further, even though we know that learning (in the sense of working with, internalizing and making information meaningful for us) requires "practice," when this practice can be controlled by the computer (for example, for learning musical notation, doing math problems, or analyzing verbal content), we become that much more convinced that learning comes from *without* rather than from *within*.[n] Moreover, learning-as-information-accumulation, the antithesis of what progressive education represents, appears much more efficient than having to learn how to think critically, to solve problems, or, most importantly, to make meaning on one's own. Such complex processes can't be as easily structured, do not occur quickly, and, despite considerable rhetoric, remain very difficult to teach (e.g., Bereiter, 2001; Coulter, Oaks, Gadbow, & Gerardi, 2000).

HOW DOES TECHNOLOGY ADDRESS THE NEEDS OF ADULT STUDENTS? THE VIRTUAL CLASSROOM

In the midst of so many changes in the university, what seems not to have changed is the college classroom. It continues to offer in microcosm what most teachers believe are the key ingredients of education: knowledge structured to facilitate comprehension, scholarship in action through faculty lectures, directions and feedback from an expert, and practice in thinking through guided discussion. No matter what the pedagogical innovation, no matter how much emphasis upon self-direction, discovery, and "active" learning, activities associated with classrooms continue to serve as an essential expression of what is meant by legitimate formal education.

Not surprisingly, then, new technologies supported by the vast resources of the Internet have been redirected to create commercially successful and readily available "virtual" classrooms. For high schools, universities, and the business world, a variety of venders have created classroom "platforms," accessible entirely from the Internet, that contain the basic components that have been common to classrooms for hundreds of years. Courses are organized by lessons, chapters, or modules, and in each lesson, sections are devoted to: (a) lectures, study guides, or teacher comments, (b) opportunities for class questions or discussion, and (c) assignments. The software too makes easily available provision for group projects and links to relevant information from the Internet, as well as tools for testing students and automatically calculating and tracking their grades. Numerous books and materials have been published to help human resource professionals (e.g., Horton, 2006), teachers (e.g., Palloff & Pratt, 2003, 2007), and "consumers" of multimedia (e.g., Clark & Mayer, 2008) adjust and work effectively in these cyberspace-based classrooms. A whole new industry of instructional technology has burst upon the scene.

In terms of progressive education, considerable academic attention has been devoted to the "discussion" component of the virtual classroom, the only appropriate arena available for student contributions, and most particularly, the "social construction" of knowledge (e.g., Campos, 2004; Jarvela & Hakkinen, 2002; Woo & Reeves, 2007). The concepts, methods, and expectations of such discussions are not new (e.g., Brookfield & Preskill, 1999), except that virtual classrooms make available written records of every student utterance so that instructors and researchers can analyze the quality of discussions and measure each student's progress in developing his or her contributions. To date, most of this research, often based upon a close study of small classes, has raised many interesting questions (e.g., Gulati, 2008; Kirscher, Sweller & Clark, 2006; Lai, 2008; Ho & Swan, 2007). Much remains to be learned about the role of the instructor in these discussions, the degree of control needed to assure that the discussions produce meaningful learning, the relationship of discussion to the acquisition of prescribed content, effective methods of teaching naive undergraduates the requisite skills of inquiry, and the impact upon learning when a discussion is controlled, required, and known to be monitored by the instructor.

A particularly intractable issue relates to the amount of time and effort demanded of an instructor to follow, respond to, and analyze the postings of *every* student in a class of, for example, 20 students. And for many adult learners, although the asynchronous nature of these discussions makes it possible to participate at convenient times, many adults have competing obligations that allow only for periodic "attendance." Moreover, there is an important attitudinal dimension: A number of these students are not interested in "listening" to their classmates and do not see these conversations as relevant to the acquisition of new information offered by an expert. Thus, whether the discussion component of virtual classrooms can effectively realize progressive or constructivist ideals is still very much open to question.

Without a doubt, technology in education could be applied in entirely different ways. Indeed, as has been creatively imagined by many a science fiction writer (e.g., Stross, 2005; Vinge, 2006), technology is a critical stimulant for revolutionizing the ways we learn and teach in the future. Even today, some claim that hypertext, for example, encourages new forms of cognition (i.e., Burbules & Callister, Jr., 2000; Jones & Spiro, 1995; McClintock, 1992) or that the application of virtual reality techniques represent qualitatively new pedagogies (e.g., Vander Valk, 2008; Stone, 1995). Yet, for most students, hypertext is seen as either a quick way to access a required resource or the rough equivalent of an optional footnote. As for virtual worlds, their instructional value is still very far from being fully understood, let alone seriously applied.[o] They remain, at this point, for many adult learners, interesting diversions.

In comparison, at least to date, it is the virtual classroom that remains the major new structure that technology has made available to higher education. And, despite the problematic nature of the discussion component, it is extremely popular with adult students, the major market for this educational experience (see Mangu-Ward, 2008). In the face of this popularity,[p] we have asked ourselves what the virtual classroom offers the adult student that was not previously available in, for example, correspondence courses. We speculate that part of the answer is related to the isolation adult students have traditionally experienced. As has been argued elsewhere (Coulter & Herman, 1994), an important reason why adult students have been marginalized is their enforced "distance" from the main point of action – the college campus. Educators, and society in general, tacitly regard any education that occurs outside the walls of the campus to be, by definition, second-rate. One of the novel features of the virtual classroom is an immediate perception by students who join such a "class" that they are now "going to school," exactly as if they might be "going to campus." And while the adult learner may not see much value in constructing knowledge with fellow learners, they do value the companionship such virtual classes afford. Instead of having to work alone with a teacher by mail (as in the traditional correspondence course) or in individual tutorials (as in independent study) or in often segregated courses (as at night or on weekends), they now have a chance to be in a "classroom," where the world agrees "real" education takes place.

Aside from offering adult students the opportunity to actually "go to college" like emergent adult students, we cannot overemphasize how much virtual classrooms are also seen as convenient and efficient. Clearly specified in the appropriate locations in the classroom template, students learn what they must read – and when, what information they must look for – and where, what topics they should discuss – and how, and the nature and timing of the required assignments. The course structure also allows faculty to know exactly when, where, and for how long a student logs on, and also to record the number of times a student participates in any given discussion with either original postings or responses. With such data available, teachers can then easily specify what level and what kind of participation in the course they expect. And with expectations laid out in advance, students can plan their schedules accordingly. The clearer the expectations, the easier it is for any given student to fully comply.[q]

Contrast this model with the kind of free-flowing face-to-face conversation between a professor and student that ideally characterizes progressive education. Working together in an open-ended study, student and teacher collaborate in specifying and clarifying what is expected (e.g.,

Daloz, 1999). The questions students must address grow, organically, out of the initial conversations about the course direction, and, equally important, as students begin to share what meaning they have made of what they discovered on their own (e.g., Mandell & Herman, 2007). This form of pedagogy makes sense in a Houlian world of inquiring adult learners. But, when education for adults focuses on the needs of a globalized economy and workplace rather than individual development, a predetermined curriculum with carefully structured and prescribed classroom activities clearly appears to be the most efficient way to provide the knowledge students expect to acquire.

CAN FACULTY ALTER THE DIRECTION OF THE VIRTUAL CLASSROOM?

Walters (2000) makes a similar distinction when discussing adult education in the globalized world. Globalized education, she argues, focuses in a competitive world on "human capital" whereas in a cooperative world, it could emphasize "human development." Dewey (1938) too identifies a similar conflict between the goals of "traditional" education that address societal needs and the goals of progressive education that focus on the individual learner. One might similarly interpret the distinction between "pedagogy" (the imposition of a prescribed curriculum in teaching children) and "andragogy" (free inquiry in education designed primarily for adults), noted by Lindeman (1927), subsequently Knowles (1976), as simply another variation on a similar theme. These writers, and certainly others, believe that dual perspectives on education have legitimacy. Indeed, these scholars, particularly Dewey, were explicitly concerned with finding ways of balancing, integrating, or even transcending these tensions, rather than claiming an inherent superiority of one view over the other.

The problem, as we see it, is that the forces influencing higher education today have coalesced to such a degree that together they are tipping the balance away from any serious attention to progressive ideals. In the end, only one view of education may prevail – the one that fits best with the assumptions of a market-driven, technology-supported world. We have already discussed many of the current circumstances that are changing the landscape of higher education: a new market of adult learners, the commercialization of the university, the focus on information as a goal in itself, and the delivery of this information in structured online classrooms. But we have so far not yet considered how these forces have impacted the individual teacher. Is it possible that college faculty can effectively resist the impact of these developments? Can they, given a renewed commitment to active, constructivist, and open-ended inquiry, restore the balance between free and controlled methods of learning?

Our personal experiences do not make us optimistic. After all, teachers are themselves adults and subject to the same overt and tacit social messages as our adult students. We too are dazzled, if not swept away, by new technologies and the increased information and communication resources that now pervade our institutions. But in the rush to wire our workplaces, efforts by faculty to deliberate on the likely educational consequences of these changes have been essentially co-opted. At the same time, we, again like our students, live very busy lives with many competing priorities. For example, in all major universities, despite considerable hand-wringing (e.g., Bok, 2006, Hersh & Merrow, 2005), faculty continue to be held accountable far more for their research activities than for their teaching prowess. Thus, time available for faculty to investigate the effectiveness of their teaching, much less resist the pressure to incorporate these new technologies, is very limited. Moreover, with research grants ever more important economically (Stein, 2004), those who teach are increasingly regarded as merely human capital (Burgan, 2006) -- interchangeable parts hired to keep the student pipelines moving smoothly from admission through graduation. Clearly this environment does not invite faculty to pay careful attention to individual learners or to thoughtfully develop courses customized around each student's personal experiences, questions, interests, and needs. The creative responsiveness associated with "the best college teachers" described so cogently by Bain (2004) requires truly Herculean efforts to practice.

A second issue for college teachers, noted earlier, is the problem of disciplinary expertise. In a progressive educational environment, such expert knowledge must be set aside in order to allow students an active role in their own learning[r]. But it's not easy for university-trained professionals to accord students, even adults, equal learning status, let alone to genuinely relinquish control. In fact, one reason the new educational technologies appear to be so positively received is that they actually *strengthen* the level of faculty control over the materials students should learn (Coulter, 2006).[s] For example, we can post our complete lectures online and thus bypass the need for student notes, often viewed by faculty as inadequate and inaccurate. At some universities, college policy now *requires* such postings. Technology now offers faculty an alluring opportunity: an easily accessible, fixed location where their thoughts can be optimally expressed and consistently available any time for everyone.

Even though it seems only a small step removed from responding personally to student inquiries, the posting of lectures, syllabi, and other course materials in the templates of virtual classrooms is, in fact, a *major* step toward reinforcing the traditional view of the teacher as a subject matter expert. Nor does this step take place in a vacuum, since our professionally-trained administrators encourage this teaching strategy. For them, particularly given the increased enrollments they have actively sought, it is much more efficient to capture online the expertise of their existing faculty and then reuse it, than to hire additional teachers who, arguably, will simply be offering multiple versions of the same material.[t] Thus, rather than a benign reproduction of the physical classroom, the virtual classroom is turning into a "learning repository" for the expert knowledge of faculty that can potentially be delivered by almost anyone. Already the stated goal of many instructional technologists is to protect students from the idiosyncrasies of individual teachers by creating what Kohn (2002, p. 94) describes as "teacher-proof" curricula. Even today the work of part-time disembodied instructors who had no part in the creation or updating of the virtual courses they teach[u] consists almost exclusively of giving feedback and assigning grades. It is probably not that far off in the future when even this work will, with sophisticated software, become fully automated. At that point, live instruction will be almost entirely obsolete.

The impact of the virtual classroom and the changing role of the teacher in higher education are mirrored by changes in the primary and secondary schools. The K-12 educational world has been radically altered by the standardization movement and the de-skilling of teachers obliged to follow the dictates of No Child Left Behind and other such legislation. Already this movement has begun to directly influence higher education. Many state legislatures, boards of regents, and regional accrediting bodies are intimately involved not only in stipulating the contents of what must be taught, but also in prescribing how the learning must be assessed.[v] The standardization of learning outcomes, the evaluation of general education requirements through examination, and other measures that undercut the academic flexibility of the individual professor, and even (in the case of larger university systems) the individual college, are every day more in evidence (e.g., U.S. Department of Education, 2006).[w]

It seems fair, then, to conclude that teachers appear poorly positioned to revolt against the direction education seems to be taking. Even if they are disposed to resist, the ability of faculty to make a difference is certainly weakened by their increasing irrelevance in the instructional process and by the development of overwhelmingly powerful management systems that dwarf any individual teacher's voice. What remains for them to do, however, is not an insignificant role, requiring as it does that they stay abreast of new developments in their fields, find ways of skillfully creating stand-alone courses that reflect this new information, and make clever use of whatever new tools the Internet increasingly offers. And, as we have pointed out above, it's not as if faculty are not prepared for, or don't enjoy the role of content expert. Certainly, once technological developments advance to the point that virtual classrooms can manage themselves with little live assistance, it's not even a question as to what teachers would do if the choice were between being a content expert or having no job at all.

WHAT IS THE FUTURE OF PROGRESSIVE EDUCATION FOR ADULT LEARNERS?

Because of their pragmatic needs, adult learners seem willing, more so even than children or emergent adults, to submit to the iron hand of standardization and expert knowledge.[x] For them, the siren call of the virtual classroom, a mode of education not only easily accessible, but universally acceptable, is irresistible. With an ever-increasing number of such students, vulnerable and often co-opted faculty, and a worldview that enshrines information above all else, what then is the future of progressive ideals within the wired realm of higher education?

One compromise has received considerable attention: blended or hybrid classes, a format popular with campus-based students that combines online information delivery with potentially progressive off-line discussions (e.g., Picciano & Dziuban, 2007; Rossett, Douglis, & Frazee, 2003). Here students participate in a virtual classroom that structures the required information and directs them to explore other, often online, resources. Then, on a regular basis, with the guidance of an instructor in a face-to-face classroom, they share what they have learned, collectively reevaluate this knowledge, and together construct new meanings. Unfortunately for the majority of adult students seeking easy, convenient and, above all, distant access, this model is not really viable. Thus, blended classrooms, attractive as they may seem, will probably remain primarily on-campus options for high school or emergent adult students who share a common physical location and have time for regularly scheduled face-to-face meetings. Most importantly, even though this model purports to meld the directive and progressive methods in one setting, the blended model still remains, in essence, a teacher-centered experience.

Why, we wonder, is this teacher-centered approach so attractive? Why have learner-centered approaches, the essence of the progressive education agenda, not taken greater hold (e.g., Chicoine, 2004) despite philosophical and empirical support, even occasional revivals of wide-spread interest (e.g., Barr & Tagg, 1995)?

Some argue that we cannot let go of the information delivery model because we do not yet fully understand the nature of human learning and therefore operate according to models of learning that are fundamentally incorrect (e.g., Bereiter, 2001; Rumelhart, 1989; Lave & Wenger; 1984, Salomon, 1993). Bereiter, for example, argues that we hold an intuitive and tacit assumption that the mind is a container, which leads us to wrongly believe that education consists of making sure the "filing cabinet" of the mind is properly filled. As he puts it:

The idea of knowledge as the contents of a mental filing cabinet is, I believe, the most stultifying conception in educational thought [even though]... it has been shared by all the major combatants in the educational debates of this century. There are traditionalists who want to make sure the filing cabinet is filled with the right things, there are child-centered and "constructivists" educators who insist the contents of the filing cabinet should be the results of the child's own inquiries, and there are the thinking skills enthusiasts who want to ignore the mental filing cabinet (whose contents they believe to be rapidly obsolescent) and to focus on developing skills in accessing various external filing cabinets and applying their contents. There is merit in all these positions but they appear unreconcilable. Moreover they all undervalue knowledge as it figures in a knowledge-based economy and the careers of experts (p. 24).

If Bereiter is correct that our current ways of teaching are informed by an inadequate "folk" theory about how the mind works, it is very unlikely that institutions of education, much less individual educators, will abandon any time soon the classroom model so perfectly in tune with our current emphasis upon information our minds need to acquire. Considerable research has already shown how difficult it is for non-physicists to discard intuitive notions of the physical world for the laws discovered by Newton, even when they know the old notions are wrong (e.g., Bain, 2004, pp 22-24, Wandersee, Mintzes, & Novak, 1994, as cited by Bereiter, 2001; p 95). Most educators are not cognitive psychologists informed about theories of mind; indeed, in higher education, faculty training about how minds learn is virtually nonexistent.[y] To challenge, much less change, our assumptions about the nature of learning, when a powerful bureaucratic apparatus exists to support our current beliefs, may require something akin to a Kuhnian revolution (1970).

Until this revolution occurs, we feel compelled to conclude that within the realm of formal education, fueled by the demands of information-hungry and degree-seeking adults (Finke, 2000), the virtual classroom will continue to reign. No doubt some dedicated faculty and educational researchers will continue to promote constructivist and progressive methods in their classrooms (e.g., Ball & Lai, 2004; Hodgkinson-Williams, Slay, & Sieborger, 2008). However, in many instances, these pockets of resistance will probably not represent a repudiation of the information delivery model so much as an attempt to improve the quality of the information transfer process. In other words, the ultimate goal of current undergraduate educators, no matter how enlightened, will still remain that of helping students ingest already preselected and vetted facts, figures, terms, concepts, theories, methods, and whatever other information we believe constitute the disciplines we teach.

The irony is that, left to themselves, adults already *do* learn in progressive and constructivist ways. Away from formal schooling, at home or at work, we watch what others do and interact with them, customize what we learn to suit our own needs and purposes, and often create new, even unique, applications–sometimes out of curiosity, sometimes with a particular purpose in mind. It was Dewey's hope that the introduction of these "natural" methods into the classroom would help children see the relevance of school learning to their everyday lives and thus find it easier to learn. But adult students already know that school learning is important or needed. Perhaps we should simply trust these students to find their own ways of absorbing what they genuinely need to know from the lessons we teach. In other words, perhaps we shouldn't worry so much about our inability to relinquish the information delivery model of learning that seems so embedded in the virtual classroom.[z]

We must add here that the Internet, the very technological innovation critical to the development of the virtual classroom, also offers unlimited possibilities for the kinds of unregulated education, outside the academy, that adults have always sought (Kett, 1994) such as in libraries, museums, archives, and grass roots organizations of all kinds.[aa] Why not look at the Internet as a kind of "de-schooled" world of education, a technologically inspired network like the "learning webs" proposed by Illich (1971) more than 40 years ago? Why not view the Internet as a world of "digital disorder" (Weinberg, 2008) where learners are free to choose topics and areas of immediate concern to them – whether the focus is on practical problems or on queries animated by a simple desire to know – and which they can then discuss with people from much more diverse cultural, national, or class backgrounds than can be found in any one campus classroom? Does not such an open-sourse environment meet many of the prerequisites for both progressive and social constructivist models? Are not surfing and blogging and downloading and text-messaging extensions of what people already do spontaneously on the job, in their communities, with their family and friends, in their myriad "learning projects" (perhaps even more vibrantly today than when Tough [1968] recognized their significance)? And unlike what undergraduate institutions expect from the majority of their students, does not the Internet have the added advantage of encouraging people, through wikis, for example (see Susstein, 2006; Tapscott & Williams, 2008), to actually contribute in a serious way to the creation and dissemination of new knowledge?

Currently most educators do not take very seriously the many social communication networks that are blossoming on the Internet; and those who do are mostly interested in finding ways of confining them within the boundaries of their formal course offerings (e.g., Rogers & Swan, 2004). And perhaps that's as it should be. The spontaneous embrace of the Internet by many thousands, some say millions, of people who happily engage in a kind of chaotic, unsupervised exploration of ideas, may turn out to be a better realization of progressive and constructivist ideals than what some dedicated teachers try to create in their classrooms. If so, we should not worry about the future of progressive education. Our speculations suggest that a rebalancing of the tensions described by Dewey and others – a corrective to our over-dependence upon the information delivery model that seems so appropriate for adult students in a globalized, market-driven society – will not come from our educational institutions, but from the freedom of inquiry offered by the very technologies educators at this moment are so eagerly trying to corral.

REFERENCES

Andersson, P., & Harris, J. (2006). *Retheorising the recognition of prior learning*. Leicester, UK: NIACE.

Arnett, J. J. (2000). Emerging adulthood: A theory of development for the late teens through the twenties. *American Psychologist, 55*(5), 469-480.

Aronowitz, S. (2000). *The knowledge factory*. Boston: Beacon Press.

Ashburn, E. (2008, May 2). After the deluge, the drought? How admissions offices are preparing for a shrinking pool of traditional applicants. *Chronicle of Higher Education* (pp. B5-6).

Bain, K. (2004). *What the best college teachers do*. Cambridge, MA: Harvard University Press.

Ball, E. & Lai, A. (2004). What's special about online education. Retrieved Sept. 20, 2008, from http://empire2.esc.edu/facultyweb/ericball.nsf/3c6bed5f67edce6185256889006cd8e8/c6faea6fc298b883852573b400548a55?OpenDocument

Barr, R. B., & Tagg, J. (1995). From teaching to learning: A new paradigm for undergraduate education. *Change*, 27(6), 12-25.

Bauerlein, M. (2008). *The dumbest generation*. NY: Tarcher/Penguin.

Bereiter, C. (2001). *Education and mind in the knowledge age*. Mahwah, NJ: Lawrence Erlbaum.

Bloom, M. (1996). Transformative moments in education. *Holistic Educational Review*, 9(4), 44-47.

Blumenstyk, G. (2008, May 9). In turbulent times, two small colleges brace for the worst: They find strategies to keep enrollments up. *Chronicle of Higher Education*, (pp. A1, A12-13).

Bok, D. (2006). *Our underachieving colleges.* Princeton, NJ: Princeton University Press.

Bowers, C. A. (2000). *Let them eat data.* Athens, GA: University of Georgia Press.

Brookfield, S. (1986). *Understanding and facilitating adult learning.* San Francisco: Jossey Bass.

Brookfield, S. (1987a). *Developing critical thinkers: Challenging adults to explore alternative ways of thinking and acting.* San Francisco: Jossey-Bass.

Brookfield, S. (Ed.) (1987b). *Learning democracy: Eduard Lindeman on adult education and social change.* London: Croom Helm.

Brookfield, S., & Preskill, S. (1999). *Discussion as a way of teaching.* San Francisco: Jossey-Bass.

Burbules, N. C. (2000). Does the internet constitute a global educational community? In N. C. Burbules & C. A. Torres (Eds.), *Globalization and education: Critical perspectives* (pp 323-355). New York: Routledge.

Burbules, N. C., & Callister, Jr., T. A. (2000). *Watch IT: The risks and promises of information technologies.* Boulder, CO: Westview/Perseus.

Burgan, M. (2006). *Whatever happened to the faculty? Drift and decision in higher education.* Baltimore: John Hopkins University Press.

Clark, R. C., & Mayer, R. E. (2008). *E-learning and the science of instruction.* San Francisco: Pfeiffer.

Chicoine, D. (2004). Ignoring the obvious: A constructivist critique of a traditional teacher education program. *Educational Studies, 36*(3), 245-263.

Compos, M. (2004). A constructivist method for the analysis of networked cognitive communication and the assessment of collaborative learning and knowledge-building. *Journal of Asynchronous Learning Network*, 8(2), 1-29.

Coulter, X. (2006). What do we know about student learning? A case study. *All about mentoring*, Issue 31, 35-39.

Coulter, X., with Oaks, S., Gadbow, N., & Gerardi, J. (2000). Technology and education. *All About Mentoring*, Issue 18, 3-10.

Coulter, X., & Herman, L. (1994). Individualization in distance education. *Proceedings: International Conference on Distance Education* (pp 190-192). Moscow, Russia: Association for International Education.

Council for Adult and Experiential Learning (CAEL) (2005). *Principles in practice: Assessing adult learning focused institutions – Case studies.* Chicago: CAEL.

Cross, K. P. (1992). *Adults as learners: Increasing participation and facilitating learning.* San Francisco: Jossey-Bass.

Daloz, L. A. (1999). *Mentor.* San Francisco: Jossey Bass.

Dewey, J. (1916/1966). *Democracy and education.* New York: Free Press.

Dewey, J. (1938). *Education and experience.* New York: Touchtone Books.

Empire State College Office of Institutional Research (2007). 2006-2007 fact book. Retrieved May 18, 2008, from http://www.esc.edu/ESConline/Across_ESC/InstResearch.nsf/db1a77fb2f6bcf2085256bfa005466b0/

88f47019d6617326852573e7004e9bc0?OpenDocument

Erikson, E. H. (1959). *Identity and the life cycle.* Madison, CT: International Universities Press, Inc.

Finder, A. (2007, Nov. 20). Adjuncts outnumber tenured professors on U.S. campuses. *International Herald Tribune.* Retrieved May 14, 2008, from http://www.iht.com/ articles/2007/11/20/america/college.php

Finger, M., & Asun, J. M. (2001). *Adult education at the crossroads: Learning our way out.* London: Zed Books.

Finke, W. F. (2000). *Lifelong learning in the information age.* Bueren, Germany: Fachbibliothek Verlag.

Friedman, T. (2000*). The Lexus and the olive tree.* New York: Anchor Books.

Freire, P. (1972). *Pedagogy of the oppressed.* New York: Herder and Herder.

Giddens, A. (2000). *Runaway world: How globalization is reshaping our lives.* New York: Routledge.

Giroux, H. A. (2007). *The university in chains.* Boulder, CO: Paradigm.

Gulati, S. (2008). Compulsory participation in online discussions: Is this constructivism or normalization of learning? *Innovations in Learning and Teaching International, 45*(2), 183-192.

Herman, L., & Mandell, A. (2004). *From teaching to mentoring: Principle and practice, dialogue and life in adult education.* New York: Routledge.

Hersh, R., & Merrow, J. (Eds.) (2005). *Declining by degrees.* New York: Palgrave/Macmillan.

Ho, C., & Swan, K. (2007). Evaluating online conversation in an asynchronous learning environment. *Internet and Higher Education, 10*(1), 3-14.

Hodgkinson-Williams, C., Slay, H., & Sieborger, I. (2008). Developing communities of practice within and outside higher educational institutions. *British Journal of Educational Technology, 39*(3), 433-442.

Horton, M. (1990). *The long haul: an autography by Myles Horton.* New York: Anchor.

Horton, W. (2006). *E-learning by design.* San Francisco: Pfeiffer.

Houle, C. O. (1961/1993). *The inquiring mind: A study of the adult who continues to learn.* Norman, OK: Oklahoma Research Center.

Hung, W., Bailey, J. H., Jonassen, D. H. (2003). Exploring the tensions of problem-based learning: Insights from research. In D. S. Knowlton & D. C. Sharp (Eds.), *Problem-based learning in the information age* (pp 13-23). New Directions for Teaching and Learning, 95. San Francisco: Jossey-Bass.

Illich, I. (1971). *Deschooling society.* New York: Harper and Row.

Jarvela, S., & Hakkinen, P. (2002). Web-based cases in teaching and learning: The quality of discussions and a stage of perspective-taking in asynchronous communication. *Interactive learning environments, 10*(1), 1-22.

Jones, D. and Kelly, P. (2007). *Mounting pressures facing the U.S. workforce and the increasing need for adult education and literacy.* National Commission on Adult Literacy. Retrieved October 10, 2008 from http://www.nationalcommissiononadultliteracy.org/content/nchemspresentation.pdf

Jones, R. A., & Spiro, R. J. (1995). Contextualization, cognitive flexibility, and hypertext: The convergence of interpretive theory, cognitive psychology, and advanced information technologies. In S. L. Star (Ed.), *The cultures of computing* (pp. 146-157). Oxford, UK; Blackwell Publishers.

Kegan, R. (1994). *In over our heads: The mental demands of modern life.* Cambridge: MA. Harvard University Press.

Kett, J. F. (1994). *The pursuit of knowledge under difficulties.* Stanford CA: Stanford University Press.

Kirscher, P.A., Sweller, J., & Clark, R. E. (2006). Why minimal guidance during instruction does not work: An analysis of the failure of constructivist, discovery, problem-based, experiential, and inquiry-based teaching. *Educational Psychologist, 41*(20), 75-86.

Knowles, M. S. (1976). *The modern practice of adult education: Andragogy versus pedagogy.* New York: Associated Press.

Knowlton, D. S., & Sharp, D. C. (Eds.) (2003). *Problem-based learning in the information age.* New Directions for Teaching and Learning, *95.* San Francisco: Jossey-Bass.

Kohn, A. (2000). *The schools our children deserve.* NY: Houghton-Mifflin.

Kuhn, T. C. (1970). *The structure of scientific revolutions.* Chicago: University of Chicago Press.

Lai, A. (2008, March 25). Adult students' perceptions of learning through online discussion. Paper present at the American Educational Research Association annual conference.

Lave, J., & Wenger, E. (1991). *Situated learning: Legitimate peripheral participation.* Cambridge, UK: Cambridge University Press.

Lore, A., & Tait, A. (2004). Time: The new commodity in lifelong learning. *All About Mentoring*(Issue 27), 14-18.

Magolda, M. B. B. (2004). *Making their own way: Narratives for transforming higher education to promote self-development.* Sterling, VA: Stylus Publishing.

Mandell, A., & Herman, L. (2007). The study and transformation of experience. *Journal of Transformational Education*, 5(4), 339-353.

Mangu-Ward, K. (July, 2008). Education for profit: Why is everyone flaming the University of Phoenix? *Reason.* Retrieved July 10, 2008, from http://www.reason.com/news/show/126856.html.

Manning, M., Manning, G., & Long, R. (1994). *Theme immersion: Inquiry-based curriculum in elementary and middle schools.* Portsmouth, NH: Heineman.

McClintock, R. (1992). *Power and pedagogy: Transforming education through information technology.* New York: Teacher College Columbia University Institute for Learning Technologies.

Merriam, S. B., Caffarella, R. S., & Baumgartner, L. M. (2006). *Learning in adulthood* (3rd ed). San Francisco: Jossey-Bass.

Mezirow, J. (1991). *Transformative dimensions of adult learning.* San Francisco: Jossey-Bass.

Michelson, E., & Mandell, A., & contributors (2004). *Portfolio development and the assessment of prior learning.* Sterling VA: Stylus.

Miyoshi, M. (1998). "Globalization," culture, and the university. In F. Jameson & M. Miyoshi (Eds.), *The cultures of globalization* (pp. 247-270). Durham, NH: Duke University Press.

Mossburger, K., Tolbert, C. J., & Stansbury, M. (2003). *Virtual inequality: Beyond the digital divide.* Washington, DC: Georgetown University Press.

Negroponte, N. (1996). *Being digital.* NY: Vintage Press.

Newfield, C. (2008). *Unmaking the public university: The forty year assault on the middle class.* Cambridge, MA: Harvard University Press.

Palloff, R. M., & Pratt, K. (2003). *The virtual student: A profile and guide to working with online learners.* San Francisco: Jossey-Bass.

Palloff, R. M., & Pratt, K. (2007). *Building online learning communities: Effective strategies for the virtual classroom.* San Francisco: Wiley.

Picciano, A. G., & Dziuban, C.D. (Eds.). (2007). *Blended learning: Research projects.* New York: Sloan-C.

Postman, N. (1985). *Amusing ourselves to death.* New York: Penguin.

Postman, N. (1993). *Technopoly: The surrender of culture to technology.* New York: Vintage Books.

Pusser, B., Breneman, D. W., Gansneder, B. M., Kohl, K. J., Levin, J. S., Milam, J. H., & Turner, S. E. (2007). Returning to learning: Adults' success in college is key to America's future. Retrieved May 18, 2008, from http://curry.edschool.virginia.edu/files/pathways/ReturntolearningApril2007.pdf

Rogers, D., & Swan, K. (2004). Self-regulated learning and internet searching. *Teachers College Record, 106*(9), 1804-1824.

Rossett, A., Douglis, F., & Frazee, R. V. (2003). Strategies for building blended learning. *Learning Circuits.* Retrieved May 14, 2008, from http://www.learningcircuits.org/2003/jul2003/rossett.htm

Roszak, T. (1986). *The cult of information: The folklore of computers and the true art of thinking.* New York: Pantheon.

Rumelhart, D. E. (1989). The architecture of mind: A connectionist approach. In M. I. Posner (Ed.), *Foundations of cognitive science* (pp. 133-159). Cambridge, MA: MIT Press.

Salomon, G. (Ed.) (1993). *Distributed cognitions.* Cambridge, UK: Cambridge University Press.

Schribner, S. (1984). Studying working intelligence. In B. Rogoff & J. Lave (Eds.), *Everyday cognition: Its development in social context* (pp. 9-40). Cambridge, MA: Harvard University Press.

Sennett, R. (2006). *The culture of the new capitalism.* New Haven: Yale University Press.

Sennett, R. (2008). *The craftsman.* New Haven: Yale University Press.

Stein, D. G. (Ed.). (2004). *Buying in or selling out: The commercialization of the American research university.* Rutgers, NJ: Rutgers University Press.

Stone, A. R. (1995). Sex and death among the disembodied: VR, cyberspace, and the nature of academic discourse. In S. L. Star (Ed.), *The cultures of computing* (pp. 243-255). Oxford, UK; Blackwell Publishers.

Stross, C. (2005). *Accelerando.* New York: Ace Books.

Susstein, C. R. (2006). *Infotopia: How many minds produce knowledge.* Oxford, UK: Oxford University Press.

Tapscott, D., & Williams, A.D. (2008). *Wikinomics: How mass collaboration changes everything.* NY: Portfolio/Penguin.

Torres, J. T. (2008, May 3). Stop dreaming and think: A 24-year-old's advice for high-school seniors. *Orlando Sentinel* (p. A19).

Tough, A. (1968). *Why adults learn: A study of the major reasons for beginning and continuing a learning project.* Toronto: Ontario Institutions for Studies in Education.

University Continuing Education Association (2006). *Lifelong learning trends: A profile of continuing education.* Washington, DC: National Center for Educational Statistics.

University of Washington (n.d.). *WOT history project*. Retrieved May 18, 2008, from http://depts.washington.edu/wtohist/about_project.htm

U.S. Census Bureau (2008). *Question & answer center*. Retrieved May 18, 2008, from https://ask.census.gov/cgi-bin/askcensus.cfg/php/enduser/std_adp.php?p_faqid=284&p_created=1077815981&p_sid=nI3o5k3j&p_accessibility=0&p_redirect=&p_lva=&p_sp=cF9zcmNoPTEmcF9zb3J0X2J5PSZwX2dyaWRzb3J0PSZwX3Jvd19jbnQ9MjYsMjYmcF9wcm9kcz0mcF9jYXRzPSZwX3B2PSZwX2N2PSZwX3BhZ2U9MSZwX3NlYXJjaF90ZXh0PWNvbGxlZ2UgZ3JhZHVhdGVz&p_li=&p_topview=1

U.S. Department of Education (2006). *A test of leadership: Charting the future of U.S. education*. Retrieved May 14, 2008, from http://www.ed.gov/about/bdscomm/list/ hiedfuture/reports/final-report.pdf

Valentine, T., & Darkenwald, G. G. (1990). Deterrents to participation in adult education: Profiles of potential learners. *Adult Education Quarterly*, *41*(1), 29-41.

Vander Valk, F. (2008). Identity, power, and representation in virtual environments. *Online Journal of Learning and Teaching*, *4*(2). Retrieved Sept. 20, 2008, from http://jolt.merlot.org/vol4no2/abstracts.htm

Vinge, V. (2006). *Rainbows end*. New York: Tor Books.

Vygotsky, L.S. (1934/1986). *Thought and language*. Cambridge, MA: MIT Press.

Vygotsky, L.S. (1978). *Mind in society: The development of higher psychological processes*. Cambridge, MA: Harvard University Press.

Walters, S. (2000). Globalization, adult education, and development. In Stromquist & Monkman (Eds.), *Globalization and education: Integration and contestation across cultures* (pp. 197-215). Lanham, MD: Rowman & Littlefield.

Weinberger, D. (2008). *Everything is miscellaneous: The power of the new digital disorder*. New York: Holt Paperbacks.

Welton, M. (2005). *Designing the just learning society: A critical inquiry*. Leicester, UK: NAICE.

Wlodkowski, R. J., & Kasworm, C. C. (2003). *Accelerated learning for adults: The promise and practice of intensive educational formats*. San Francisco: Jossey-Bass

Woo, Y., & Reeves, T. C. (2007). Meaningful interaction in web-based learning: A social constructivist interpretation. *The Internet and Higher Education*, *10*(1), 15-25.

ENDNOTES

[a] We work at SUNY Empire State College, which offers an array of study modes including individual tutorials and individualized curricula that include strong emphasis upon the assessment of learning through lived experience.

[b] Currently older students represent more than 40% of all undergraduates (University Continuing Education Association, 2006).

[c] Arnett (2000) also importantly acknowledges that this stage is not universal and occurs only in societies that can afford to offer young people time off from social obligations and (although he does not make this second point explicit) in which life expectancies are long enough that they will have sufficient time afterward to make the contributions to society for which this long period of education prepares them.

[d] Using the description developed by The National Center for Education Statistics (NCES), the Council for Adult and Experiential Learning (CAEL, 2005) identifies seven characteristics of nontraditional students: "Have delayed enrollment into postsecond-

ary education; attend part-time; are financially independent of their parents; work full-time while enrolled; have dependents other than a spouse; are a single parent; lack a standard high school diploma (p.1)."

e John Dewey, the principal theorist of progressive education, upon whose ideas we so strongly depend, developed its tenets almost entirely with only children in mind.

f Since about 85% of adult students (at least at our institution) are actually "returning" students, their view of school as a repository of knowledge is the reality they have already experienced not only in high school, but in their earlier aborted attempts in college.

g A frequent response to this concern is, ironically to offer students courses in "time management," which in its instrumental focus seems somewhat incompatible with a progressive or constructivist view of education.

h Few adults return to college without having already engaged in serious thought about the direction of their lives. Nor have their personal experiences been exactly devoid of perspective changes or transformational moments.

i More than 70% of adult Americans still do not have college degrees (U.S. Census Bureau, 2008).

j It is thus not by chance that the "National Commission on Adult Literacy" describes "those individuals who started, but did not complete a college education" as "low-hanging fruit" (Jones and Kelly, 2007).

k These younger students too are increasingly pragmatic (e.g., Torres, 2008). Escalating tuition rates alone discourage them from taking extra time, even if time were available. Moreover, they too look for efficiencies, and for two quite different reasons. First, they, like adults, have competing priorities -- typically their college work versus many of the extracurricular activities offered by the college itself. The less time spent on studies, the more that is available for sports, student government, and part-time jobs (all activities intended to pad their young resumes) to say nothing of socializing, and just hanging out. But secondly, it may be simply part of human nature to seek the shortest route (for a classic example, see Scribner, 1984).

l We refer here, of course, to the *potential* for universal access, reminding ourselves that a "digital divide" continues to serve as a barrier for large numbers of underserved populations (e.g., Mossburger, Tolbert, & Stansbury, 2003). In considering the provision of college study for adult learners, the underprivileged in our world still remain largely outside serious consideration, raising, as they do, abiding questions not only of access, but of class and status.

m By focusing upon the Internet, we clearly include the personal computer, upon which the ability to access the Internet depends, and through which many educational innovations were originally developed. We can point, for example, to computer software developed to simulate various scientific experiments that, once available, swiftly migrated to the Internet.

n Postman (1993) makes a similar point as he imagines the impact of the then new technology of writing several thousand years ago when people then came "to rely upon external signs instead of their own internal resources [p. 12]."

o Some commentators point out that technological developments in education are still only at an early, primitive stage, that is, at a period roughly equivalent to that of the "model-T" when automobiles still strongly resembled a horse carriage. However, we note that long-standing models, like the classroom, are not always easy to transcend. The functional components of the modern car are still, 100 years later, recognizably the same as in the model-T.

p That the virtual classroom has been eagerly accepted by the adult student can be easily illustrated by enrollment figures from our institution. Since its inception, Empire State College consisted of a variety of geographically dispersed learning centers that offered individualized instruction, plus a small program of distance education that provided pre-designed courses for self-study delivered by mail and expanded by telephone contact with a tutor. The distance program at that time, and for many years thereafter, accounted for about 10% of the college enrollments (including students from the other locations in the college who supplemented individual tutorials with an occasional distance course). However, with the advent of the virtual classroom, the once ancillary distance program began to grow exponentially, and now accounts for almost a third of the college undergraduate enrollments (79,610 out of a total 228,440 credits). More startling, during the past 5 or 6 years as the virtual classroom became fully institutionalized, the distance program itself, which typically grew – as did the rest of the college – a few percentage points a year suddenly doubled in size (Empire State College Office of Institutional Research, 2007).

q It is interesting to follow an instructor's learning curve in operationalizing the requirements for a meaningful "discussion." It begins when the instructor specifies that each student is expected to participate at least three times in a given discussion. When she then observes that some students meet this requirement by making three different comments on, say, the last day of the discussion, she then changes the instructions to specify "one main posting" and "three comments to your classmates" *per week*, and then later to prevent all this activity from still occurring on one day, "one main posting, several responses to other classmates, and *several follow-up comments*" in a given discussion. Finally, a teacher may learn to be prepared to write private notes to students in order to ask them to participate at least three or four or five *different days* during the discussion period. Students demand this level of specificity even though the instructor may also emphasize that an important goal of class discussions is the co-construction of knowledge and explain various ways by which this can be accomplished.

r Indeed, "expertise" in a constructivist educational world has to be redefined. No longer based upon the disciplinary knowledge represented by the Ph.D., or conventional scholarly pursuits, it must refer to a whole new variety of complex skills related to teaching -- skills of mentoring and facilitation that are totally ignored in most graduate programs, acquired almost entirely by experience, and practiced only when this special kind of teaching is seen by the faculty as worth the extra effort involved.

s At the first State University of New York Conference on Instructional Technologies (SUNY CIT) in Oneonta in 1992 (see http://www.cit.suny.edu/) – in a presentation designed to demonstrate the usefulness of technology in education -- several science teachers could not contain their excitement over the ways in which instructional software would now allow them to totally control the results and consequences of various procedures in classical experiments from their disciplines.

t Currently, administrators seek to streamline education through the use of a "reserve army" of adjuncts, a huge body of floating, highly skilled academics upon which so many institutions now depend (e.g., Finder, 2007). We suggest below, however, that even this army may soon become obsolete.

[u] Our institution is still quite flexible in that, at least to date, it allows instructors leeway to change up to 20% of a virtual course they are hired to teach.

[v] Sennett (2008) provides a sensitive and detailed discussion of the training, tools, skills, consciousness, and even the "hand" of the craftsman, and above all, "the drive to do good work (p. 267)." One could argue that it is exactly such "quality-driven work" (quality-driven work Chapter IX) that is being tested if not completely undercut by the recent changes in higher education just described.

[w] Think too about the practical and ideologically-tinged implications of the term, "accountability," and the development of nationwide databases that describe and compare the achievements of our higher education institutions.

[x] Although it might seem ironic that these same students are also perceived as quite capable of charting their own course and of being open to the uncertainties and challenges of complex inquiry, we should also remember that they are fully aware of the importance of following the rules of society. And why should they question the offerings and available methodologies of higher education when they purportedly offer exactly the information that they need?

[y] This lack of training reflects in yet another way how strongly we assume that our minds are containers waiting to be filled with expert knowledge. We may need help in making those containers more receptive to what we want students to learn, but we don't even imagine that the container metaphor itself may need to be challenged.

[z] Clearly, the vast majority of us did not experience a constructivist or progressive education either as children, teenagers, undergraduates, or, in many cases, even as graduate students.

[aa] We refer here to the social or political movements exemplified in the writings and practices of such important critical educators as Friere (1972), Lindeman (Brookfield, 1987b) and Horton (1990). Also relevant today are new experiments in alternative forms of unregulated education as, for example, the WTO History Project connected to the 1999 Seattle protests against the World Trade Organization (University of Washington WTO History Project, n.d.) and other such Internet-based digital initiatives.

Section II

Teaching and Learning with Information Technology:

Constructivist Approaches

Chapter VIII
Online Learning Conversations:
Potential, Challenges and Facilitation

Jakko van der Pol
Utrecht University, The Netherlands

ABSTRACT

This chapter aims to perform a thorough analysis of students' online learning conversations. Although offering a high potential for collaborative learning, successful online learning conversations are not easy to realize. After discussing the specific challenges of conducting conversations in general, conversations-for-learning and learning conversations online, the author uses this investigation to discuss ways to effectively facilitate them. Van der Pol demonstrates, then, that the context-creating effect of anchored discussion can effectively address some of these difficulties by turning opinion-oriented exchange of ideas into a more meaning-oriented processing of material, while increasing communicative efficiency.

INTRODUCTION

Imagine a teacher in higher education: a teacher with a progressive state of mind and an enthusiasm for social constructivist theories who is continuously trying to improve the learning of his or her students, but is nevertheless still struggling to create successful and sustainable social constructivist teaching practices in his or her everyday life. Especially when trying to fulfill the promise of Computer Supported Collaborative Learning (CSCL), this teacher may have trouble to deepen students' learning processes and to maintain students' initial levels of motivation and engagement after a fading first wave of technology-related enthusiasm. The needs and problems of this imaginary teacher are at the basis of the theoretical and practical issues I will report on in this chapter and they have determined its approach in four different ways.

First, in this chapter the use of CSCL is regarded as a means to reach a certain goal, and it studies CSCL with deep or effective learning processes of students in mind. As indicated by Säljö (2003), introducing technologies in institutionalized forms of learning has not always lived up to the initial expectations in reaching these goals. In response to this observation, Säljö proposes to abandon the question of whether new technologies can improve learning altogether, because learning does not become better or more efficient, just different. However, I think it is still worthwhile to investigate whether it is possible to enhance learning through CSCL and how this can be done. Not only do I think that educational research is not neutral when it comes to learning, I feel it also has a certain responsibility. As the positive expectations about the advantages of CSCL for educational practice have been partially responsible for its widespread introduction, I think educational science should also investigate whether and how these expectations can be realized. I continue to believe in the potential of CSCL for student learning, albeit depending on many related variables and conditions. I therefore propose to direct research towards studying how to shape and develop the different (specialized) uses of CSCL in such a way that their learning potential can be better realized.

Second, seeing CSCL as a means and not as a goal in itself, I will be aiming for something more than the mere presence of online interaction. Andriessen, Baker, and Suthers (2003) distinguish between learning from a certain dialogue and learning to conduct that dialogue. Although the second learning goal will usually need to be reached to a certain degree as a necessary condition for reaching the first, by itself it may not be sufficient. In order to justify the investments in students' and teachers' time and the efforts that are required, I feel that it is important that students' efforts are directed as much as possible to activities that lead to and deepen learning, in relation to their content-related learning goals. As stated by Kirschner, Paas, and Kirschner (Submitted), it is important in collaborative learning to maximize the students learning-generating effort (or "germane cognitive load") while minimizing the required interaction costs. In this way, it is possible to let students invest as much of the available effort as possible in the production of rich interactions.

Third, our concerns for educational practice have led to the choice of a form and area of CSCL where I expect to achieve the greatest gains. The area I will primarily focus on is students' processing of academic texts, using *online asynchronous discussion* in blended learning situations. As stated by Lapadat (2002) asynchronous online discussion is particularly suitable for the collaborative construction of meaning and presents a great potential for conceptual change. I will argue that online discussion is especially useful for the appropriation of abstract knowledge, which can still be seen as an essential activity in higher education and as forming the basic ingredient for the development of many required competences. The medium presents students with an open learning environment that allows interaction on a conceptual level. This "discursive" interaction offers the possibility for education to connect to the way students (initially) make sense of the material (Laurillard, 1993), while at the same time interactively stimulating them to expand this understanding to a higher level (Petraglia, 1998). By offering the potential for students to engage in an active, thoughtful and personally meaningful processing of content, it can facilitate the appropriation of new ideas and the deepening of existing understanding in a self-discovering way. Thus, the kind of online discussion I wish to realize is not so much a critical and opinion-centered debate, but a more constructive conversation that is aimed at the processing of content and a deepening of understanding. I will refer to this kind of discussion as "learning conversations," described by Bellamy (1997) as conversations that lead to learning because they allow participants to make connections between previously unrelated ideas, see old ideas in a new way, and lead to conceptual change. However, this great potential to support the essential deep learning processes of students also comes with a great challenge and the reflective learning conversations described can also be the most difficult to realize. Being in a learning situation means that students are not only expected to share the knowledge they already have, but also collaboratively construct new knowledge and understanding. Collaboratively developing new meaning and creating new understanding that surpasses that of the individual participants, may be paradoxical in itself, and will be difficult and complex to say the least. As such, the development of semantic understanding presents a fragile process that can only progress in a gradual way (Alargamot & Andriessen, 2002). This challenge is further increased by the fact that asynchronous online learning conversations can also be hindered by certain medium-related constraints, such as the limited degree of interactivity (being asynchronous and producing fewer "turns") and its natural tendency to diverge. This divergent of character of online discussion, as found by Andriessen and Sandberg (1999) and by Mäkitalo, Häkkinen, Salo, and Järvelä (2001) seems to be the result of a greater freedom to engage in new topics. As stated by Romiszowski (1995), this makes it more difficult to keep an asynchronous discussion on topic than is the case face-to-face.

Fourth, our approach to tackling these challenges will be to investigate and manipulate the characteristics of different tools that can be used to support students' online learning conversations. Like Kirschner, Sweller, and Clark (2006), I believe students do need to be guided in creating successful constructivist and discovery-based learning practices, especially when they only possess low levels of prior knowledge and experience. However, I will not investigate how to provide this guidance by means of direct instruction, but how to do this by using the "affordances," or "aspects of an artifact's design that suggests how it is to be used" (Johnson, 1995, p. 219), of different tools. Although I will make use of the vast amount of knowledge that already exists on how to successfully create and implement online learning conversations (see for instance Pilkington, 2004; Salmon, 2002; Van der Pol & Admiraal, 2003), it is important to focus on how the affordances of different tools can be of further use. As stated by Pilkington, "It is also important to meet students 'where they are at' and not where we want or expect them to be if we are to effectively facilitate online discussion" (p. 163). In our case this means that, as several ways to create a sufficient quantity of responses already exist, I will direct our attention to further enhancing their content and quality. This focus on affordances of tools means I will investigate how the functional characteristics of a tool can facilitate and stimulate the production of rich and constructive interactions. Providing explicit additional training and instruction on how to interact constructively presents an additional burden on a teacher's available time and, while known to be effective, this effect may be temporary and fade over time. In addition, an advantage of using the affordances of a tool to facilitate students' interaction is that it does not directly steer or constrain students' actions too explicitly or strongly and preserves an open learning environment. As expressed by Innes (2004), using criteria to direct students' collaboration too explicitly is not desirable, as "they sometimes shape the students' behavior without expanding their understanding" (p. 247). As stated by Dillenbourg (2002) and demonstrated by Beers (2005), constraining students' collaboration too much by "overscripting" it, can have disruptive effects.

In the next sections, I will examine the notion of online learning conversations in more detail to understand better why they can be difficult to create in academic education. First, I describe how people in conversation aim to reach mutual understanding based upon the available common ground and grounding. Next, I will identify the "effort after shared understanding" (Schwartz & Lin,

2000), or "grounding" (Clark & Brennan, 1991) as an essential aspect of learning conversations. Especially in a university setting, grounding-for-learning, or "semantic grounding", (see Baker, Hansen, Joiner, and Traum, 1999) is suggested to differ from grounding as found in regular, everyday conversations. Then I will discuss the additional affordances and constraints of conducting learning conversations online. Finally, using the insight gained about the difficulties and demands of successful online learning conversations, I describe a possible way to effectively facilitate them based on the affordances of a specialized tool for online discussion.

CONDUCTING CONVERSATIONS AND THE ROLE OF COMMON GROUND

Before exploring the specific characteristics of conversation for learning and conversation online, it is useful to look at some of the basic mechanisms of conversation in general.

Conversation can be seen as a specific form of communication, in that it is generally interactive, in large part language based, and conducted by a limited number of participants. This paragraph will mainly cover dyadic conversations, as this is often the focus of traditional face-to-face conversation theories which I will start with. In most kinds of conversation, the goal usually lies in reaching a certain degree of mutual understanding, meaning that both partners understand to a certain degree what the other means by his or her utterances, in order to communicate effectively. Two main elements that play a role in achieving this are grounding and common ground (see Clark & Brennan, 1991; Clark, 1992). Roughly, grounding can be described as an active process which takes place during the conversation itself, whereas common ground can be seen more as an available resource on which the conversation is based. As explained by Clark, common ground lets speakers use the information they have about the knowledge, ideas, intentions, or (shared) history of one's communication partner to omit redundant information from their utterances and communicate more efficiently. In instances where relying on the available common ground is not sufficient, people can resort to the use of grounding activities. In a broad sense, grounding can be identified as anything a person does to increase the degree of (mutual) understanding. In a more limited sense, grounding can be identified as certain explicit actions one can undertake to reach this goal, such as checking, acknowledging, or repairing (see for instance Poesio & Traum, 1997).

In sum, I interpret common ground as something that conversation partners start with and base their grounding efforts on, in order to achieve a certain degree of mutual understanding. Using the terms offered by Akkerman et al. (in press), I view common ground as an "overlap" in the range of meaning that is available to the different partners in a certain conversational situation, and mutual understanding as a degree of "similarity" in the meaning that these partners attribute to specific utterances during the conversation itself. To clarify this interpretation, I have schematically depicted it in Figure 1.

In Figure 1, arrows indicate the respective relations between common ground, grounding and mutual understanding. As indicated by the positive arrow from mutual understanding back to common ground, it is possible for the mutual understanding to become part of the common ground of certain participants for future conversation. Another self-reinforcing relation, (indicated by the negative arrow from common ground to grounding), is formed by the functional nature of grounding, which means that grounding activities are driven by mechanisms of efficiency and will occur (only) to the extent they are needed (Clark & Wilkes-Gibbs, 1986; Wilson and Sperber, 1995). Thus, the less common ground there is to start with, the more grounding activities will occur (as confirmed in Van der Pol, 2002).

Figure 1. Schematic view of common ground, grounding, and mutual understanding and the relations between them

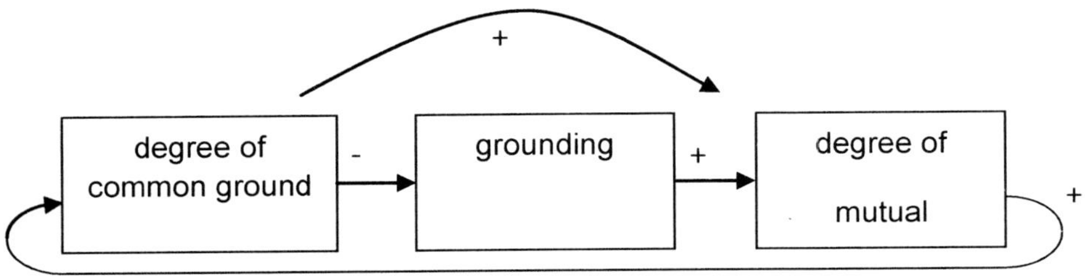

Finally, the success and ease of grounding itself also depends on the available common ground, as indicated by the positive arrow at the top of Figure 1. Kraus and Fussell (1991) express this reciprocal relation between grounding and common ground as follows: "It is hard to find some if you don't have some already and you don't have any unless you find it" (p. 4). Thus, although it is possible to (partially) correct and prevent misunderstanding by performing grounding activities, the available amount of common ground also plays an important role in determining the efficiency of communication. Having explained our basic model of common ground, grounding and mutual understanding, two additional comments should be made regarding the fact that common ground is not a fail-proof resource that simply accumulates and leads directly to mutual understanding, and regarding the context-sensitive nature of common ground.

First, the fuzzy and ambiguous nature of common ground makes using common ground less straightforward than it may seem. While in Clark's (1992) reasoning piece of information X is either "known" or "unknown" to person A or B, there can also be many stages in between, and many different ways of knowing piece of information X. Presenting an alternative perspective on common ground, Wilson and Sperber (1995) view it rather as a "mutual cognitive environment", which they define as the overlap in the participants' *manifest* meaning. Thus, we can identify a person's manifest meaning as the range of possible meaning that is evoked or triggered by the presented evidence, in a particular situation. As stated by Croft and Cruse (2004), this body of conceptual content (also known as the purport), does not correspond to any specific interpretation but presents a wide range of meaning which, as demonstrated by Billig (1988), can even contain contradictory meanings as well. Viewing common ground as an overlap in people's manifest meaning means that a high degree of common ground does not ensure people will attribute the same meaning to a certain utterance, but that they will be more likely to do so. This perspective on common ground is also more in line with the "mutual knowledge paradox" as described by Clark and Marshall (1981). As someone's knowledge, intentions or experiences can never be accessed directly and can only be inferred from their communicative interactions (Draper & Anderson, 1991), mutual understanding can never be reached completely but only to a certain degree.

Second, because a person's range of triggered manifest meaning depends greatly on his/her perception of the collaborative situation, so does the amount of common ground among multiple conversation partners. As emphasized by Bereiter (2002), the individual process of meaning-making is intrinsically connected to a particular context and cannot be seen in isolation. Because I have defined common ground as overlap among individuals' ranges of manifest meaning, this degree of overlap is strongly influenced by the strength or the focusing effect of the conversational context. As expressed by Wilson and Sperber, "a mismatch between the context envisaged by the speaker and the one actually used by the hearer may result in a misunderstanding" (1995, p. 16). Consequently, because participants' perceptions of their conversational context are not stable but dynamic, so is the available common ground (see Koschmann & Le Baron, 2003). As stated by Baker et al. (1999), participants' perceptions of the communicative situation (such as the collaborative goal and the medium that is used) also influence their grounding process. Next I will further investigate the nature of conversation with a learning goal and of conversations which are conducted online.

THE SPECIAL NATURE OF LEARNING CONVERSATIONS

In conversation for learning, one of the main goals is to engage students in the collaborative processing of meaning in order to deepen their levels of understanding. Therefore, as expressed by Schwartz and Lin (2000), the process of grounding itself, which they describe as the effort after shared understanding, plays a central role. However, three additional remarks bearing on this observation should be made: First, although it is the activity of grounding itself that is closely related to learning, not *all* or *any* kind of grounding will automatically produce the desired results. Second, the kind of grounding that is most strongly associated with learning cannot be expected to be as self-regulatory and taken for granted as the grounding-for-conversation we know from conversation in everyday life. Third, in learning situations where participants have only a limited understanding of the content they are talking about, successful semantic grounding becomes even more difficult.

Our first remark, that not any kind of grounding will suffice, applies especially to the setting and the goal that I focus on in this chapter. In university education, learning conversations will often concern abstract knowledge which cannot be experienced directly but only through the descriptions of others, making learning essentially a mediated phenomenon (Laurillard, 1993). Especially when the goal lies in collaboratively developing a better understanding of abstract subject matter, it is mainly conceptual or semantic grounding (see Baker et al., 1999), that is most strongly associated with learning. Besides the semantic level of grounding, there also exists a pragmatic level of grounding, which is more directly concerned with recognizing other students' communicative intentions and with ensuring the continuation of the collaboration. As stated by Baker et al. "Semantic level grounding however, relates to attaining mutual understanding of what is meant by certain terms and expressions; it thus relates more closely to learning in a specific knowledge domain by means of interpersonal interaction" (1999, p. 42).

Regarding the second remark, we think that students' semantic grounding processes may often not progress as far as we would want them to, as this may not be necessary to continue the conversation at a seemingly acceptable level. As a primal mechanism in everyday communication, the process of grounding is driven by a principle of "least collaborative effort" (Clark & Wilkes-Gibbs, 1986), meaning that participants will limit the efforts they invest in grounding activities to what is sufficient to continue the conversation. Similarly, Wilson and Sperber's (1995) relevance-theoretic comprehension procedure indicates that a hearer will also follow a path of least effort in deriving the meaning and implications of a certain utterance. Wilson and Sperber also mention that "a communicative intention can be fulfilled without the corresponding informative intention being fulfilled" (p. 30). In conversations in general, a high degree of semantic grounding may not always be necessary to continue the conversation (which can, for example, be the case in a competitive debate). Kuhn and Goh (2005, p.

347), state that one of the mistakes teachers make when conducting classroom discussions is

... to allow the activity to relapse into nothing but consecutive self-expression, first on the part of one student, then another. It does not matter much what each student says, and no student need listen to another. In this worst-case scenario, the only attention the next student pays to the speaker is to wait to observe a signal that this speaker is about to finish, so that he or she can begin. As long as everyone gets their share of turns to speak and no one speaks too long, there is a wealth of opportunity for self-expression. Yet, no further purpose is fulfilled.

Similarly, Schwarz and Glassner (2003) conclude that people are generally highly skilful in challenging, counterchallenging, justifying or agreeing during everyday conversation, but that the arguments are often mediocre according to analytical criteria. In sum, as concluded by Baker et al. (1999), what is sufficient to continue the conversation, might not be the same as what is sufficient for learning, and instead of striving for a *minimal* collaborative effort, participants in learning conversations should strive for an *optimal* collaborative effort (see Dillenbourg, Traum, & Schneider, 1996). Finally, even when people do aim for a high degree of semantic grounding and mutual knowledge, they may succumb to the "illusion of shared knowledge," which, as shown by Ross, Greene and House (1977), means that people are likely to overestimate their amount of mutual knowledge.

Our third remark on the additional difficulty of semantic grounding for learners refers to the influence of students' understanding of the subject matter on the effectiveness of their communication. As identified by Innes (2004), language has a double function in collaborative learning, as both a means to conduct the collaboration itself and as its object in the goal of coming to understand and appropriate certain scientific discourse and concepts. According to Innes, these two functions of dialogue "are enmeshed because students' inability to communicate with each other impacts the inquiry process and consequently affects their ability to acquire useful knowledge" (p. 192). Or, in the words of Baker et al.: "Learning from grounding - collaborative learning - can thus be viewed as appropriation of semiotic tools, mediated by those very tools" (1999, p. 12). Thus, similarly to the reciprocal relation between grounding and common ground that I described above, the ability of students to construct knowledge together also largely depends on the amount of prior knowledge they have available. Especially in learning situations where students are still wrestling with the subject matter, it may become more difficult for them to properly understand others and explain themselves, which in turn affects the collaborative deepening of understanding.

CONDUCTING LEARNING CONVERSATIONS ONLINE

Thus, as a high degree of semantic grounding may not be required to continue the conversation, the effectiveness of students' learning conversations cannot always be taken for granted and even if this is the case, the process may be hindered by students' low levels of prior knowledge. In this section, I will show that conducting learning conversations online (focusing on the use of asynchronous discussion boards), can present some additional problems.

First, compared to communication face to face, the distance and freedom that is associated with online forums seem to present users with a reduced expectancy of conversational coherence and a preference for expressing and exchanging existing opinions and experiences. Online discussion often can be seen to resemble a sequence of monologues rather than a true dialogue (Fay, Garrod, & Carletta, 2000). Although, as expressed by Herring (1999), this reduced interactional coherence is usually not a serious impediment to users' recreational use of computer-mediated communication (CMC), it can hinder them in conducting effective online learning conversations. As indicated by Andriessen (2006), constructive forms of argumentation have more to offer to education than the aggressive forms that can be found in talk shows or the political sphere. Elaborating on this distinction, I think that a further subdivision can be made. When students interact in a constructive and collaborative way, two different kinds of argumentation can emerge. First, students may argue about what they think of certain statements in a text ("argumentation for opinion") and second, they may argue about the meaning they attribute to it ("argumentation for interpretation"). The latter requires students to stand back from their personal viewpoints and to look at the material from the author's perspective, trying to develop new understanding. Although this argumentation-for-interpretation is more difficult for students (as demonstrated by Van der Pol, Admiraal, & Simons, 2006a), I believe that it also offers the most possibilities for learning. This preference is also shared by Klemm (1998, p. 3), who expresses it in the following way:

Don't settle for just opinions. Everybody has opinions. They are like knee jerk reflexes, occurring with little thought once they have been formed. Thus, it is not surprising that many classroom discussion groups online are dominated by opinion messages, rather than rigorous analysis and creative thought.

Second, the reciprocal relationship between grounding and common ground and the double function of language are even more emphasized when conducting learning conversations online. The context-reduced nature of communicating at a distance (Smith, 2003). may leave students with a greater range of manifest meaning, presenting a less focused area of common ground. Thus online conversations may leave more room for miscommunication than communication face-to-face. In addition, the reduced level of interactivity and absence of non-verbal communication make miscommunication or misalignment both harder to detect and to repair. While we have seen that people will usually invest as little effort as necessary to continue the conversation and repair any miscommunications after they occur, asynchronous conversation places more emphasis on preventive strategies, like perspective taking (Järvelä & Häkkinen, 1999) and audience design (Clark, 1992). As for these processes the existence of a largely shared frame of reference is an important prerequisite, because asynchronous conversations depend less on grounding and more on common ground. A complicating factor in collaborative learning conversations is that these processes of audience design and perspective taking become more complex in "many-to-many" communication, as it then becomes more complex to decide who to design for, or whose perspective to take.

The following example (see Figure 2) shows two messages that were collected in a study of using asynchronous discussion groups for the collaborative processing of academic texts (Van der Pol, 2002). The data collected in this study often showed messages that in form resembled successful online discussions with successive messages containing new knowledge and argumentations. Although on a pragmatic level students′ actions were sufficiently grounded, their mutual understanding on a semantic level was insufficient to ensure a successful collaborative construction of knowledge. In the given example we see that is can be very difficult to determine how the different messages relate to one another and what the logical implications of a reaction is for the messages that precede it, which makes it difficult for students to build upon each others' contributions towards a deeper understanding of the subject matter.

When we look more closely at the second message, we see that it aims to provide "some more clarity" in response to the first message.

Figure 2. Fragment of an online learning conversation

Discussion statement: "According to Laurillard, phenomenography is a research method that focuses on task specific characteristics"

Date: January 12, 2000 05:06 PM
Subject: wrestling with the statement

I find the start statement of this discussion not as clear than it is at first sight. If I read Laurillard I can only partly confirm the statement. De question is, however, whether phenomenography is a suitable research method for task specific characteristics.

Date: January 19, 2000 02:49 PM
Subject: some more clarity

In our class of 12/1/2000 my group has posed a 'reading question' about phenomenography. In the discussion that followed a few things obviously came forward, after which a joint
conclusion was drawn, which is the very thing that: Laurillard prefers the phenomenographical approach (p.35). The drawbacks that she mentions after that are aimed against old forms like observation of behaviour and in favour of phenomenography. Only by asking about it and not by observing, can one know how student experience a certain concept.

However, this reply presents a gradual shift in topic as it discusses the question whether Laurillard is in favor of phenomenography or not, instead of the question whether, according to Laurillard, phenomenography focuses on task specific characteristics. Although both messages are related, the subtle change in topic of the second one means it presents no direct implications for the first, which hinders the drawing of a new conclusion. This short fragment makes clear that students not only have to guess the meaning of Laurillard's book (see Flower, 1990 on text comprehension), but also the meaning of the discussion statement and the meaning of each other's messages. This makes collaboratively trying to develop new understanding of subject matter that is unknown and unfamiliar a difficult and gradual process (Alargamot and Andriessen, 2002). As demonstrated by Webb and Mastergeorge (2003), this fragile collaborative process requires students to ask precise questions and to provide each other with detailed and relevant reactions, based on a thorough and iterative study of each other messages and the material at hand.

In the previous sections I have looked at conversations in general, conversations for learning, and (learning) conversations online, for each section identifying obstacles that can impede the creation of successful online learning conversations. This examination has underlined the importance that online learning conversations are meaning-oriented, constructive, and interactionally coherent. I have demonstrated that these characteristics will usually not arise naturally and, thus that they need to be facilitated.

FACILITATING ONLINE LEARNING CONVERSATIONS

A possible way to increase the meaning-orientedness, the constructive nature, and the interactional coherence of students' online learning conversations, is to use a tool for "online anchored discus-

Figure 3. The IVLOS Annotation Tool (www.annotatiesysteem.nl)

sion" (see Bernheim Brush, Bargeron, Grudin, Borning, & Gupta, 2002). As claimed by Hunt (1998), the fact that students in regular online bulletin boards appear to respond immediately and without much reflection is caused by a lack of *context*. It is exactly this integration (or "anchoring") of discussion in a shared context that distinguishes anchored discussion from regular forum discussion. As can be seen in Figure 3, a tool for anchored discussion displays a document or artefact next to the threaded discussion. In addition, messages (or "annotations") can simply be linked (or "anchored") to specific selections of the document. These two functional differences with regular online forum discussion are hypothesized to have two different effects. Firstly, the online presence of the learning material may serve as a context for collaboration that can set students' collaborative orientation and strengthen the link between discussion and study material (Herrmann and Kienle, 2003; Gay, Sturgill, Martin, and Huttenlocher, 1999; Guzdial and Turns, 2000). Second, the frame of reference for individual messages that is created by anchoring them may facilitate interpretation and reduce the "space for misunderstanding" (Dillenbourg, 1999). Combined, the two effects are expected to make anchored discussion a more effective tool for conducting theory-oriented online learning conversations (Herrmann and Kienle, 2003; Gay, Sturgill, Martin, and Huttenlocher, 1999; Guzdial and Turns, 2000).

Van der Pol, Admiraal, and Simons (2006a) have conducted a experimental comparative study on the affordances of anchored discussion and regular online discussion to support students' online learning conversations. In their study, 1181 messages generated in a university course on General Pedagogy were analysed using a newly constructed coding scheme (with message as unit of analysis and an average interrater reliability across all variables of Cohen's $\kappa = 0.82$). Results of this analysis are shown in Table 1, showing the relative number of messages containing the different mentioned activities.

These results confirm both hypothesized effects. First, they show that dialogue in the system for anchored discussion was indeed more directed at reconstructing the meaning of the text than discussion in the traditional forum (which was more oriented towards the exchange of personal opinions and experiences). This increased percentage of meaning-oriented discussion was combined with a significantly higher level of referring to content of other messages (5 % in the anchored discussion versus 0 % in the regular forum discussion) and of the text (12 % in the anchored discussion versus 5 % in the regular forum discussion). Students also reported reread relevant passages from the document before replying to a message significantly more frequently (means on a 5 point scale were 3.58 for the anchored discussion and 2.77 for the regular forum discussion). These three results all point to an enhanced link between discussion and study material in the system for anchored discussion, which may be caused by the tool's specialized design or affordance for this purpose. This difference in orientation also produces a more constructive dialogue in the system for anchored discussion, versus a more debate-like dialogue in the forum discussion. Second, the system for anchored discussion seems to indeed offer students less 'room for miscommunication' and an increased communicative efficiency. With briefer referrals (demonstrative rather than comprehensive) and messages containing fewer self-clarifications than the system for regular forum discussion, participants need fewer words to express their ideas. As expected, these referrals point to the existence of a shared frame of reference.

The results also show a third effect, with conversations in the tool for anchored discussion devoting less attention to establishing social relationships and regulating the collaborative processes in the regular forum discussion and being more straightforward and "to-the-point" (with shorter messages also translating in a greater number of messages). Figure 4 and 5 show two threads from both experimental conditions, where this difference can be observed. Starting with the regular discussion forum (Blackboard), messages can be seen to resemble email correspondence in more than one respect, most of them carrying more information than just the student's actual question or answer about the subject matter (section 3). First of all, most open and close with a *social* introduction and a departure phrase (section 1). They also contain *regulative* statements about what it is the author is going to do in this mail and what subject he or she wants to talk about, requesting for replies, or justifying their reaction (section 2). As described by Hammer and Veronesi (1999) This kind of message structure (with an opening, a core section, a directive speech act, and a closing) serves to construct a social relationship with the purpose of keeping in contact and increasing the changes of a response (increasing the level of "answerability").

Table 1. Activities found in anchored and regular online discussion

	Percentages of activities:	Regular forum discussion (n=462)	Anchored discussion (n=719)
1)	Reconstructing meaning	58 %	77 %
	Social comments	42 %	2 %
	Regulation of the process	25 %	4 %
2)	Self-clarifications	35 %	20 %
	Demonstrative referring	5 %	19 %

Figure 4. Thread from a traditional discussion forum

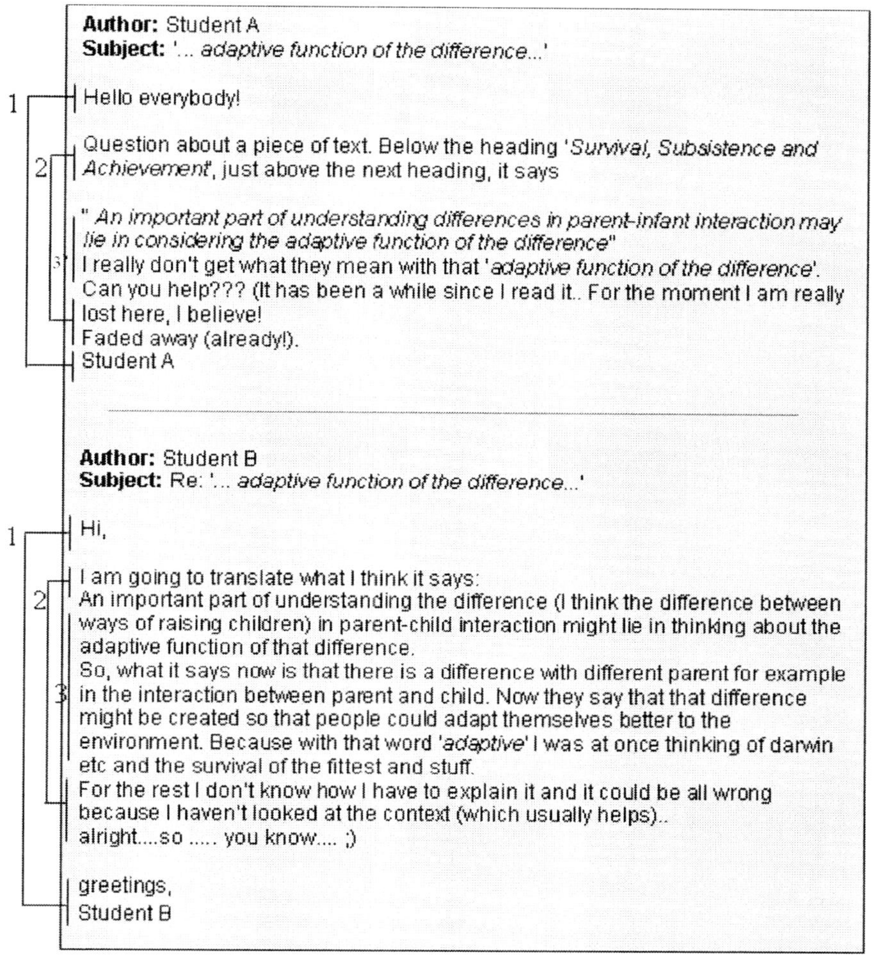

Author: Student A
Subject: '*... adaptive function of the difference...*'

1 Hello everybody!

2 Question about a piece of text. Below the heading '*Survival, Subsistence and Achievement*', just above the next heading, it says

3 " *An important part of understanding differences in parent-infant interaction may lie in considering the adaptive function of the difference*"
I really don't get what they mean with that '*adaptive function of the difference*'.
Can you help??? (It has been a while since I read it.. For the moment I am really lost here, I believe!
Faded away (already!).
Student A

Author: Student B
Subject: Re: '*... adaptive function of the difference...*'

1 Hi,

2 I am going to translate what I think it says:

3 An important part of understanding the difference (I think the difference between ways of raising children) in parent-child interaction might lie in thinking about the adaptive function of that difference.
So, what it says now is that there is a difference with different parent for example in the interaction between parent and child. Now they say that that difference might be created so that people could adapt themselves better to the environment. Because with that word '*adaptive*' I was at once thinking of darwin etc and the survival of the fittest and stuff.
For the rest I don't know how I have to explain it and it could be all wrong because I haven't looked at the context (which usually helps)..
alright....so ..... you know.... ;)

greetings,
Student B

When looking at a typical sequence from the system for anchored discussion (see Figure 5), the short and to-the-point nature of the messages immediately stands out. We do not find any social introductions or closings and only one regulative comment about the reason for posting a message (section 2). The messages consist, for the most part, solely of the actual question or remark about the content (section 3), making the conversation more communicatively efficient.

Interpreting these results takes us back to our model of common ground and grounding, as depicted in Figure 1. This model indicated that grounding is functional, meaning that in human communication active grounding will usually occur to the extent that it is needed. From this perspective, we can interpret the absence of social and regulative coordination in the system for anchored discussion as a possible reduced need for these activities. By focusing students' collaborative intentions (suggesting to them that "We are here to conduct a constructive meaning-oriented conversation"), the system for anchored discussion may have presented students with higher levels of regulative common ground, reducing the need for explicit pragmatic grounding (negotiating the collaboration goals) and social grounding (establishing social ties to ensure continuing participation towards these goals). Instead, more effort can be directed into the semantic grounding, constituting the primary collaborative learning process. The model by Van der Pol, Admiraal, and Simons (2006b) that is shown in Figure 6 depicts this the parallel existence of different sub area's of grounding (adding social grounding as a third form of grounding) and their possible competition or interference.

This competition effect between different forms of grounding seems to be supported by other studies. Dewianti (2005) concluded that stimulating the regulation of group processes does not enhance the process of knowledge co-construction, and Lockhorst (2004) found that when students' social communication in online discussions increased, communication about task content decreased. Similarly, Mäkitalo et al. (2004, p. 312) found a negative effect of coordinative and social communication on content-oriented behavior as their uncertainty reducing script increased the amount of discourse but decreased information seeking. Thus, although a certain level of common ground on both social

Figure 5. Thread from a tool for anchored discussion

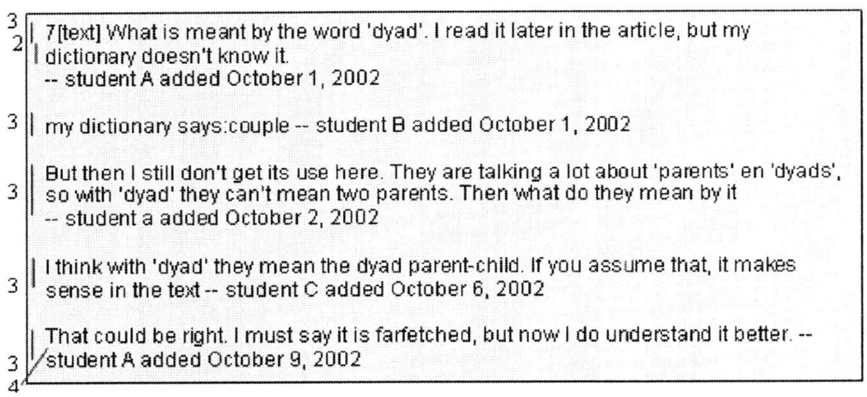

Figure 6. Interrelated sub areas of grounding and mutual understanding

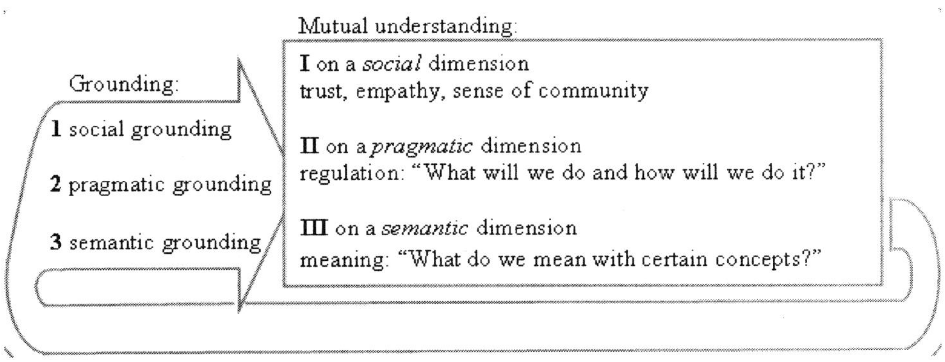

and regulative aspects seems to be necessary to "grease" the collaborative process, they may also direct part of students' efforts away from their content-processing activity. The optimal division of students' collaborative efforts, then, may lie in allowing them to direct as much effort as possible into semantic grounding and collaborative meaning-making, as long as sufficient levels of social and regulative common ground are maintained. Naturally, this means that the optimal division of students' grounding efforts over social, semantic, and pragmatic grounding will always differ per situation, with for instance blended learning situations (combining online learning conversations with face-to-face meetings) usually offering more social and regulative common ground and allowing for a bigger part of students' available effort to be directed towards semantic grounding and collaborative meaning making, than education that is purely provided online.

CONCLUSION

In the course of this chapter, I have described several obstacles to conducting successful online learning conversations. We have also seen that it is possible to facilitate this paradoxical process of collaboratively developing new understanding that surpasses that of the individual participants. The main implication for educational practice is that using anchored discussion can help teachers to create effective online learning conversations. Returning to our struggling teacher from the introduction, this means that using specialized tools can partly reduce the need for additional explicit training or instruction in order to achieve the desired results. With regard to the broader theory and research on CSCL, the results imply that the affordances of tools can be an effective means to evoke certain activities in an open learning environment that does not explicitly direct or constrict students' actions. In addition, we have seen the success of an approach that aims to directly facilitate students' primary process of conducting effective learning conversations, instead of facilitating this indirectly by focusing on interaction-regulating processes such as social and pragmatic grounding.

REFERENCES

Akkerman, S., Van den Bossche, P., Admiraal, W., Gijselaers, W., Segers, M., Simons, R. J., & Kirschner, P. (in press). considering group cognition: From conceptual confusion to a bouRendary area between cognitive and socio-cultural perspectives? *Educational Research Review.*

Alamargot, D., & Andriessen, J. (2002). The "power" of text production activity in collaborative modelling: Nine recommendations to make a computer supported situation work. In M. Baker, P. Brna, K. Stenning, & A. Tiberghien (Eds.), *The role of communication in learning to model* (pp. 275-302). Mahwah, NJ: Lawrence Erlbaum Associates.

Andriessen, J. (2006). Arguing to learn. In K. Sawyer (Ed.), *Handbook of the learning sciences* (pp. 443-459). Cambridge, MA: Cambridge University Press.

Andriessen, J., & Sandberg, J. (1999). Where is education heading and how about AI. *International Journal of Artificial Intelligence in Education, 10,* 130-150.

Andriessen, J., Baker, M., & Suthers, D. (2003). Argumentation, computer support, and the educational context of confronting cognitions. In J. Andriessen, M. Baker, & D. Suthers (Eds.), *Arguing to learn: Confronting cognitions in Computer-Supported Collaborative Learning environments* (pp. 1-25). Dordrecht: Kluwer Academic Publishers.

Baker, M., Hansen, T., Joiner, R., & Traum, D. (1999). The role of grounding in collaborative learning tasks. In P. Dillenbourgh (Ed.), *Collaborative learning: Cognitive and computational approaches* (pp. 31-63). Amsterdam: Pergamon/ Elsevier Sience.

Beers, P. J. (2005). *Negotiating common ground: Tools for multidisciplinary teams.* Heerlen, The Netherlands: Open University of the Netherlands.

Bellamy, R. (1997). *Support for learning conversations.* Abstract retrieved November 11, 2002, from http://kmi.open.ac.uk/seminars/abstract.cfm?id=10.

Bereiter, C. (2002). *Education and mind in the knowledge age.* Mahwah, NJ: Lawrence Erlbaum Associates.

Bernheim Brush, A. J., Bargeron, D., Grudin, J., Borning, A., & Gupta, A. (2002). Supporting interaction outside of class: Anchored discussion vs. discussion boards. In G. Stahl (Ed.), *Computer support for collaborative learning: Foundations for a CSCL community. Proceedings of CSCL 2002* (pp. 425–434). Hillsdale, NY: Lawrence Erlbaum Associates.

Billig, M. (1988). *Ideological dillema's; a social psychology of everyday thinking.* London: Sage Publications.

Cherubini, M., Van der Pol, J., & Dillenbourg, P. (2005, July). *Grounding is not shared understanding: Distinguishing grounding at an utterance and knowledge level.* Paper presented at CONTEXT'05, the Fifth International and Interdisciplinary Conference on Modelling and Using Context, Paris.

Clark, H. H. (1992). *Arenas of language use.* Chicago: University of Chicago Press.

Clark, H., & Brennan, S. (1991). Grounding in communication. In L. Resnick, J. Levine & S. Teasley (Eds.), *Perspectives on socially shared cognition* (pp. 127–149). Washington, DC: American Psychological Association.

Clark, H. H., & Marshall, C. R. (1981). Definite reference and mutual knowledge. In A. K. Joshi, B. L. Webber, & I. A. Sag (Eds.), *Elements of discourse understanding* (pp. 10-63). New York: Cambridge University Press.

Clark, H., & Wilkes-Gibbs, D. (1986). Referring as a collaborative process. *Cognition, 22,* 1-39.

Croft, W., & Cruse, D.A. (2004). *Cognitive Linguistics.* Cambridge: Cambridge University Press.

Dillenbourg, P. (1999). What do you mean by collaborative learning? In P. Dillenbourg (Ed.), *Collaborative-learning: Cognitive and computational approaches* (pp.1–19). Oxford, UK: Elsevier.

Dillenbourg, P. (2002). Over-scripting CSCL: The risks of blending collaborative learning with instructional design. In P. A. Kirschner (Ed.), *Three worlds of CSCL. Can we support CSCL?* (pp. 61-91). Heerlen, The Netherlands: Open University of the Netherlands.

Dillenbourg, P., & Bétrancourt, M. (2006). Collaboration load. In J. Elen, & R. E. Clark (Eds.), *Handling complexity in learning environments: Research and theory.* Amsterdam: Elsevier.

Dillenbourg, P., Traum, D., & Schneider, D. (1996) Grounding in multi-modal task-oriented collaboration. In P. Brna, A. Paiva, & J. Self (Eds.), *Proceedings of the European Conference on Artificial Intelligence in Education* (pp. 401-407). Lisbon, Portugal: Fundacao Calouste Gulbenkian.

Draper, S. W., & Anderson, A. (1991). The significance of dialogue in learning and observing learning. *Computers & Education, 17*(1), 93-107.

Fay N., Garrod, S. C., & Carletta J. (2000). Group discussion as interactive dialogue or as serial monologue: The influence of group size. *Psychological Science, 11*(6), 481–486.

Flower, L. (1990). Negotiating academic discourse. In L. Flower (Ed.), *Reading to write: Exploring a cognitive and social process*. Oxford, UK: Oxford University Press.

Gay, G., Sturgill, A., Martin, W., & Huttenlocher, D (1999). Document-centered peer collaborations: An exploration of the educational uses of networked communication technologies. *Journal of Computer-Mediated Communication, 4*(3).

George, S., & Labas, H. (in press). E-learning standards as a basis for contextual forums design. *Computers in Human Behavior.*

Guzdial, M., & Turns, J. (2000). Effective discussion through a computer-mediated anchored forum. *Journal of the Learning Sciences, 9*(4), 437–469.

Herring, S. (1999, June). Interactional coherence in CMC. *Journal of Computer-Mediated Communication 4*(4). Retrieved March 4, 2004, from http://jcmc.indiana.edu/vol4/issue4/herring.html

Herrmann, T., & Kienle, A. (2003, January). *Integration of communication, coordination and learning material. A guide for the functionality of collaborative learning environments.* Paper presented at the 36th Annual Hawaii International Conference on System Sciences, Hilton Waikoloa, Hawaii.

Hunt, R. A. (1998). Electronic discussions in learning and teaching: Why they don't work, and how they might. *Connexions: The Newsletter of the International Society for the Exploration of Teaching Alternatives, 10*(2), 1–7.

Innes, R. B. (2004). *Reconstructing undergraduate education: Using learning science to design effective courses.* Mahway, NJ: Lawrence Erlbaum Associates.

Järvelä, S., & Häkkinen, P. (1999, March). *Web based cases in teaching and learning - reciprocal understanding and perspective taking in conversation.* Paper presented at the conference for Computer Assisted Learning, London.

Johnson, J. A. (1995, June). *A comparison of user interfaces for panning on a touch-controlled display.* Paper presented at the Conference on Human Factors in Computing Systems, Denver, CO.

Kirschner, P. A., Sweller, J., & Clark, R. E. (2006). Why minimal guidance during instruction does not work: An analysis of the failure of constructivist, discovery, problem-based, experiential, and inquiry-based teaching. *Educational Psychologist, 41*(2), 75–86.

Kirschner, F., Paas, F., & Kirschner, P. A. (Submitted). Effects of individual and group-based learning.

Klemm, W. R. (1998). Eight ways to get students more engaged in online conferences. *T.H.E. Journal*, August. Retrieved, March 27, 2007, from http://www.thejournal.com/articles/14054.

Koschmann, T., & Le Baron, C. (2003). Reconsidering common ground: Examining Clark's contribution theory in the OR. In K. Kuutti, E. Karsten, G. Fitzpatrick, P. Dourish, & K. Schmidt (Eds.), *ECSCW 2003: Proc. Eighth European Conference on Computer-Supported Cooperative Work.* Amsterdam: Kluwer.

Krauss, R. M., & Fussell, S. R. (1990). Mutual knowledge and communicative effectiveness. In J. Galegher, R. E. Kraut, C. Egido (Eds.), *Intellectual teamwork: Social and technological foundations of cooperative work.* Hillsdale, NJ: Erlbaum.

Kuhn, D., & Goh, W. (2005). Arguing on the computer. In: T. Koschmann, D. Suthers & T. W. Chan (Eds.), *Proceedings of the 2005 conference on Computer support for collaborative learning: the next 10 years!* (pp. 346 - 352). Mahwah, NJ: Lawrence Erlbaum Associates.

Lapadat, J. C. (2002). Written Interaction: A Key Component in Online Learning. *Journal of Computer Mediated Communication (JCMC),*

7(4). Retrieved April 6, 2007, from http://jcmc.indiana.edu/vol7/issue4/lapadat.html

Laurillard, D. (1993). *Rethinking university teaching*. London: Routeledge.

Mäkitalo, K., Salo, P., Häkkinen, P., & Järvelä, S. (2001, August). *Analysing the mechanisms of a common ground in Web-based interaction*. Paper presented at the JURE pre-conference of the 9th European Conference for Research on Learning and Instruction, Fribourg, Switzerland.

Paas, F., & Van Merriënboer, J. J. G. (1994). Instructional control of cognitive load in the training of complex cognitive tasks. *Educational Psychology Review, 6*, 51-71.

Petraglia, J. (1998). *Reality by design: The rhetoric and technology of authenticity in education*. Mahwah, NJ: Lawrence Erlbaum Associates.

Pilkington, R. (2004). Developing discussion for learning. *Journal of Computer Assisted Learning, 20*(3), 161-164.

Poesio, M., & Traum, D. (1997, November). *Representing conversation acts in a unified semantic/pragmatic framework*. Paper presented at Working Notes: AAAI Fall Symposium on Communicative Action in Humans and Machines, Cambridge, MA.

Riva, G. (2001). Communicating in CMC: making order out of miscommunication. In L. Anolli et al. (Eds.), *Say not to say: new perspectives on miscommunication* (pp. 203–233). Amsterdam: IOS Press.

Romiszowski, A. (1995). Use of hypermedia and telecommunications for case-study discussions in distance education. In: Lockwood, F. (Ed.) *Open and distance learning today*. New York: Routledge.

Ross, L., Greene, D., & House, P. (1977). The false consensus phenomenon: An attributional bias in self-perception and social perception processes. *Journal of Experimental Social Psychology, 13*, 279-301.

Säljö, R. (2003). Representational tools and the transformation of learning. In B. Wasson, S. Ludvigsen & U. Hoppe (Eds.), *Designing for change in networked learning environments: Proceedings of the international conference on CSCL*. Dordrecht: Kluwer Academic Publishers.

Salmon, G. (2002). *E-tivities: The key to active online learning*. London: Kogan Page.

Schwartz, D. L., & Lin, X. D. (2000). Computers, productive agency, and the effort after shared meaning. *Journal of Computing in Higher Education, 12*(2), 3-33.

Schwarz, B. B., & Glassner, A. (2003). The blind and the paralytic: Fostering argumentation in everyday and scientific issues. In J. Andriessen, M. Baker, & D. Suthers (Eds.), *Arguing to learn: Confronting cognitions in computer-supported collaborative learning environments* (pp. 227-260). Dordrecht: Kluwer Academic Publishers.

Smith, B. (2003). The use of communication strategies in computer-mediated communication. *System, 31*, 29-53.

Van der Pol, J. (2002). *Identifying and modelling variables in complex CSCL-situations. Case study: The use of asynchronous electronic discussions*. Paper presented at the Conference on computer support for collaborative learning (CSCL2002), Boulder, CO.

Van der Pol, J., & Admiraal, W. F. (2003). Het succesvol inzetten van asynchrone elektronsiche discussies [Succesfully implementing asynchronous online discussions]. *Onderzoek van Onderwijs, 32*(2), 26–31.

Van der Pol, J. Admiraal, W. & Simons, P. R. J. (2006a). The affordance of anchored discussion for the collaborative processing of academic texts.

International journal of Computer Supported Collaborative Learning, 1(3), 339-357.

Van der Pol, J. Admiraal, W. & Simons, P. R. J. (2006b). Context Enhancement for co-intentionality and co-reference. *Journal of artificial intelligence & society, 20*(3), 301-313.

Veerman, A. L. (2000). *Computer-supported collaborative learning through argumentation*. Unpublished doctoral dissertation, Utrecht University, Utrecht.

Webb, N. M., & Mastergeorge, A. M. (2003). Promoting effective helping behavior in peer-directed groups. *International Journal of Educational Research, 39*, 73-97.

Wilson, D., & Sperber, D. (1995). *Relevance: communication & cognition*. Blackwell publishing, Oxford.

Chapter IX
Virtual Constructivism: Avatars in Action

Laura M. Nicosia
Montclair State University, USA

ABSTRACT

Contemporary educators have been reassessing pedagogical frameworks and reevaluating accepted epistemologies and ontologies of learning. The age-old debate whether knowledge is gained or constructed seems drawn to a consensus in the 21st Century: those who seek knowledge are active participants in the learning process and they have uniquely 21st Century attributes. Web 2.0+ technologies, various social media (Facebook, MySpace, Blogger, YouTube) and online virtual reality environments (Second Life, World of Warcraft, Sims) have influenced today's students in ways that constructivists should explore, embrace and exploit. This essay explores how Second Life (SL) effectively employs and distills the principles of educational constructivism. SL offers endless opportunities for immersion within user-constructed environments and activities. Educational use of SL may facilitate learner-led activities and yield learning that is prompted by desire and curiosity rather than learning for learning's sake. By exploiting these qualities with constructivist pedagogies, educators create environments that challenge and enable students to engage in the deepest kinds of learning.

INTRODUCTION

Teaching within Second Life—one of the most successful three-dimensional online virtual worlds—characteristically enables the deployment, application and distillation of several key principles of educational constructivism. Second Life (SL) offers its participants virtually endless opportunities for immersive experiences within user-constructed environments, communities and quests. By exploiting these qualities inherent within the SL platform, educators may create mediated, distributed learning scenarios and environs that challenge and enable students (as

3-D graphical avatars) to participate in their own learning. Such learning is motivated by curiosity and may provoke deep thinking and the construction of new knowledge.

The use of SL as a constructivist *milieu* ultimately rouses learner-led and learner-centered activities. My experiences inworld have shown that hands-on learning (even virtual hands-on) is constructivist at its core due to the intense engagement of immersive cognitive responses and the application of scaffolded learning from user-participants. Utilization of such pedagogical practices is promising, especially considering how many neomillennials seem to live their wired lives. The advent of Web 2.0+ technologies, various social media (e.g. Facebook, MySpace, Blogger, YouTube) and online virtual reality environments (e.g. Second Life, World of Warcraft, Webkinz) have appealed to and acted upon today's youths in ways that constructivists should explore, embrace and exploit for educational purposes—across the disciplines.

We have all seen our students juggle hypertexts, collaborate on synchronous and asynchronous storytelling, produce PowerPoint presentations and iMovies, send photos via their cell phones, update their romantic status on their web pages, while they simultaneously text and instant message each other. Despite questions of access and the injustices of the economic and digital divides, this generation of students—in general—is comfortable in the digital world.

Consequently, as SL and other virtual environments continue to grow in their capabilities, and as their technologies become more elegant and more ubiquitous, the academy may find itself on the wrong side of the educational digital divide. Unless academicians learn to appropriate, adopt and adapt various digital and virtual reality environments to our special disciplinary needs and to our students' multi-modal, multi-tasking learning styles, we may lose this opportunity to stay on or ahead of the technological curve. We must do so however, with pedagogical authenticity and constructivist validity.

Based on the work of several foundational pedagogical theorists and educational technologists (such as Dede, Gee, Kuhn, Vygotsky and Yee, among others) this essay discusses how educators may exploit these constructivist characteristics of SL and enable avatars to:

- journey through the learning environment's unfolding episodes and processes
- foster and nurture community between and among classmates and the instructor
- engage in collaborative knowledge-building
- distribute cognitive capital among and between the engaged groups and individuals
- forge identity formation strategies
- communicate with each other
- strengthen their own sense of agency
- engage with digital and virtual artifacts
- construct and learn content in context and in application
- activate meaning-making strategies

While theoretical in principle, this essay is the outgrowth of nearly two years' worth of personal and academic, inworld experiences and gleanings from my time within SL exploring ways to supplement my upper-level university literature courses.

BACKGROUND

This is the age of millennial and neomillennial learners, two subsets of digital natives whose learning styles have affected shifts in the skills and knowledge that our contemporary society prizes and which higher education demands. While we educators are well aware of these multi-tasking, collaborative, SMS-sending students who populate our classes, there are distinctions between millennials and neomillennials. At the risk of sounding reductive, it is desirable to distinguish between these students since the premise of this chapter is founded upon the argument that SL is a pedagogically valid constructivist tool for today's and tomorrow's student population.

Chris Dede of Harvard identifies several key characteristics that distinguish these two groups of students. Millennials typically use a limited number of media that are comfortably familiar and that appeal to users' individual learning styles. Additionally, millennials have found success with using Internet searches and various software programs and juggle these tools nimbly. Neomillennials, on the other hand, have a fluency and comfort with numerous media and deploy all for the empowerment they offer both individually and collectively. They manipulate media and cobble together various data from these deployments in collaboration with their peers and via a process of accumulation (Dede, 2005, p. 7).

Consequently, we have students in our classes who are as facile with certain technologies as we are, if not more so. In addition to being introduced to the skills, dispositions and contents of our disciplines, today's students expect to acquire skills and tools that are innovative, challenging, relevant, and useful. These learners are in our college courses now; we must, therefore, find and utilize appropriate pedagogies to produce deep thinking and to foster respect for both the content of our respective disciplines and how we go about doing our disciplinary work in this wired era.

To learn effectively and deeply, today's students expect cutting-edge technologies and hands-on experiences to engage their learning strengths and styles. Dede asserts that these styles and skills are a form of "mediated immersion," which include, virtual settings, communal learning, distributed knowledge, "experiential learning, guided mentoring, and collective reflection," and "co-design of learning experiences personalized to individual needs and preferences," among other traits (2005, p. 7).

As a corollary to this list of immersive learning styles and skills, I offer that virtual reality environments (VREs) such as SL, create, foster, and engage those very immersive traits that Dede exhorts us to address. These environments become virtual *learning* environments (VLEs) in the process. The proliferation of these VLEs has prompted numerous researchers and educators to investigate whether they are an optimal medium for constructivist pedagogies. Educators may embed constructivist activities within SL to instigate an educational techtonic shift—drawing students away from the passive to the active construction of new knowledge via engagement with digital and virtual artifacts, and by the scaffolding of prior knowledge. Such SL praxis is limited by only an instructor's imagination, willingness to experiment, and the time required to create such environments.

WHAT IS SECOND LIFE?

In short, SL "is a 3-D virtual world created by its Residents. Since opening to the public in 2003, it has grown explosively and today is inhabited by millions of Residents from around the globe" (www.secondlife.com/whatis/). It started with the equivalent of 64 acres of virtual land and stands now at 65,000 acres and is growing still. Residents may buy virtual plots or islands of land, own buildings and furnishings, and customize their avatars to suit their tastes and whims. The servers that house SL are managed by Linden Lab Corporation, but the virtual environ "is owned and built by its resident participants" (www.secondlife.com). Most important, while people sometimes say they "play" SL (referring to being inworld), SL is not a game—there are no rules or pre-established goals as there would be for a true game and no points to gather along predetermined courses of action. There may be games within SL, but the grid itself is user-constructed to be what the resident creators desire their environment to be.

Renting space or purchasing virtual land in SL is akin to renting space on a server for web hosting. The website creator builds the site, but pays for the maintenance and upkeep of the web space. All costs are, however, optional; SL offers free accounts and owning virtual property is not required to be inworld. It is only when an avatar opts to own land that payments are made for the maintenance of space on Linden Lab's servers. Each SL parcel of virtual land is a blank canvas awaiting the touch of a creator. M.D. Dickey (2005) points out that SL does offer "at least three things: (1) a 3D space or environment; (2) avatars that represent the individual user; and (3) interactive chat, either using text or voice or both" (as cited in Berge, 2008, p. 27). These conditions bear out why SL is not a game and why some first time visitors become either quickly overwhelmed with the sense of expansive openness or quickly bored with what appears to be so little "to do."

CONTEXTUALIZING LEARNING WITH VIRTUAL CONNECTIONS AND LINKED NARRATIVES

Humans love narratives; we may best be described, perhaps, as *homo narrans,* or in SL, *avatar narrans.* We thrive on stories, yearn for contextualizations, and are most comfortable when we sense causality. This is so, whether the stimulus is a piece of literature, an object, an idea, an image, or a digital artifact. The linkages we create form a narrative. Any time we verbalize or describe conceptualizations and relationships between objects, even digital objects, a narrative is created and we become *homo* or *avatar digitalis.*

For instance, when I am teaching a traditional face-to-face course in American literary fiction at my university, Montclair State in New Jersey, I strive to establish a sense of historicity, provide a literary dialog between texts and/or authors, and examine the myriad anxieties of influence from which texts spring. These connections lead to more than a chronology of literary linearity when students realize that art and artistic movements are not solely or always cause/effect—but very well may be synchronous or parallel with other art and movements. Such temporal and spatial understandings are exciting, and when students take preexisting knowledge and synthesize it with new knowledge, the creation of this linked narrative is deep learning. In fact, powerful scholarship occurs when my students understand that, for example, Jean Toomer's *Cane* was written in both metaphoric and literal dialog with Sherwood Anderson's *Winesburg, Ohio,* or that some of Willa Cather's and Flannery O'Connor's works may be discussed as sharing several stylistic devices. This type of lesson is the basis for many English major courses, delivered traditionally and supported by taking notes and lists.

Consider how effective class discussions would be if students could also encounter images, documents, and sites from the texts by also experiencing them? What would happen if they could take a field trip to visit a representation of Anderson's city of Winesburg and move through the locale he created? How powerful would it be to have students engage with artifacts from and about the stories? What kinds of learning would occur if our college readers could recreate or simulate a character's home (or other key site from the text) in three dimensions to enable others to walk into and through the space?

Bryan Carter (2006) suggests that teaching in VREs "addresses several important learning objectives . . . active learning (versus passive learning), interactivity with the subject matter, assisting visual learners who may find lecture formats difficult, and contextualizing the information resources" (p. 52). Such "eyes-on" experiences and vigorous connections with the "stuff" of novels is one of the key components of constructivist pedagogies. In real life (RL) this kind of activity, however, is not always possible or even desirable (taking students to tour the setting for Dante's *Inferno,* is impossible and highly undesirable, for instance). In SL, however, such field trips and immersions are likely.

Inworld, avatars engage with objects that are selected, designed and mediated by the educator to provide a virtual sense of place and space for a particular text or specific concept that is deemed integral to the content. This is what Sosnoski and Portlock (2006) refer to as "narrative architecture" (p. 67). As such, avatars are immersed in the text as they move through key scenes and experience, visualize and even hear events (mediated by the educator) to evoke comprehension, provoke deep thinking and foster collective reflection.

In all honesty, however, the following must be addressed: there is a steep learning curve to acquire the basic skills an avatar requires to successfully navigate within the SL metaverse (for example, walking, flying, chatting, teleporting, IM-ing, searching and using aerial maps). Skills such as these are transferable, however. If these college students have ever played computer or console video games such as Madden NFL, Tomb Raider, The Legend of Zelda, Mario Brothers, Grand Theft Auto, and so on, they are used to navigating through three-dimensions and understand the vagaries of using both joysticks and arrow keys. However, even students who have never engaged with video, computer or Massively Multi-User Online Role-Playing Games (MMORPGs) will acquire those basic skills of navigation and maneuverability, albeit at varying speeds and levels of success. The acquisition of these start up skills are necessary for students (as avatars) to achieve success and enjoyment within SL.

In my undergraduate-level English elective, The Art of Fiction, students read novels, short stories and collateral critical and theoretical readings over a traditional 15-week semester. Students read Gloria Naylor's novel, *Mama Day,* and while doing so, worked on their SL skills and abilities. To augment our face-to-face class meetings with traditional large and small discussions, one of their assignments was to pay particular attention to the extended members of the Day family as recorded in the Family Tree (provided as a paratext at the start of the novel). This record of family births and deaths is a key element to understanding the central mysteries of the novel and my students were assigned to keep track of these records.

They gathered information for each family member and recorded events for that character throughout the course of the narrative. Then, they entered SL and found a gravesite for the original founder of the island community, Bascombe Wade, that I hid within a forest; quests and scavenger hunts are popular tools within SL. Wade had been murdered by one of his slaves (the founding matriarch of the Day clan). Her subsequent flight after the murder and her seemingly absolute disappearance inform the central mysteries of Willow Springs. Death and the graveyard are central to the plot, mysteries and conflicts within this novel. Once they found the grave, each student "touched" the tombstone with his or her cursor and received the formal task and the next step in the hunt through the novel's setting.

They were next sent to a well, where they were each given the name of a character. They were to write an epitaph for that character with the salient information culled from their reading. Once they wrote the epitaph, they built, purchased or traded for a headstone, inscribed the character's name on it, placed the headstone within the graveyard and embedded their epitaphs into each grave. This, by the way, is done with click-and-drag techniques, akin to copying or moving folders from within a word processing program. By interacting with their environments, creating new knowledge and leaving a virtual footprint or artifact behind—in this case a virtual annotation—they grew active in their learning, "thus making meaningful connections both between prior knowledge, and new knowledge and among disciplines" (Carter, 2006, p. 53). This graveyard exercise helped students concretize their understanding of the content of this novel rife with mysterious deaths and births of ex-slaves and their progeny.

This assignment poses several high-level, intellectual challenges. First, students must understand the content of the novel. That is, learning about the various characters, their sometimes mysterious deaths, and the socio-cultural events of the novel becomes primary among other conditions. Second, they must write an acceptable epitaph (a sample was provided for modeling). Third, they must acquire a headstone or create one using SL's basic building tools. In short, to successfully complete this task avatars needed to: read and comprehend the text, implement social strategies, engage with other avatars, and create a digital artifact embedded with new knowledge and appropriate to the geographic and chronologic specifications constructed by the author, Naylor. Additionally, what is pertinent to this study, these avatars actively cooperate with others within small- and large-group collaborations, build content and engage with the virtual environment based on the novel.

These avatars are immersed within the text in ways that utilize group strategies and engage in what Sosnoski calls "configuring...a cognitive process whereby persons understand other persons by recalling analogous past experiences" (p. 35). Working with other avatars and joining together to a common end, lead to the establishment of community between and amongst the participants. Concomitantly, students' emotional investment to the assignment is heightened. While it may appear that these avatars are playing a role-playing game, they are not simply pretending. Nicolas Yee discusses the depth of user engagement within virtual environments: "The emotional investment that these environments derive from users is one way of countering that assumption. Users in fact take these environments very seriously" (2006, p. 14). In what may seem paradoxical, student avatars learn while having fun and while seriously engaged to accomplish a task. Learning within SL is serious play or rather, playful work. These avatars are active—not passive. They build knowledge and immerse themselves within the text in ways that utilize constructivist pedagogies in authentic methods.

FOSTERING COMMUNITY AND COLLABORATIVE TIES

As a VRE, SL is a social medium—there is no evading this point. There is a strong social element to the SL experience. Without it, the metaverse would be devoid of meaning—no different than sitting in a warehouse with only boxes lining the walls. It would be like going to a theme park with no one else in the park and no one to run the rides.

SL is meant to be a shared, collaborative experience. The most effective class tasks and assignments take advantage of the social nature of the medium and exploit it by constructing challenges that necessitate collaboration, foster community-building, and the distribute cognitive capital among and between groups. When inworld, I teleport the avatars to my location—my home or one of my university's islands—and have them landmark the location as a "home base." Once they are with me, I give them objects they can use inworld (clothes, shoes, household objects, furnishings and landmarks of interest).

Additionally, I distribute a few Linden dollars (the inworld currency which at the time of this writing is 264L = $1.00 US) to each avatar for the purchase of incidentals—like tombstones, for instance. Generally, 20L per student avatar is sufficient for the entire semester. My total outlay costs me under $2.50 US for the whole dispersal to 30 avatars. My objective is to get student avatars accustomed to sharing with each other and to fostering a sense of generosity—I model the behaviors I want to see manifested. Avatars, in general, are very willing to share with each other when they realize that the cost is non-existent to minimal and when the payoff is so high. Friendships develop, and avatars have fun sharing their Gucci bags and Porsches. While SL is a for-profit platform much of what is inworld is free and open access.

Over time, students develop an attachment to their avatars and invest much of their real life personalities into the graphical representations they have created for themselves. However, when a person first creates an avatar (a representation of his or her choice—humanoid or not) several default physiques are available. The user may choose from among skin colors, and may select any body weight, height, hair style, eye color, and so on—all of which are modifiable. Choices for one's body are only as limited as one's willingness to experiment with appearance.

Once the user is satisfied with the avatar's appearance, the avatar ventures out into the SL metaverse and engages with other avatars. This is especially so at Orientation Island—SL's designated way station for skills development—though other orientation sites are proliferating. Avatars traverse an island, learning how to perform basic functions such as walking, turning corners, navigating stairs and ramps, flying, and so on. In order to succeed at these oftentimes awkward maneuvers, avatars read interactive signs, ask questions of other avatars, and rely on social interactions for answers to their queries: "How do I pick up that item? How do I teleport to CHSS Island?" and so on. The best way to steer one's avatar through this metaverse is by building relationships and working collaboratively.

In RL, when I assign a task, I direct my students to work either independently or in groups; I create the task to be appropriate to the content and skills I want them to develop, and toward the specific learning outcomes I have identified as desirable for this unit or text. In face-to-face class assignments student achievement can be affected by: the length of the class session before the next class takes use of the room (generally 75 minutes); attendance issues for absent group members who missed the bus or are stuck in traffic; pre-conceived ideas and biases about a classmate based on appearances, among other variables.

These factors are ameliorated when using SL. Student avatars can (and tend to) remain inworld as long as they can sit at their keyboards; they are not limited by the end of a class period or constrained by room furnishings. While I can cite anecdotal data about how many hours my college students spent inworld, and about how many hours I spend in SL, Yee reports that avatars "spend on average 22.72 hours (n = 5471, SD = 14.98) each week [inworld]. The lower quartile and upper quartile boundaries were 11 and 30 respectively. The distribution showed that about 8% of users spend 40 hours per week or more in these environments" (2006. p. 18). These numbers point to the attraction such VREs have for a large slice of technology-oriented learners.

Students inhabit their avatars even when they are tired or not feeling well enough to drive to campus. Additionally, avatars cannot see the person behind the keyboard when they interact with other graphical representations. Their inworld relationships are based on performance and intrinsic value—not preconceived notions about a fellow student. In a face-to-face class setting, when groups self-select, it is common to see students gravitate to those they feel are peer members from similar backgrounds and ages. Despite this real-life tendency, Yee's research has determined that "seemingly disparate demographic groups would oftentimes be collaborating and working together to achieve the same goals.... This finding is particularly striking given that these disparate demographic groups seldom collaborate in any real life situation" (2006, p. 9). Avatars engage in heterogeneous interactions and forge bonds across boundaries rather than remain within their real-world comfort zones.

This exposure to other avatars leads to an increasing comfort level when encountering an "Other," and facilitates building community and fostering collaboration. It is common to have humanoids, furries, cyborgs and fantastic creatures populating an inworld group (in any number of combinations). This lack of cultural baggage or back-story is termed "null experience" by Sosnoski (2006, p. 36) and it is a position of pure possibility and liminality. Berge (2008) acknowledges that SL "is very *culturally* diverse, with people found from around the world, and avatars that expand culture to other levels.... With such cultural diversity in an environment where anything is possible, the impossible sometimes happens. Discovery and exploration are encouraged" (p. 29). However, as with any group project, the level of success is directly related to a number of factors. Among these are: the delicate social interactions between group members; the varying levels of skills among members; the work ethic of individual group members; and the effectiveness of the instructor to encourage members in unity toward one goal—the triumphant completion of the assignment or task.

Through the process of creating an avatar, experiencing a willing sense of disbelief in the immersive qualities of the act, and engaging with other avatars, the human behind the keyboard may choose to project an aspect of his or her personality that he or she wishes to develop. The level of anonymity provided by inhabiting an avatar may activate the development of agency and may facilitate the progress of social skills that might otherwise be hampered by face-to-face experiences.

FORGING IDENTITY FORMATION STRATEGIES AND STRENGTHENING AGENCY

Creating an avatar may be an invitation to role-play or to take on a new personality. Despite this opportunity, avatars tend to be an extension of one's real-life personality—although enhanced and perhaps more freely wielded. Recent research has shown that since identity is unfixed, we (and our avatars) may change and transform in myriad ways when we feel secure enough to do so. Angela Thomas' research (2007) has identified that in the cyber world there "are multiple layers through which we mediate the self and [those] include the words we speak, the graphical images we adopt as avatars to represent us, and the codes and other linguistic variations on language we use to create a full digital presence" (p. 5). Sherry Turkle's research has pursued this avenue by elaborating upon and critiquing the "fluid and decentralized nature of identities" in virtual environments (as cited in Yee, 2006, p. 6).

The lack of personal risk, the non-existent physical danger, the potential to interrupt the laws of physics and the curtain of anonymity are ingredients to cultivate avatar identity formation. Avatars often explain, albeit anecdotally, that they feel more free to assert themselves, to ask for help, to take risks inworld. Knowing that one may teleport from any perceived danger or discomfort with a simple keystroke or pull-down menu command, and knowing that mistakes are not fatal or irreversible, free many avatars from the tyrannies of embarrassment or fear. Being an avatar is empowering in ways that may become habitual.

Vygotsky's (1978) concept of self-mediation and of the psychological tool is applicable in this instance: students create avatars (the tools) as mental signs (projections) to represent themselves and the concepts for which they stand. These tools or avatars are then used to mediate their exchanges with others and with the environments within which they find themselves. The avatars observe themselves in action, and use or construct knowledge by assessing and evaluating their exchanges within those new relationships and relations. This is a strong tenet of both radical and social constructivism and it is inherent in the fabric of being in a VRE or in SL, in particular (Doolittle, 1999).

What seems more important, perhaps, than acknowledging inworld personality development and the fostering of inworld agency is clarifying whether any such identity formation has real, empirical transference to the real world and whether these new forms of social interactions and personal projections have any lasting effects in one's RL. Recent studies by Nakamura (2000), Bushman (2002), Anderson (2000), and Ferguson (2002) all critique the effects of video games, media and online role-playing; however, these studies focus on aspects of violence. Nakamura, however, is the exception; she has studied racial identity perception and formation in cyberspaces. What seems clear is that more scholarly social research needs to be done to examine the possible carryover of other derived experiences (both social and psychological) and content acquisition from virtual environments such as SL to RL.

COLLABORATIVE KNOWLEDGE-BUILDING AND THE FREE DISTRIBUTION OF COGNITIVE CAPITAL

In traditional learning environments, to provoke deep thinking and to engage active learning, instructors plan their class meetings ahead of time and work backward from their desired learning outcomes toward the process that is most likely to reach that goal. Similarly, in a non-face-to-face environment, if an instructor wishes to utilize constructivist pedagogies for activating avatar learning, there must be an effective application of procedures and tasks. Lessons must be planned, sites must be constructed and outcomes must be identified well ahead of the actual immersive encounter. Bain (2004) suggests that in order to prepare for meaning-making rather than knowledge-transmission, instructors should ask themselves the following: "How will I create a natural critical learning environment in which I embed the skills and information I wish to teach in assignments . . . that students will find fascinating—authentic tasks that will arouse curiosity, challenge students to rethink their assumptions and examine their mental models of reality?" (p. 60). Effective planning and narrative architecture are *de rigueur* in both RL and in SL.

If an avatar enters a space where the lesson is not well conceived or where the virtual constructions are haphazard or poorly considered, then the learning outcomes will be as unsuccessful and inauthentic as any poorly designed traditional class. When avatars enter teacher-mediated learning spaces that are well-designed and effectively deployed, however, the immersive qualities and sense of social presence intensify the engagement factor and increase learning in visceral and synesthetic ways. Assignments such as these deploy constructivist pedagogies in several ways—even in a virtual environment.

While it can be safely assumed that Bruner (as cited in Huitt, 2003, ¶ 1) wrote about constructivist teaching and learning in face-to-face environments, the main principles of constructivism that he identified are applicable in the virtual realm, too. Three of Bruner's principles speak to the kinds of experiences that avatars engage with whenever they enter the virtual realm. These acknowledge that "instruction must be concerned with the experiences and contexts that make the student willing and able to learn (readiness)," that it "must be structured so that it can be easily grasped by the student," and that all teaching "should be designed to facilitate extrapolation and or fill in the gaps (going beyond the information given)". (1990, p. 1) These qualities point to a mode of learning that is a dynamic process. Through using VLEs like SL, instructors make learning engaging (and I dare say, fun). Avatars create new knowledge, synthesize data by combining and scaffolding various schema, makes choices, and share that knowledge with others.

Take for instance, another assignment I designed for my undergraduates while studying *Mama Day*. After they completed their epitaph task, avatars were to consult the map (found at the front of the novel) of the setting, Willow Springs, and to find the magical herb garden (located next to the graveyard). This garden provides the elixirs and herbals that the protagonist, Miranda "Mama" Day, uses to heal the sick, counter the evil spells cast by her nemesis (Ruby) and to cure her niece, Cocoa. This garden is the key geographical and thematic element to resolve the major conflict of the text. During their reading, students took notes regarding the ways that Mama used this herb garden. Avatars researched plants that the author mentioned and those that might be indigenous to an island off the coast of Georgia and South Carolina. Second, they were to identify several (3-4) of those plants that Mama could have used. As a corollary, no plants were to be duplicated within the entire class collection. Avatars had to cooperate with each other in order to avoid recurrence. Third, they were to find examples of those plants in SL (whether these were actual 3-D renderings of those plants or 2-D pictures) and "plant" them in the graveyard. They also had to write short, descriptive summaries of how each plant would be used and what ailments they would treat. Of course, they were required to cite their sources (using MLA formatting). Such writings, by the way, were delivered to me inworld via a simple note-card email process (similar to sending an email with an attachment in RL).

Building upon their classroom discussions on the novel, the avatars were excited to teleport to other places in SL and rummage through inworld gardens or stores. Some avatars purchased their researched plants. Others bartered for desired flora. Still others received those plants as gifts from friendly avatars who were not students in our class. The goal for this task was to immerse students in one of the central geographic sources for Mama Day's magical powers.

Additionally, this assignment pushed avatars to go beyond the information given in the novel and to extrapolate from the text to fill in readerly gaps. To create an authentic, aesthetic, and useful medicine garden, the avatars created knowledge, shared that knowledge with each other, and actively built a virtual herb and plant patch. (As American poet Marianne Moore might have said, they built "virtual gardens with real herbs in them.") Problem solving, collaboration and engagement became the norm. Most importantly, perhaps, is that such constructive meaning-making was distributed across the small groups and between classmates rather than delivered from the instructor to the avatars. The construction of knowledge from this exercise embraced the tenets of social constructivism—where avatars created learning from "the result of social interaction.... between people collectively searching for truth, in the process of their dialogic interaction" (Doolittle, 1999, p. 4). Constructed truth becomes the product of a series of social negotiations that distribute knowledge between avatars. Such truths are not delivered or transmitted from the teacher. Rather, they are dispersed across and among peer groups.

AVATARS ENGAGING WITH DIGITAL AND VIRTUAL ARTIFACTS

To complete their task for the above-mentioned garden assignment, avatars performed various research functions. They could have completed the task via any number of avenues: searching on the web with a browser to find images on Flickr or Google Images; drawing accurate images using tools within their word processor; taking a picture of an herb and either downloading it to their computer or scanning it; purchasing a virtual plant inworld, and so on. Regardless of the method they chose to use, they engaged with digital or virtual artifacts in an active and methodical manner. Ultimately, avatars built upon their own comfort and skill levels—if they were most comfortable with using a search engine to download and copy images, they chose that method. If they preferred taking photos with their digital cameras or cell phones, then that was their method of choice. Clearly, "in the modern world," says James Paul Gee (2004) print literacy is not enough. "People need to be literate in a great variety of different semiotic domains.... the vast majority of domains involve semiotic (symbolic, representational) resources besides print and some don't involve print as a resource at all" (p. 19). This inworld exercise requires student avatars to be literate in various domains and to use those domains to create new knowledge.

In this age, students bring many skills and talents to their classes. When I allow them to use their pre-existing strengths to engage their own learning schema, I enable them to succeed. By encouraging students to employ their own knowledge, to engage their technology-oriented skill sets and to tap into higher-order thinking and learning, they gain confidence in their abilities to do scholarly research and to acquire contextual information both in SL and in RL.

Through such immersion, college students, even if they are not English majors, learn that novels can affect them in ways that are salient and pertinent. Most avatars enjoyed this task and found that researching herbalism and reading about Gullah agricultural practices was aesthetically pleasing and practically useful. Such learning becomes real to the student avatars because "the learning takes place within a meaningful...context. What you must learn is directly related to the environment in which you learn and demonstrate it; thus, the learning is not only relevant but applied and practiced within that context" (Van Eck 2007, p. 56). Even the most esoteric and aesthetic text may become visceral, understandable and real when avatars engage with and create digital artifacts in constructive and immersive situations.

Additionally, engaging with digital media develops a set of transferable tech media skills that students deem useful in their lives and in their pursuits for a career. Such relevancy "is likely to lead to an increase in motivation... the individual comes to understand the need for certain knowledge" (Doolittle 1999. p. 5). Moreover, Thomas (2007) notes that engaging with digital media aids in the acquisition of new media literacies. She has asserted, "[o]ne of the most important factors is that with every new form of community, children are participating in new forms of literacy. Literacy is being transformed and is evolving with every new set of social practices" (p. 4). Student avatars recognize that learning how to import photos, to engage with data in both 2- and 3-D formats, and to navigate through alternate virtual environments are both useful and pleasurable.

WHAT'S THE HITCH? ISSUES AND CONTROVERSIES

Teaching in SL is not a panacea for student disengagement or for a weak work ethic. It is neither simple to build sites, nor is it second nature to deploy constructivist activities inworld. Building requires a significant amount of time, training and a certain amount of funding for persistent land sites and plots. As for the activities, instructors must learn how to translate face-to-face activities to the virtual world—recognizing that just as in RL, avatars require ample space within which to move and navigate. If space is not considered, avatars might very well bump into each other and chaos may ensue.

Teaching in SL is not "quiet," despite the ability to limit our conversations to textual chats and IMs if we choose to abandon the use of audible voice. Consequently, instructing in SL requires the teacher to juggle multiple chat boxes or long, scrolling dialogue screens. This may all seem cacophonous at first. However, this kind of multi-tasking becomes more comfortable over time and with practice. Look at our students. They talk to their friends, text others, and listen to their iPods all at the same time! So can we. It merely takes a sense of adventure and a willingness to practice.

This speaks to the steep learning curve for using SL. It takes time to become facile with the interface. Using arrow keys, pull-down menus and right clicks are not the norm for many in the academy. Berge writes that "[n]othing seems very intuitive in-world" and that it took over ten hours for him to effectively navigate in the environment (2008, p. 30). All this is true. There is an intensive, immediate need to acquire specific skills in order to succeed inworld. Developing these skills takes time—valuable time that could be spent teaching content and disciplinary skills.

The demand for up-front skill acquisition, however, might be offset by the intense immersive qualities that SL affords. Avatars tend to enjoy their time in SL and I can anecdotally offer that college students will oftentimes spend more hours inworld than they do in face-to-face class periods or doing homework in the traditional sense. Immersion yields increased engagement. And, once avatars become facile with the few skills needed inworld they are able to succeed at the tasks I assign.

Some critics assert that teaching in "SL cannot accomplish anything that they could not accomplish in regular Websites" (Berge, 2008, p. 30). While this has merit, my point about taking students to see and maneuver through exotic, foreign or dangerous environs still holds true. While students may gain endless information and data about Dante's *Inferno* by searching through web sites, scouring various scholarly indices, and perhaps flying to Italy to tour museums, I can take them to a 3-D rendering of the levels of Hell and have the avatars add content from the text and from their research. *That* cannot be accomplished in regular websites.

Somewhere along our schooling years, we came to the understanding that learning should not be fun or pleasurable—that being engaged with materials and content in sensory and visceral ways is not appropriate. Using VLEs like SL provokes deep learning, promotes engagement and prompts the occasional laugh. While it may not be content rich to trade a designer-inspired pocketbook for a medicinal herb, and while disciplinary core content curriculum standards may not require collaboration and group work, these are useful and sound outcomes.

I do not advocate the use of VLEs in place of content. Content and disciplinary skill building must always be primary and must always drive the lessons toward specific and clear learning outcomes. Lessons must be planned with both content and outcomes foremost in the instructor's thoughts—working backward to the mechanics of the delivery inworld. This is, however, no different than traditional face-to-face teaching. An instructor who attends class unprepared or with no clear goals in mind creates a class bereft of meaning, purpose and validity, just as it would in SL. Likewise, using any technology for technology's sake is unsound practice.

This begs the question: just as in RL, issues of assessment must be addressed. How will instructors measure the successes, failures, and the achievements of each lesson inworld? Are students to be assessed on the level of their technology skills? What would happen if a student just does not become facile with the SL or VLE platform, but *does* understand the content of the course? Will the grade be affected? How will the avatars' achievements be evaluated and measured? Will empirical data be gathered? And, how will the instructor be evaluated? Will peers and administration understand the vagaries of teaching in such a situated and mediated environment? These questions require serious consideration and further study.

Additionally, there are other contentious topics regarding technology glitches and flubs. We know that servers go down and that the Internet access slows down always just when we need it. Teaching with the aid of any technology necessitates back-up plans B and C at all times—just in case.

The use of Web 2.0+ technologies also brings into question issues regarding the technology gaps (between various economic haves and have-nots, and between instructors who use outdated equipment despite the cutting-edge requirements for running some of these programs). Schools will need to designate specific labs for running SL (for both class sessions and for student body use) and/or will have to provide students with jump drives with the program installed.

Access issues for economically or technologically challenged students are also important considerations, as are issues regarding students with visual disabilities. These students may not (at the time of this writing) be able to use SL since it is a visual medium. Students with severe physical disabilities who may not be able to manipulate a keyboard, joystick or mouse may need extra support and modifications. Alternate methods of simulating immersion may have to be addressed for students who have legally recognizable disabilities and varying levels of abilities. Ultimately, administrative policies will need to be set in place for many of the aforementioned (and not mentioned) issues.

It must also be acknowledged that virtual vandals and trouble-makers (called "griefers") are present in SL (as are "hackers" in the web-based cyberworld). These avatars take pleasure in disrupting normal social engagements and while they are generally not violent, *they* play malicious pranks with viruses, may disrupt sites and may threaten an avatar's sense of security. There are mechanisms to eject and prevent these griefers from entering educational (or private) sites. But these mechanisms are at the advanced level of skill acquisition. One must know a bit about how SL works and a bit about how to set certain scripts into action to use these inworld tools. There are surveillance tools inworld and there are methods to track and record avatar activity. Ultimately, there will always be griefers, just as there will always be real-life vandals and computer hackers.

These inworld malcontents who disrupt discussions are annoying, angering and occasionally threatening, but their actions do not damage any real property or physically endanger any real humans. It is true that a griefer may cause a site to crash, But sites can be rebuilt or reset fairly easily—not like a building that has had a fire set in it. And, while it is possible that griefers may cause an avatar to become uncomfortable or frightened, an avatar always has the ability to teleport to a safe space (home or a class, or a campus) and can quit SL if the encounter is untenable. Very little is irreplaceable in SL, unlike in RL. Regardless, some educational institutions are exploring how to set policies into place that offer varying levels of protection to students and faculty (and the university from legal responsibility) as we venture into this brave new world.

Finally, issues surrounding intellectual property ownership and rights, along with standards and procedures for copyright need to be clarified. If a faculty member uploads lessons, supporting materials and assignments to SL (or to any web-based server, for that matter, including Blackboard) who owns and controls the rights to those products? Do digital and virtual materials fall under the same scrutiny and regulations as print or visual documents? Additionally, administration, faculty and students need to be made aware of and to adhere to the regulations of Fair Educational Use and attribution (both from the standpoint of copyright and Creative Commons Licensing).

These issues are pressing and there are no easy answers. Directors of Information Technology, librarians and media specialists, human resources, legal teams and perhaps even professional unions need to investigate policies and procedures for all web- and Internet-based technologies, whether they are generation 1.0 or 2.0+.

WHERE DO WE GO FROM HERE?

Second Life is growing rapidly and does not appear to be slowing down. In fact, it is expanding at a tremendous rate. In his article on SL in higher education, Kelton (2007) reports that from August 2006 to August 2007, the number of SL residents rose from 350,000 to over 8.5 million. Numbers like these are telling. It is unlikely that membership will continue to grow at such an exponential rate; however, SL is not going to simply go out with a whimper. Unless one works at the upper echelon of Linden Lab, one cannot predict whether SL will continue as is in perpetuity. The next step would be for SL to facilitate and encourage users to work across several platforms using various VREs. This cross-platform exchange has already been accomplished between SL and IBM's OpenSim, successfully demonstrating virtual world interoperability. Increasing instances of free-flow between VREs would expand the possibilities for creating new learning environments and for more social interactions between larger numbers of avatars/learners.

A recent article in *Virtual World News* (January 2008) reported that a cross platform educational grid is in the planning stages and would permit educational systems to work across various virtual world platforms. The article reported that the importance of the education grid "is not just that it's a 3D environment, but that it's a part of a larger system. The education grid is a repository. . . . Some of it will be pre-made for teachers to just use and some that they can create. It's both a repository and the server-side architecture. It houses the servers that allow students and teachers and corporate learners to get together." This means that a multi-platform server system, devoted to teaching and learning is in the works. Perhaps this will solve some of the aforementioned controversial issues and will signal a new evolution in VLEs.

Recent research and development are even looking into creating an interface that will allow avatars to be controlled without the use of a mouse or keystrokes. Movements would be controlled by eye and/or body movements similar to the ways players engage with the Nintendo Wii. This would take SL into an entirely new type of immersive experience. The most recent Sun Services White Paper (2008) predicts that "a user experience similar to that of Nintendo Wii may become common for Internet users. And graphical realism will vary depending on the purpose and focus of the virtual world" (p. 3). What comes next insofar as the future for VREs and VLEs is limited by the human imagination and the human quest for technological experiences.

FINAL THOUGHTS

It is apparent that virtual worlds are not going to go the way of the VHS tape—at least not for the foreseeable future. My experiences within Second Life have shown that students actively participate in the construction of their own learning as they co-journey with their avatars through mediated learning spaces and activities. In wending their way through these various sites, my students encounter artifacts placed there for them, explore websites, examine primary documents, synthesize their reflections on their experiences through writing, and become partners in the construction of their own learning, leaving virtual artifacts as proof of their accomplishments. This is constructivism at its core. I believe it holds promise for the future—not as a utopian teaching environment—but as an extension and an enhancement for my face-to-face class sessions. As Second Life and other virtual environments continue to grow in their capabilities, the technology will become more ingeniously neat and user-friendly.

Millennials and neomillennials will rise up through the education system and will be in our college classes and in our workforce. Academicians and educators must continue to explore the possibilities for authentic constructivist learning using Virtual Learning Environments. These students, who are growing up playing Wii and Webkinz, will be fully familiar with having an avatar. They will be in our classes. We should adopt their tools and adapt their technologies to our ends—teach them using the skill sets and comfort zones they already possess. Then, we will speak to them in their own language, using their own tools, to create new knowledge, to foster active learning, and to nurture collaborative networking skills. This is the best kind of constructivist learning.

REFERENCES

Bain, K. (2004). *What the Best College Teachers Do.* Cambridge, MA: Harvard University Press.

Berge, Z. L. (May-June 2008). Multi-User Virtual Environments for Education and Training? A Critical Review of Second Life. *Educational Technology* (pp. 27-31).

Bruner, J. (1990). *Acts of meaning.* Cambridge, MA: Harvard University Press.

Carter, Bryan. (2006). Virtual Harlem in the Beginning: Retrospective Reflections. In J. J. Sosnoski, P. Harkin, & B. Carter (Eds.), *Configuring History: Teaching the Harlem Renaissance through Virtual Reality Cityscapes* (pp. 47-57). New York: Peter Lang.

Dede, C. (2005). Planning for Neomillennial Learning Styles. *Educause Quarterly, 28*(1), 7-12.

Dickey, M. D. (2005). Three-dimensional virtual worlds and distance learning: Two case studies of active worlds as a medium for distance education. *British Journal of Educational Technology. 36*(3), 439-451.

Doolittle, P. E. (1999). *Constructivism and Online Education.* http://edpsychserver.ed.vt.edu/workshops/tohe1999/pedagogy.html

Gee, J. P. (2004). *What Video Games Have to Teach Us About Learning and Literacy.* New York: Palgrave Macmillan.

Huitt, Caroline. (2003). Constructivism. *Educational Psychology Interactive.* Valdosta, GA: Valdosta State University. http://chiron.valdosta.edu/whuitt/col/cogsys/construct.html.

Interview: Media Grid to Take Education Across the Virtual World--and the XO. January 29, 2008. http://www.virtualworldsnews.com/2008/01/interview-media.html.

Kelton, A. J. Second Life: Reaching into the Virtual World for Real World Learning." *EDUCAUSE Center for Applied Research: Research Bulletin, 27*(17), 1-13.

Second Life. http://www.secondlife.com

Second Life--North American User Statistics. SL Wikispaces. https://secondter.wikispaces.com/Second+Life--North+American+User+Statistics?responseToken=206937f26492c9a5fbe6b2046cdc01f6

Sosnoski, J. J. (2006). Configuring African-American Culture as Virtual Experiences of History. In J. J. Sosnoski, P. Harkin, & B. Carter (Eds.), *Configuring History: Teaching the Harlem Renaissance through Virtual Reality Cityscapes* (pp. 31-41). New York: Peter Lang.

Sosnoski, J. J., & Portlock, T. (2006). Design for Narrating History in Virtual Reality Scenarios. In J. J. Sosnoski, P.Harkin, & B. Carter (Eds.), *Configuring History: Teaching the Harlem Renaissance through Virtual Reality Cityscapes* (pp. 61-68). New York: Peter Lang.

Sun Microsystems. (2008). Current Reality and Future Vision: Open Virtual Worlds. A Sun Services White Paper. http://www.sun.com/service/projectdarkstar/index.jsp.

Thomas, Angela. (2007). *Youth Online: Identity and Literacy in the Digital Age.* New York: Peter Lang.

Van Eck, Richard. (2007). Digital Game-Based Learning: It's Not Just the Digital Natives Who are Restless. In J. J. Hirschbuhl & J. Kelley, (Eds.), *Annual Editions: Computers in Education. Twelfth Edition* (pp. 55-63).

Vygotsky, L. S. (1978). *Mind in Society: The Development of Higher Psychological Process.* Cambridge, MA: Harvard University Press.

Yee, N. (2006). The Psychology of Massively Multi-User Online Role-Playing Games: Motivations, Emotional Investment, Relationships and Problematic Usage. In R. Schroeder & A. Axelsson (Eds.), *Avatars at Work and Play: Collaboration and Interaction in Shared Virtual Environments* (pp. 187-207). London: Springer-Verlag.

Chapter X
The Power and Promise of Web 2.0 Tools

G. Andrew Page
The University of Alaska Anchorage, USA

Radwan Ali
Kennesaw State University, USA

ABSTRACT

The key idea that sets constructivism apart from other theories of cognition was launched about 60 years ago by Jean Piaget. It was the idea that what is called knowledge does not and cannot have the purpose of producing representations of an independent reality, but instead has an adaptive function (Von Glasersfeld, 1996, p.3). In this chapter, a variety of Web 2.0 applications and their affordances are presented and discussed in relation to constructivism in higher education. The aim is to explain how these applications can be used in higher education to promote interactive and engaging learning environments. Recommendations for harnessing the potential of these tools along with practical examples will assist facilitators of higher education with creative means to design their courses and thus promote Learning 2.0.

WHAT IS WEB 2.0?

The Internet has presented great opportunity for global human participation by transcending geographical, cultural, religious, social classification, and political barriers. As the proliferation of learning with technology increases, there is also amplification in the array of technological possibilities for a variety of asynchronous and synchronous interactions. Therefore, it becomes necessary to provide insight into the effective use of these technologies and the facilitation of e-learning. According to Schrum and Hong (2002) the goals of teaching with technology should include facilitating higher-level, thinking skills, such as analysis, synthesis, and evaluation. Alexander describes this phenomenon:

Web 2.0 is defined as a way of creating webpages focusing on microcontent and social connections between people. It also exemplifies that digital content can be copied, moved, altered, remixed, and linked, based on the needs, interests, and abilities of users—quite against the grain of both traditional and recently expanded copyright (Alexander, 2008, p.151).

There are many creative Web 2.0 applications, tools, and services available online (See Appendix A for a list of digital resources and links). These tools or web-based learning objects have the potential to engage and involve the learner with technology as opposed to having a student learn from a computer module or digital lecture. There are many Web 2.0 tools that can be found at different sites on the web. For the purpose of this chapter, we will discuss blogs, wikis, widgets, nings, plugins, social networking (MySpace and Facebook), and virtual environments (Second Life). What issues will Web 2.0 solve? What are the potential learning outcomes from using Web 2.0? It is the promises of positive educational outcomes that web-based tools can potentially produce through quality interaction at a meta-level that are intriguing many educators and trainers.

This chapter has implications for instructors, students, instructional designers, and administrators involved with e-learning in higher education. This chapter also provides a synthesis of e-learning issues and an overview of Web 2.0 tools for promoting a constructivist environment.

The 5th edition of *The Principles of Instructional Design* (Gagne, Wager, Golas, Keller, 2005), states a very important instructional design question which is often overlooked. The question is "For what problem is instruction the solution?"(p.23). This question is especially relevant considering the numerous challenges and limitless potential of Web 2.0 tools. By using Web 2.0 tools the theory of constructivism can be applied to reach new levels of digital creation and e-learning.

Constructivism is the process of linking new understanding to old, modifying and enriching existing knowledge, and enhancing depth of comprehension about a topic. McFedries refers to Web 2.0 as functioning as a platform (p.68) because of the ability to delete, edit, and add content and work collaboratively with others in a synchronous approach.

"Constructivism affords a knowledge building process that engages active learners with the physical and social world" (Twomey-Fosnot, 1996, p.30). These interactive online tools include portals, blogs and video blogs, widgets, plugins, wikis, conferencing, games, survey instruments and games. Using these Web 2.0 tools in an educational setting builds on Vygotsky's (1978) view of interactive learning within the zone of proximal development (ZPD) and how the social process is crucial to the development of thought and behavior patterns. The application and integration of quality educational content is an important part of the constructivist framework. This is the adaptive function of constructivism alluded to by Von Glasersfeld (1996).

When interactive technology is applied in accordance with constructivist principles, it has been called "learning 2.0" (Brown & Adler, 2008). Learning 2.0 is about actively using technology. The multiliteracies of education (e.g., critical thinking, creating content, collaboration) are fostered due to the constructivist nature and exchange of knowledge that is part of the Web 2.0 phenomenon (Alexander, 2008). Learners taking the initiative to create content and learning by being engaged and involved is at the heart of constructivist practice. Brown and Adler (2008) acknowledge the impact of Web 2.0 and how it can add a new dimension to communication and participation:

The latest evolution of the Internet, the so-called Web 2.0, has blurred the line between producers and consumers of content and has shifted attention from access to information toward access to other people. New kinds of online resources—such as social networking sites, blogs, wikis, and virtual communities— have allowed people with common interests to meet, share ideas, and collaborate in innovative ways. Indeed, the Web 2.0 is creating a new kind of participatory medium that is ideal for supporting multiple modes of learning. (p.18)

THE USE OF WEB 2.0 APPLICATIONS TO PROMOTE CONSTRUCTIVIST LEARNING

This section will consider the benefits of constructivism and match them with the characteristics of a few of Web 2.0 applications. Constructivism, as a learning theory, provides an appropriate fit for the use of Web 2.0 applications. Constructivism as a learning framework can be served immensely by using the tools and affordances (Jonassen, 1999) presented by the Web 2.0 applications. Constructivism suggests that learners can enjoy their experience by being more involved in their learning than by being listeners (Franklin, 2000; Ge & Tok, 2003). The idleness or boredom aspects of passive listening prevent learners from realizing the potential of the subject matter being discussed. However, by engaging in relevant learning activities, the learner will have the opportunity to apply the concept learned. Web 2.0 applications can be very engaging for different reasons: the novelty of the technologies, the aesthetic appeal, and their interactive modes.

Web 2.0 applications present many affordances for constructivist learning. Some of these applications include social networking, blogs, podcasting, media sharing, and virtual worlds. The common trait of these applications is shareability. These applications and similar others allow for global interaction where participants contribute and use others' contributions. The contribution can be as simple as a few paragraphs posted to a blog or as complex as creating avatars on a Second Life (SL) island. The main premise is that these applications represent great potential for learning. In this section, a sample of five Web 2.0 applications is used to illustrate the use of Web 2.0 applications as contexts for constructivist learning. This sample includes social networking and bookmarking, blogs and wikis, podcasts, and media sharing, and virtual worlds

Blogs have been around for several years and we are seeing the impact of their use for positive learning outcomes. Basically, blogs are web-based journals that allow participants to contribute to or create discussion threads. The threads can discuss different subjects separately or concurrently. There are possibly millions of blogs on the web whose subjects include religion, politics, medicine, and car repairs in addition to others. Blogs are characterized by their asynchronous nature and ease of use. However, they also provide opportunities for audio and video files. Blogs can be used for constructivist learning in many ways. First, a blog can serve as an individual learning journal. A learner may start a blog to keep inventory of his/her learning. This inventory will help learners build from one phase of their learning to another. By doing so, learners use previous knowledge to generate new insight. Another example of a blog being used for educational purposes is a class group project. Learners can be engaged with discussion topics and debates in a flexible learning space that allows for the incorporation of videos, graphics, and audio and the enhanced creativity. Blogs are user-generated web-based journals that offer opinions and information and that may include text, images, and links to other blogs and web pages. Some blogs are confined to personal expressions, but others make provision for reactions and comments from readers. In higher education, blogs have been used as a means by which students can collaborate asynchronously. A blog can also serve as a group knowledge builder. A group of learners can join forces in order to complete tasks or projects. Each user is usually assigned a section of the task to complete and contribute to the blog. Through this collaboration, useful content is usually produced as a solution for the task or project. The produced content is presented as the fruition of the learning experience. By collaborating with others, a learner can be benefit by adding to his/her knowledge through the different members' contributions.

A wiki is a collaborative website that allows visitors to add, remove, edit, and change content, typically without the need for registration. It also allows for linking among any number of pages. This ease of interaction and operation makes a wiki an effective tool for mass collaborative authoring. (e.g., Wikipedia.org). Similar to blogs in many aspects, Wikis can serve numerous constructivist learning opportunities such as content production. Wikis are made of many web pages that allow contributors to add and modify content, theirs and others'. The modification opportunities make it possible for all wiki members to share their knowledge in addition to assessing that of other members. By assuming the editor (modifier) role, a wiki contributor is improving his/her writing and critical thinking. Through this practice, new knowledge may be found and added to the learner's previous knowledge. The aggregation of faculty publications in a wiki is described by Conner (2007) as a means to socially construct an online library resource and promote research productivity.

An example of an educational wiki in higher education would be Ask Dr. Wiki (http://www.askdrwiki.com/). This is a medical wiki devoted to creating a free source of medical information. Individuals can publish clinical notes, pearls, X-ray images, angiograms and more on the site. Using this wiki, anyone with a medical background can contribute or edit medical articles.

Media Sharing and Construction

Podcasts and media sharing include video, audio, photos, slides, and others to present other constructivist learning opportunities through producing and sharing digital content. Media sharing is appealing because of its interactive and personal nature. The content can be audio or video files. The main premise here is that learners can generate new knowledge by creating these podcasts and learn from the creation process. Through podcasts, learners can share information and broadcast many useful materials. Learners, then, can use podcasts as tools to learn from and teach others and provide follow-up on previous discourse. Dearstyne (2007) notes the trends in collaboration such as the increase in blogs, mashups, and wikis. Mashups are websites, or other applications, that integrate and aggregate content from more than one source into an integrated application (e.g., combining data on a topic of interest with geographical data such as Google Maps).

Virtual worlds applications such Second Life are interactive environments where participants assume personalities (avatars) to interact with other individuals and objects such as rooms, documents, presentations, and videos. By interacting with the different objects, a learner will gain a bit of knowledge from each object and then connect these different bits from the different objects to form the new experience. Similarly, the virtual world setting provides group collaboration contexts, and consequently, opportunities for constructing content. The application of collaborative and social constructivist-oriented activities, such as virtual spaces and a "user-centric approach" that promotes higher level learning are described by Sommerville and Nino (2007). The social synchronicity achieved through these media allows distant learners to connect in real-time and promote connectedness and community.

The group experience and the community membership provide positive confines that embrace collaboration and individual creativity. While the participants are making products, they will have opportunities to acquire and gain feedback from other community members (Jonassen, 1999). That will provide the different participants with new skills and knowledge as benefits of the feedback. These applications present opportunities where a learner can physically actively engage through the human computer interface (HCI) technology, virtually, and mentally active. An example of how a learner can be physically engaged in a virtual environment (and arguably a web 3.0 phenomenon) is in the work of Mitch Kapor and Philippe Bossut (2008) who have developed a hands-free navigation device for SecondLife. A video of this prototype is available at: http://www.youtube.com/watch?v=2t52gkAwJq8

Consequently, by being mentally active, learners will expand their cognitive abilities to move beyond memorization. Within that practice, long-term critical thinking becomes the norm for these learners. Furthermore, as the learners' critical thinking evolves, a few positive habits such as planning and organization are gained. These habits become second nature for these learners in their potential careers. Web 2.0 activities are similar in those regard. For example, learners who use blogs to track their learning (writing) and to collaborate with others (building camaraderie) will develop better writing and interpersonal skills to help in their respective careers. The main point is that many of these learning experiences are transferrable, and can be potentially helpful in many different contexts.

Perhaps the most important aspect of constructivism is that learners are charged with their learning. That charge will breed a sense of ownership and accordingly, learners become more interested in the process. Similarly, Web 2.0 applications encourage all participants to create their own work that can be seen by millions around the world and have the potential for cross-hemispheric collaboration. Because they own that work and the fact that it is in the public spotlight, those participants become more motivated to do work of high quality. There is an interesting dynamic and digital responsibility that is inherent to content creation on the web. The participants will seek and explore new venues and resources from which to learn to make better products. These results contribute to personal involvement. Furthermore, the experience will generate a long-term memory of the new knowledge and skills that are gained (Doolittle, 1999).

Another significant aspect which relates Web 2.0 applications to constructivist settings is that designing and implementing authentic activities to which the learners can relate can be a great motivation for the learner to participate (Zaulkernan, 2006). That authenticity can pique the learners' interest and curiosity. For example, the use of podcasting and vodcasting (video casting) by music students enhances the learning experience and enables the transmission of content to mobile devices (m-Learning) such as iPods and cell phones.

How will a facilitator use the different Web 2.0 applications as contexts for constructivist learning experiences? This question should be considered in relation to learning objectives, activities, and outcomes.

Implementing Web 2.0 technologies, while establishing learning objectives, is the beginning in creating a constructivist approach to learning. To accommodate for the potentially different levels of motor skills among the learner, the facilitator must include learning the technology as one of the learning objectives of the learning experience. By doing so, the learner will understand the importance of these technologies for completing their tasks. For example, a facilitator may create a list of learning objective as shown below in Table 1 to promote Learning 2.0.

Accordingly, the activities in the same learning experience should be planned to help the participants use the respective Web 2.0 tech-

Table 1. An example of implementing Web 2.0

The learners will be able to	
1.	*Identify what a learning strategy is.*
2.	*Explain what a lesson plan is.*
3.	*Create a lesson plan.*
4.	*Explain what a blog is.*
5.	*Use a blog to post his/her lesson plan.*
6.	*Provide and receive feedback.*

nologies while completing their tasks toward achieving the learning outcomes. In order to apply an activity to achieve Learning Objective 5, a facilitator may require the learners to find a blog and monitor the interaction or to set up an RSS feed to manage the flow of created content. Furthermore, the learners are asked to explore how to participate in the respective blogs they found. Since participation is multi-layered the teacher could also encourage the participants to use other Web 2.0 technologies on the blog such as widgets or avatar creation (meez.com). Assessments of the participating learners should include measure to assess whether they learned and used the respective Web 2.0 technologies. Using survey instruments such as Zoomerang or SurveyMonkey can provide feedback to faculty about the level of learning.

WHY ADOPT WEB 2.0 TOOLS?

There are numerous benefits of using Web 2.0 tools: access to social networking tools, knowledge sharing, digital publishing, and cost effectiveness, are just a few. *Computerworld* magazine (2008) reports that Web 2.0 tools such as wikis, blogs, and social networks are being used by many corporations (e.g., Intel Corporation) to solve complex problems, to promote training, and to build an online community. Higher education is also using these emerging tools in creative ways to promote a higher quality of interactivity and collaboration among learners. Palloff and Pratt (2005) discuss the idea that teaching and learning involve more than taking old teaching models of lecture and transferring them to a different medium. They describe an engaged learner who collects, create, and recreates knowledge. The definition of dialogue is enhanced to included Web 2.0 tools such as blogs, synchronous audio and video conferencing. Social constructivist theory emphasizes the negotiation of meaning and construction of shared understandings through dialogue (Jonassen, Davidson, Collins, Campbell, & Haag, 1995; Bonk & Kim, 1998).

There are numerous issues with the quality of online or blended learning courses, including how course objectives are accomplished, and the levels of interaction with content, peers and the instructor (Legon, 2007). *Can Web 2.0 tools increase the Zone of Proximal Development (ZPD)?* Research centered on making the transition to the Web 2.0 world and the use of podcasting audio technologies to promote collaborative learning uncovered some interesting findings. Lee, Mcloughlin and Chan (2008) found that students were empowered by the creation of rich content, as was evident in their own reflections and analysis of constructivist thought. However, Sandars and Schroter (2007) surveyed 3000 medical students and found that while there was a high familiarity and interest with a wide variety of Web 2.0 technologies, there was hesitancy in how to use these technologies. The authors conclude that there needs to be additional training in the instructional approaches for achieving desired educational outcomes through the power of technology.

The increase of online interactivity through these tools offers solutions to communicating ideas collaboratively not only effectively and efficiently, but also with a high degree of creativity. Interactivity is complex and important to the learning process. There is interactivity between the learner and the teacher, the learner and peers, and the learner and the technology. Woo and Reeves (2007) deconstruct the notion of interaction in online learning through the social constructivist lens and show how it can promote meaningful learning. They conclude:

The bottom line is that to increase the learning effects of online interaction, we should, first of all, understand clearly the nature of interaction within the framework of social constructivist learning theory. Once we gain such an in-depth understanding, we should be able to engage in productive research and development to identify the necessary design principles for implementing more effective interaction activities within web-based learning environments. (p.23)

Likewise, according to Jonassen (1995), constructivists use technologies for purposes that are (1) applied to real-world situations, (2) problems and constructs, and (3) authentic and appropriate social context. There are numerous benefits to integrating the Web 2.0 tools as learners are empowered through the creation of content and the many forms of communication. What are some of these Web 2.0 tools and how are they being used in higher education?

WEB 2.0 AND HIGHER EDUCATION

Thompson (2007) reasoned that "Today's institutions of higher education (IHEs) need to consider how they can move from being Education 1.0 institutions because their competitors are revising how they provide services and coursework using Web 2.0 applications" (p.40). However, Eijkman (2008) believes that ". . . the effective educational use of Web 2.0 will in due time lead to a radical reframing of our educational thinking and practices and a redesign of digital learning spaces around interdependent acculturation . . ." (p. 102). From the standpoint of the learner, Karpinski (2008) found that a majority of learners surveyed were unfamiliar with Web 2.0 resources. While students may be familiar with the Web 2.0 tools there was hesitancy on their part to adopt them because of (1) difficulty of use (2) skepticism about the quality of content, and (3) the lack of perceived need. Clearly there needs to be more of an acculturation of Web 2.0 tools so that current and future educators can be more confident in effectively using them in their practice. In some cases there will be a need for an epistemic instructional transformation for faculty who are in various stages of adopting new technologies.

Campus Technology magazine published an article in April 2008 about creative ways that college instructors have integrated these tools into their practice. Examples include the use of a wiki that allows students to submit exam questions for consideration. A portal is a system that has many functions including serving as an online community and a means to catalog or organize different digital creations and provide them to members in an effective manner.

Alexander (2006) describes Web 2.0 tools and how this "heterogeneous mix of relatively familiar and emergent technologies" (p.33) is used in a social networking context. Another similar example is the University of Alaska Anchorage's Web 2.0 wiki which not only defines what a Web 2.0 tool is but how this online learning instrument can be used in an educational setting. Graduate students enrolled in an instructional design course collaborated to create the content on the wiki. Each learner was given editor status so that they could post their content defining and describing the technology but also sharing their creativity by sharing how they used the tool for a positive learning outcome.

In addition to the most popular (blogs and wikis), other Web 2.0 tools include social bookmarking (de.li.ci.ous), collaborative webpage design (jot.com), and (Gnosh.org). Wix.com is a free online tool that helps the web designer with the development of a Flash interface website. This intuitive tool offers "what you see is what you get" (WYSIWYG) design that is dynamic and fluid.

Conferencing software is a tool that allows for the synchronous sharing of ideas. Commercial products include HorizonLive, Elluminate, and Adobe Breeze. WizIQ is a free online conferencing software which provides the interactivity of a whiteboard, webcam, direct messaging, and more, (http://wiziq.com). Understandably, Alexander concludes: "Meanwhile, academic implementations are bubbling" (2006, p.44) with the advent of Web 2.0.

While there have been studies on interactivity and collaborative learning through community building (e.g. Davies, Ramsay, Linfield, & Couperthwaite, 2005; Palloff & Pratt, 2005; Sherry, 2000), Hung (2001) contends there still remains much research on *how* the social interactions that are an inherent Web 2.0 component support the Vygotskian claim that cognition begins at the social level. What does this mean in terms of utilizing Web 2.0 technologies for facilitation of Learning 2.0? What is the educational utility of the Web 2.0 tools?

THE OPPORTUNITIES OF WEB 2.0

The purpose of this section is to convey how Web 2.0 applications represent and provide opportunities for constructivist learning. The capacity to apply and construct new knowledge, to encourage critical thinking, promote collaboration, and to use guided scaffolding to support higher level thinking and learning is at the very center of the constructivist paradigm. It has been argued that higher education needs to move "beyond the individual mind to include learning that is built up by mediated conversations among members of peer groups, local learning communities, and broader cultural systems" (Sherry, 2000, p. 21). Conrad and Donaldson (2004) also take this point of view: "Engaged learning stimulates learners to actively participate in the learning situation, and thus gain the most knowledge from being a member of an online learning community" (p.7). Lave and Wenger (1991) state: "A learning curriculum is not something that can be considered in isolation, manipulated in arbitrary didactic terms, or analyzed apart from the social relations that shape legitimate peripheral participation" (p. 97). But what is meant by "actively participate" and which technologies and approaches support positive learning experiences and promote quality interaction or legitimate peripheral participation?

Social Networking

Social networking Sites (SNS) applications allow participants to belong to "friends" communities where individuals are invited to be friends. Two well-known applications are MySpace and Facebook. On the other hand, social bookmarking applications are used to help users organize information by saving different web sites in one location because the user deems these web sites important for learning and research. Tufekci (2008) did a study on social networking sites (SNS) by college students and draws an important distinction between the "expressive Internet" with an emphasis on the social interactions and networking and the "instrumental" Internet that focuses on e-commerce and news. Users of SNS are also much more frequent users of the expressive Internet.

Global access and democracy can easily intertwine to create new possibilities. For example, what if classrooms in Tasmania were paired with their counterparts in Alaska to study polar science from an Antarctic and Arctic perspective? The learners could collaborate to write and illustrate a report that could involve topics such as global warming, environmental, or ecological science. The learners could communicate through web-based synchronous and asynchronous mediums.

The Internet has made digital space and the ZPD much larger and the world much smaller. Two video game players can enjoy the same game via separate individual web access points without regard to their location or time zone; that is the larger space. These games can be embedded within a blog, webpage, or wikis (See, for example, http://onlinegames.net for the extractable and embeddable code). With the possibility that one of these players may be in the USA and other may be in China, the world reach becomes smaller. The world of Web 2.0 applications such as MySpace (http://www.myspace.com/) and FaceBook is also small; through them, a person may have hundreds, if not thousands, of friends around the world. These and similar social networking applications make global interaction and cross-hemispheric collaboration easier and provide settings for global communities.

Another example of a Web 2.0 tool that promotes interactivity and community is the Ning. A Ning aggregates Web 2.0 tools such as a webpage discussion forum, file sharing, within its multi-functional framework. While MySpace and Facebook have been used for several years, a Ning allows the user to create their own network and not just be a participant in a network that has already been created. It is a means by which private or public groups with a shared interest can construct digital space and interact through a variety of mediums online. The Ning at http://Classroom20.ning.com is an example of constructed digital space for educators who can share how they are integrating technology into their practice. It is a dynamic community that is changing because of the content constructed by participants. Nings have been created for a myriad of groups including those focused on action research, Scholarship of Teaching and Learning (SoTL), computer enthusiasts, history buffs, etc. It takes only a few minutes to construct a Ning and create a viable network that aggregates the virtual social presence of learners, their thoughts and discussions, and digital artifacts. The Ning has enormous potential for providing structured community and while it is in its early stages of development it has been widely adopted by higher education departments, workgroups, ESL classes, significant interest groups, and so on.

Web 2.0 tools support many forms of social networking and collaboration, including those which a more revolutionary sense of emancipatory Freirean education, such as Flash mobs, groups that get together physically to demonstrate or protest using digital technologies (e.g. LiveJournal.com). There are countless possibilities.

Widgets and Plugins

Other tools include widgets or interactive add-ons to supplement the learning experience. These widgets can be embedded into course blogs to achieve a variety of results. There are numerous widgets to help with the learning of language and testing of skills on a variety of topics, from tracking the financial markets, to creating artwork like Matisse. A good example of a widget used in education is to have it deliver course content that the students can embed in their individual web sites. Various widgets include embedded videos, photos, podcasts, maps, all aggregated on one page (pageflakes.com) (*The Chronicle of Higher Education*, 2008).

Web 2.0 tools may fill a specific need or address a certain problem, such as the management of the enormous amount of digital information. For example, a pragmatic plug-in tool for browsers is Real Simple Syndication (RSS). RSS feeds allow the user to aggregate various resources (e.g. blogs, wikis libraries, periodicals, websites) when new content has been added. Another example is Zotero, a free plug-in for anyone who needs assistance when managing reference data from online database searches. It helps organize the citations of research articles and makes the writing experience flow easier because the knowledge can be arranged, classified, and shared. It also facilitates the importing and exporting of information. For example, you can capture citations from webpages, store documents, images, and other indexed citations from large databases. The user has the ability to co-create digital libraries and be notified via Really Simple Syndication (RSS) when a relevant research publication is tagged for retrieval from similar articles. Filters can also be set up for advanced searches. Users can also write annotations that will be saved with each reference.

FUTURE TRENDS

The question "What issues will Web 2.0 address?" was asked earlier in this chapter. The new approaches to learning with digital media have the potential to enhance the higher education learning experience. We are at the nexus of a historic transformation in how learning is shared, transmitted, collected, and constructed. The social web has seen a profound shift in higher education and the building of community through virtual social networking sites. As the facilitation of online learning evolves (artificial intelligence and non-human facilitation, web 3.0?) it becomes necessary to adopt Web 2.0 tools in the design of effective and efficient e-learning.

CONCLUSION

This chapter discussed the promise of using Web 2.0 technologies to extend the classroom. Some of the advantages associated with implementation of Web 2.0 were listed and discussed. The potential for a revolution of learning that involves constructivism will inherently involve these tools. "Constructing an understanding requires that the students have opportunities to articulate their ideas, to test those ideas through experimentation and conversation, and to consider connections between the phenomena that they are examining and other aspects of their lives"(Julyan & Duckworth, 1996, p.58). We can use these Web 2.0 tools to transform our practice and enhance the reality of learning. The dialectic between the individual and society through the use of Web 2.0 tools is expanding dramatically as these new possibilities suggest that learning is a constructive building process of meaning-making. Faculty should adopt these constructivist tools or run the risk of "engage or enrage" (Prensky, 2005). Faculty cannot continue to teach as they were taught but rather they must employ the interactive Web 2.0 tools to engage learners. This is the power and promise of Web 2.0 tools.

REFERENCES

Addison, C. (2006). A new chapter in web development? *Development in Practice, 16*(6), 623-627.

Alexander, B. (2006). A new wave of innovation for teaching and learning. *Educause* (March/April) 33-44. Retrieved April 5, 2008 from http://www.educause.edu/ir/library/pdf/ERM0621.pdf

Alexander, B. (2008). Web 2.0 and emergent multiliteracies. *Theory Into Practice, 47*(2), 150-160.

Bonk, C. J., & Kim, K. A. (1998). Extending sociocultural theory to adult learning. In C. M. Smith & T. Pourchot (Eds.), *Adult learning and development: Perspectives from educational psychology* (pp. 67-88). Mahwah, NJ: Lawrence Erlbaum.

Brown, J. S., & Adler, R. P. (2008). Minds on fire: Open education, the long tail, and learning 2.0. *Educause Review, January/February,* (pp. 17-32).

It's a Web 2.0 world: Are you ready for it? (2003, April). *Campus Technology* (pp. 43-50).

Computerworld. (2008). *IT will "take head out of sand" on Web 2.0 in 2008, 42*(6), 5.

Connor, E. (2007). Using Wiki technology to build a faculty publications database. *Journal of Electronic Resources in Medical Libraries, 4*(4), 11-25.

Conrad, R. M., & Donaldson, J. A. (2004). *Engaging the online learner: Activities and resources for creative instruction.* San Francisco: Jossey-Bass.

Davies, A., Ramsay, J., & Linfield, H. (2005). Building learning communities: Foundations for good practice. *British Journal of Educational Technology, 36*(4), 615-628.

Dearstyne, B. W. (2007). Blogs, mashups, & wikis. T*he Information Management Journal, 41*(4), 24-33.

Dede, C. (1996). Distance learning: Distributed learning: Making the transformation. *Learning and Leading with Technology, 23*, 25-30.

Doolittle, P. (1999). *Constructivism and online education.* Virginia Polytechnic Institute and State University. Retrieved July 2008 from: edpsychserver.ed.vt.edu/workshops/tohe1999/text/doo2.pdf

Doll, W. (1993). *A post-modern perspective on curriculum.* New York: Teachers College Press.

Eijkman, H. (2008). Web 2.0 as a non-foundational network-centric learning space. *Campus-Wide Information Systems, 25*(20), 93-104.

Franklin, G. (2000). Augmenting print-based distance learning with other technologies can enhance its effectiveness for adult learners. *Journal of Educational Technology Systems, 28*(4), 327-334.

Gagne, R. M., Wager, W. W., Golas, K. C., & Keller, J. M. (2005). *Principles of instructional design.* (5th ed) Belmont, CA: Wadsworth Publishing.

Ge, S., & Tok, M. (2003). Enhancing online education using collaboration solutions. *Journal of Educational Technology Systems, 31*(4), 361-380.

Glogoff, S. (2005). Instructional blogging: Promoting e-learning, student-centered learning, and peer input. *Innovate, 1* (5). http://www.innovateonline.info/index.php?view=article&id=126

Hung, D. (2001). Design principles for web-based learning: Implications for Vygotskian thought. *Educational Technology, 41*(3), 33-40.

Jonassen, D. (1999). Designing constructivist learning environments. In Reigeluth, C. M. (1999). *Instructional Design Theories and Models: A New Paradigm of Instructional Theory, 2*, 215-241. Mahwah, New Jersey: Lawrence Erlbaum Associates, Publishers.

Jonassen, D., Davidson, M., Collins, M., Campbell, J., & Haag, B. B., (1995). Constructivism and computer-mediated communication in distance education. *American Journal of Distance Education, 9*(2), 7-26.

Julyan, C., & Duckworth, E., (1996). A constructivist perspective on teaching and learning science. In C. Twomey-Fosnot (Ed.), *Constructivism: Theory, perspectives, and practice*. (pp. 55-72). Columbia University. Teachers College Press.

Karpinski, J. L. (2008). Disconnected in a connected world: Knowledge and understanding of Web 2.0 tools at the University of Pennsylvania Medical Center. *Medical Reference Services Quarterly, 27*(1), 53-72.

Lave, J., & Wenger, E. (1991). *Situated learning: Legitimate peripheral participation.* Cambridge, MA: Cambridge University Press.

Lee, M. J., Mcloughlin, C., & Chan, A. (2008). Talk the talk: Learner-generated podcasts as catalysts for knowledge creation. *British Journal of Educational Technology, 39*(3), 501-521.

Legon, R. (2007). Online courses lead the way in quality assurance. *Sloan-C View: Perspectives in Quality Online Education 6*(1). Retrieved July 20, 2007 from http://www.sloan-c.org/publications/view/v6n1/viewv6n1.htm

McFedries, P. (2006). The web, take two. *IEEE Spectrum, 43*(6), 68-68.

Palloff, R., & Pratt, K. (2005). *Collaborating online: Learning together in community.* San Francisco: Jossey-Bass.

Prensky, M. (2006). Engage me or enrage me: What today's learner's demand. *Educause* October, 60-64. Retrieved September 17, 2008 from http://net.educause.edu/ir/library/pdf/erm0553.pdf

Sandars, J., & Schroter, S. (2007). Web 2.0 technologies for undergraduate and postgraduate medical education: an online survey. *Postgraduate Medical Journal, 83*(986), 759-762.

Schrum, L., & Hong, S. (2002). Dimensions and strategies for online success: voices from experienced educators. *JALN, 6(1).* Retrieved August 11, 2002 from http://www.aln.org/alnWeb/journal/jaln-vol6issue1.htm#schrum

Sherry, L. (2000). The nature and purpose of online conversations: A brief synthesis of current research. *International Journal of Educational Telecommunications, 6*(1), 19-52.

Somerville, M. M., & Nino, M. (2007). Collaborative co-design: A user-centric approach for advancement of organizational *learning. Performance Measurement & Metrics, 8*(3), 180-188.

The Chronicle of Higher Education (2008, September 11). *Professor uses web 'Widgets' to share course content.* Retrieved October 10, 2008, from http://chronicle.com/wiredcampus/article/3307/professor-uses-web-widgets-to-share-course-content.

Thompson, (2007). Web 2.0 *takes on colleges and universities: The dawn of education 2.0.* Retrieved April 29, 2008 from http://www.masternewmedia.org/news/2007/04/20/Web_20_takes_on_colleges.htm

Tufekci, Z. (2008). Grooming, gossip, Facebook and Myspace. *Information, Communication & Society, 11*(4), 544-564.

Twomey-Fosnot, C. (1996). Constructivism: A psychological theory of learning. (pp. 8-33). In C. Twomey-Fosnot (Ed.), *Constructivism: Theory, perspectives, and practice.* New York: Teachers College Press.

Von Glasersfeld, E. (1996). Introduction: Aspects of Constructivism. In C. Twomey-Fosnot (Ed.). *Constructivism: Theory, perspectives and practice* (pp. 3-7). Teachers College, Columbia University.

Woo, Y., & Reeves, T. C. (2007). Meaningful interaction in web-based learning: A social *constructivist* interpretation. *Internet & Higher Education, 10*(1), 15-25.

Zualkernan, I. A. (2006). A framework and a methodology for developing environments. *Educational Technology & Society, 9*(2), 198-212.

APPENDIX

Tool	Example	Link
Open Source Tools	Open Office Linux The Open Source Initiative Source Forge	http://Openoffice.org http://www.linux.org http://www.opensource.org/ http://sourceforge.net/
Open CourseWare	Massachusetts Institute of Tech (MIT) Carnegie-Melon University Utah State University	http://ocw.mit.edu http://www.cmu.edu/oli/ http://ocw.usu.edu/
Portals	Merlot (Multimedia Educational Resources for Learning and Online Teaching) UPortal	http://merlot.org http://www.uportal.org/
Collaborative design & Project Management	Jot Gnosh Project Zoho	http://jot.com http://gnosh.com http://projects.zoho.com/jsp/home.jsp
Survey Instruments	Zoomerang Survey Monkey	http://zoomerang.com http://surveymonkey.com
Video Blogs	Video Blogs	http://www.vblog.com/
Directory of Web 2.0 Tools	http://www.go2Web20.net/ Centre for Learning & Performance Technologies	http://www.allthingsWeb2.com/ http://c4lpt.co.uk/handbook/elearning20.html
Social Networks	Ning.com	http://education.ning.com http://www.classroom20.com/
Blogs	Blogger.com	http://asdtechnology.blogspot.com
Wikis	PBwiki.com wikispaces.com	http://pbwiki.com http://wikispaces.com
Video Blogs	Vobbo	http://www.vobbo.com/
Webpage design	Weebly FreeWebs Google websites	http://Weebly.com http://webs.com http://sites.google.com/
Conferencing Software	HorizonLive Elluminate Live! Breeze, WizIQ;	http://Elluminate.com www.wimba.com http://www.adobe.com/resources/breeze/ http://www.wiziq.com
Social bookmarking	De.li.ci.ous	http://delicious.com/
Widgets	Widget Box	http://www.widgetbox.com/tag/education http://pageflakes.com
Online Answers	One Big U	http://www.onebigu.com/user/homepage.php
Games	Online Games	http://onlinegames.net

Chapter XI
IT-Enabled Strategies for Mental Modeling in E-Learning

Shalin Hai-Jew
Kansas State University, USA

ABSTRACT

This chapter examines some ways information technologies (IT) are deployed in higher education courses to help learners create robust mental models. In e-learning, mental models are created through the following: digitally mediated (1) information visualization work; (2) virtual immersion, and (3) human interactivity. For (1) information visualization work, information technologies afford multi-sensory learning channels: texts, visuals, slideshows, screen casts and animations, audio, video, interactivity, immersiveness, and simulations. IT supports the archiving of digital learning artifacts through eportfolios, digital gallery shows, and informational multimedia databases. (2) Virtual immersion has been enabled in 3-D interactive spaces where learners may experience multi-faceted learning. More complex simulations have also been created with animations and long-term continuous learning. To promote (3) human interactivity, IT reifies human intellects and perceptions, and social software uses "swarm intelligence" to support consensus-driven realities. IT creates contexts for co-learning and intercommunications. Collaborative pedagogical strategies—online teaching case studies, mock trials, facilitated games, and various assignments and assessments—highlight the multiple-views, multiple-realities of constructivist learning.

BACKGROUND

With the recent popularization of constructivist ideologies in e-learning, practitioners have found that information technologies may be used to help learners build efficacious mental models. Information technologies (IT) offer three main strengths for creating mental models: information visualization work, virtual immersion and human interactivity. This chapter will open with a brief definition of constructivism and mental models. Then five main IT affordances or "enablements" supporting the use of constructivist mental models will be introduced and supported with technological examples. Constructivism essentially posits that individuals learn when they self-create mental models and meanings, by visualizing certain roles and relationships, by understanding larger systems, and by engaging in human interactions. Individuals must be motivated to learn with authentic and applied learning situations.

A brief look at constructivism. Moshman (1982) suggests that there are three interpretations of constructivism: *endogenous, exogenous* and *dialectical:* The Endogenous focuses on the individualistic and internal nature of knowledge construction, unique to each learner. Here, an instructor facilitates the experiences which may lead to changes and maturation of existing mental models. Exogenous constructivism advocates formal instruction along with active exercises to help learners develop practical knowledge representations. Dialectical constructivism emphasizes the importance of realistic experience for the creation of sophisticated mental models (Dalgarno, 2001, p. 185). Learning contexts are viewed differently based on varying constructivist approaches: *"Ontological.* Contexts are constructed socially, in interaction with other agents in the world, and psychologically, in making sense of sense data. *Epistemological.* Interpretation of context is always constituted within a frame of reference. *Pragmatic.* Instead of labeling contexts, computers can provide resources for people themselves to create and maintain contexts in their action" (Oulasvirta, Tamminen, & Höök, 2005, p. 195).

Some tenets of mental modeling. A basic definition of a mental model is that it is the learner's internal conceptualization of a particular system or paradigm, the functions of equipment, or phenomena. A mental model involves the respective learner's sense-making and may involve both implicit and explicit ideas, both unconscious and subconscious beliefs, and internalized and externalized knowledge. To borrow a concept from Argyris and Schön, a mental model would involve the "theories in use" vs. rather than "espoused theories" (1974, as cited in Argyris, 1993, p. 20). A mental model captures the substructure or underlying theory by which a thing or phenomena may be understood. Making meaning from a mental model involves "balance, comparison and the reorganization of existing knowledge and the input of information," assert Lei, Yang and Zhang (2006, n.p.). This incremental and cumulative process of building schemas is context-dependent. "Mental model construction involves mapping active memory objects onto components of the real-world phenomenon, then reorganizing and connecting those objects so that together they form a model of the whole situation" (Derry, 1996, p. 168).

Mental models affect decision-making; therefore, a mental model is also "a hypothetical knowledge structure that integrates the ideas, assumptions, relationships, insights, facts, and misconceptions that together shape the way an individual views and interacts with reality" (Steiger & Steiger, 2007, p. 1). Shared mental models strengthen the work of teams with "the ability to successfully navigate unforeseen situations or problems as they arise" (Thomas & Bostrom, 2007, p. 3). Knowledge structures show relationships among knowledge components and knowledge objects. A symbolic reasoning (instructivist) view would focus more on stored representations, and a situated learning (constructivist) view would focus on "connections, potentials" (Jonassen, et al., n.d., n.p.). "Mental models are particular organizations of memory objects that constitute a specific event interpretation," writes Derry (1996, p. 169).

The figure "Some Mixed Elements of Constructivism" introduces the central types of constructivism—endogenous, dialectical and exogenous; the ontological, epistemological, and pragmatic. These concepts have implications for different types of e-learning because each high-

Figure 1. Some mixed elements of constructivism

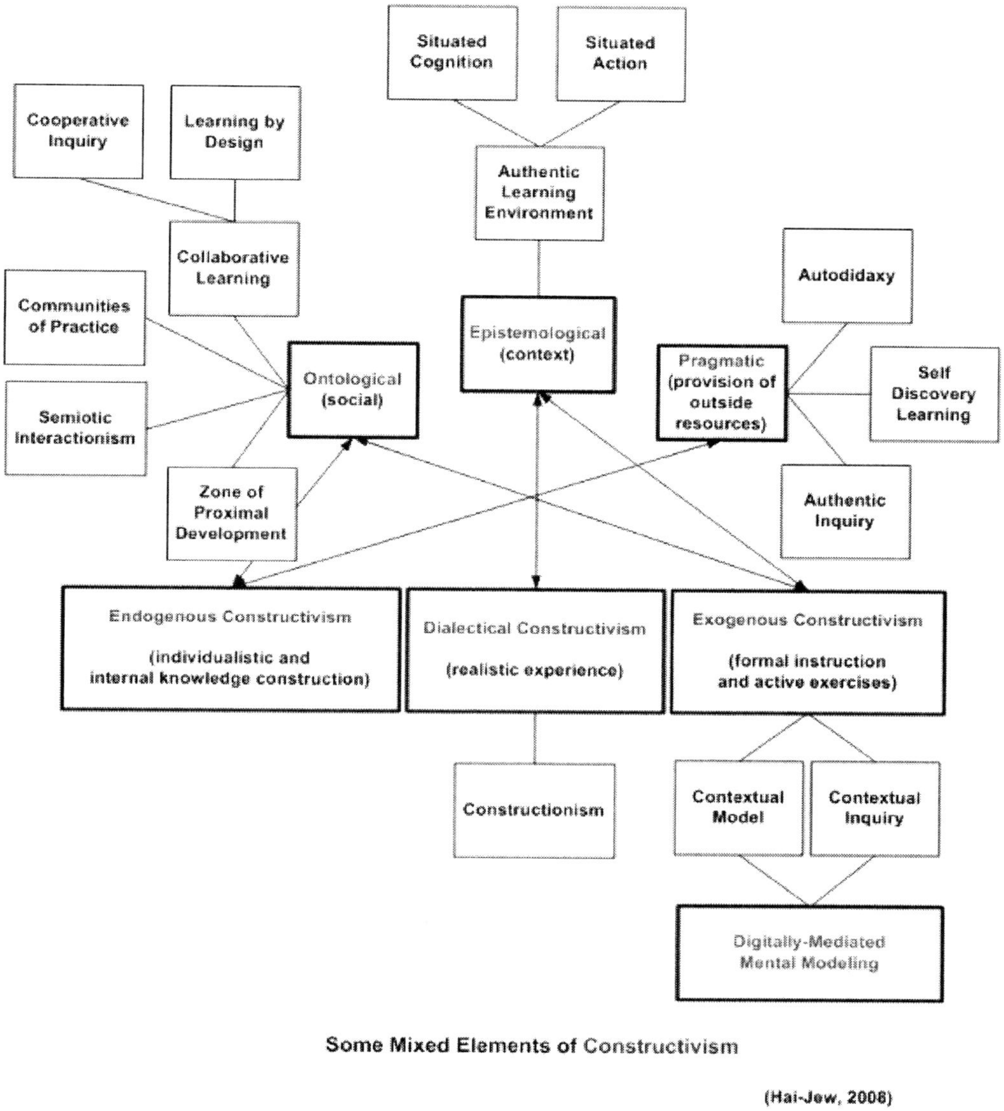

lights different aspects. Each model introduces concepts about how constructivist learning occurs, and there are related pedagogical concepts tied to the models. Mental modeling is achieved through active and contextual learning.

The tie between mental models and computers goes back at least three decades, with a particular focus on human-machine systems. Another angle has been the use of computers to mimic what are thought to be the cognitive structures of the mind through the creation of knowledge structures. A knowledge structure is a form of schema; it represents the information (knowledge components and the relationships between them) that is required if a learner is to be able to solve problems (Merrill, 2000, p. 245).

Most mental models are built around a particular topic within a knowledge domain. Such areas may be organically defined, while others may be human-made and artificial. A knowledge domain may rely on the traditional academic categorizations, such as plant species labels or chemical typology. Still other mental model areas may be built around the identification and diagnosis of a problem (Crapo, Waisel, Wallace, & Willemain, 2000, p. 219). Various activities in the context of these domains enhance mental constructions, such as reading (Anderson, 1984; Schank & Obelson, 1977); mathematical problem-solving (Newell & Simon, 1972; Greeno, 1978; Larkin, 1983), and scientific modeling (as cited in Olssen, 1996, p. 284).

The figure "Core Elements of a Mental Model" provides a simple depiction of objects and entities, and their interrelationships (also known as data "values" and "structures"). A mental model includes the larger context and environment in which these relationships exist, and the objects interact and change. Mental models may focus on small aspects of a system in depth or portray more abstract, higher-level realities.

Obtaining content for conceptual models. Conceptual models are those built by subject matter experts in a domain to represent the particular aspects of a knowledge domain. Conceptual models rely on substantial background information even if only a portion of a knowledge domain is depicted. Designing learning around conceptual models requires defined learning objectives and outcomes. All relevant terminology and nomenclature need accurate expression. Objects (characteristics and behaviors) have to be defined as well as their interrelationships. Processes need clear explication. Learning scenarios need to be designed and created in multimedia Web-enabled spaces. The model then is tested with experts and novices, often against real-world, measureable outcomes. Design standards for the mental model build are defined and adhered to for consistency.

Getting information from subject matter experts (SMEs) to build into digitized mental models may involve a range of processes. One would be the verbal protocol analysis (Ericsson & Simon, 1980), for higher-level user-oriented

Figure 2. Core elements of a mental model

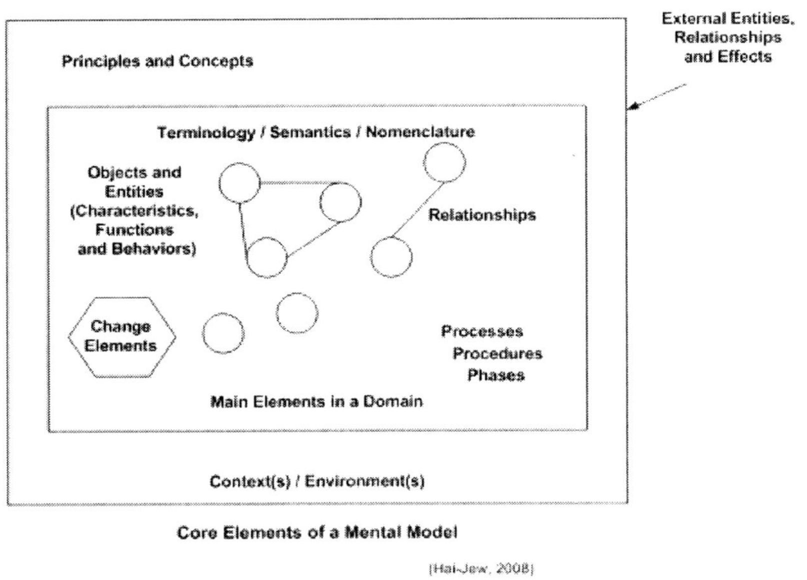

requirements via open-ended questioning. Sequential discovery (O'Bannon, 1987) involves having current users "identify specific values and procedural steps related to objects, events, conditions, and activities in their domain of expertise." Concept discovery surfaces major ideas and objects of interest. Cognitive network analysis is used to see how objects and concepts are "grouped and related" (Snyder, Paap, Lewis, Rotella, Happ, Malcus, & Dyck, 1986, as cited in Jones & Popken, n.d., p. 814).

No tabula rasa*: Naïve mental models.* Naïve mental models that are maintained by novices tend to be weak, with elusive, illogical, incomplete, or poorly formed details. The mental conceptualizations may be invalid. Important distinctions may be missing (Nardi & Zarmer, 1991, p. 487). Yet prior mental models are foundational for new learning. Learners need to be able to articulate their existing mental models and have the metacognitive ability to guide them in recreating a stronger mental model, and not merely have their misconceptions pointed out to them (Ben-Ari, 1998, p. 258). Metacognition involves the awareness of thinking and learning processes; acquiring and analyzing new data; reflecting on the impact of the new information on former understanding, and ultimately constructing meaning.

Mental models and pedagogical design. Some writers advocate complex learning to match the complexity of the world. A central goal is to increase individual cognitive flexibility, which Spiro, et al. (12, p. 165) define as "'*the ability to spontaneously restructure one's knowledge, in many ways, in adaptive response to radically changing situational demands'*" (Chieu, Milgrom & Frenay, 2004, p. 222). This capability is especially important in ill-structured domains where examples are "diverse, irregular and complex". Being able to simultaneously hold multiple perspectives, multiple metaphors and analogies and multiple interpretations of the same information (Feltovich, et al.) strengthens cognitive flexibility. Analogies help learners compare and contrast features and enrich their knowledge base (Brandt, 1997, p. 115). Through an assortment of mediated environments, instructors may revisit the learning materials "at different times, in rearranged contexts, for different purposes, and from different conceptual perspectives" for advanced knowledge acquisition, according to Spiro, et al. (Chieu, Milgrom & Frenay, 2004, p. 221) and cognitive flexibility. Offering diverse ideas and problem-solving assignments may create perturbations, which lead to the accommodation of new ideas (Götschi, Sanders, & Galpin, 2003, p. 347).

Driscoll suggests that (1) multiple modes of learning (different representations of contents, different methods for exploring contents) and (2) multiple perspectives ("expression, confrontation, and treatment of multiple points of view") may improve the conditions of learning for cognitive flexibility. Information technology allows multiple modes of learning, and human-facilitated learning may improve the richness of a multiplicity of perspectives. Multiple modes involve the use of a variety of multimedia and communications tools for interactivity (Chieu, Milgrom & Frenay, 2004, p.222). Situated learning proponents suggest that learners need to draw on real-world situations and particularly issues from their own communities (Freire, 1974; Ladson-Billings, 1995; Lave & Wenger, 1991; Greeno, 1998, as cited in Abrahamson, et al., n.d., p. 1015). There needs to be a porous representational exchange between the e-learning conceptual modeling and the external real.

Standards for the viability of mental models. Mental models may be representational, describing a physical reality as an abstraction. They may be predictive or anticipatory of the future. They may be proscriptive prescriptive in describing how something should optimally be. They may be speculative or purely theoretical, possibly even proposing an untestable thesis. Mental models may be used for troubleshooting current systems; they may build related knowledge further. Human cooperation often requires shared mental model-

Figure 3. The viability of a digital mental model

Mental models need to be viable before they are adopted and internalized by individual learners or groups. The viability of a digitized mental model stems from a variety of factors.

APPLICABILITY

1. **Informational Accuracy:** How accurate is the information or depictions in a mental model?
2. **Representativeness and Fidelity:** How representative is the mental model of the real world? Is the fidelity high or low?
3. **Coherence:** How aligned and clear is the model?
4. **Predictiveness:** How predictive is the mental model of the real world?
5. **Efficacious:** How efficacious is the model used in conjunction with other tested models in a complex environment?
6. **Timeliness:** How timely or current is the mental model?
7. **Completeness:** How complete is the mental model? Is all relevant information included?
8. **Foundational Assumptions:** Are the foundational assumptions of the mental model valid and sound? Are there inherent biases of this mental model?
9. **Refinability for New Information:** How refinable is the mental model in terms of incorporating new information and fine-tuning it? Does it flexibly incorporate new data?
10. **Usefulness:** Is the digital mental model consistently useful for its defined purpose or purposes?

TECHNOLOGICAL FLEXIBILITY

11. **Portability:** How portable is the mental model between technological systems (like databases, immersive environments, learning management systems, and others)?
12. **Scalability:** How adaptive is this digital mental model to usage by larger numbers of people?
13. **Integration with other Mental Model Systems:** How easily does this mental model integrate with other mental modeling systems in the same domain knowledge? In cross-domain knowledge?
14. **User Interface Alignment:** How aligned is the user interface with the conceptual mental model behind the digital mental model? How navigable is the user interface? How user-customizable is the interface?
15. **Accessibility**: How accessible is the system to those with visual acuity, hearing, mobility or cognitive processing challenges?
16. **Multimedia Richness:** How full-sensory and multimedia rich is this digitized mental model? How full-sensory is it?

LEARNING VALUE

17. **Expressibility and Transferability:** How expressive or transferable is the mental model? (Note: The higher the abstraction, the fewer the constraints in terms of expressibility or transferability.)
18. **Learning Value for Diverse Learners:** What is the learning value of this mental model for a wide range of learners?
19. **Non-existent or Minimal Negative Learning:** How much unintended negative learning or inaccurate learning (or noise) is involved in the mental model? Is it easy to address and defuse negative learning?
20. **Learning Augmentation:** How easy is it to create effective front-end learning pieces, effective debriefing, and localized applications of the mental model?

The Viability of a Digital Mental Model

(Hai-Jew, 2008)

ing. This is especially so in situations that may be high-risk, intense, or high-reliability ones. Shared mental maps may enhance consensus among members of a team; these understandings enhance teamwork, especially in "high criticality" or time-critical settings.

While constructivism and mental modeling theoretical frameworks allow for greater ranges of "truth," not all viewpoints are equally viable or defensible: ". . . coherence and viability are valued over an abstract notion of truth." A mental model is seen as viable when it works—e.g. for troubleshooting code (Ben-Ari, 1998, p. 260); diagnosing faults through pattern recognition (Sanderson & Murtagh, 1989, p. 1238), and accurately predicting the realities represented by the model. One risk of building mental models is doing it in such a way as to accidentally support learner misconceptions or negative transfer. Learning through mental modeling should involve ways to surface and address misconceptions, in part because of the persistence of "established misconceptions go, even in the face of compelling evidence" (Jackson & Fagan, 2000, p. 85).

The figure "The Viability of a Digital Mental Model" stems from a range of factors, which are based in three main areas: applicability, technological flexibility, and learning value. Depending on the knowledge domain, some of the sub-factors within these categories may have more importance than others.

Five main IT affordances or enablements support the use of constructivist mental models in e-learning.

I. Information Visualization Work
   1. The support of learner self-discovery
   2. Knowledge construction
   3. Digital content building
II. Virtual Immersion
   4. Digital context building, and
III. Human Interactivity
   5. Mediated community for constructivist learning.

Information visualization work. IT affordances include the rich multi-sensory multimedia objects that may be built with authoring tools or Learning / Course Management Systems (L/CMSes). Knowledge structures may be built to represent information to enhance human abilities to visualize systems and domain realities. Information visualization work supports learner self-discovery, knowledge construction, and digital content building. Autonomous learners may be empowered to seek out particular mediated experiences and information to build their own mental models. *Virtual immersion* has been enabled in 3D interactive spaces where learners may experience rich multi-faceted, multi-sensory learning. More complex simulations have also been created with immersive and complex learning. The uses of augmented reality through global positioning systems (GPSes), handhelds, laptops, and other portable devices extend the digital delivery of learning content well into physical embodied spaces for ambient "mixed reality" experiences and the benefits of physical, embodied motions. Virtual immersion is created through digital context building and experiential constructions. *Human interactivity.* The mediated interactivity made possible by synchronous and asynchronous mediums of communication enables community-building, partnering, and group collaboration. Social software captures "swarm intelligence" knowledge, attitudes, processes, and other "intelligence." This information may be used for building further learning contents or learner trails. Fresh collaborative pedagogical strategies have also emerged--online teaching case studies, mock trials, facilitated games and digital sims, and diverse assignments and assessments—to highlight the "shimmer" (multiple views, multiple realities) that is the result of constructivist learning. IT supports building mediated community for constructivist learning. The remainder of this chapter offers further elaboration on these points.

The figure "Building Learner Mental Models via IT Affordances" shows how naïve mental models may be shaped through information visualization work, virtual immersion, and human interactivity. These modalities for learning may all be supported by existing IT.

Figure 4. Building learner mental models via IT affordances

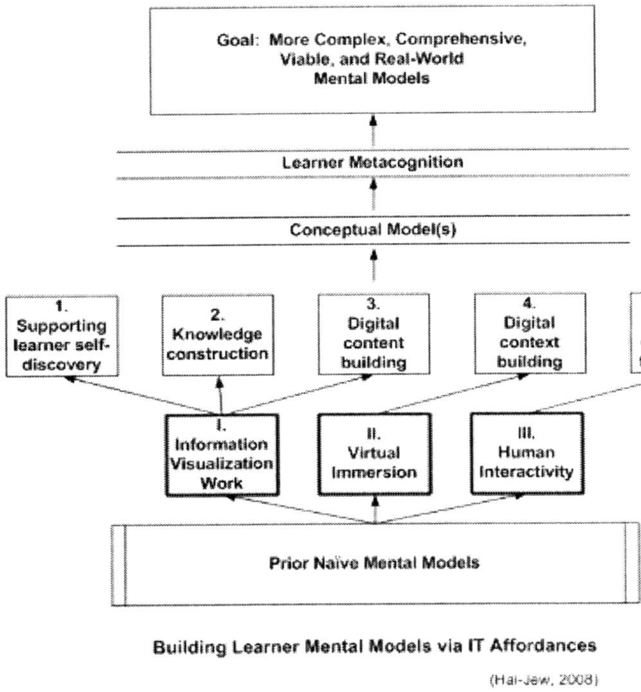

INFORMATION VISUALIZATION WORK

When learners build mental models, they need to be able to visualize particular systems, structures, objects, and interrelationships. Their exposure to the conceptual models of the experts may lead to their realization of inconsistencies in their own models, which then leads to accommodation, which leads to changes in internal representations (Dalgarno, 2001, p. 184). Driscoll offers five major teaching strategies consistent with constructivism: support of learning conditions for learners for "(1) reasoning, critical thinking, and problem solving; (2) retention, understanding, and use; (3) cognitive flexibility; (4) self-regulation; and (5) mindful reflection and epistemic flexibility" (Chieu, Milgrom & Frenay, 2004, p. 221). Evocative multimedia visualization tools, both in terms of user creation (of digital artifacts) and their ease of consumption, enhance the experience of alternate realities and the building of more sophisticated mental models.

Supporting Learner Self-Discovery

In the long term, individuals need to be able to initiate their own learning and to regenerate their knowledge bases as they gain expertise. Some thinkers have observed a shift in learning from supply-push to a demand-pull model (Brown & Adler, 2008, p. 30). The previous teaching and learning of degree-granting institutions, as "supply-push" entities, may provide e-learning for those who want the formalized structures. Many people, as they move into the workforce, will go to other e-learning providers for their own internally-driven interests. Supporting this learning means providing access to rich learning communities with plenty of digital learning resources in discovery learning spaces.

IT innovations encourage active learner initiative (autodidaxy) in seeking out new learning.

Endogenous constructivism is enhanced through learner controlled browsing of contents and their active exploration in simulations and microworlds (Dalgarno, 2001, p. 186). Learners need support in selecting relevant learning materials and paths at different stages. "The use of *cognitive tools*, to assist with knowledge construction and articulation during instruction, including concept mapping tools and hypertext editing tools is consistent with exogenous constructivism. Practice modules, for example quizzes and problem solving exercises, which allow the learner to obtain feedback on their own construction of knowledge, are also consistent with this view" (Dalgarno, 2001, pp. 187 - 188). Problem-solving through mental models may help learners to explain, encode, evaluate, communicate, troubleshoot, plan and predict actions in complex situations.

Active learner self discovery. The relative longevity of digital objects with caching by search engines and archiving in digital databases means that the learning may aggregate and continue over time. Investigative learning tasks often involve long-term and complex learning activities (Davydov, 1988, as cited in Stary & Totter, n.d., p. 41). Indeed, constructivism applies to learning over a long time horizon, and having digital resources in the various stages of learning and for various learning purposes over lifetimes is a technological enablement. Constructivism envisages a learning process which does not lend itself to quick fixes: internal cognitive models of the problem domain are constructed and refined over a period of time and in the light of experience and reflection" (McKenna & Laycock, 2004, p. 170). This aspect may be even more critical for irregular domain situations with fast-changing information and practices, where the learning must continue well beyond the confines of a formal classroom or training.

Learner goal setting is an integral part of constructivism because learning occurs in learners' goal-directed consciousness (Olssen, 1996, pp. 282 – 283; Dark & Winstead, 2005, p. 29). The empowerment of learners to take the initiative and chart the course of some of their learning is a core constructivist assumption. Learning opportunities should "provoke reflective experimentation, discourse, and negation of conceptual conflicts should be provided. Such environments promote assimilation and accommodation and lead to the construction and reconstruction of knowledge" (Derry, 1996, p. 165). Online spaces, if facilitated well, may offer learners "choice, diversity, and congeniality," according to Shang, Shi and Chen (Summer 2001, p. 3). These designed digital spaces will encourage learners' own senses of internal agency towards learning.

User-created-content publication. Constructivism suggests that each learner theoretically brings something to the shared online learning experience (Zauder, Lazić & Zorica, 2007, p. 437). The reification of user-generated contents in so-called Web 2.0 phenomena supports this valuation. In a "convergence of literacies," learners create their own digital contents with technologies that offer lower learning thresholds for entry. Multimedia creations are tools that enable the easy sharing of such literacy artifacts. "Reflective practice is an important part of this learning process. Students are encouraged to document their own learning through personal portfolios, sketches, photography, videotape and contributions to the class' newsletter. The social culture of the CLL (constructionist learning laboratory) values collaboration, idea sharing and a sense of community" (Stager, 2002, para. 14). In even more protected spaces like online classrooms, these technologies have enabled a more private sharing of user-created learning objects.

These contents then become part of their student and professional portfolios (which may be private, semi-public or fully public); the digital contents may migrate to the WWW based on the precept that information wants to be free. Such student-created materials have become central parts of the university learning experience (Lippincott, 2007, p. 16). A variety of higher education projects and curriculum focus on preparing students to be content creators in their disciplinary areas, with e-portfolios, digital gallery spaces, poster sessions and virtual conferences to showcase their work. Numerous publicly-available sites offer specialty aggregation of such digital artifacts.

Efficient searchability of digital contents. Automated 'bots and software programs navigate the Web to collect relevant finds for users, including search engines and derived data aggregators. Automated union catalogs with metadata harvesters collect open access information for human use. Screen scraping, a technique for extracting information from a Web page, enables more focused searches. The new focus on proper labeling of resources, via metadata standards for machine interoperability, has also enabled easier self-discovery of learning materials for the building of mental models (Al-Khalifa & Davis, 2007, n.p.). While "low semantic" searches are quite easily achieved, "high semantic" ones that require more in-depth knowledge of the contents will require more back-end complexity (Hasenblas & Nack, 2007, p. 5). More customized searches are now being made possible with user profiling and "stereotyping" (MacManus & Porter, 2005, n.p.). The creation of *folksonomies* (collaborative tagging) place information into a hierarchy of popularity and topical categorizations of contents. *Folksonomies* or user-created taxonomies refer to the totality of tags on a certain subject matter (Zauder, Lazić & Zorica, 2007, p. 437). Tag clouds offer a weighted list of user-generated tags dependent on user-determined value. "And semantic tagging technologies that exploit metadata functionality can greatly increase the navigation of internal and external information overload, as well as influence information-based product consumption and use" (Knights, 2007, p. 35). Really Simple Syndication (RSS) feeds pull multimedia information to a user's browser, laptop, desktop computer, cell phone, or mobile device for convenient consumption. These endeavors "facilitate organization, information retrieval and knowledge discovery" (Zauder, Lazić & Zorica, 2007, p. 438).

Knowledge Construction

Knowledge construction, which involves the creation of knowledge and its structuring in IT spaces, relies on a range of technologies. Traditional ones captured expert knowledge on intranets and databases. Wikis capture users' contributions. Such open-source endeavors are said to quicken knowledge management with fewer mistakes than closed-source knowledge management environments (Watson & Harper, 2008, p. 2). Online survey systems make primary data sources more readily accessible (Maia, Machado, & Pacheco, 2005, p. 220). Informational visualizations of complex data sets, with archived or live information, may enhance human understanding and decision-making in diverse fields from helicopter piloting in treacherous wind conditions (Aragon & Hearst, Apr. 2005); bioinformatic gene expression microarrays in the life sciences (Saraiya, North, & Duca, July – Aug. 2005, p. 443), and immersive archaeological data visualization (Acevedo, Vote, Laidlaw & Joukowsky, Oct. 2001, p. 493).

Articulating knowledge structures. Knowledge may be constructed using a variety of multimedia modes. Knowledge may be built up in ontologies or taxonomies. These may be reflected spatially to show the related objects and interrelationships in a subject domain. Knowledge structures (operationalized mental templates) are imposed on an information environment to give it "meaning and form" (Riggins & Slaughter, 2006, p. 4). "Cognitive psychology suggests that a mental model consists of two major components: knowledge structures (schemata) and processes for using this knowledge (mental operations). A major concern of instructional design is the representation and organization of subject matter content to facilitate learning" (Merrill, 2000, p. 244).

Four types of knowledge objects may be seen as representing all cognitive subject matter content. "*Entities* are things (objects). *Actions* are procedures that can be performed by a learner on, to, or with entities or their parts. *Processes* are events that occur often as a result of some action. *Properties* are qualitative or quantitative descriptors for entities, actions, or processes" (Merrill, 2000, pp. 244-245). Another system divides knowledge into declarative and procedural types (Derry, 1996, p. 165), with declarative knowledge as encompassing factual information and procedural content covering how-to actions required to achieve a task.

Knowledge structures are built into different types of databases or repositories to emulate cognitive brain structures with node-link structures. "Consequently, it has been argued that an effective way for learners to articulate their knowledge representation is to construct their own hypermedia databases (Jonassen, 1992b) or through computer-assisted concept mapping (Dalgarno, 2001, p. 189). Understanding such structures, relationships and objects may provide a holistic understanding. Such knowledge structures not only present schemas but may express the information experientially. "Learners should build organizational patterns of association between primary concepts and affiliated parts through experience" (Dark & Winstead, 2005, p. 29).

Ontologies (rigorous often-comprehensive hierarchies of defined knowledge domains) define the various components of a system and the interrelationships between them in a knowledge domain. *Taxonomies* (classifications or arrangements of related objects in ordered systems) are the classifications of various objects (with their respective descriptions, identifications and naming). These tend to be formal structures created by subject matter experts instead of novices. Live real-time modeling involves the use of live data streams; these streams may be observed for anomalous patterning as part of monitoring, inspection, exploration and projection of potential risks (D'Amico & Kocka, Oct. 2005, p. 110). Those data streams may be combined with artificial or simulated elements for "mixed reality" interactions and experiences. Interactive live knowledge structures may be built through interactive role plays, strategic games, group simulations, and game theory decision-making. Some decision support systems are knowledge structures with convenient user interfaces to tap into those systems for information and directions. Computer interfaces (and equipment or system "dashboards") should align with the mental model. For example, a user interface of a particular piece of equipment should reflect conceptual model of that equipment. A digital simulated dashboard of a virtual vehicle should emulate that of the physical one. How the information is portrayed affects how people perceive, learn and remember it. Multimedia involves an "integration of metaphors of different medium" (Lei, Yang & Zhang, 2006, n.p.). Disjunctures in interface design may confuse the conveyance and acquisition of the learning in the model.

Digital Content Building

Digital learning objects form critical pieces in the building of mental models—especially given their interchangeability and reusability among technological systems. Learners both consume digital contents in the experiencing of conceptual models and also create / co-create these to express their own mental models.

Hypertext. Hypertext files, as sign systems and forms of speech, play an important foundational role in the building of individual mental models. "Vygotsky argued that the higher mental functions rely on the mediation of behavior by signs and sign systems, the most important of which is speech. Vygotsky saw signs as a special type of stimuli that are used as 'psychological tools,' tools that are 'directed toward the mastery of control of behavioral processes—either someone else's or one's own (1981b, p. 137, as cited in Hung, May – June 2001, p. 34). Hypertext consists of chunks of textual information (nodes) with groups of words acting as automatic links to other chunks (McKnight, Dillon & Richardson, 1991, as cited in Dalgarno, 2001, p. 186). Learners prefer a variety of textual representations for their learning (Fountain, 2007, pp. 44 - 45). New electronic writing forms have emerged, with the wide proliferation of IT—in collaborative spaces and in e-books (Nelson, 2008, p. 1). Some "adaptive hyperbooks" are collated based around metadata, and the books' pages are automatically aggregated web pages. Textual diversities may be helpful in supporting learner creativity (Surowiecki, 2004, as cited in Fountain, 2007, p. 45).

Visuals and informational graphics. Digital visuals—which may be static or dynamic and interactive—enhance learners' abilities to conceptualize complex ideas. Static images include still photos, sketches, blueprints, diagrams, timelines, tables, icons, photo montages, and screenshots. Images may be representational and descriptive. Dynamic visuals involve movement—like image maps, screen casts, fractals, animated agents, and interactivity—involve user interaction with the image for deeper understanding. Live images may include webcam feeds, acoustical imaging, satellite imageries, and other such captures; the informational streams may be evolving over time. The visual information may exist both in real-time, the past, and potentially continuing longitudinal time. The visual information may be holistic and macro, or it may be partial and focused. Images tend to be visceral and affective. They may have low, medium or high fidelity. They may be realistic or theoretical, or conceptually mixed (ranging from photo-realistic to fantastical). The visuals may be plain or aesthetically stylized. Sequence diagrams show complex narrative sequencing. 3D visualizations allow users "to see the created VE" (virtual environment) as the end user would (Winterbottom & Blake, 2004, p. 69).

Audio files may involve speeches, foreign language drills, musical files, and sounds that authenticate an experience. Rare sounds may be made more widely accessible. Sounds, built on differing tracks, may be mixed to evoke a more complex environment. A rich range of digital video types have been captured for e-learning. These may be videos of live labs, lectures, panels, events, demonstrations, simulations, and field trips. Live interactive television may be captured. In immersive digital game and simulation spaces, *machinima* (machine animation and cinema) may be captured for analysis, learning, digital archival and re-packaging.

VIRTUAL IMMERSION

IT enables the creation of full contexts for the building of mental models. Persistent immersive worlds offer 3D engagements with the virtual environment and live real-time interactivity with human-controlled avatars. Simulations combine powerful visuals and sound with artificial intelligence (AI)-driven non-player agents for social re-creations; they use complex programmed robots to imitate ever-more-complex environments. The virtual world flows into the real with wireless fidelity (wi-fi) connectivity into ambient spaces, with multimedia augmentation through mobile devices, laptops and head-mounted sets. GPS-enabled functions enable the location of individuals within a full three-dimensional real space and the activation of particular effects and digitally-mediated experiences overlaid over real-space. Effective mental modeling in the context of constructivism should take place in "authentic and real-world environments" (Doolittle, 1999, n.p.). In some cases, there's nothing quite like being physically in designed real-space.

Digital Context Building

A digital learning context must be persistent over a period of time or into practical perpetuity. These environments must have proper augmenting cues to aid in the reinforcement of important learning points. A constructivist context for mental modeling may involve the simulated environment: outdoors or indoors, photo-realistic or not, with selective fidelity highlighting the relevant aspects of the environment. The context may involve the presence of others acting in particular roles or identities. It may involve the sharing of values, ideas, practices, and cultural assumptions. The actions in the context may be both planned and improvised for learning value. The environment offers social cues for intercommunications, interactivity, exploration, and actions. IT, with the 3D immersive spaces, multi-sensory communications channels, and wrap-around learning / course management systems (L/CMSes), may offer a variety of field-specific learning, with situated cognition, situated actions and decision-making. Many systems even allow individual user tracking to follow individual user behaviors, to further enhance the customized support for their respective learning.

Virtual learning environments. Virtual learning environments, often using virtual reality (VR) technologies, LMSes and immersive spaces, promote learner-initiated self-discovery learning as well as high interactivity among mediated learners. Zeltzer (Winter 1992) suggests that "autonomy, presence, and interaction" are the main affordances of VR (Jackson and Fagan, 2000, pp, 84 – 85). Virtual reality may support visual metaphors, which may be an important part of mental modeling. "Other affordances include: (1) Allowing students to gain a greater understanding of abstract concepts through the creation of visual metaphors; (2) Allowing students to directly manipulate and scale virtual objects or environments for clearer understandings; and (3) Allowing students to visit places and interact with events that distance, time, or safety concerns would normally prohibit" (Jackson & Fagan, 2000, pp. 84 – 85). Other such environments may involve mock conferences and tradeshows, with both automated (slideshows, voice files) and live (audio-video presentations) information delivery.

Immersive simulations. Simulations and games often offer iterative experiences; they may be experienced again and again, and they are scalable for various numbers of users. Modifications (mods) may be made to the digital experience for variance and updating. Simulations provide operational models of some aspects of the real world. Beedle and Wright identify four higher-order thinking areas from the research literature in which gaming may have powerful effects on learners: motivation, problem-solving, communication and creativity (2007, pp. 153 – 155), elements which support constructivist co-learning. These digital experiences may simulate situations, devices, interactions, processes or environments. Sims may feature changing sets of relationships among various variables that show authentic causes and effects, with algorithmic modeling. Some simulations use live multivariate data feeds from remote sensors, cameras, labs, radar, scanners, satellites or people to feed information to those inside the "emergent situativity" or evolving simulation. Others have designed stochastic ranges, so each experience with the simulation is variable. Many of the complex ones are full-sensory and immersive.

Higher-end simulations have focused on workplace environments, organizational processes, problem-solving, fault diagnoses of machines, foreign language interactions, job interviews, and leadership scenarios. Epidemiological human or animal infection paths are expressed as fractals. Molecular interactions are modeled at an atomic level. Numerous system types have been emulated: electrical, organizational / bureaucratic, natural / ecological, cultural, societal, architectural, and historical. On the simpler lower-cost ones, there are digital "wetlabs" (for investigative learning) and human-mediated role-plays and case studies and dramatic recreations of historical events. Learning here may be cognitive, affective and psychomotor; it happens when participants internalize the processes within the system based on their interactions and experiences. Simpler simulations have involved guided tours of various locales and experiences—based on the real and the fantastical.

Immersive Games. Some digital games fall into this realm of immersive digital environments. Here there are full-sensory digital surrounds. Online games may be designed as microworlds with social learning in the interactions between live individuals and artificial agents (non-player entities). Immersive games often deploy artificial intelligence 'bots (robots) or agents "with its own goals, sensory inputs, reasoning, reaction, and social ties (Li, Musilek & Wyard-Scott, 2004, p. 734), and embedded beliefs, desires and intentions that guide their online actions (p. 734). Such games are about exploration, socialization, achievement, and dominance / control over others, according to Bartle's game motives. Others enjoy the challenges of "hard fun," the tasks, digital sensations and discoveries. "Gameplay" refers to "all the *doing, thinking* and *decision-making*" that makes a game fun or not, according to Prensky (2002, p. 9). "Meaning that is derived from experience is powerful because it is fundamentally self-referent; that is to say it is deeply rooted in personal identity and viewing life from the inside out in the context of social systems" (Dark & Winstead, 2005, p. 29).

Augmented reality. E-learning technologies are moving into real space with the use of ambient intelligence, mobile devices (handhelds, laptops, personal digital assistants, and VR devices). Here real spaces are mediated with the use of additional information provided through various wireless devices. Ambient intelligence refers to electronic environments that respond to the presence of people. These may offer dynamic representation. Spatial nodes may be activated when users arrive at particular geographic regions. These technologies offer richer learning opportunities and ubiquitous (ubi-) learning environments. Some examples of augmented reality learning involve trainings for close-body combat in war zones using a complex mix of laptop backpacks with wearable head-mounted sets in specially designed four-walls environments that combine full-sensory details. Others involve GPS-enabled treasure hunts and real-world tours with agricultural or natural environmental overlays. These ambient spaces may involve three-way relationships among humans, robots and the world for "the construction of shared grounds" for efficiency, collaboration and an "affective relationship" (Hwang, Lee, & Kwon, 2005, p. 623). Interfaces now may involve speech or gesture ones for future mobile or ubiquitous situations (Krum, Omoteso, Ribarsky, Starner, & Hodges, Oct. – Nov. 2002, p. 411). Multimodal interfaces—for example, those that combine speech and gesture—may be employed.

Haptics. Various devices now are used with 3D virtual interfaces to create real-space experiences. One example is through the use of haptics or touch-devices. Braille devices include haptics keyboards. Software makers have created interfaces which emulate the wielding of surgical devices to simulate the grinding of bones for hip replacement surgery or a scalpel for surgical endeavors.

Larger-motion devices may enable the creation of embodied mental modeling, with the building of muscle memory through movement-detection mobile hardware and software devices in wi-fi spaces.

HUMAN INTERACTIVITY

For all the focus on technologies, human collaboration plays a central role in implementing constructivism and some types of mental modeling. Various activities, such as "negotiation, mediation, and experimentation," promote the creation of individual mental models (Moreno, et al., 2007, p. 105). The focus of learning is not on regurgitating objective knowledge but "the creation of knowledge that reflects their comprehension and conception of the information"

(Makkonen, 2000, pp. 1-2). Learning should involve social negotiation and mediation (Doolittle, 1999, n.p.) among co-learners as novices and experts. Collaborative learning emerges as learners interact (Shen, Hiltz, & Bieber, 2006, p. 1045). Mediated communications technologies enhance virtual collaborations—that may be totally user-driven or facilitator-guided. In a sense, in cooperative inquiry, there may never be a final point of arrival. Some projects may be open long-term collaborations without clear end points (Fountain, 2007, p. 45). The human mediation of learning experiences in IT-enabled e-learning adds a competitive advantage and flexibility in terms of learning activities and focuses. As socio-technical systems, virtual communities' successes may be measured in multi-dimensional ways: the quality of transactions, the benefits to individuals in terms of "information, support, friendship (and) distraction"; the recruitment of members; the retention of members, and the production of a shared product (Kraut, n.d., p. 10). Long-term virtual community planning must identify user motives and encourage their altruistic participation (Wagner & Prasarnphanich, 2007, p. 5).

Mediated Community for Constructivist Learning

Mediated communities for learning focus on the role of facilitator leadership and follower-ship in such environments. Both must play active roles in maintaining a productive and mutually beneficial virtual community. These are communities of practice (CoP), in which people interact through dialogue and collaborations. Such communities are described as "self-organizing networks of people dedicated to sharing knowledge" (Breu & Hemingway, 2002, as cited in Watson & Harper, Feb. 2008, p. 4), or "networks of practice". Through their engagements, the members of the communities develop new ways of seeing and immerse in the community "beliefs, norms and dispositions" (Hung, 2001, p. 33). Knowledge sharing lies at the heart of such collaborative communities. New members do not have to attain expertise to be enculturated into the respective professions but are made welcome early on (Brown & Adler, 2008, p. 20). The internalization of the interactivity in such zones hones the varied mental models of the participants.

"Versions of constructivism all focus to some extent on interaction, either within an individual in terms of the interaction between existing knowledge structures and new information, or between individuals in a society, jointly constructing knowledge" (Winterbottom & Blake, 2004, pp. 67-68). Dialogue not only helps people process mental models internally by "explaining, clarifying, elaborating, and defending our ideas" but also to co-build understanding (by "integrating, elaborating, and defending" ideas) with others in social spaces" (Pena-Schaff & Nicholls, 2004, as cited in Dron, 2007, p. 33). Control is negotiated through dialogue, which serves as self-expression, negotiation and collaboration. For far-transfer learning, abstraction (as compared with specificity) tends to make the learning more transferable (Hung, 2001, p. 35).

Social computing makes possible a rich variety of human interactivity (Wang, et al., 2007, p. 79). Software tools enhance user abilities at creating individual and group telepresence (digitally-expressed identity and being); these support the expression of user self-construal. These tools enhance learner inter-communications, which may enhance the enlivening of curriculums. Avatars, as digital *doppelgängers*, have special meanings to their humans. "...your character there is not merely a collection of screen art and software subroutines; it is, in large part, you" (Ludlow & Wallace, 2007, p. xv).

Facilitator roles. Instructors clearly need to stay cognizant of learners' emotional and cognitive states in order to adjust the learning environment, learning activities, and interactions. They need to design proper setups and debriefings of synchronous online experiences. They need to create a safe learning environment and tolerance for multiple points-of-view. They need to be able to elicit tacit learner mental models. Instructors would do well to use their own conceptual models (the teacher's representation of knowledge to be transferred) to define the anchors for the learning. "This conceptual model gives context and meaning to the target which further allows facilitation of knowledge through connection with mental models" (Brandt, 1997, p. 114). They need to direct learners to test their theories to see what happens and then to draw out the proper learning from those experiences. They may smooth relationships between people in order to promote more effective learning. They may highlight analogical relationships and metaphors. They need to enhance the sense of social trust in the group (Yang, Zhang, & Chen, 2007, n.p.). The discomfort implied in the mental dissonance—inconsistency, disagreement, contradiction, disjunction, and discrepancy in a situation relating to beliefs, perception, attitude and behavior / action—plays a key role in helping learners adjust their own internal mental models to the external simulations.

Small group work for co-construction. Small group work may be enhanced with a clear setup of the small group tasks. "In order to achieve collaboration, it is essential that the activity include: opportunities for face-to-face interaction; joint attention to ideas or materials, positive interdependence (each student's contribution is needed for the group to succeed), individual responsibility, interpersonal and small group skills, and group processing (Johnson, Johnson & Smith, 1991; Holubec, 1999; Cohen, 1994, Barron, 2003, as cited in Roschelle, Rosas & Nussbaum, n.d., p. 521).

Online tutors—human-facilitated and automated. Online tutors may support learning globally. Indeed, a number of colleges and universities outsource the tutoring to live human tutor services, which often train their tutors to follow clearly defined guidelines for engaging with learners. One central requirement is to understand learners' levels of understandings. It may require human finesse to facilitate "the process of capturing and describing the important features of an individual's mental model in a domain-specific arena, including the assumptions, key factors (conceptual, logical, physical) and their interrelationships (temporal, causal, spatial)" (Steiger & Steiger, 2007, p. 2). A human facilitator "assumes a major role in modeling the meta-cognitive thinking associated with the problem-solving process" (Savery & Duffy, 2001, p. 10). Scaffolding assumes considerations both for novices and experts, to keep both of these groups of outliers engaged.

Automated teaching agents and tutors enhance the scalability of the tutoring. Such agents may use natural language (artificial intelligence or AI) communications with pre-coded responses. Methods to profile the learner (based on their submitted information, test results and online behaviors) have enhanced the abilities of automated systems to respond to the individual and changing needs of learners. "From a course agent, each teaching agent obtains course material and course-specific teaching techniques and then tries to teach the material in the most appropriate form and pace based on the background and learning style of the student," write Shang, Shi and Chen (Summer 2001, pp. 5 – 6).

Case-based learning environments. Cases, based on situated action theory, may offer transferable learning. Real-world may enhance complex knowledge transfer with "(a) the use of multiple knowledge representations, (b) linking abstract concepts in cases to depict knowledge-in-use, (c) demonstrating the conceptual interconnectedness or web-like nature of complex knowledge, (d) emphasizing knowledge assembly rather than reproductive memory, (e) introducing both conceptual complexity and domain complexity early, and (f) promoting active student learning. These features have been used prescriptively to specify design elements for complex, multidimensional, and nonlinear environments such as hypertext and hypermedia" (Makkonen, 2000, p. 4). Both social negotiation and internal negotiations must occur for learning (meaning making) to happen, and some tools through which this may be achieved involve articulation and reflection (Jonassen, et al., n.d., n.p.).

Open-ended problem-based learning. One way that instructors have used e-learning for creating mental modeling is by engaging problem-based learning. This may be deployed for both individual learners and distributed groups. The problem-solving needs to be valuable and real-world. Problem-based learning is "a method that sits well within the ethos of constructivism that is learning as an active acquisition of ideas and an assimilation of those ideas into a framework that the learner already possesses…Students are not simple vessels to be filled up with facts; they need to be able to apply their knowledge in the real world and in circumstances that are not always predictable" (Newman, Daniels & Faulkner, 2002, n.p.). The way the projects are assessed then will require creative and variable approaches.

Collaborative wiki and blog techno. The adoption of wikis and blogs for higher education has been low, with percentage estimates in the single digits for their adoption (Du & Wagner, 2005, p. 1). Message board functionalities bring learners together to share different multiple viewpoints. Wikis allow individuals to share knowledge, "invite critique, present multiple points of view, and seek to change others' ideas" (Wagner & Bolloju, 2005, as cited in Watson & Harper, 2008, p. 3). Many use wikis in endeavors to keep documents live and updated (p. 4). Wikis have been used to provide learning-rich assignments, including the building of a town (as a shared mental model) (Lund & Smørdal, 2006, p. 42). Wikis have been used for research projects, enhanced analytical reading, knowledge management, concept mapping, presentation tools, and group authoring (Duffy & Bruns, as cited in Chao, 2007, n.p.). The use of blogs for the creation of shared collaborative mental models around a particular subject has also gained traction in higher learning. Blogs allow for the sharing of textual and multimedia objects publicly. Participants may receive feedback about their work from their peers, and blogs and wikis transform learners into "authors and publishers" in an empowering way (Driscoll, 2007, p. 10).

Synchronicity. Synchronous interactivity has become more multi-channeled with information technology advancements. Individuals may now connect with live voice chats, text chats, virtual audio-video classroom interactivity and live video via webcams. Archived digital materials may be brought into synchronous play. Interactive performances and collaborations on different continents occur using online platforms (Neumark, 2007, pp. 8 – 10). Guest speakers present to individuals, small groups, or even thousands in the audience, via mediated distances. Virtual team members collaborate live. Subject matter experts cooperate with apprentices in live, real-time and focus on shared digital objects. Live and transient data streams may be microcast or broadcast for information-sharing, collaboration and decision-making.

User-centered software designs focus on learner needs (Soloway, et al., 1996, p. 189). The ability to "mash-up" various interface companies allows for the sharing of data, services, resources and interface components (Floyd, et al, 2007, p. 2) and empower users with more flexible tools (Miller, 2005). Service oriented architectures allows for more collaboration and the "coupling of remote resources or services" for flexibility (Schroth, 2007, n.d.). Traditional educational technology values of "replicability, reliability, communication, and control" (Heinich, 1984) contrast with "the seven primary constructivist values of collaboration, personal autonomy, generativity, reflectivity, active engagement, personal relevance, and pluralism" (Lebow, 1993, p. 5, as cited in Savery & Duffy, 2001, p. 3).

CONCLUSION IT AND MENTAL MODELING

The e-learning and social technologies exist today to enable powerful development of mental modeling for particular domain fields and some learning objectives. Technologically, the most complex to build would be immersive simulations, with only well capitalized companies and rare institutions of higher learning able to pursue their development. However, the other types of mental model building are likely within reach of most line-level educators and instructional designers. With growing pedagogical sophistication and simpler technologies, they may create rich mental models with constructivist underpinnings on the ever-more sophisticated IT of today.

Information technologies have strengthened instructional abilities for mental modeling through enhanced abilities for information visualization, virtual immersion, and human interactivity. New technological systems strengthen learner self-discovery, knowledge construction, and digital content building. Virtual immersions promote full-sensory context building for complex learning. Human interactivity has been enhanced manifold for the building of learning communities. Numerous endeavors in the IT realm are moving forward to provide greater connectivity, multimedia richness, mobility, the interchangeability of digital learning objects, and standards-setting for digital portability. The push for anywhere, anyplace, anytime technologies will enhance the ability to deliver mental modeling through various situations in virtual and real spaces. The leveraging of mediated social networks strengthens collaboration around shared and divergent mental schemas.

IT has supported the redefinition of community around shared interests; it has empowered people to collaborate as cyborgs, a cybernetic organism that is a human being augmented by embodied and embedded technology. The value of qualitative research has expanded, with automated ways of collecting information and online surveys. Quantitative findings may be captured and presented in ever more enriched and complex ways for easier integration into mental models. A greater spirit of sharing, diversity of thinking and more open intellectual property practices have permeated the public sphere. Higher education itself has shown a greater acceptance of born-digital information and e-journals as valid, and many faculty members have demonstrated a higher comfort level with emergent curriculums. Open source sharing of knowledge through wikis, blogs, "deep Web" tagging and information surfacing, and born-digital publishing have challenged people to think about the evolution of knowledge and what information is reliable or important, and why. IT has enabled mental modeling in e-learning and expanded real-space modeling of our shared universes.

ACKNOWLEDGMENT

Thanks to Scott Finkeldei, Phyllis Epps, and Dr. Rebecca Gould at Kansas State University for running a creative shops. My colleagues Joshua Works, Andrea Mendoza, and Brent Anders, bring thoughtfulness and massive talents to their work, one project after another. Thanks to Julie Bell for her gift of helps. I am glad for all my faculty and administrative clients, who help take this work to a higher level. My gratitude to R. Max.

REFERENCES

Abrahamson, D., diSessa, A. A., Blikstein, P., Wilensky, U., Uttal, D. H., Amaya, M. M., Mariulis, L. M., & Collins, A. M. (n.d.). *What's a situation in situated cognition? – A constructivist critique of authentic inquiry* (pp. 1015-1021).

Acevedo, D., Vote, E., Laidlaw, D. H., & Joukowsky, M. S. (2001, Oct.). *IEEE Visualization 2001* (p. 493).

Al-Khalifa, H. S., & Davis, H.C. (2007). Replacing the monolithic LOM: A folksonomic approach. *Seventh IEEE International Conference on Advanced Learning Technologies (ICALT 2007).*

Aragon, C. R., & Hearst, M. A. (2005, Apr.). Improving aviation safety with information visualization: A flight simulation study. *CHI 2005* (pp. 441-450). ACM.

Argyris, C. (1993). *Knowledge for action: A guide to overcoming barriers to organizational change.* San Francisco: Jossey-Bass Publishers. 20.

Beedle, J. B., & Wright, V. H. (2007). Perspectives from Multiplayer Game Players. In D. Gibson, C. Aldrich & M Prensky (Eds.), *Games and simulations in online learning: Research and development frameworks* (pp. 153-155). Hershey: Idea Group.

Ben-Ari, M. (1998). Constructivism in computer science education. *SIGSCE 98* (pp. 257-261).

Brandt, D. S. (1997, Oct.) Constructivism: Teaching for understanding of the Internet. *Communications of the ACM, 40*(10), 112-117.

Brown, J. S., & Adler, R. P. (2008, Jan. – Feb.). Minds on fire: Open education, the long tail, and learning 2.0. *Educause Review. 20 & 30.*

Chao, J. (2007). Student project collaboration using wikis. *20th Conference on Software Engineering Education & Training.*

Chieu, V. M., Milgrom, E., & Frenay, M. (2004, Sept.). Constructivist learning: Operational criteria for cognitive flexibility. *Proceedings of the IEEE International Conference on Advanced Learning Technologies (ICALT '04)* (pp. 221-225).

Crapo, A. W., Waisel, L. B., Wallace, W. A., & Willemain, T. R. (2000). *Visualization and the process of modeling: A cognitive-theoretic view* (p. 219). Association for Computing Machinery, Inc.

Dalgarno, B. (2001). Interpretations of constructivism and consequences for computer assisted learning. *British Journal of Educational Technology, 32*(2), 183-194.

D'Amico, A., & Kocka, M. (2005, Oct.) *Information assurance visualizations for specific stages of situational awareness and intended uses: Lessons learned* (p. 110). Workshop on Visualization for Computer Security. IEEE.

Dark, M. J., & Winstead, J. (2005, Sept.). Using educational theory and moral psychology to inform the teaching of ethics in computing. *Information Security Curriculum Development Conference '05* (p. 29). Association for Computing Machinery, Inc.

Derry, S. J. (1996). Cognitive schema theory in the constructivist debate. *Educational Psychologist, 31*(3/ 4), 163-174.

Doolittle, P. E. (1999). Constructivism and online education. Virginia Polytechnic Institute and State University.

Driscoll, K. (2007, May – June). Collaboration in today's classrooms: New web tools change the game. *Multimedia & Internet @ Schools* (pp. 9-12).

Dron, J. (2007). *Control and constraint in e-learning: Choosing when to choose,* 33. Hershey: Idea Group Publishing.

Du, H. S., & Wagner, C. (2005). Learning with weblogs: An empirical investigation (pp. 1-9). IEEE.

Ebner, M. (2007). E-Learning 2.0 – e-Learning 1.0 + Web 2.0? *Second International Conference on Availability, Reliability and Security (ARES'07)* (pp. 1-5).

Floyd, I. R., Jones, M. C., Rathi, D., & Twidale, M. B. (2007). *Web mash-ups and patchwork prototyping: User-driven technological innovation with Web 2.0 and open source software* (p. 2). IEEE.

Fountain, R. (2007). *Conceptual enhancement via textual plurality: A pedagogical wiki bow towards collaborative structuration* (pp. 43-46). Association for Computing Machinery, Inc.

Götschi, T., Sanders, I., & Galpin, V. (2003, Feb.). Mental models of recursion. *SIGCSE '03* (pp. 346-350).

Hendler, J. (2007). The dark side of the semantic web. A letter from the editor. *IEEE Intelligent Systems* (pp. 2-4). IEEE Computer Society.

Hung, D. (2001, May – June). Design principles for Web-based learning: Implications from Vygotskian thought. *Educational Technology* (pp. 33-41).

Hwang, J.-H., Lee, K. W., & Kwon, D.-S. (2005). The role of mental model and shared grounds in human-robot interaction. *2005 IEEE International Workshop on Robots and Human Interactive Communication* (pp. 623-628).

Jackson, R. L., & Fagan, E. (2000). Collaboration and learning within immersive virtual reality. CVE2000 (pp. 83-92). Association of Computing Machinery, Inc.

Jonassen, D., Davidson, M., Collins, M., Campbell, J., & Haag, B. B. (n.d.) Constructivism and computer-mediated communication in distance education.

Jones, P. H., & Popken, D. A. (n.d.). *Interactive prototyping methodology for a domain dependent user interface* (p. 814). US Government.

Kraut, R. (n.d.). Analyzing & improving online communities. *Educause*, (pp. 1-45).

Knights, M. (2007, Feb. – Mar.) Web 2.0. *IET Communications Engineer* (pp. 30-35).

Krum, D. M., Omoteso, O., Ribarsky, W., Starner, T., & Hodges, L. F. (2002, Oct. – Nov.) *Evaluation of a multimodal interface for 3D terrain visualization* (p. 411). IEEE.

Lawrence, K. F., & Schraefel, M. C. (2006). *Freedom and restraint: Tags, vocabularies, and ontologies* (pp. 1745-1750). IEEE.

Lei, T., Yang, Y., & Zhang, Y. (2006). The usability of multimedia interface based on user's mental models. *Proceedings of the 16th International Conference on Artificial Reality and Telexistence—Workshops (ICAT '06).*

Li, Y., Musilek, P., & Wyard-Scott, L. (2004). Fuzzy logic in agent-based game design (pp. 734-739). IEEE.

Lippincott, J. K. (2007, Nov. / Dec.). Student content creators: Convergence of literacies. E-Content. *Educause Review* (pp. 16-17).

Ludlow, P., & Wallace, M. (2007). *The Second Life Herald: The Virtual Tabloid that Witnessed the Dawn of the Metaverse.* Cambridge: The MIT Press.

Lund, A., & Smørdal, O. (2006, Aug.). Is there a space for the teacher in a WIKI (pp. 37-46). Association for Computing Machinery, Inc.

MacManus, R., & Porter, J. (2005, May). Web 2.0 for designers. *Web 2.0 Design: Bootstrapping the Social Web.*

Maia, L. P., Machado, F. B., & Pacheco, A. C. (2005, June). A constructivist framework for operating systems education: A pedagogic proposal using the SOsim. ITiCSE '05 (pp. 218-222).

Makkonen, P. (2000). Do WWW-based presentations support better (constructivistic) learning in the basics of informatics? *Proceedings of the 33rd Hawaii International Conference on System Sciences.*

McKenna, P., & Laycock, B. (2004). Constructivist or instructivist: Pedagogical concepts practically applied to a computer learning environment (pp. 166-170). Association for Computing Machinery, Inc.

Merrill, M. D. (2000). Knowledge objects and mental models (pp. 244-246). IEEE.

Moreno, L., Gonzalez, C., Castilla, I., Gonzalez, E. J., & Sigut, J. (2007, May). Use of constructivism and collaborative teaching in an ILP processors course. IEEE Transactions on Education: 50 (2), 101-111.

Nardi, B. A., & Zarmer, C. L. (1991). *Beyond models and metaphors: Visual formalisms in user interface design* (p. 487). IEEE.

Nelson, M. R. (2008, Jan.). E-books in higher education: Nearing the end of the era of hype? *Educause Center for Applied Research: Research Bulletin.*

Newman, I., Daniels, M., & Faulkner, X. (2002). Open ended group projects a 'tool' for more effective teaching. Australian Computer Society, Inc.

Olssen, M. (1996, Sept.) Radical constructivism and its failings: Anti-realism and individualism. *British Journal of Educational Studies, 44*(3), 275-295.

Oulasvirta, A., Tamminen, S., & Höök, K. (2005). Comparing two approaches to context: Realism and constructivism (pp. 195-198). Association for Computing Machinery, Inc.

Prensky, M. (2002). The motivation of gameplay, or the REAL 21st century learning revolution. < http://www.marcprensky.com/writing/Prensky%20-%20The%20Motivation%20of%20Gameplay-OTH%2010-1.pdf>.

Riggins, F. J., & Slaughter, K. T. (2006). The role of collective mental models in IOS adoption: Opening the black box of rationality in RFID deployment.

Roschelle, J., Rosas, R. & Nussbaum, M. (n.d.). *Towards a design framework for mobile computer-supported collaborative learning* (pp. 520-524).

Sanderson, P. M., & Murtagh, J. M. (1989). Troubleshooting with an inaccurate mental model (pp. 1238-1243). IEEE.

Saraiya, P., North, C., & Duca, K. (2005, July – Aug.) An insight-based methodology for evaluating bioinformatics visualizations. IEEE Transactions on Visualization and Computer Graphics, *11*(4), 443.

Savery, J. R., & Duffy, T. M. (2001, June). *Problem based learning: An instructivist model and its constructivist framework* (pp. 1-17). CRLT Technical Report No. 16-01. Center for Research on Learning and Technology.

Schroth, C. (2007). Web 2.0 versus SOA: Converging concepts enabling seamless cross-organizational collaboration. *The 9th IEEE International Conference on E-Commerce Technology and The 4th IEEE International Conference on Enterprise Computing, E-Commerce and E-Services (CEC-EEE 2007).*

Shang, Y., Shi, H-C., & Chen, S-S. (2001, Summer). An intelligent distributed environment for active learning. *ACM Journal of Educational Resources in Computing, 1*(2), 3, 5-6.

Shen, J., Hiltz, S. R., & Bieber, M. (2006, Nov.). Collaborative online examinations: Impacts on interaction, learning, and student satisfaction. *IEEE Transactions on Systems, Man, and Cybernetics—Part A: Systems and Humans, 36*(6), 1045-1053.

Soloway, E., Jackson, S. L., Klein, J., Quintana, C., Reed, J., Spitulnik, J., Stratford, S. J., Studer, S., Eng, J., & Scala, N. (1996). *Learning theory in practice: Case studies of learner-centered design* (pp. 189-196). Association for Computing Machinery, Inc.

Stager, G. S. (2002, July). Computationally-rich constructivism and at-risk learners. Australian Computer Society, Inc. *CRPIT '02: Proceedings of the Seventh World Conference on Comput-*

ers in Education conference on Computers in Education, 8.

Stary, C., & Totter, A. (n.d.). On learner control in e-learning. *Trust and Control in Complex Socio-Technical Systems* (pp. 41-48).

Steiger, N. M., & Steiger, D. M. (2007). Knowledge management in decision making: Instance-based cognitive mapping. *Proceedings of the 40th Hawaii International Conference on System Sciences* (pp. 1-2).

Thomas, D. M., & Bostrom, R. P. (2007). The role of a shared mental model of collaboration technology in facilitating knowledge work in virtual teams. *Proceedings of the 40th Hawaii International Conference on System Sciences* (pp. 1-8).

Wagner, C., & Prasarnphanich, P. (2007). Innovating collaborative content creation: The role of altruism and wiki technology. Proceedings of the 40th Hawaii International Conference on System Sciences 2007 (pp. 1-10). IEEE.

Wang, F-Y., Zeng, D., Carley, K. M., & Mao, W. (2007, Mar. – Apr.). *Trends and controversies: Social computing: From social informatics to social intelligence* (pp. 79-83). IEEE.

Watson, K., & Harper, C. (2008, Feb.). Supporting knowledge creation: Using wikis for group collaboration. *Educause Center for Applied Research: Research Bulletin: 2008*(3), 2-13.

Winterbottom, C., & Blake, E. (2004). Designing a VR interaction authoring tool using constructivist practices (pp. 67-71). Association for Computing Machinery, Inc.

Yang, S. J. H., Zhang, J., & Chen, I. Y. L. (2007). Web 2.0 services for identifying communities of practice through social networks. *2007 IEEE International Conference on Services Computing (SCC 2007).*

Zauder, K., Lazić, J. L., & Zorica, M. B. (2007, June). Collaborative tagging supported knowledge discovery. *Proceedings of the ITI 2007 29th International Conference on Information Technology Interfaces* (pp. 437-442). Cavtat, Croatia.

Section III
International Approaches to Constructivism:
Case Studies

Chapter XII
Transformative Potential of Constructivist Blended Problem-Based Learning in Higher Education

Roisin Donnelly
Dublin Institute of Technology, Ireland

ABSTRACT

This chapter critically explores the design and implementation of a blended problem-based learning (PBL) module for academic professional development in higher education. A core aspect of the chapter is the overview of the design framework used for the application and specific use of learning technologies in the PBL module. As it would appear that E-Learning courses are often lauded on the basis of their constructivist approach to learning, but in reality sustained inter-student contact and discussion can be difficult, an underlying purpose of the chapter is to show how interactional analysis helps in understanding the potential of transformative pedagogy within blended PBL. This chapter aims to highlight how emerging constructivist theories of learning may be applied to the blend of PBL and E-Learning. It addresses the need for an analysis of the interactions taking place in blended PBL with a specific focus on academic staff who are engaged in professional development in higher education in Ireland. It applies the relevant constructivist theories to the face-to-face PBL tutorials, online discussions, focus group interview texts and reflective papers generated over two years in a professional development module involving 17 academic staff. The chapter concludes with an analysis of the difference that the blended delivery made to both tutors and participants and discusses the design implications of a blended PBL model for the practice of academic development.

INTRODUCTION

Some of the most innovative E-Learning applications used today tend to be among advocates of a constructivist approach to learning who have high levels of information and communication technology (ICT) skills, often in higher and adult education. Influenced by constructivist philosophy and new learning technologies, there is increasing interest among higher education faculty/academic staff in authentic activities as a basis for learning. In recent years, the integration of constructivist learning theory and problem-based learning (PBL) strategies with online scenarios and virtual role playing have taken student activities to the heart of the curriculum. This chapter explores the constructivist nature of blending face-to-face and online PBL to offer opportunities to educators in higher education seeking to confront pedagogical challenges in their on-the-ground practice.

It is useful here for the context of the study discussed in this chapter to make explicit definitions of the key principles involved: transformative pedagogy, problem-based learning and blended learning. From a definitional perspective, descriptions of transformative pedagogy originated in the adult education literature and Myers (2006) believes it has been regarded as an approach to teaching that encourages students to grapple with disorienting dilemmas, critically examine their assumptions related to the contradictory information, seek out additional perspectives, and ultimately acquire new knowledge, attitudes and skills in light of these reflections – all in order to experience personal and intellectual growth.

Many definitions currently exist in the literature for blended learning; however, the definitions seem to converge around the idea of synthesizing E-Learning with the more traditional forms of teaching and learning, drawing together the "e" with the classroom, the laboratory, the seminar and the tutorial setting. PBL is an educational strategy that involves the presentation of significant, complex and "real-world" problems to students that are structured in such a way that there is not one specific correct answer or predetermined outcome. The blended learning used in this study has been described as a form of complex blending in that it combines face-to-face and online PBL. Savin-Baden (2006) has concluded that the objective of combining PBL and E-Learning is in itself complex. She also recognises that this terminology is problematic since it offers little indication "about the ways in which technology is being used, areas where students interact, which tools are used, how learning materials are selected and applied and the extent to which any of these fit with PBL" (p. 4).

Issues, Controversies, Problems

Blended PBL is a growing field of study whose impact is becoming increasingly relevant in higher education in a period of continuing rapid change and within a set of global challenges. These challenges affect the teaching staff in higher education today, and as a result, the faculty/academic development experts who are charged with easing their transition through the sea-change in how education is delivered. There is currently little evidence of any comprehensive studies into the transformational potential of constructivist blended PBL in faculty/academic development; critical research to date has been sporadic at times but widespread and plentiful in addressing such issues as what is blended learning or PBL and how online PBL occurs.

Although the PBL tutorial is the central and key learning encounter in PBL curricula, and the integration of technology has increased in practice in recent years, surprisingly very little research has been done on what actually happens in blended PBL tutorials. The purpose of this study is to provide research-based information about the realities of delivering a constructivist PBL programme using technology. Central to this study is the role of interactivity and technology in supporting it in the contemporary PBL tutorial classroom, both real and virtual. In view of the constructivist approaches to education now increasingly coming to prominence, it is appropriate to enquire into the perceptions of the academic staff in order to further the effectiveness of interactivity in a blended environment. It is hoped that the presentation and exploration of issues faced by those working with blended PBL tutorial groups will provide fellow researchers and practitioners in the field with a rich text from which they can form their own ideas and research-based practice.

Purpose of the Study

In Ireland, and indeed in the host Institute where the study was conducted, as elsewhere, the use of online technologies has become an increasingly important challenge in academic staff development. There has been a demand for higher education institutions in Ireland to put E-Learning initiatives and the accompanying academic staff training and development firmly on their agendas. Informally enthusiasm among academic staff for E-Learning continues to grow and where explicit institutional policies are lacking, pressure on lecturers to engage with new technologies is coming from students and from their own peers. Alongside this, new pedagogical approaches emerge on the educational scene to support complex, flexible and integrated learning and the development of professional competencies. Although not new, PBL is one which appears to have captured the imagination and support of teachers; there has been a growing interest in the last few decades, particularly in the collaborative construction of knowledge through active learning and the importance of higher-order skills such as problem solving. Given our increasingly networked society, interest has grown in innovative conceptualizations of pedagogies such as problem-based learning and in where and when to use them.

Kirkup & Kirkwood (2005) reported that over the past 15 years the discourse about higher education teaching has been couched in terms of transformation of learning and revolution in technology. However, there has been a sense of disappointment that the transformative potential of technology is being missed or resisted. Transformative learning theory is being proposed in this study as a means to understand the complexities of education in an age where information and communication technologies (ICTs) are constantly reshaping and redefining our accepted notions of what it means to teach and learn in a higher education environment. This study contributes to the gap in the literature by illustrating how academic staff can implement transformative pedagogy such as PBL in their blended classes.

What perpetuated the need to have the participants engage in the blended PBL activities was the fact that many academic staff still lack the online experience of the Internet generation, and so do not feel as confident in an online environment as they do in a traditional classroom setting. In this context, the problem is a social rather than a pedagogic one and lecturers may need to experience being online students themselves in order to gain the necessary confidence to move to facilitating an online environment. Increasingly it is also becoming important not just to make such training more accessible, but explicitly designed to produce qualitatively improved pedagogy (Ham & Davey 2005, p. 263). It is important to ensure that the lecturer's first experience of an online environment is positive, one that will allow them to see the pedagogic possibilities at their disposal.

A blended rather than a fully online approach was chosen. While there are a number of successful fully online degrees reported on offer today such as those through 'Ultraversity', a distance learning subsidiary of Anglia Ruskin University, where all students are taught fully online. generally fully online qualifications suffer from low numbers of enrolled students and /or low retention rates, perhaps indicating that this is not always what the majority of students seek for their university education (Hoare, 2006). Arguably blended approaches have emerged as an alternative to fully online programmes.

The emphasis in this next section of the chapter is on an examination of the blended environment and PBL tutorial used on this postgraduate programme. It will explore the impact of blended PBL on the learners, particularly the impact of shifting back and forth between face-to-face (f2f) and virtual environments. The blended PBL approach to learning in this study was chosen because it is both varied and flexible and introduces questions about what it means to be a problem-based learner and tutor in both a classroom and online setting. Finally, consideration is given to whether the course design and not the technology caused both the benefits and the problems, or whether it was a blend of both.

Introduction to the Blended PBL Module

Issues related to the design and implementation of blended learning environments have increasingly surfaced in recent years, as technological advances continue to blur the lines between distributed learning and traditional campus-based learning. This has raised questions about advances in technology during that last decade that have brought challenges and opportunities to the ways in which individuals are educated and trained, in particular through blending online and f2f instruction.

The focus of the research reported here is a postgraduate programme in higher education learning and teaching which was developed in 2001, and has over 100 graduates today. The 'Designing E-Learning' module at the heart of this study is one of a number of ten week modules which carry ten ECTS (European Credit Transfer and Accumulation System) credits. The 17 participants on the module in this study were all either lecturers or educational support staff teaching in varied subject disciplines in higher education.

The programme grew out of identification of a specific need. The nature of these academic staff's varied work responsibilities today is complex, with demands on their time (ranging from lesson preparation, student support and research, to staff meetings, curriculum development) pulling them in many directions. In my experience as an academic developer, over a number of years, there have been a significant number of academic staff who have been inducted and trained to set up the basics of operations in technologically mediated learning environments across the Institute, but few who have been developed further in terms of pedagogical training. Indeed, of those staff who have expressed a wish to engage with E-Learning, on average more than half of those who attend an initial day-long introductory session do not proceed immediately to employ it as part of their practice, citing time constraints as the main inhibiting factor (Donnelly & O'Rourke, 2007). As a result of all the pressures academic staff face in today's higher education environment, Donnelly & O'Farrell (2006) have argued that for their own professional development, they need to be provided with streamlined learning experiences which deliver essential topics and learning materials in readily accessible formats. A central challenge was to create and sustain quality learning environments of enduring value for these teachers.

Motivation of the participants in the module was important. Consideration of motivational factors could have a large impact on module completion and work levels of participants. A pre module questionnaire had been distributed to all participants at the outset of the study to gauge their reasons for wanting to undertake a blended PBL approach to their learning. It is suggested here that the need to encourage engagement amongst academic staff with regard to their E-Learning and pedagogical professional development op-

Figure 1. Activities in the blended PBL module

Features of a Blended PBL Environment	Duration of Activity
Face-to-face PBL tutorials	10 x 3 hours
Between tutorials: researching, reading, planning, designing ideas	Over 10 weeks
Online reflective journal entries	1 per week x 10 weeks
Video conferencing session	3 x 1 hours
Asynchronous discussions	5 per week x 10 weeks
Synchronous chat sessions	10 x 30-60 mins
International guest tutor collaboration	3 x 1 week

portunities has never been greater. By so doing, participants would be enabled to experience, discuss and reflect on issues related to teaching and learning in a blended environment. This could thereby enable them to relate their understanding and practice to appropriate educational principles and key institutional policies. Whilst the first stage of the programme, the Postgraduate Certificate in Third Level Learning and Teaching is required to get tenure, all participants who proceed to the postgraduate diploma are self-selecting and choose to pursue this professional development opportunity themselves for personal learning.

Sharpe et al. (2006) identified transformative course level designs as one of the characterizations of successful blended E-Learning. A review of the literature showed that studies repeatedly acknowledged that engaging in course design or redesign was critical to success. The PBL approach in the module at the centre of this study claims to incorporate such transformative potential and can be summarised as including stages of problem identification, deconstruction, seeking and using knowledge and experience, understanding, thinking, choosing a strategy, acting and then critically evaluating and reflecting on the action.

However, there is no specific recipe for mixing up the ingredients of the blend of problem-based learning adopted; the amount of face-to-face, synchronous and self-directed work is prescribed by the learning outcomes of the module itself. Online communication plays a vital role and it is blended with event-based activities, the main ones being the PBL tutorial itself and self-paced learning. However, one of the most exciting opportunities afforded by blended learning experiences and which is central to PBL, is the creation of learning communities. As the development and availability of online tools for communication has led to a concomitant rise in the concept of an online community, Donnelly (2008) advocates that inherent within this is a discussion regarding the consideration of suitable technologies and media choices available in a fluid field such as academic development today. Figure 1 provides a breakdown of the ingredients in the blended PBL module and estimated time for completion of each activity.

In the module design, online components were communication, participation and activity completion and each needed to be considered equally as important as attending any of the live, face-to-face PBL tutorials. Otherwise, there may be a possibility of participants waiting for the live events to obtain the important information. This blended PBL module was designed based on a model originally developed by Oliver (1999) (illustrated in Figure 2).

From a pedagogical perspective, for an academic developer and a designer and tutor in blended PBL, it is important to seek best practices for how to combine instructional strategies in face-

Figure 2. The three learning design elements of the blended PBL module

Learning Design Elements	Description
Learning tasks	The activities, problems and interactions used to engage the learners, on which learning is based
Learning resources	The content, information, and resources with which the learners interact, upon which learning is based
Learning supports	The scaffolds, structures, encouragements, motivations, assistances and connections used to support learning

to-face and computer-mediated environments that take advantage of the strengths of each and avoid their weaknesses. It is acknowledged that the conjunction of PBL and new technologies which have the potential to facilitate the development of novel pedagogical solutions, implies a challenge for academic staff. For greater transformation of teaching and learning to occur, it is argued that lecturers need appropriate support to make the conceptual shift toward the generation of more sound technology-based learning activities in their practice.

The module used the WebCT course management system, which provided both asynchronous and synchronous interaction tools. For the former, the module had discussion forums where the participants posted their messages and its own email system that enabled the participants to exchange private emails. For synchronous communications, the module had chat rooms where individuals exchanged instant messages at the same time. By virtue of its online setting, the module web site was accessible from anywhere and at any time. Marratech software was used for video conferencing with guest tutors, and on one occasion, with a peer from the PBL group from abroad.

Along with course design, the tutor's role is readily recognized as central to the success of both PBL and E-Learning. In the E-Learning literature especially, the tutor is referred to as an e-moderator (Salmon, 2000). A hugely important area in any form of instructional delivery is the role of the tutor; in an online environment, it is even more crucial. Each year, two guest tutors from abroad are invited to join the online discussions over a three week period in the module. They are also fellow academic developers internationally, and for many years have been designing and delivering forms of academic professional development, using a combination of online and face-to-face strategies, some of which include problem-based learning.

It can be argued that the impact on the tutor is analogous to the impact a faculty member would find in a classroom setting. A tutor who values a cohesive, supportive and productive blended PBL class can accentuate exchanges of positive affect in learners; they can encourage collective and achievement orientations toward learning in students; they can show appreciation for the uniqueness of each particular learner; and they can facilitate open and diffuse discussions about the problem in a virtual and face-to-face learning environment.

LITERATURE REVIEW

In the past several years there have been studies conducted into the effectiveness of PBL in a virtual environment in higher education in a number of differing contexts, for example in Ireland (Donnelly 2006b; Jennings, 2006)), in the UK (Juwah, 2002; Valaitis *et al.*, 2005) and in Finland (Portimojärvi, 2006); there has also been recent work conducted into the traditional PBL tutorial in a staff development context (Barrett, 2008). Panda & Juwah (2006, p. 207) note that the increased use of the web for learning and teaching has "necessitated a re-examination of some of the issues with E-Learning and the professional development of academic staff engaged in an online facilitation role".

Masie (2006) contends that blended learning has always been a major part of the landscape of training, learning and instruction; however, in reality it has only become a noticeable feature of the higher education sector in Ireland since the beginning of this new millennium. There has been a community of practitioners studying and working in blended courses for many years, nevertheless research in the main has been inadequate.

A central concept of the literature review in blended PBL is how students shift between online and f2f environments and how they engage in learning and collaborate within the two. This is particularly important to explore within the framework of supporting transformation.

As PBL is consistent with a social constructivist learning pedagogy, one of the challenges faced is to integrate E-Learning technologies into PBL where it is appropriate, and where it supports the learner. Within PBL, Ronteltap & Eurelings (1997) classified PBL activities for learners into two sets and it is based upon these that E-Learning technologies were integrated into the module. Information-related learning activities benefit greatly from the vast array of resources available on the Internet to support the research and resource-based searching, selecting, collecting and presentation aspects of the PBL problem. It is argued that the skills of analysis and synthesis still need to be developed within the individual learner and cultivated in the PBL group. Communication and collaboration activities included peer, tutor and international guest expert communications and all support the participants in their questioning, challenging and constructing of knowledge. A third set of activities deliberately integrated into this module are reflective in nature. In a knowledge construction process with learner intention and self consciousness, Jonassen & Land (2000) suggest that learning activities need the support of reflection and self-regulated learning and the module requires that participants have both the consciousness and capability for both.

Hanson & Clem (2006), Hofmann (2006), and Owston et al. (2006) all provide evidence for learners preferring live interactions over those that are computer-mediated in a blended learning experience. It is suggested that these findings are credible because there is a pervasive argument in the literature that there is a synergy that is manifest in face-to-face contact that the computer cannot replicate. Donnelly (2006a) has suggested that discussion forums, chat rooms and email are an adjunct to the PBL group experience by providing the spontaneity and momentum achieved face-to-face in the classroom. However, in contrast to this, Offerman & Tassava (2006) make the claim from their research that the live components are unnecessary and primarily used for socialisation reasons. This begs the questions, when and why should academics be considering human interaction such as collaboration and learning communities, and how live interaction versus what has been coined by Graham (2006) as *low-fidelity* (asynchronous) interaction affects the problem-based learning experience.

This study recognises that there is still confusion about the models, media and environments used to support PBL that use technology in some way, and is particularly concerned with illuminating current knowledge on PBL group-oriented interaction. Interaction between participants in the PBL group is the key element to a successful blended learning experience for all involved. This is based on a sociological understanding of one of the dimensions of interaction for describing groups, coined by Wagner (2006) as "interactions as transactions". Interaction has been and continues to be one of the most hotly debated constructs in the realms of distance and E-Learning, instructional design and academic transformation, to name but three. The ability to interact - with tutors, students, content interfaces, features, code, channels and environments - can be argued to be analogous to being connected. Whilst this may appear simplistic, for technology-mediated learning, interaction is undoubtedly a key value proposition. It continues to be perceived as the defining attribute for quality and value in a blended learning experience. Interactivity is the core of learning, and is evident at all levels of engagement. However, the term *interactivity* is used so loosely that in the fields of E-Learning and blended learning, it has become almost synonymous with the notion of learning itself. By bringing the concept into sharper focus here, real insight will be gained into the nature of blended PBL.

Interaction has been explored in the literature at three levels: interaction with concepts, tasks and people (peer learners and tutors). These three levels have been previously represented in a popular framework for interactive learning by Mayes & Fowler (1999). However, it is suggested that a case can be made for proposing a new dimension of interaction that focuses on the blended PBL interaction activity experience. This was based upon recognition that transformative learning is a complex process of interaction between people, the tools they use and the context in which they are embedded. Today, educators have more choice than ever when it comes to selecting types of interactions to include in their blended courses; unfortunately they often do not even know the potential of the tools that are available to them or how to use then effectively. To further confound the situation, the use of interactive technologies alone does not ensure that meaningful interactions will occur in a blended course. All decisions regarding types of interaction in a blended course should be driven by pedagogical principles and grounded in research.

Research Study: Solutions and Recommendations

There were two research objectives in this study:

1. To establish, in a PBL tutorial setting the factors that govern the success of blended problem-based learning in order to establish participant learning as transformational;
2. To examine the participants' perspectives on how an E-Learning approach to PBL impacts their own learning and practice.

THEORETICAL FRAMEWORK

Interpretivism was both the paradigm and the methodology for this study. Interpretivism seeks to understand the complex world of lived experience from the perspectives of the participants. As a methodology, it has the goal of analysing, understanding and interpreting the participants' discussions in the face-to-face and online PBL tutorials from their own perspective. The interpretivist emphasis on meaning and the relationship between language and meaning is addressed in the concept of discourse, which in this study is a web of statements, categories, beliefs and practices. The aim of this approach was to explore the research objectives in ways that made connections between the words from the videos and online discussion forum transcripts, the social functions these words perform and wider social practices.

The analysis of the literature indicated that qualitative data was needed as evidence of the changes that occurred in the nature of blended PBL in the context of academic development in higher education. This decision was arrived at from recognition that transformative learning would be the outcome of a complex process of interaction between people, the tools they use and the context in which they are embedded. I was aware of the difficulty in finding specific examples of participants' behaviours or learning that could be identified as transformative. Instead, I was looking for fundamental shifts in participants' orientation towards new knowledge which I expected might manifest itself in changes in their engagement in activities with their peers and tutors.

Interpretivism was chosen to identify the "essence" of human experiences concerning the phenomenon of blended problem-based learning as a model of academic development, as described by the participants in this study. It involved a small number of subjects in extensive and prolonged engagement designed to provide a basis on which to understand their lived experiences and develop what Creswell (1998) calls patterns and relationships of meaning. Consequently, the motivation of this research was to gain a rich insight into the problem-based learning nature of the blended environment rather than focus on the statistical analysis of quantitative responses by participants.

Participants

The sample for this study was the total population (17 participants) of blended PBL groups undertaking the PBL module in the two years the data was collected, in order to explore the lived experience of a heterogeneous population of academic staff in higher education; there were three PBL groups in this study, and all participants completed the ten week module. Figure 3 shows the breakdown of the background of the participants.

METHODS

The general research design was to observe the process of learning on the module in some depth. In order to ensure that sufficient observational data had been collected for a thorough analysis of what was occurring in the module, three complementary methods were chosen to provide the data relating to the experiences of the participants, in addition to my more obvious tutor role in the module delivery. The research methods employed to

Figure 3. Sample for the study

Number	Attribute
17	part-time postgraduate learners in total
9	had competed a PBL module previously
5	had prior experience of blended learning
8	males
9	females
15	subject disciplines in higher education represented: Group 1 (psychology, social science, culinary arts, information literacy, adult literacy); Group 2 (Biology, Apprentice Plumbing, Apprentice Joinery, Apprentice Metalwork, Adult Literacy); Group 3 (Architecture, Marketing, Culinary Arts, Refrigeration, Printing, Fine Art, Chemistry).

collect face-to-face and online observational data were participant observation, online discussion logs, open-ended focus group interviews and self-reflective papers to capture the participant's own thoughts and experiences. Participant observation was the primary method of data collection and occurred with the three blended PBL groups in the module, using video recordings of the face-to-face PBL class tutorials and video recordings of the participant/guest tutor synchronous video conferencing sessions. The PBL tutorial observations for each group were transcribed. The face-to-face classroom and online observation was complemented by two focus group interviews for all three groups. It was important to observe the groups over the complete ten week period of the module in order to examine how the groups negotiated the problem face-to-face and online, how the group dynamics worked in the blended environment, how the life of the group unfolded thereafter and what influenced the participants reaching a transformation in their learning.

Each method was chosen for the opportunity it could offer to explore interactions and dimensions of transformation, both of which were central to this study. A main concern has been to provide meaningful and accessible insights into the practice of blended PBL based on the analysis of real-life situations. There were two levels in the analysis of the data. Level One was descriptive in nature and through video observations explored the interactions between the peers, the tutors and the content of the blended PBL tutorial. Level Two was a thematic analysis of transformative learning in blended PBL.

Through the combination of methods, categories and themes emerged to inform the findings of the study and implications for practice. It was important for the researcher to be engaged with the events as they happened in the field and to bring holistic attention to the practices as constitutive of a distinct culture. As suggested by Hine (2000, p. 20), this study has examined those enduring practices through which the blended PBL groups have become meaningful and perceptible to participants.

DISCUSSION OF FINDINGS

Validity was considered through triangulation of data from the multiple sources (participant observation, textual analysis and focus group interviews) and comparison with other models of academic development; this provided support for assumptions and conclusions. The multiple sources of data provided a rich description of the situation, the factors that impacted on success in a blended PBL tutorial setting and on transformation of learning and practice, along with identification of areas for possible improvement to module design. Presentation of data from all sources allows readers to focus on aspects of particular interest and to draw their own conclusions about the validity of findings.

Whenever possible by using the words of the participants themselves (in line with the ethical considerations adopted for the study, pseudonyms are used throughout), key issues will be highlighted. For inclusion of participant quotes, the following applies: RP = reflective paper.

Module design (illustrated in Figure 4) emphasises what Uden & Beaumont (2006) have called the close alignment between PBL and ICTs, specifically in the areas of resources, interactive tasks and support. Kiser's research (2002) identifying five core elements of blended learning (use of scenario-based exercises, integration of learning objects, early use of knowledge and skills, access to live mentors and assessments mimicking real world tasks) maps onto PBL and Uden & Beaumont (2006) believe that this reinforces the argument that PBL and E-Learning technology can indeed be integrated successfully to the benefit of learners. Furthermore, to summarise, they assert that "the student-centred, task-focused, research-based, collaborative learning characteristics of PBL make it a particularly suitable environment in which to blend technology" (p. 207).

Figure 4. A design of the blended PBL module

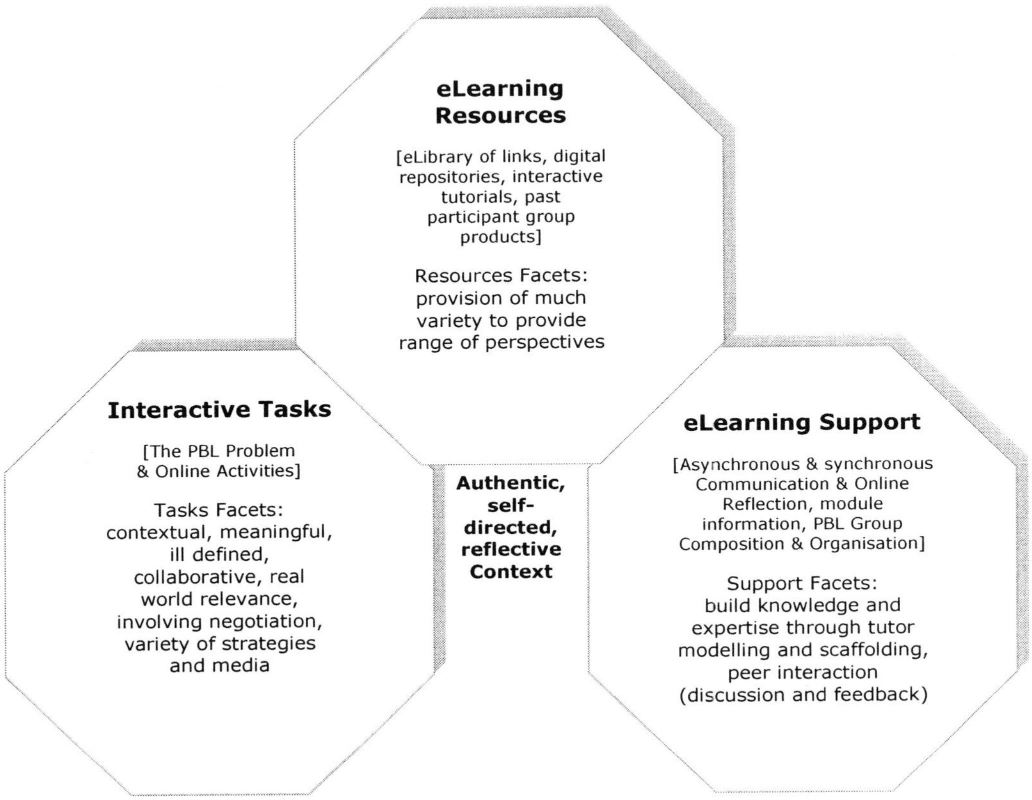

A critical factor in the success of PBL is the provision of appropriate resources for active research and location of information for the solving of the PBL problem. Organisational support information about the module includes: explanation of PBL and links to some key PBL sites; roles and responsibilities of the participants; the group composition; expectations of time to be spent online individually and in the PBL group; milestones, deliverables and deadlines for the PBL problem; negotiated group ground rules based on netiquette; a database of frequently asked questions; links to relevant interactive tutorials on web researching, interpersonal communications, conflict resolution, self and peer assessment strategies within PBL; and participant home pages. The intention is to provide a dynamic website, to complement the f2f PBL weekly tutorials and adapt to the participant group experience and progression of the PBL Problem.

It is the essentially rich environment of synchronous and asynchronous communication tools that truly support distributed (virtual) PBL. Group working and associated communication is a vital component of PBL and it is dependent on participant collaboration for its success. Again, it is argued that E-Learning technology should only be used if it adds value to the existing f2f tutorial sessions. It is interesting to note that a review by Ronteltap et al. (2004) at the University of Maastricht, a higher education institution famed for its use of PBL, prioritised the use of communication tools because planned f2f meetings "appeared to be insufficiently supportive of the learning process" (p. 274). This enthusiasm has been tempered by the realisation that one of the most challenging tasks is to design and facilitate an effective communication space for any form of online learning, including PBL.

The interactive use of E-Learning in this module is the process-supporting kind, making use of the technologies of discussion forums, chat rooms, video and audio conferencing to link to live international experts, blogging software and what Buzan & Buzan (1996) have termed mind mapping techniques, with supporting software. Building semantic links in this way represents information more visually and also uses more than one dimension. In the past, Schank et al. (1995) have reported other software explored by researchers which was specific to social and cognitive process support. One such example was by Koschmann *et al.* (1996), who worked on developing a comprehensive computer-supported environment for PBL. What is interesting about their work is their suggested use of technological supports needing to subtly change pedagogy - a theme that has since been enhanced and propagated in E-Learning research ever since.

Bereiter & Scardamalia (2000) picked up on ways that E-Learning technology could truly add-value to the process of PBL. A particular benefit occurs in the directive tutor's role which Donnelly (2006a) reported as having a three-fold impact: the tutor encouraging a more equal relationship with the participants; the tutor-dependent students' role; and disparities in student participation and involvement. In this blended PBL model, the tutor is no longer at the centre of the communications web. The asynchronous nature of the online discussions means that participants do not have to capture a conversational turn to contribute to discussions and so it becomes less likely that a few participants will dominate. Conversely, in this same medium, the more reticent or less verbal participant may also benefit from having more time to formulate a response. Of course, such a rosy picture in itself can be misleading.

Donnelly (2004) reports from a survey of the literature, that there appears to be significant opportunity in PBL for the use of asynchronous forums. Based on this, threaded discussion postings are used on the module for posing questions, arguing positions and challenging statements, achieving consensus and co-constructing meanings about E-Learning. At the same time, Donnelly & O'Rourke (2007) contend there is a plethora of research reporting difficulties generally with participation in online discussions. Particularly with campus-based students, the use of discussion forums can be regarded as an imposition by students unless they can see that the discussion relates to their study and that the discussion forum is an appropriate location and medium for the interaction. Once students are convinced of the benefits of sharing references and sources, it can galvanise the group into getting started. Palmer (1998) believes a teaching and learning space needs to be more than a form for individual expression; it must also be a place in which the group's voice is gathered and amplified, so that the group can affirm, question, challenge and correct the voice of the individual. Sustained asynchronous collaboration over time remains to this day a problematic area. Whilst learning can demand solitude, it also demands community, a dialogical exchange in which our ideas can be tested, our biases challenged, our knowledge expanded and even our ignorance aired; all of which is best served with others, rather than in an exchange in which we are simply alone with our thoughts. Alongside this, it is recognised that there is difficulty entailed in such endeavours in an online forum. Bereiter & Scardamalia (2000, p. 193) rightly argue that "utilising an educational approach with the high aspirations of PBL means looking for ways to make participation cognitively more rewarding to the students".

There is a growing emergence of the use of synchronous forums for supporting PBL (Portimojärvi, 2006). Each year newer technologies are ensuring that synchronous communication is becoming even more ubiquitous amongst students and arguably teachers need to follow suit. Some claims of preference for one medium over the other have been too hastily accepted and in their research, Chew & Beaumont (2004) found that an integrated set of synchronous and asynchronous tools were important within a distributed PBL environment.

Resolving the tension over which medium to use seems to be a perpetual task for the educator. The fissure is particularly acute in this debate over synchronous vs. asynchronous delivery. At the turn of the century, McFerrin (1999) reported on studies highlighting the benefits of online communication by extending classroom discussions, improving interaction between student and teacher (Collins, 1998) and increasing time-management ability, self-directive behaviour, self-confidence and self-discipline. Yet today a common problem for some academic staff is that they find asynchronous facilities a hindrance rather than a help to learning. Students too are reluctant to use them in a formal academic setting for a number of reasons, in contrast to the growing popularity of social networking sites such as those provided by, for example, "Bebo" and "Facebook"; asynchronous interaction can inhibit spontaneous development of ideas; in collaborative problems, a student may also make significant progress down the wrong path through research or practice before his or her group-mates can correct an improper understanding of that student's role in the group for that particular assignment. In addition, Garrison & Anderson (2003) have reported that asynchronous interaction inhibits the quick allocation of tasks and formation of schedules to get problem-solving activities completed. Furthermore as traditional face-to-face group dynamics can still tend to be the benchmark by which the value of the learning-teaching experience is judged, online pedagogies are frequently valued by academic staff only in proportion to how well they seem to reproduce or simulate an equivalent face-to-face experience. Where this fails (as it often does) lecturers may revert to using the virtual learning environment as a method for distributing lecture notes, or may simply abandon using it altogether.

Figure 5 shows the implementation of the blended PBL model of academic development proposed in this study. Colour is used to show at a glance those components that are face-to-face (outlined in green), those that are fully online (outlined in blue) and the blended components (outlined in purple). The WebCT courseware management system is highlighted in grey. The discourse in the PBL tutorial is supported with activities such as guided reflection, guest seminars, demonstrations of the technologies and peer evaluation. The online components of the blend include some course management standard features such as student homepages, online timetables and a set of personal individual progress statistics. These are augmented by virtual PBL tutorials which encompass online activities such as discussions, presentations, formative assessments and delivered using a set of tools such as video conferencing, podcasting, discussion forums, chat rooms and interactive tutorials.

Drilling down to explore levels of interaction spanning the f2f and online environments is key on this study. Figure 6 represents the different levels of activity and interaction in the blended PBL tutorial at three key stages in the ten week module at week 1, at the half way point (week 5), and at the close of the module (week 10). The PBL tutorial is shown in a face-to-face setting and also in its online environment. At week 1, the interaction is fully between the participants and their tutor separately in the face-to-face and online tutorial. At this stage, there is little crossover of interaction between the two environments. By week 5, there is much more cross-over of interaction between the face-to-face and online tutorials. Guest international tutors are introduced to participants and the deepest and most prolonged online discussions of the entire module occur at this juncture. Towards the close of the module, at week 10, the focus of the participants is on the

Figure 5. The implementation of the blended PBL module

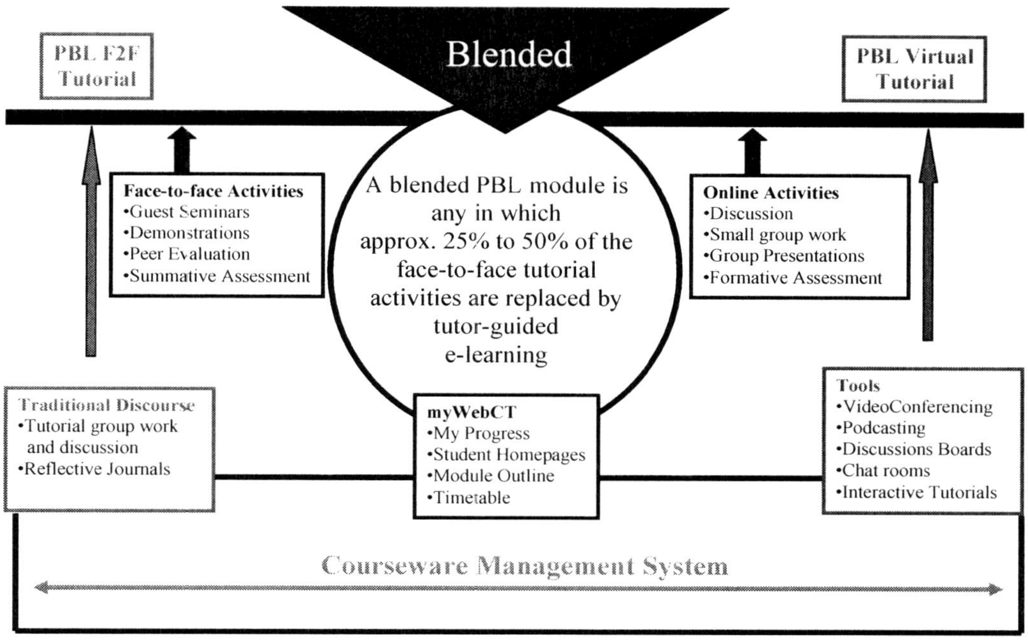

PBL content and their end product and once again there is little intersection of interaction types in the blend.

Deep learning often involves the element of time, but the findings suggest a theoretical possibility for deep learning in 10 weeks. These professional practitioners made strides towards transformative learning through change in their cognitive structure and reflection. The participants transformed their frames of reference so that they could best appreciate and understand their own experience on the module.

Change does not come easy to individuals or institutions for that matter. We are all creatures of habit. Our institutions will have to initiate and adapt training programs in order to prepare lecturers for the increased expectations of the 'new' student that is emerging. This is essential as learning is becoming more self directed. (Declan, RP15)

For transformative learning to take place for me, new learning needs to challenges parts of the existing framework of assumptions, beliefs and thinking processes. To facilitate this transformation in my learning it is necessary to reflect on experiences on the module and undergo an ordered, systematic, structured, deliberate, logical analysis of events and situations. (Eimear, RP6)

This research has suggested that blended PBL might lead to an enduring transformation in teaching practice. However, change in departments proceed in slow, linear, incremental steps - improving, expanding, contracting, and reforming - without altering its fundamental institutional mission, approach, or structure. Whilst this observation on change is neither new nor the preserve of the academic institution alone, it points to the importance of never losing sight of the state of affairs on the ground. Of course, transforming a

Figure 6. Levels of activity within the blended PBL approach

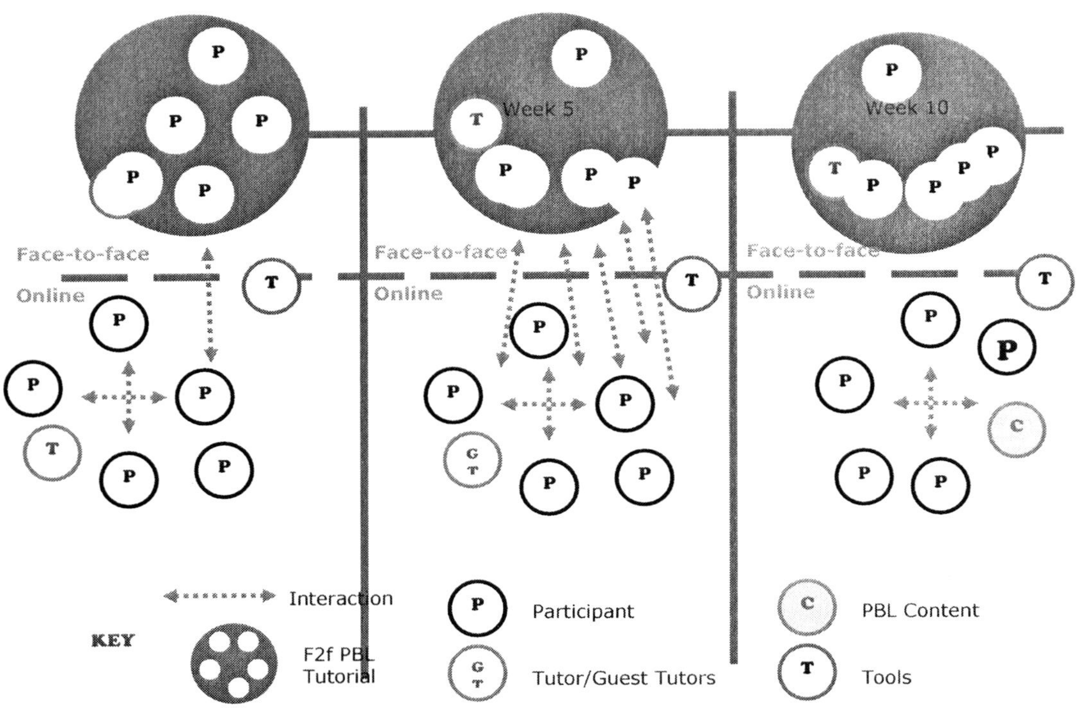

departmental approach to learning and teaching in such a complex phenomenon as the higher education institution is neither linear nor predictable. Transformation is an iterative process, since as an institution proceeds, experience leads to learning that can modify the transformation process.

As we would expect, questions about pedagogy and technology are inseparable in this study. The pedagogical innovations and module design are as integral to the investigations on transformative learning as the technologies being drawn upon. It is not the purpose of this study to examine the impact of technology in isolation from these larger transformations in teaching and learning. What this research wanted to track is the interaction and interdependence of innovative pedagogies such as PBL and new media technologies.

In looking at the nature of transformation on the module, several overlapping categories of questions emerge. First, there are questions about shifting the time and space for learning. Looking at the impact of online communication tools to connect participants on the module to experts in other countries, shifting time and space implies the possibilities of connecting the classroom to resources in the community. Second, there are questions about getting the participants to slow down and document the process of learning through critical reflection. Third, there are questions about how academic developers can best support academic staff in adopting a learning approach and a change to teaching practice that differs from what they have previously experienced.

It was important for all participants to step back from the module in order to reflect upon and understand the impact of contextual issues that may have been cultural, social or even economic on thinking and professional practice. If transformative learning is to take place and a more holistic approach to academic development cultivated, these most ingrained assumptions have to be recognized and questioned. These may not have been fully and universally borne out but an insight provided by Nobel Laureate Herbert Simon (1982, p. 46), does bring us to a place where we are better able to witness a transformation in learning: "for the computer to bring about a revolution in higher education, its introduction must be accompanied by improvements in our understanding of teaching and learning."

Bach et al. (2007) suggest that the continuing rapid pace of change and evolution is unlikely to lead to a singular and idealistic outcome where technology inspires a very large transformation of the quality of learning that empowers students. However, in their view, technology will continue to offer some real opportunities to increase the quality of learning and teaching, and in my view transform the learning experience for the individual lecturer within a problem-based learning context. Giroux (1988, p127) coined the phrase "teachers as transformative intellectuals" and has suggested that we use forms of pedagogy that treat students as critical agents, make knowledge problematic and utilise critical and affirming dialogue. This he argues in part suggests that transformative intellectuals take seriously the need to give students an active voice in their learning experiences. This voice and the ongoing commitment of the module participants is evidenced by two of the groups having published a chapter on their transformative experiences of the module. Instilled confidence from the module was evidenced by a number of participants from each of the three groups presenting their ongoing work at a number of recent national and international conferences including Educational Technology of Ireland 06 and International Technology, Education and Development (INTED) 07, all under blended learning streams.

In implementing a constructivist learning approach in blended problem-based learning on this module, some aspects of constructivist learning may be directly stimulated by technology, notably an increased level of collaboration, heterogeneity of roles and greater complexity and authenticity in assigned tasks. Other aspects, such as involvement with content that incorporates multiple academic disciplines may not be influenced by technology per se, but are often reinforced by technology use.

CONCLUSION

This final section provides the conclusions of this study, a contemplation of what will happen after the study, including what other avenues are open for exploration, along with critical reflection about the ultimate application or use of the research. This study was an attempt to provide insights into blended PBL in a higher education academic development context in order to encourage in-depth discussion and open debate on the field. One of the main challenges faced by today's academic developers is the development of E-Learning technologies to support constructivist and social constructivist teaching and learning approaches among the academic staff with whom they work so that this is, in turn, is carried forward into their own classrooms and subject disciplines (Donnelly, 2006a).

Figure 7 shows which elements of the blend are best conducted face-to-face, and which work best online.

The findings in this study indicated a preference for live tools in the blending of PBL and eLearning, specifically the face-to-face PBL tutorial. Of the online tools utilized in the module, the online reflective journals were the most favoured. The blending of delivery mechanisms (online and face-to-face), instructional approaches such as PBL and technologies such as asynchronous and synchronous discussions, video conferencing and online reflection have shown potential in evolving to support collaborative and interactive

Figure 7. Blended ingredients that proved effective

Face-to-face	Online
Induction involving ICT skills training, icebreaker and socialization sessions and introduction to PBL	Reflective journals
Brainstorming	Sharing resources
	Utilising the online library and ebooks
Critiquing peer ideas	Interdisciplinary international experts dialoging through video conferencing and asynchronous forums
Sharing peer experiences	Positive peer feedback (both seeking and giving)
Resources: involving past graduates of the module	Online peer tutoring and support with technical problems

learning. From a design perspective, it is useful to employ a holistic approach to blending media and activities. Developing technical competencies and pedagogical skills in tandem and via authentic tasks, provides the learners with technological skills to succeed in both face-to-face and computer-mediated environments.

Worldwide, higher education institutions are using formal, accredited courses for academic staff as a vehicle for diffusion of staff development in new learning technologies. It is still arguable that a new cadre of teachers in higher education need an understanding of the dynamics of online communication and interactions; this study has suggested that alongside needing to learn effective ways of facilitating and teaching online, a change in attitude, values and beliefs will develop confidence for ongoing learning.

FUTURE TRENDS

To fully understand the journey of transformation, there are some key points worth remembering; there is no set timetable for this journey as some participants move through some stages

Figure 8. Current and future considerations of blended PBL

Blended PBL Today	Considerations for Academic Development Tomorrow
Tutors facilitate; students do, present, think, construct Working together facilitates learning and problem solving Subjects are integrated into a learning whole Learning is problem-centred There are many rich resources for learning Concepts are explored using a variety of media Assessment is based on each student's developing abilities to solve problems to communicate ideas, present information, and learn how to learn Technology connects the world to the classroom vice versa	To help teachers: - find ways to exploit the capacities offered by new technologies; - to acknowledge the ever-changing nature of the e-landscape; - to be aware of a more robust body of theory and practice on blended learning; This will require a careful process of ensuring collaboration between tutors and experts, gaining successful experience in teaching with the technologies and participating in a community that provides continuous support.

more quickly than others; comfort level, interest, technology access and time are important determinants of any individual's timeline. When institutions in higher education recognize the need to learn technology and pedagogy as an opportunity to transform teaching and learning, this can open up new opportunities for all involved. Figure 8 summarises blended PBL today and what needs to be considered for academic developers in the future, when designing and tutoring in this field.

Although generalisability was not an intention of this study and no groups are the same, for future research, an intention is to replicate the study with other groups to see if the conclusions are sustained. This additional research across other groups could be used to determine how widespread the interpretations are. The study focused on how the participants in the groups interacted with one another in a blended environment and to an extent the learning that was transferred to professional practice. Applying Activity Theory as an analytical tool for representation to support the analysis of the differences/similarities across different learning groups in blended PBL is also a promising area for future research, as is a focus on the interplay between the levels of transformative learning reached in blended PBL academic development and its actual impact on students' learning in different disciplines.

REFERENCES

Bach, S., Haynes, P., & Smith, J. L. (2007). *Online Learning and Teaching in Higher Education*. Maidenhead: Open University Press.

Barrett, T. (2008). *Students Talk About Problem-based Learning in PBL Tutorials: Illuminative Concepts*. Unpublished doctoral dissertation, Coventry University, UK.

Bereiter, C., & Scardamalia, M. (2000). Commentary on Part 1: Process and Product in Problem-based Learning (PBL) Research. In D. H. Evensen & C. E. Hmelo (Eds.), *Problem-based Learning. A Research Perspective on Learning Interactions* (pp. 185-195). Mahwah, NJ: Lawrence Erlbaum Associates.

Buzan, T., & Buzan, B. (1996). *The Mind Map Book: How to Use Radiant Thinking to Maximise your Brain's Untapped Potential* (reprint ed.). New York: Plume Books.

Chew, S. C., & Beaumont, C. (2004, June). *Evaluating the effectiveness of ICT to support globally distributed PBL teams*. Paper presented at the ITiCSE Conference, ACM/SIGCSE, (pp. 47-51).

Collins, M. (1998). The use of email and electronic bulletin boards in college-level biology. *Journal of Computers in Mathematics and Science Teaching, 17*(1), 75-94.

Creswell, J. (1998). *Qualitative Inquiry and Research Design: Choosing Among Five Traditions*. Thousand Oaks CA: Sage Publications.

Donnelly, R. (2004). The effectiveness of teaching 'Online Learning' in a problem-based learning classroom environment. In M. Savin-Baden & K. Wilkie (Eds.), *Challenging Research into Problem-based learning*. Buckingham: Open University Press.

Donnelly, R. (2006a). Blended problem-based learning for teacher education: lessons learnt. *Journal of Learning, Media and Technology, 31*(2), 93-116.

Donnelly, R. (2006b). The Academic Developer as Tutor in PBLonline in Higher Education. In M. Savin-Baden & K. Wilkie (Eds.), *Problem-based Learning Online* (pp. 79-97). Maidenhead: Open University Press.

Donnelly, R. (2008). Virtual Problem-based Learning Communities of Practice for Teachers and Teacher Educators: An Irish Higher Education Perspective. In C. Kimble & P. Hildreth (Eds.),

Communities of Practice: Creating Learning Environments for Educators. USA: Information Age Publishing.

Donnelly, R., & O'Farrell, C. (2006). Blended E-Learning for Continuous Professional Development of Academic Staff. In J. O'Donoghue (Ed.), *Technology Supported Learning and Teaching: A Staff Perspective.* Hershey PA: Information Science Publishing.

Donnelly, R., & O'Rourke, K. (2007). What now? A critical evaluation of E-Learning CPD practice in Irish third level education. *Journal of Further & Higher Education.*

Garrison, D. R., & Anderson, T. (2003). *E-Learning in the 21st Century. A Framework for Research and Practice.* London: RoutledgeFalmer.

Giroux, H. A. (1988). *Teachers as Intellectuals. Towards a Critical Pedagogy of Learning.* Critical Studies in Education Series. New York: Bergin & Garvey.

Graham, C. (2006). Blended Learning Systems: Definition, Current Trends, and Future Directions. In C. J. Bonk & C. R. Graham (Eds.), *The Handbook of Blended Learning. Global Perspectives, Local Designs* (pp. 3-21). San Francisco: Pfeiffer.

Ham, V., & Davey, R. (2005). Our first time: Two higher education tutors reflect on becoming a 'virtual teacher'. *Innovations in Education and Teaching International, 42*(3), 257-264.

Hanson, K., & Clem, F. (2006). To Blend or Not to Blend: A Look at Community Development Via Blended Learning Strategies. In C.J. Bonk & C.R. Graham (Eds.), *The Handbook of Blended Learning. Global Perspectives, Local Designs* (pp. 136-149). San Francisco: Pfeiffer.

Hine, C. (2000). *Virtual Ethnography.* London: Sage.

Hoare, S. (2006). *The Guardian.* Tuesday June 20 2006. Retrieved September 24, 2008, from http://www.guardian.co.uk/education/2006/jun/20/elearning.technology15

Hofmann, J. (2006). Why Blended Learning Hasn't (Yet) Fulfilled Its Promises: Answers To Those Questions That Keep You Up at Night. In C. J. Bonk & C. R. Graham (Eds.), *The Handbook of Blended Learning. Global Perspectives, Local Designs* (pp. 27-40). San Francisco: Pfeiffer.

Jennings, D. (2006). PBLonline: A Framework for Collaborative E-Learning. In M. Savin-Baden & K. Wilkie (Eds.), *Problem-based Learning Online* (pp. 105-123). Maidenhead: Open University Press.

Jonassen, D., & Land, S. (2000). (Eds.) *Theoretical Foundations of Learning Environments.* Mahwah, NJ: L. Erlbaum Associates.

Juwah, C. (2002). *Using communication and information technologies to support problem-based learning.* HEA Academy. Retrieved February 28, 2007, from http://www.heacademy.ac.uk

Kirkup, G., & Kirkwood, A. (2005). Information and communications technologies (ICT) in higher education teaching: A tale of gradualism rather than revolution. *Learning, Media and Technology, 30*(2), 185-199.

Kiser, K. (2002). *Is blended best?* E-Learning Magazine. Retrieved July 31, 2006, from http://www.elearningmag.com/elearning/article/articleDetail.jsp?id=21259

Koschmann, T. D., Kelson, A. C., Feltovich, P. J., & Barrows, H. S. (1996). Computer-supported Problem-based Learning: A Principled Approach to the Use of Computers in Collaborative Learning. In T. D. Koschmann (Ed.), *CSCL: Theory and Practice of an Emerging Paradigm* (pp. 83-124). Hillsdale, NJ: Lawrence Erlbaum Associates.

McFerrin, K. M. (1999, February). Incidental Learning in a Higher Education Asynchronous Online Distance Course. In J. D.Price, J. Willis, M.

Jost, & S. Boger-Mehall (Eds.), SITE 99: *Proceedings of the Society for Information Technology & Teacher education International Conference*, San Antonio TX (Charlotteville VA, Association for the Advancement of Computers in Education (AACE)), (pp. 1418-1423).

Masie, E. (2006). The Blended Learning Imperative. In C. J. Bonk & C. R. Graham (Eds.), *The Handbook of Blended Learning. Global Perspectives, Local Designs* (pp. 22-26). San Francisco: Pfeiffer.

Mayes, J. T., & Fowler, C. J. H. (1999). Learning technology and usability: a framework for understanding courseware. *Interacting with Computers, 11*, 485-497.

Myers, S. A. (2006). Can I use transformative pedagogy when teaching online? *College Teaching, 2*(1), 82-94.

Offerman, M., & Tassava, C. (2006). A Different Perspective on Blended Learning: Asserting the efficacy of Online Learning at Capella University. In C. J. Bonk & C. R. Graham (Eds.), *The Handbook of Blended Learning. Global Perspectives, Local Designs* (pp. 516-528). San Francisco: Pfeiffer.

Oliver, R. (1999). Exploring strategies for online teaching and learning. *Distance Education, 20*(2), 240-254.

Owston, R., Garrison, D., & Cook, K. (2006). Blended learning at Canadian universities: issues and practices. In C. J. Bonk & C. R. Graham (Eds.), *The Handbook of Blended Learning. Global Perspectives, Local Designs* (pp. 338-350). San Francisco: Pfeiffer.

Palmer, P. (1998). *The Courage to Teach: Exploring the Inner Landscape of a Teacher's Life*. San Francisco, CA: Jossey-Bass.

Panda, S., & Juwah, C. (2006). Professional Development of Online Facilitators in Enhancing Interactions and Engagement: A Framework. In C. Juwah (Ed.), *Interactions in Online Education. Implications for Theory and Practice*. London: Routledge.

Portimojärvi, T. (2006) Synchronous and Asynchronous Communication in Online Problem-based Learning. In Verkkovirityksiä – ongelmaperustainen oppiminen ja projektioppiminen verkossa.

Ronteltap, C. F. M., & Eurelings, A. M. C. (1997). POLARIS: The Functional Design of an Electronic Learning Environment to Support Problem-based Learning. In T. Müldner & T. C. Reeves (Eds.), *Educational Multimedia and Hypermedia*. Charlottesville: Association for the Advancement of Computing in Education (pp. 1802-1807).

Ronteltap, C. F. M., Goodyear, P., & Bartoluzzi, S. (2004). A pattern language as an instrument in designing for productive learning conversations. *World Conference on Educational Multimedia, Hypermedia and Telecommunications, 1*, 4271-4276.

Salmon, G. (2000). *E-Moderating. The Key to Teaching and Learning Online*. London: Kogan Page.

Savin-Baden, M. (2006). The challenge of using problem-based learning online. In M. Savin-Baden & K. Wilkie (Eds.) *Problem-based Learning Online* (pp. 3-13). Maidenhead: Open University Press.

Schank, P., Ranney, M., & Hoadley, C. (1995). Convince Me [Computer Programe and Manual]. In J.R. Jungck, N. Peterson & J.N. Calley (Eds.), *The BioQuest Library*. College Park, MD: Academic Software Development Group, University of Maryland.

Sharpe, R., Benfield, G., Roberts, G., & Francis, R. (2006). *The undergraduate experience of blended E-Learning: a review of UK literature and practice*, Retrieved December 12, 2006, from http://www.heacademy.ac.uk/4884.htm

Simon, H. (1982). *Sciences of the Artificial.* Cambridge, MA: The MIT Press.

Uden, L., & Beaumont, C. (2006). *Technology and Problem-Based Learning.* Hershey, PA, USA: Information Science Publishing.

Valaitis, R. K., Sword, W. A., Jones, B., & Hodges, A. (2005). Problem-based learning online: perceptions of health science students. *Advances in Health Sciences Education, 10*, 231-252.

Wagner, E. D. (2006). On Designing Interaction Experiences for the Next Generation of Blended Learning. In C. J. Bonk & C. R. Graham (Eds.) *The Handbook of Blended Learning. Global Perspectives, Local Designs* (pp. 41-55). San Francisco: Pfeiffer.

Chapter XIII
Critical Survey of Information Technology Use in Higher Education: Blended Classrooms

James G. R. Cronin
University College Cork, Ireland

John Paul McMahon
University College Cork, Ireland

Michael Waldron
University College Cork, Ireland

ABSTRACT

Reception and use of information technology by lifelong learners within a "blended" learning environment needs to be articulated within a constructivist paradigm. Increasingly, the term reflective practice is appearing in the vocabulary of adult education discourse. Educators have become familiar with the concept of reflective practice through Donald Schön's writings. Schön's work is founded on a tradition of learning supported by Dewey, Lewin, and Piaget. As a learning group, lifelong learners are receptive to constructivist learning interventions where facilitated activities provide learners with opportunities to enact and collaboratively construct meaning as interventions unfold. This case study reviews learning enactments through an online discussion forum in an evening diploma in European Art History, University College Cork, Ireland.

INTRODUCTION

In Ireland, the majority of lifelong learners are part-time students trying to maintain work and family commitments. Ireland's participation in the European Union's Bologna Declaration (1999) is establishing a higher educational qualification framework that predicts flexible career progression paths and greater educational opportunities by 2010. If course providers are to fully integrate non-traditional learners within higher education they need to consider opportunities afforded by learning technologies to enhance self-directed inquiry.

Studies in the application of learning technologies in higher education continue to focus on an 18-30 age group. The reception and use of information technology by lifelong learners still needs to be articulated within a constructivist paradigm. The Internet facilitates student and teacher access to many inquiry-based learning experiences. Online social networking, as mediated through discussion boards and chat rooms, allows for information collaboration and knowledge construction outside of the formal educational institution. Judith V. Boettcher (2007) argues that online tools are particularly valuable in this context because they provide a public forum in which the cumulative, step-by-step process of concept formation, refinement, application, and revision is fully visible to student peers as well as their mentors. Boettcher suggests that discussion forums, blogs, journals, and small group work are all excellent strategies for allowing learners to enlarge their mental models, to clarify concepts, and to establish meaningful links and relationships.

Malcom S. Knowles (1970, 1975 & 1984) acknowledged the importance of prior experience and self-direction in adult education. Increasingly, the term reflective practice is appearing in the vocabulary of adult education discourse. Broadly speaking, reflective practice is a mode that integrates thought and action with reflection for the benefit of improving one's professional practice. Over the past two decades, educators have become familiar with the concept of reflective practice through Donald Schön's writings about reflective practitioners (Schön, 1983, 1987). Schön's work has an historical basis in a tradition of learning supported by Dewey, Lewin, and Piaget, each of whom advocated that learning is dependent upon the integration of experience with reflection and of theory with practice. Lee S. Shulman's paper on signature pedagogies in the professions (2005) implicitly owes a debt to the work of Donald Schön. Signature pedagogies are composed of three elements: surface structures, deep structures and implicit structures. The surface structure, applies to the acts of teaching and learning. The deep structure refers to the values of the discipline (e.g., what makes a lawyer or a doctor?). The implicit structure or hidden curriculum is revealed through performances of student understanding. In his seminal paper, Shulman mentions in passing the transformative potential of learning technologies upon a traditional curriculum design, but he does not expand on this. (Shulman, 2005, pp. 52–59).

This case study reviews learning enactments, through an online discussion forum on an evening diploma in European Art History at University Cork, as a means of demonstrating that lifelong learners, as a learning group, are receptive to constructivist learning interventions where facilitated activities provide learners with opportunities to enact and collaboratively construct meaning as interventions unfold.

Since 2004, the Centre for Adult Continuing Education, University College Cork, has seen the potential for using online resources to enhance disciplinary understanding within "blended" learning environments whereby online is integrated with class-based teaching and learning (Cronin, 2005, 2008; Young & Cronin, 2005).

The two-year, part-time diploma in European Art History aims to provide lifelong learners with a sound base in the historical, critical and contextual study of European painting, sculpture and architecture. Teaching and learning is enacted through lectures, case studies, tutorials, workshops, field trips, and web-based media supported on the university's Blackboard Virtual Learning Environment. This "blended" model supports social interaction appreciated by adult learners. A recent shift to a modular structure is facilitating greater student self-direction than previously available. In future, students will be able to exit the course after the first year with a certificate qualification. Prospective students are expected to demonstrate a genuine interest in the visual arts when making their application. The diploma is examined through continuous assessment. There is no final examination. Students take four modules in first year and four in second year. In order to achieve the award of diploma all modules must be passed in both years. There are opportunities for students to matriculate onto the degree in Art History, Department of History, University College Cork.

The impact of art history education in Cork has been dramatic over the past decade. The two-year diploma in European Art History, co-ordinated by the Centre for Adult Continuing Education, University Cork began in 1991 with 25 students and recorded 72 students in 2008. In the past decade the age profile has widened, but two-thirds of the diploma participants remain female. A degree in art history was established in University Cork in 2001. In the first four years of being offered this degree accounted for 10% of the overall first year undergraduate intake to the Humanities in University Cork. The success of both programmes reflects deep interest in the visual arts by sections of the wider community: professionals seeking to enhance their careers and those in retirement seeking to fulfill a lifelong interest. The diploma in European Art History strives to engage with the cultural life of both the college and the wider community. Applicants decide to pursue the diploma for various reasons. For some it is for personal ambition, while for others it is to develop their chosen career or to embark on a career change. Most applicants have spent years engaged with the visual arts through reading art history and visiting galleries, but wish to engage more formally with the discipline. In Cork employment opportunities in the visual arts have increased dramatically in the past decade. This is reflected in the careers of adult learners on the part-time diploma. A decade ago the majority of students on the course were retired professionals and those in employment were mainly secondary school teachers. Now a myriad of careers feature: gallery administrators, artists, picture framers, medical professionals, art teachers, graphic designers, publishers, interior designers, and local architects. In turn, the diploma provides teaching and administrative opportunities to graduates from the degree. This gives them an opportunity to develop and refine their skills: vital for the competitive academic jobs market. Currently, six graduate tutors comprise the core teaching team. Two graduated from the diploma and then went on to read art history at degree level. Their example provides a tangible progression path for lifelong learners.

The initial cycles of the diploma in European Art History were taught over two nights a week. However, in 2004, due to a demand, the diploma was scaled back to one evening a week. While this model is extremely popular with the public, accounting for large application numbers, nevertheless, the disadvantage is that there is reduced contact time. The decision to adopt an online discussion forum in the second year of the diploma was a decision made jointly between students and course instructors. Because the class only met one night a week it was felt that there was a real need to carry on debate and reflection beyond the tutorial room, lecture theatre and gallery space.

TEACHING THE CRAFT OF THE ART HISTORIAN

The diploma in European Art History asks students to engage with the following fundamental questions: What is art history? How do art historians and critics engage with the visual arts?

Art history is commonly understood as the study of art objects within their historical and cultural context. However, this definition is too narrow to embrace the plurality of art histories currently studied in higher education. Art history, as an academic discipline, was initially introduced to universities on continental Europe during the mid-nineteenth century and so is relatively new. However, during the past twenty-five years, the nature and purpose of the discipline has been questioned. James Elkins, a leading art historian and art critic, currently teaching at the School of the Art Institute of Chicago, has classified art history according to the following broad category titles: normal art history, visual theory, serious art criticism, and commercial art writing (Corcoran, 2000). For Elkins, normal art history applies to studies in the history of certain canonical sequences mostly within the context of Western culture. Its journals include, broadly speaking, *Art Bulletin*, *Burlington Magazine* and *Zeitschrift für Kunstgeschichte*. Visual theory also called new art history is seen by Elkins as more experimental than the first category. It is interdisciplinary, drawing on such diverse areas as film theory, psychoanalysis, and semiotics. This group is represented by the journal *October*. Serious art criticism, says Elkins, is largely ignored by followers of the first two categories, but is read by curators and artists. Commercial art writing is vast in comparison to the tiny productions of art history and serious criticism and is a product of the mass media (Corcoran, 2000).

Donald Preziosi, formerly art history professor in California, argues that new art history is engaged in playing out multi-discourses (research, teaching, curatorial issues, criticism, and the art market). These discourses are, partially, focused on professional goals. Today, art history education (either at university or in art school) needs to acknowledge the plurality of discourses within the discipline many of which have begun as academic concerns, but have been shaped through professional practice (Preziosi, 1992).

What characterises an apprentice art historian? On the diploma in European Art History this is expressed through learning outcomes where students should be able to perform the following learning competencies:

1. Identify appropriate vocabulary for analysis and interpretation of works of art in their historical contexts.
2. Discriminate a variety of art historical methodologies that include the use of biography, stylistic analysis, iconography and social history.
3. Employ library and online resources to the preparation and presentation of short and long papers that critically examine and/or compare works of art, or exhibit knowledge in development of a theme.
4. Reflect upon the appreciation of art and architecture directly through visits to museums, churches, galleries, and historical sites.
5. Organise written and oral presentations in a critical and original manner.
6. Evaluate the values (social, political, religious, artistic, intellectual) of Western culture through their thematic study of the visual arts (painting, sculpture and architecture).

The course examines the changing intellectual traditions, institutions, social practices and issues of taste governing the canon: the body of works of art considered at any one time to represent great art. These learning competencies slot the diploma into Elkins' normal art history category.

SCAFFOLDING THE LEARNER THOUGH A "BLENDED" MODEL

A decision to adopt a "blended" teaching and learning model, comprising both direct in-class contact and remote online communication, was motivated by a desire to give the class an opportunity to post their comments on a topic of art history, covered either at a lecture or in a tutorial, over a longer time-period. It was hoped, in turn, that this process would consolidate in-class teaching and learning.

The potential of technology to enhance teaching and learning is now well established. That said, however, it is still worth briefly outlining some salient focuses of opinion in the scholarly literature as well as predicting a future concern.

Gerard Delanty, a British sociologist, has argued that a major cognitive shift is currently taking place in society brought about by information technology, resulting in a dissolving of the divisions between professional (in the sense of expertise) and lay knowledge (Delanty, 2003). The Internet has resulted in a cognitive revolution creating new forms of information literacy and online social networking activities now expressed by the term Web 2.0. As a concept it refers to a perceived second generation of web-based communities and hosted services — such as social networking sites popular in Ireland like Facebook and Bebo — which aim to facilitate creativity, collaboration, and sharing between users. Ease of access to information via the World Wide Web has changed the role of teachers from being information providers presenting in front of a class to becoming facilitators in guiding learners through information access and processing, problem-solving, and creative learning projects, using learning technologies in a learner-centric classroom. (Leu, et al, 1997; Reynolds, 1999). Focus is placed on assisting learners to construct knowledge both as independent self-directed enquiry and communally in peer groups in order to demonstrate their knowledge attainment through enactment and application (Tishman and Palmer 2005; Boettcher 2007). Rhizome theory, a term borrowed from Deleuze and Guattari (1980/1987), is gaining currency in educational research as a metaphor to articulate the fluid nature of unmediated knowledge construction especially through social networking sites (Cormier, 2008). Wikis are examples where online communities supply and regulate content. Here, authorship is collective, difficult to attribute, or authenticate. In principle, the collective regulates itself. What role does higher education now play in knowledge construction? Conversations concerning the authority of authorship will become pronounced as boundaries between professional and lay knowledge construction continue to blur in cyberspace. The increasing speed and greater availability of information technology means that students have more access to information than ever before, but, to paraphrase Paul Virilio (2000), is it also creating information fallout from the exploded "information bomb"? Can educators assume that students know how to pick their way through the mass of content in a discerning, critical, and ethical manner?

The application of technologies to teaching and learning has created a new term – cybergogy. One of the central elements of cybergogy is the intent to combine fundamentals of both pedagogy and andragogy (the methods of teaching adults) to arrive at a new approach to learning. Cybergogy focuses on helping adults and young people to learn by facilitating and technologically enabling learner-centric, autonomous and collaborative learning in a virtual environment (Carrier & Moulds, 2003). Current research stresses the centrality of the learner and the need to actively engage the learner so as to scaffold independent thinking. Studies by Goody and Malone (1999), Laurillard (2002), Carrier and Moulds (2003), Tishman and Palmer (2005), Boettcher (2007), and Wang (2007) broadly identify features of good practice for cybergogy: placing the learner at the centre of the teaching and learning experience; fostering an engaged learning environment; and reflecting on practice. To accomplish these we need to consider building online Communities of Practice in the disciplines and across disciplines; cyber ethics (e.g., copyright issues, computer hacking, privacy and plagiarism) and an awareness of socio-economic barriers to technological inclusion – the so-called digital divide discussed by Carrier and Moulds (2003), Ó Fathaigh (2002, 2003), Gurstein (2003), and O'Brien and Ó Fathaigh (2007).

The ability to effectively use information technology is fast becoming a basic literacy skill. By incorporating **information literacy skills** into the learning process the curriculum designers were hoping to enhance student confidence and develop socially transferable skills. Encouraging the use of websites can give a student the background they need to gain a deeper understanding of a topic. The Internet offers many credible and expansive resources for students to do research. Jos Boys (2001) has engaged with models of creative learning to suggest that the Internet's form allows for juxtaposition of information via hyperlinks, thereby helping to foster the creative and independent learning process. Students should be shown how to effectively search the web on their own for resources in a structured way. Applications such as QuickTime image tools have allowed more architectural case studies to be included in the current diploma cycle. In the first year of the diploma, case studies on the sixth-century Ravenna churches and Renaissance Vatican complex have benefited from the inclusion of these web tools. Conveying the three-dimensional sense of a building or piece of sculpture is something which is extremely difficult to convey through the static slide format but can be achieved through an enhanced web tool.

The Internet facilitates student and teacher access to many inquiry-based learning experiences. Discussion boards and chat rooms allow for collaboration outside of the educational institution. As mentioned, Judith Boettcher (2007) argues that online tools are particularly valuable in this context because they provide a public forum in which the cumulative, step-by-step process of concept formation, refinement, application, and revision is fully visible to student peers as well as their mentors. Boettcher suggests that discussion forums, blogs, journals, and small group work are all excellent strategies for allowing learners to enlarge their mental models, to clarify concepts, and to establish meaningful links and relationships.

UNCOVERING UNDERSTANDING: TWO E-MODERATED CASE STUDIES

Just as Web 2.0 technologies are clearly reshaping the Web, innovations in instructional practice and academic technology are moving higher education in new directions. E-moderation of online discussion forums is a growing aspect of the Instruction 2.0 paradigm. The term "e-moderation" describes the process of managing the communication of others online (Coghlan, 2001). Michael Coghlan's critique of e-moderation shows that focus has moved from content to process in online courses, and that this shift of focus means that many online teachers will need to develop e-moderation skills. Coughlan argues that the skills of an e-moderator are the same as those of a good classroom teacher when managing face-to-face discussions: listen; ensure that all have their say; guard against dominators; encourage respect for different opinions while encouraging people to challenge their beliefs; resolve disputes; bring discussions to meaningful closure, etc. Coughlan (2001) reviews models of knowledge construction proposed by Guy Bensusan, Gilly Salmon and Edward DeBono. In his analysis, all three paradigms partially intersect as each is concerned with a staged theory of knowledge construction where the learner gradually appropriates and personalises acquired disciplinary knowledge. The stages of Salmon's five-stage model are as follows:

1. Access and Motivation: Exploring the technology and motivation building are key issues. The e-moderator helps learners navigate the environment.
2. Socialisation: Building on the first stage, this stage focuses on social processes and "community building".
3. Information Exchange: Information is exchanged and co-operative tasks can be achieved. Interaction occurs between content, other participants and the e-moderator who assists exploration activities.
4. Knowledge Construction: Knowledge development and discussion activities become important. Participants start to take control of knowledge construction.
5. Development: Participants become responsible for their own learning and that of their group. Ideas are applied to individual contexts. This stage is characterised by reflection and assessment.

Coughlan argues that Salmon's five-stage model of e-moderating is comprehensive in that it caters for all levels of CMC (Computer Mediated Communication), and affords the e-moderator with some clues about the type of discourse to be involved in at each level (Salmon, 2002). This model informed our forum design.

Initially, the objective for the Blackboard discussion in the diploma in European Art History evening diploma was to serve as a virtual learning environment to allow learners increase their limited contact hours beyond those provided by lectures and tutorials as the diploma cycle is currently only one night a week. It was anticipated that the discussion board would support and enhance discussion beyond face-to-face tutorial contact. The extent to which this was achieved in an academic year, from late September to early June, is what follows.

AD2803: *Art in the Age of Enlightenment Neoclassicism and Romanticism*

This second year module examines the re-interpretation of Antiquity during the eighteenth-century Enlightenment. Topics examined include: Rococo art and architecture in Italy and France, neo-Classical art and architecture, and 18th Century British portrait painting. This module includes visiting collections in the Crawford Art Gallery, Emmet Place, Cork. This city gallery boasts a fine collection of Irish 18th and 19th Century holdings.

According to Meyer and Land (2003, 2006), a threshold concept represents a transformed way of understanding, or interpreting, or viewing something, which must be understood before a learner can progress. This transformation may be sudden or it may be protracted over a considerable period of time. The transition to understanding may prove troublesome. One of the elements of troublesome knowledge for students taking art history is trying to understand the historical context of artistic production. AD2803: *Art in the Age of Enlightenment Neoclassicism and Romanticism*, is concerned with art produced during a period of great social and political upheaval in the wake of the Enlightenment. Jacques-Louis David (1748-1825) is one of the central artists discussed in the module, as he was a key exponent of Neoclassicism and his work is important in debates concerning the genesis of Modernism and the Modernist aesthetic in the 20th Century. David's career was intertwined with the events of the French Revolution (1789-92). In order to understand David's work one must understand the context in which he was working. The rapidly changing fortunes of the revolutionaries and counter-revolutionaries are complex. Therefore, the e-moderated forum for this module was thematically designed with this difficulty in mind and expanded upon face-to-face tutorial discussion. Initially, hyperlinks were posted to relevant supplementary material aimed to enhance student understanding of the historical and cultural context. Links were provided to a documentary series on the French Revolution, produced for the History Channel (http://www.thehistorychannel.co.uk), and broadcast on YouTube (http://www.youtube.com) This resource demystified many of the complex historical details that directly impacted David's painting during this period.

Course statistics indicated that the forum received 895 views, but a mere 21 posts. Although 67% of the class logged on to view discussions on the forum it is interesting that less than one-third of the 40 students taking the diploma through assessment contributed to the online discussion. Lack of confidence accounted for this online silence. When discussions did take place they were concerned with topics and themes to be covered in the tutorials supporting the module. The forum, therefore, served mainly as a virtual message board between the tutor and the students. Six students contributed continuously to the discussions. These students were high achievers performing well in assessments and also contributing well to lectures, tutorials and field trips.

With the support and agreement of the class Michael Waldron, module tutor and moderator, conducted a feedback session in period one (September to December, 2007) of the second year of the diploma cycle. This session was very informative with regard to how the class (which is entirely comprised of mature students) approached, viewed, and used the online discussion board. The following four questions were put to them as a class during this session:

"Question: In what ways did the discussion forum promote 'deep learning'?" The class unanimously stated that had this resource been available from the beginning of the two-year diploma cycle then it would have made more of an impact on how they use it. They felt that the positive energy and good will that came hand-in-hand with beginning the course in year one could have translated into using Blackboard and the discussion forum more effectively. The notion of "deep learning" through the discussion forum therefore, in their eyes, was hampered by its introduction in year two. However, they were in agreement that, as an additional resource, it had considerable potential to promote better understanding and knowledge.

"Question: In what ways did it support students?" As an aid to the course the class agreed that Blackboard as a whole had definite merits with regard to course information being uploaded, such as timetable information, lecture guides and slide lists as well as announcements. In addition, they found that Waldron's tutorial summaries on the discussion board were a helpful support to their own notes and also for focusing and clarifying their own perception of the class discussion. The discussion forum also enabled them to keep up-to-date with topics being discussed, access useful online materials posted by tutors and to actively participate in the selection of each week's discussion topic.

"Question: What were the barriers to online participation?" It was discovered that many more students viewed the discussion forum than actually contributed to it. Initial feelings of inadequacy, lack of self-confidence, constraints of time, and issues of user-friendliness in site navigation were factors contributing to online silence. A number of students expressed a feeling of intimidation or embarrassment with regard to posting comments, even brief questions, for fear of not appearing to be profound or informed enough of the subject. When assured that any and all comments would be welcome there was still a sense of reticence among them. They felt that, as they were new to many of the topics that were being discussed from week-to-week, they lacked the confidence and knowledge to be able to give a valuable contribution to the discussion. Their own perceptions of the level of the discussion were unnecessarily high. Added to these concerns was the fact that many members of the class worked full-time and had children and therefore time was a major issue regarding their availability to participate online. A number perceived that full-time students would use the resource much more frequently. This perception was unfounded. Cronin, who also moderated two undergraduate online discussion groups, found that undergraduate forum activity was less frequent than with the evening diploma class.

"Question: How could it be improved?" Ease of navigation and overall user-friendliness was a factor considered necessary for forum improvement. In the initial start up of each module's discussion forum some students expressed feelings that a single, more coherent and more easily accessed forum would be preferable to a number of sub-forums. The class felt anonymity in posting would be beneficial. The consensus was that if they could select usernames rather than use their real names then they would be more inclined to contribute to the discussion forum without fear of embarrassment. The idea of being named, and therefore accountable for a posting, even a superficial comment within a deeper discussion, was seen as a significant impediment to participation. Feedback revealed self-censorship as the class also requested an edit function to remove unwanted posts.

AD2805: *Birth of Contemporary Art*

AD2805: *Birth of Contemporary Art* examines the various diverse strands of the Modernism from 1900 to 1945 and the Modernist aesthetic during the early 20th Century. Topics examined include: Symbolism, Cubism, Expressionism, Dada, and Surrealism.

The questions asked at the end of the first term informed the forum design for the last module of the second term. John Paul McMahon, who taught this module, acted as forum moderator. Instead of allowing the discussion forum to stand alone as a tutorial support it was embedded in the module assessment.

An Andy Warhol retrospective happened to coincide with this module. Entitled, *The Eternal Now,* the exhibition showed pieces produced at the Factory studio in New York, between 1963-1968, at the zenith of Warhol's career. As a learning resource for the teaching programme, it engaged with four fundamental aspects of Modernism: authorship, originality, collaboration, and documentation. Students were invited to post on the online discussion forum, short, but critical statements, no longer than ten sentences, revealing connections between the themes of the exhibition and the module.

Although nearly 80% of the class logged on to view discussions on the forum, over a six-week period, interestingly, again, less than one-third of the 40 students taking the diploma through assessment contributed to the online discussion. This forum received 918 views, but merely 53 posts. Participation marginally increased from 6 to 10 students. Those who participated were enthusiastic and reflective.

Troublesome knowledge in this module coalesces on approaches to Modernism: How is it defined? When does it begin? How does it break with earlier artistic traditions? How is it made manifest? Does Modernism change our notions of the art historical canon? The forum was embedded in the lectures so as to assist students with tackling these issues. John Paul Mc Mahon, acting as both module teacher and forum moderator, urged the class to engage with Elkins' categories of art history. Broadly speaking, forum participants were less concerned with debate on definitions of the disciplinary canon more than on seeking out assistance on specific topics they were completing for terminal assessment. The general group response, as revealed through online postings, told us more about general learner behavior than it did about specific online enactments. Learners, no matter what age group, have a tendency to be strategically motivated, whereby, they focus on particular targets or goals, principally, assessments.

From his experience of teaching the group and moderating the forum, John Paul McMahon answered the original four survey questions at the end of the six-week module.

"Question: In what ways did the discussion forum promote 'deep learning'?" McMahon believed that if the resource had been available from the beginning of the two-year diploma cycle and had been better integrated into the course then it would have made more of an impact on how the class used it. Approximately half of the class (those who were being assessed) engaged with the discussion forum in a focused and thoughtful manner.

"Question: In what ways did it support students?" McMahon stated that the moderated forum had given the students a sense of real-time engagement through synchronous and asynchronous communication. McMahon's comments generated an on-line dialogue of question, response, and further questions being posed by the students. The forum empowered the students to begin to discuss ideas regarding art history outside the essay and exam context. It made the experience of writing about art history more authentic and empowered contributors to develop their individual critical voices.

"Question: What were the barriers to online participation?" Previously, a number of students had expressed a lack of confidence in posting online. Confidence grew in the second term because the class began to see the forum as a safe and supportive learning environment. This was achieved through enthusiastic moderation and encouraging postings in a supportive, open and non-judgmental manner.

"Question: How could it be improved?" Again, McMahon believed that if the resource had been available from the beginning of the two-year diploma cycle then it would have made more of an impact on overall participation. However, the forum suited some learning styles and did not suit others. McMahon suggested that mentoring on how to use the discussion forum should be encouraged.

BARRIERS TO FULL PARTICIPATION: IRELAND'S DIGITAL DIVIDE

A 21st Century learner is required to acquire diverse skills-sets: multi-tasking, critical thinking, problem solving, information discernment, ethical concern and global awareness. Learning technologies, when applied within an intentional pedagogical framework, have the potential to enhance disciplinary understanding. However, in Ireland a digital divide remains to be overcome. Digital divide, as a manifestation of digital inequality, is marked by the following structural barriers to full educational participation: equipment, location of access, autonomy of use, technical skills, social support, purpose for using information technology, gender divide, and age divide (O'Brien & Ó Fathaigh, 2007, pp. 59, 225). O'Brien and Ó Fathaigh have explained the digital divide using Pierre Bourdieu's knowledge capital conceptual lens. For Bourdieu, capital -- economic, social, cultural and symbolic -- operates as an instrument of cultural reproduction thereby helping to explain inequalities in educational achievement (O'Brien & Ó Fathaigh, 2007, p. 227). The majority of diploma students continue to be female, over 45 years of age. In this study, those with limited keyboard skills initially felt uneasy accessing and using information technology. This was a potential barrier to full participation. While confidence grew during the year, nevertheless, the class felt that they would have initially benefited from elementary information literacy skills.

At national level, network infrastructure needs to be fully integrated if learning technologies are to be fully exploited in assisting to widen educational participation. Despite Ireland's sustained economic progress there are still isolated pockets of the Republic, particularly in the countryside, where broadband access is limited or non-existent. In addition, domestic network speeds need to be upgraded to equate with European standards.

CONCLUSION

Lee S. Shulman (2005) hinted that embedding Web resources require an imaginative curriculum modification. For us, this study validated this assumption. In both case studies the e-moderated forum was integrated so as to remove any anxiety surrounding prior skill-sets for the group. 67% of the class regularly logged onto the Blackboard site to update themselves on course information. The site received 4,000 hits between November 2007 and April 2008. Course statistics show that access was most frequent on Tuesday evenings, when the weekly lectures took place, and on Wednesdays with traffic heaviest at 10am, 5pm and 10pm.

Attempts to encourage debate on the forum were initially slow to develop. Online silence, or lack of forum participation, resulted from lack of student confidence in publicly posting written responses. Although beyond the scope of this study, this silence poses questions about perceived power-dynamics between moderators and contributors in online discourse. Participation improved during the second term due to systematically embedding the forum within the module design and by encouraging the use of the forum as a "virtual sandbox" where disciplinary understanding could be played out prior to assessment submission. Gradually, student postings became more assured and reflective. Although more postings took place in the Modernism forum, surprisingly, this did not represent a wider distribution of forum participants. Generally, the students who posted from the beginning continued posting and these were the most enthusiastic in the class. Tutors reported that they had seen an overall improvement in the quality of assessments in the second year particularly regarding visual analysis and critical awareness. A student who had posted regularly on the forum improved by 10% during the year. Students began to confidently critique art historical discourse within their assignments in a manner that was encouraged and supported through e-moderation. Students valued synchronous and asynchronous modes of communication provided by the online forum thereby validating Boettcher's observations, the e-forum provided a public platform where the cumulative, step-by-step process of concept formation, refinement, application, and revision were fully visible to peers and mentors.

ACKNOWLEDGMENT

We wish to thank History of Art, Centre for Adult Continuing Education and *Ionad Bairre*: Centre for Teaching and Learning, University College Cork. Thanks to our colleagues at University College Cork especially Bettie Higgs (*Ionad Bairre*), Marian McCarthy (*Ionad Bairre*), Stephen O'Brien (Education), and Sheila O'Driscoll (Adult Continuing Education), whose insights greatly helped us refine our understanding of reflective practice. Sincere thanks to the students of the diploma in European Art History, Adult Continuing Education, University College Cork, without whose contribution this work could not have been undertaken or completed. We remember former students Ann-Marie Hourihane (deceased 2008) and Suzanne Crosbie (deceased 2007) who share in our work.

REFERENCES

Boettcher, J. (2007). Ten core principles for designing effective learning environments: Insights from brain research and pedagogical theory. *Innovate,* 3(3). Retrieved November 5, 2007, from http://www.innovateonline.info/index.php?view=article&id=54

Boys, J. (2001). Learning creativity via the web. In R. Shaw & McKay (Ed.), *Digital creativity: crossing the border. Proceedings of CADE 2001: The 4th Computers in Art & Design Education Conference* (pp. 61-64). Glasgow: Glasgow School of Art.

Carrier, S. I., & Moulds, L. D. (2003). Pedagogy, andragogy, and cybergogy: Exploring best-practice paradigm for online teaching and learning. *Sloan-C 9th International Conference on Asynchronous Learning Networks (ALN)*, Orlando, USA [PowerPoint slides]. Retrieved May 15, 2004, from http://www.sloanc.org/conference/proceedings/2003/ppt/1471.ppt /

Campbell, J., Oblinger, D., & Colleagues (2007). Creating a culture of evidence tops the list of important issues as the academic technology profession moves to an "Instruction 2.0" world. *Educause* Quarterly, 30(3). Retrieved March, 11, 2008 from http://connect.educause.edu/library/abstract/TopTenTeachingandLea/44831

Coghlan, M. (2001). Emoderation: Managing a new language? Paper presented at the *Net*Working 2001 Conference - from Virtual to Reality*. Brisbane. Retrieved January, 12, 2008 from http://www.chariot.net.au/~michaelc/nw2001/emod_newlang.htm

Corcoran, M. (2000). Interview with James Elkins. *CIRCA,* 94. 26-29. Retrieved January, 10, 2008 from http://www.recirca.com/backissues/c94/jelkins.shtml

Cormier, D. (2008). Rhizomatic education: Community as curriculum. *Innovate,* 4 (5). Retrieved August, 15, 2008 from http://www.innovateonline.info/index.php?view=article&id=550

Cronin, J. (2005, July). Joining the digital dots: Promoting resource integration and autodidactic learning in teaching art history to adult learners (Towards a global understanding of lifelong learning: making a difference). Paper presented to the Forum for the Advancement of Continuing Education (FACE) Conference, University Cork.

Cronin, J. (2008). Scaffolding the learner: Teaching for Understanding (TfU) and the adult learning experience, University College Cork. [Poster session]. In N. Ryan (Ed.), *International perspectives on teaching and learning in higher education.* National Academy for Integrating Research & Teaching & Learning (NAIRTL) Conference Proceedings, November 2007. Cork: NAIRTL.

Delanty, G. (2001). *Challenging knowledge: The university in the knowledge society.* Buckingham: Open University Press.

Delanty, G. (2003). Ideologies of the knowledge society and the cultural contradictions of higher education. *Policy Futures in Education,* 1(1), 71-82. Retrieved December, 10, 2007 from http://www.wwwords.co.uk/pdf/viewpdf.asp?j=pfie&vol=1&issue=1&year=2003&article=4_Delanty_PFIE_1_1&id=86.41.151.59.

Deleuze, G., & Guattari, F. (1987). *A thousand plateaus: Capitalism and schizophrenia,* 2. (B. Massumi, Trans.). Minneapolis: University of Minnesota Press (Original work published 1980).

Goody, A. E., & Malone, V. M. (1999). Cybergogy and adult learning: a case research study used as a decision making tool in higher education. In H. E. Klein (Ed.), *Interactive teaching and the multi media revolution: case method & other techniques.* World Association for Case Method Research and Application (WACRA), Needham: MA.

Gurstein, M. (2003). Effective use: A community informatics strategy beyond the digital divide. *First Monday,* 8(12) Retrieved April, 2, 2007 from http://firstmonday.org/issues/issue8_12/gurstein/index.html

Knowles, M. S. (1970). *The modern practice of adult education: Andragogy versus pedagogy.* Follett: Association Press.

Knowles, M. S. (1975). *Self-directed learning: A guide for learners and teachers.* Englewood Cliffs: Prentice Hall Cambridge.

Knowles, M. S., & Others. (1984). *Andragogy in action: Applying modern principles of adult education.* San Francisco: Jossey Bass.

Laurillard, D. (2002). *Rethinking university teaching: A conversational framework for the effective use of learning technologies* (2nd ed.). London & New York: Routledge.

Leu, D. J., Leu, D. D., & Len, K. (1997). *Teaching with the internet: Lessons from the classroom.* Noorwood, MA: Christopher-Gordon Publishers, Inc.

Meyer, E., & Land, R. (2003). Threshold concepts and troublesome knowledge. ETL Project Occasional Paper 4 Retrieved March, 17, 2008 from http://www.ed.ac.uk/etl/docs/ETLreport4.pdf

Meyer, E. & Land, R. (Ed) (2006). *Overcoming barriers to student understanding: Threshold concepts and troublesome knowledge.* London: Routledge.

O' Brien, S., & Ó Fathaigh, M. (2007). *Learning partnerships for social inclusion: Exploring lifelong learning contexts, issues and approaches.* Cork: Oak Tree Press.

Ó Fathaigh, M. (2002, March). E-learning and access: Some issues and implications. Paper presented to UACE conference, University of Bath, UK. Retrieved May, 18, 2008 from http://www.ucc.ie/en/ace/Publications/DocumentFile,19890,en.pdf

Ó Fathaigh, M. (2003, November). E-learning and access: Some European/Irish issues and implications. *Sloan-C conference 9th International Conference on Asynchronous Learning Networks (ALN)*, Orlando, USA. Retrieved May, 18, 2008 from http://www.sloan-c.org/conference/proceedings/2003/pdf/1610.pdf

Preziosi, D. (1992). The question of art history. *Critical Inquiry*, 18(2). 363-386.

Reynolds, C. L. (1999). Computer technologies and teaching and learning: What do the students and teachers say? In T. H. Murphy (Ed.), *Proceedings of the western region agricultural education research conference.* Texas A&M University Retrieved December, 10, 2007 from http://pubs.aged.tamu.edu/conferences/WRAERC1999/pdf/wr-1999-178.pdf

Salmon, G. (2002). *E-tivities. the key to active online learning.* London: Tayler and Francis.

Schön, D. (1983). *The reflective practitioner.* New York: Basic Books.

Schön, D. (1987). *Educating the reflective practitioner: Toward a new design for teaching and learning in the professions.* San Francisco: Jossey-Bass.

Shulman, L. S. (2005). Signature pedagogies in the professions. *Daedalus*, 134(3), 52–59.

Smith, M. K. (2002). Malcolm Knowles, informal adult education, self-direction and anadragogy, the encyclopedia of informal education. Retrieved December, 10, 2007 from http://www.infed.org/thinkers/et-knowl.htm

Tishman, S., & Palmer, P. (2005). Visible thinking. *Leadership Compass*, 2-3. Retrieved November, 30, 2006 from http://www.pz.harvard.edu/Research/ArtThink.htm

Virilio, P. (2000). *The information bomb.* (C. Turner, Trans.). Verso: London and New York.

Wang, M., & Kang, M. (2006). Cybergogy for engaged learning: A framework for creating learner engagement through information and communication technology . In M. S. Khint & D. Hung (Eds.), *Engaged learning with emerging technologies.* (pp. 225-253). Springer: New York.

Wang, M. (2007). Designing online courses that effectively engage learners from diverse cultural backgrounds *British Journal of Educational Technology,* 38(2), 294-311. Retrieved August, 10, 2007 from http://www.blackwellsynergy.com/doi/pdf/10.1111/j.1467-8535.2006.00626.x

Young, G., & Cronin, J. (2005). Using digital images in an art history department: the experience of University College. Cork' TASI (Technical Advisory Service for Images) Institute for Learning and Research Technology, University of Bristol Retrieved August, 20, 2008 from http://www.tasi.ac.uk/resources/cork.html

Chapter XIV
Structure of a Blended University Course:
Applying Constructivist Principles to Blended Teaching

M. Beatrice Ligorio
University of Bari, Italy

Nadia Sansone
University of Bari, Italy

ABSTRACT

In this chapter, the case of a blended university course will be described in detail. The main focus of this description will be on how some constructivist principles – such as knowledge building, active and self-directed learners, collaborative learning, communities of learners and practice - can be applied to compose the architecture of a blended university course. The course carefully integrates online activities with face to face meetings. Several educational models are also combined to guide the design of individual, small-group and collective activities able to exploit issues such as digital identities, E-Tutor, online role-play, and E-Portfolio. Principles of constructivism were always followed when setting activities and meetings. The description provided is mainly useful for teachers and educators interested in implementing a blended course with clear references to constructivist pedagogy. In addition, theoretically founded roles, tasks, and activities are outlined. The thoughtful mix of pedagogical models, online and face to face activities, individual-dyads-small group and collective learning contexts is the strongest point of this course.

INTRODUCTION

When applied to education, constructivism calls for a radical redesign of educational goals. Rather than supporting the increase of students' knowledge, the focus should be on the activity the person can perform in a content domain. Students should not only acquire information, but they should be encouraged to put in practice what they learn. In this way, they can develop respect and confidence in the power of their mind and extend that power to think more generally about themselves as cultural agents and about their relationship with the environment in which they are surrounded by (Bruner, 1996; Cole, 1996). To obtain such results, it is necessary to improve the awareness of connections between human activities and the sociocultural contexts within which these activities take place. This means also becoming more aware of the cultural power technology plays which should not be considered merely as a content delivery mechanism, but rather as artefacts able to support human capacity for developing new culture (Engeström & Escalante, 1996; Wartfosky, 1973). Our contemporary culture is based – at least in developed countries – on advanced technology (i.e., web-based environments, digital objects, etc.) and therefore academic and educational contexts are called upon to develop positive, cultural models of how to use this technology. The natural disposition of youth to be excited about new and creative ways of using the Internet should be utilized as leverage to empower educational models based on recommendations constructivism offers.

We believe constructivism does not necessarily imply the complete replacement of previous educational models. On the contrary, well established models can be renewed and can offer hints to constructivism especially when new technologies and new forms of peer interaction – also face to face interactions - are introduced. The ultimate result is an architecture for teaching where constructivism has been developed into a rich and well designed pathway, which blends face to face with online meetings, where students performe many types of activities and build a wide array of products. This chapter offers a detailed description of such a course, which we believe would be particularly useful for teachers interested in implementing a blended course with clear reference to constructivist pedagogy.

THEORETICAL BACKGROUND

Constructivism and the Blended Approach

The constructivist view of education stresses how important it is to have active learners, able to self-direct and, eventually, re-direct their own learning processes. Active learners understand new information by doing something with it and are keen to enjoy group work because this enables them to compare different points of view and to reflect upon the multiple aspects of reality. In fact, constructivism sees individuals treating information and experiences by inevitably assigning meanings; therefore, learners are constantly involved in a process of sense making (Bruner, 1986; Cole, 1996; 1991; Gergen, 1994; Vygotsky, 1978). In short, the goal of education should be to equip students with an adequate and sophisticated apparatus for sense making and sustain learners' self-perceptions as active knowledge builders. Technology seems to be able to support and empower this type of learner (Crook, 2002; Salmon, 2002; Scardamalia & Bereiter, 1994).

The general guidelines and goals defined in the light of socioconstructivism need to be detailed when applied practically to a course. In the case of the course we are going to describe here, we found the blended approach (Alvarez, 2005; Bonk & Graham, 2006) a useful way to apply constructivism to higher education. In other words, the general principles inspiring the architecture of the course are indeed based on a constructivist vision of learning, but in order to apply constructivism, suggestions from the blended methodology have been employed. "Blended", in our interpretation, does not involve only mixing face to face and computer supported interactions. Instead, an extensive vision of the blended approach is embraced, which proposes to blend pedagogical methods, individual study and group activity, and a large variety of tasks and end products. Different teaching and learning strategies are also structured and many options and activities are offered to the learners. The specific mix of all these elements is based upon the specific needs, constraints, and potential of the particular setting for which the course is planned. We believe the blended approach has some advantages over what is labelled as e-learning. Compared to e-learning (Clark & Mayer, 2007; Horton, 2006), the blended approach maintains the relevance of face to face meetings and activities as contexts within which the educational mission can be effectively delivered, and as able to empower computer based resources. Blogs, wiki, YouTube, MySpace, and so on, are often included in e-learning courses (Dron, 2007; Holmes & Gardner, 2006), but students do not always have a clear idea of how to use these tools in an educational way; they do not easily see how or why this type of technology can support their learning processes. In these cases, the blended approach can be a successful procedure to sustain the encounter between learners and technology.

In the paragraphs below we will explained and detailed the rationale of the blended approach used in this course.

Blending Pedagogical Models

The main pedagogical inspiration for this course comes from the Progressive Inquiry Model (PIM) (Hakkarainen, Lipponen & Järvelä, 2002). This model suggests considering learning as an inquiry process starting from general and broad questions and proceeding toward fine critical and scientific thinking. A set of steps are proposed, starting from the negotiation of a general research question that will guide all the subsequent activities and continuing with argumentation based upon reading and confrontation among discussants.

Students are not so much required to find answers, but rather to generate new and more sophisticated questions. For example, in the case of the course on e-learning we are describing here, a good starting question is "What type of learning does e-learning support?" Once students start to read relevant materials about e-learning, they should start proposing more specific questions such as "Is e-learning constructivist? Under what conditions is e-learning constructivist? What is a suitable technology?"

Another pedagogical model inspiring this course is the Jigsaw (Aronson & Patnoe, 1997). The Jigsaw model is based on the idea of specialised groups working on subparts of a wider common task. Further specialisations are possible within the groups; in fact, each component can deepen a specific aspect of the subpart of the task assigned to the group. Personal and group specialisations are later combined by cross talks and peer teaching. In this way, a community of learners can be developed where "teachers" and "learners" are no longer traditional roles, but rather are temporary positions occupied during the group or individual performances (Brown & Campione, 1990). An adapted version of the model was developed for this course. Through their individual learning, students became experts on a small part of the task; later, they brought into the group their expertise to accomplish the group task. A specific educational material (for instance a paper, a chapter or a set of slides) was assigned to each student; subsequently, students were required to enter the discussion as "experts" on the material assigned to them, reporting the "voice" of the author(s) they read. The discussion within the groups was aimed at orchestrating the various "voices" and therefore forming a complete picture of the group's task. Cross-group discussions (online and face to face) were also proposed with the aim of comparing groups' outcomes and integrating different perspectives on the same task. During the cross-group discussions, a sense of community was supported because the perception of working on the same general task emerged. The concept of "voices" and of the consequent orchestration was inspired by the theory of Bakhtin (1930). We found this theory useful to stress the dialogical nature of a web-forum discussion, where students can report their own voice as well as other voices, gaining an extension of the space of debate and a perception of a positive multiplicity of perspectives (Koshmann, 1999; Wegerif, 2007).

Reciprocal teaching (Palincsar & Brown, 1984) is an additional pedagogical model used in this course. The reciprocal teaching model is particularly useful to foster students' responsibilities and to sustain group dynamics. The teacher initially models the educational activity and later students, in turn, take the lead in activities such as reading, asking questions and figuring out how reading will continue. In this way, students can profit from many examples of how to play the teacher role.

Finally, the lecturing model is also integrated into this course. Lecturing is completely the role of the teacher and it is performed face to face. Through benchmark lessons (diSessa & Minstrell, 1998), the teacher offers definitions of concepts, main ideas and critical issues for each unit.

All these models are intended to enhance higher levels of thinking and argumentation, critical reading, academic writing, and representational skills. We believe a well designed combination of these models should allow an even greater development of the skills.

Indeed, all the skills are crucial for active learners – which is the type of learner constructivism aims to train - in general, but also for experts on e-learning which is the desired final result of this specific course. In the next paragraph, we will describe in detail how the course sustains these skills.

THE COURSE: ORGANISATION AND CONTENTS

The University course described here is offered at the specialised level of Work Psychology curriculum, within the faculty of Psychology at the University of Bari (IT). The course was introduced as: (a) innovative, (b) following constructivist principles, and (c) based on a blended approach. The first innovation is to have students organized in small groups. Each student is required to be an active learner within his/her group and to take responsibility for achieving common goals by contributing to the definition of a joint enterprise (Wenger, 1998). As we know (Blumenfeld, Marx, Soloway & Krajcik, 1996; Dillenbourg, 1999), small groups are the engine of collaborative learning. Activities performed by larger groups are only possible when small groups tune their processes and dynamics. Therefore, in this course, we first propose small group activities and discussions. Considering that the recommended group-size is about six-eight individuals, the number of the groups to be formed depends on the number of students attending the course. In our case, 23 students – three males and 20 females, with an average age of 24 - attended the course; hence, three groups were formed randomly.

The course lasted 13 weeks and was divided into seven units. The first six units covered the educational content of the course. The last unit was devoted to the collaborative building of a checklist that functioned as an assessment tool. The "content" units were composed by items of different format, i.e., a chapter, a journal article, a website or a set of slides. The digital version of this material was posted online in a folder created in each unit called "Educational Material." The material was carefully selected by the teacher and fulfilled several criteria. Specifically, the material had to (a) be relevant for the unit, (b) provide different and sometimes even contrasting points of view on a certain topic, and (c) vary as regards difficulty and length so that students could differentiate their effort depending on contextual elements that will be later clarified.

The six educational content units were titled as follows: (1) Online Educational Models, (2) E-Tutor, (3) Digital Identity, (4) Learning Objects, (5) Open Source, and (6) New Trends. The ensemble of these units was aimed at giving a fairly good knowledge of what e-learning means, its main issues and its problematic aspects.

THE RATIONALE FOR BLENDING ONLINE AND OFF LINE

In this course, face to face meetings were held once a week with about two hours allotted for each meeting. During the meetings, participants had no access to computers.

Two types of meetings were planned: (1) Meetings with a standard routine, and (2) Discussion meetings.

The standard routine meetings were considered as the starting point for the unit. During the first standard meeting, the teacher presented the course and described the activities. The other standard meetings were composed of three steps: (a) start-up, (b) teacher lecturing, and (c) close down. The discussion meetings, on the other hand, were considered as a consolidation of the work performed previously. Three discussion meetings were scheduled in total during the course, with two types of discussion taking place: groups discussion and plenary discussion. Details about the face to face meetings are given in Table 1.

Table 1 is particularly useful for teachers and educators because it gives an easy-to-read overview of the structure of the face to face meetings and the difference between the two meeting types.

From the educational point of view the discussion meetings allowed students to accomplish many of the recommendations constructivism offers. They could improve their collaborative strategies by following-up and finalising the online discussions; they could improve the knowledge building process by comparing the points of view emerged on a topic; metacognition was advanced by reflecting upon the interactive processes involving the groups. The discussion meetings also served research aims. In fact, the analysis of such discussions made it possible to understand similarities and differences between online and face to face discussions. Furthermore, by comparing the three discussion meetings, it was possible to understand the development over time of the discussion strategies, group dynamics, and the general effects of the course.

To perform the group discussion, each group was located in different corners of the room or, when possible, in different rooms. The plenary discussion, by contrast, involved all the students, and groups could compare their points of view on the same topic.

Basically, the orchestration of small group discussions and plenary discussions mirrored the structure of the online discussions. In fact (see next paragraph) inter- and intra-group discussions occurred online also.

The discussion topic concerned the unit active at that moment and it was always introduced by a 30-40 minute teacher lecture.

By describing the online activities in the following paragraphs, we will clarify the connections between the online and face to face activities. However, first we illustrate the platform used for this course.

The Platform

The selected platform - Synergeia (bscl.fit.fraunhofer.de) – is well suited to constructivist principles because it is designed using the Progressive Inquiry Model. To reiterate, this model considers

Table 1. Synopsis of the face to face meetings

Standard meetings (about 10 in total)	Start-up: 30-40 minutes for teacher introduction or critical friends report
	Teacher Lecturing: 30-40 minutes on basic concepts and relevant issues about the new unit; negotiation of a new research question
	Close-down: 30-40 minutes' collective discussion about the organisation of online activities (roles, timing and specific contextual matters)
Discussion meetings (three in total)	Group Discussion: 30-40 minutes
	Plenary Discussion: 30-40 minutes
	Teacher Lecturing: 30-40 minutes

learning as an inquiry process that allows learners to go from general and broad questions to fine critical and scientific thinking. Synergeia is a free platform developed during a European project (www.euro-cscl.org/site/itcole). It is provided with many tools able to support knowledge building and critical, reflective thinking. It also allows students to share documents and ideas to create and present knowledge artefacts.

Synergeia's main tool is the discussion forum. Three types of discussion forum are possible within the platform: informal, problem-oriented, and knowledge building-oriented. Depending on the type of forum selected, different labels for notes are available. These labels refer to *thinking types*. Thinking types are scaffolds meant explicitly for knowledge building. In fact, before entering a note, writers are required to choose a specific label for their notes and therefore they are forced to reflect upon the contribution of their notes to the existing discussion (Is it proposing a working theory? Is it deepening the knowledge already built? Is it stating a problem?).

Synergeia offers other interesting tools supporting critical and creative thinking. For instance, Maptool is a graphics tool that combines a chat and a shared white-board. Students can work with it synchronously to construct concept maps. The chat window allows coordination of the task and discussion about the drawing of maps. To build the maps, students can use tools such as simple shapes (triangles, squares, circles), arrows and text.

Another interesting tool offered by Synergeia is the so called virtual post-it that makes it possible to add comments to a document in the form of a yellow icon, similar to a post-it note. Within the post-it, replies are possible so that specific discussions on a document are traceable.

Many tools are also available to check and supervise students' activities within the platform, as well as the history of each document posted on Synergeia. Footprints, stars and glasses are icons representing tools containing information about all the objects stored in Synergeia: who created, read, deleted or edited them. Furthermore, a tool is available for queries about each individual student within each single folder. Finally, a daily report is released by Synergeia's server, containing all the activities occurring within the online course.

Online Activities

In between the face to face meetings, students were required to perform a set of online activities. Once they have completed the initial steps regarding registration and familiarisation with the platform, students gradually fill in the digital space made available for the course. The process of appropriation of the digital environment and, at the same time, of building and sustaining a virtual community is reinforced when a set of meaningful and interesting activities is proposed by the teacher and partially negotiated with the students. The online activities are described below.

- *Role-play.* A consolidated tradition of studies has dealt with the use of the face to face role-play in educational contexts (Blatner, 2002; Bonnet, 2000; Van Ments, 1983). When playing a role, students are cognitively and emotionally involved, as they feel the action they are performing concerns not only their skills, but also who they are and how they will be perceived by the others. These aspects make role-play an effective educational tool (Linser 2004). Recently, role-play as a didactic practice has seen widespread use in virtual environments (Bell, 2001; Freeman & Capper, 1999). Virtual role-play, being free from rigid temporal boundaries, offers time to elaborate information while new mental models can be formed about what a particular role implies (Fannon, 2005). A few roles were planned in this course. All of them were designed around abilities and professional skills relevant for e-learning and each role was aimed at encouraging students to become active learners. Students covered the roles in turn so, at the end of the course, every student could play each role at least once. The online roles played by the students included: (a) tutor, (b) summariser, and (c) critical friend.

a. *Tutor.* This role is focused on group management and support to group discussion. The tutor is required to invite everyone to participate to the discussion, to read the educational material assigned and to use it during the discussion. Furthermore, the tutor keeps the discussion focused on the research question guiding the unit. This role is assigned from the second unit on. During the first unit an apprentice, who had previously participated in the course, modelled the role.

b. *Summariser.* The main task of this role is to summarize what has been discussed during the unit. This task is carried out in pairs. The students covering this role produce a brief summary (500 words) representing the answer to their research question, and often they refer to their summary during the group discussion. The summarisers start their task as soon as the tutor-supervised discussion is over, usually having three to four days to accomplish the task.

c. *Critical friend.* This role was designed to favour cross-group collaboration. One member of a group reads and comments on the summary produced by the summarisers of a different group, then writes a short report about the group's summary. The objective is to improve the quality of the summary and to sustain critical reading. The critical friend works at the end of the unit, with two to three days usually allotted to accomplish the task.

All these roles are first played by the teacher – with the only exception being the tutor who models how to play the role; afterwards, students cover the roles.

This set of roles is meant to support constructive social interaction and knowledge building as well as a sense of the students challenging themselves. In fact, students found themselves acting in ways they would not normally, and the impact of role-play could be observed on self-representation and development of new identity positioning (Hermans, 2004).

- *Reading papers and writing individual reviews.* Each unit contains a number of educational materials corresponding to the number of students composing each group. The reading material has different degrees of difficulty. Students not covering any particular role for that unit, read the most difficult material, whereas tutors, summarisers and critical friends read easier material. Each student had to read the document the teacher assigns to him/her and write a short critical review (three hundred words maximum) about it. In writing the review, students should cover the following points proposed by the teacher: (a) report the main points of the material read, (b) outline the contribution of the material to the research question of the unit, (c) describe unclear and underdeveloped points, (d) give a personal opinion, and (e) compare the paper with the previous materials read.

- *Reading the critical reviews.* Students post their individual critical reviews in a virtual folder placed in their group space. These reviews should be read by each of the members of the group. This is an important starting point for the group discussion, because it allows students to enter the discussion by reporting both the document's point of view and their own point of view. In this way, naïve discussions are avoided, and personal points of views can be compared and enriched by the scientific perspectives.

For each unit, the teacher read and commented on two individual reviews for each group. They were selected so that at the end of the course each student had at least two reviews with comments. The commented versions of these reviews were placed in a special folder called "Reviews commented on by the teacher." All students in the course were required to read the commented reviews. Teachers' comments were therefore socialised and discussed. Through this practice, the teacher modelled how to read and comment on a paper. This skill is useful for the critical friend, and supports an understanding of the type of reading and writing skills needed in higher education. To check that students read the reviews, the teacher simply clicked the virtual glasses (see the description of Synergeia).

Students could track their own progress in writing by comparing the first corrected reviews with their most recent one (see the paragraph on Self-evaluation).

- *Searching new materials.* Students were encouraged to search for new materials to better address the unit. The material selected could be posted online accompanied by a short justification reported on a digital post-it note (see Synergeia description). The justification was to contain information about author and/or website credibility, why the material was considered relevant for the unit, and how it could contribute to the inquiry on the unit's research question. This activity had a twofold aim: On the one hand, it supported the students' sensation of being active, by contributing to the selection of educational materials; on the other hand, students could reflect on the criteria of how to recognise valuable information obtained on the Internet. Students appreciated this practice and increasingly selected interesting educational material, although this meant an extra load for them in terms of reading and reporting to their colleagues on what they read.
- *Online discussions.* Three types of discussions were possible: informal, organisational and topic-specific. This latter type formed the core of the unit and it was guided by the Progressive Inquiry Model. The main purpose of the topic-specific discussion was to reach a common answer to the initial research question. Each student participated in the discussion by reporting the point of view of the material read and, of course, his/her personal processing of it. Reading other students' reviews enabled everyone to better understand the point of view that their colleagues brought into the discussion.

Each unit had its own topic-specific discussion, but the process of discussing and the performing activities was repeated for each unit. The unit's recursive structure made it possible to improve the discussion style and the appropriation of the Progressive Inquiry Model. About one week was allocated for the topic-specific discussion during which the groups were active in the forum monitored by the tutor. Immediately afterwards, a plenary discussion was introduced involving all the groups. The main objective of this cross-group discussion was to outline practical indicators about the unit. In other words, students were required to reflect on how the main points, discussed during the current unit, could be recognised when observing an e-learning experience. Therefore, the plenary discussion accomplished several purposes. First, it allowed more extensive discussion through which groups' points of view could be compared. Moving from small group discussion to larger group discussion is considered an important step for collaborative learning and community building. This type of shift was also organised during the face to face meetings and the comparison between online and face to face highlighted interesting differences and similarities, some of them reported in the conclusions section of this chapter. Second, the plenary discussions supported the transition from theoretical thinking (occurring during the small group discussion) to more practical thinking afforded during the plenary discussion which was devoted to outlining the indicators. Last, the list of indicators was aimed at building a final common tool, a sort of observational grid finalized during the last unit that students used to observe and analyze e-learning courses. This activity will be described in detail below.

Informal and organizational discussions were possible throughout the course. These were important spaces in which the sense of community was built up and various matters were faced and solved. Students were allowed to open up new discussion forums whenever they liked. For instance, forums were opened up to discuss the experience of covering a certain role. These spaces represented interesting opportunities for students to express their thoughts and feelings about their participation in the course.

- *Building concept maps*. Before moving to a new unit, students were required to represent the topic-specific discussion by building a map using the Maptool (see Synergeia description). Specific thinking and argumentative skills were supported through this activity. In fact, concept maps are considered to foster learning through the recognition of primary concepts of knowledge and the relationships between concepts (Novak, 1977; Novak & Gowin, 1984). Maptool made it possible to build maps by chatting in synchronous mode. Students (especially those covering the role of summarisers) set a day and time to meet online. Preparatory face to face meetings were encouraged to prepare draft maps before logging in. The draft maps could be uploaded into the shared whiteboard provided by Maptool and edited, changed and reorganised during the session. The chat-logs and the final maps were stored in a folder and students could discuss and comment on them. In this way, reflection was fostered on the process of building a concept map, and on the differences between composing a text and a map.
- *Building a collaborative artefact*. Students were required to build a checklist aimed at guiding e-learning course observation. As mentioned previously, during the plenary discussion at the end of each unit, students

had to agree on a list of indicators concerning the unit. The final unit of the course was devoted to defining the checklist that will actually be used to observe a set of e-learning courses selected by the teacher. The construction of this list was meant to be a plenary work, although students spontaneously tended to coalesce themselves into groups. Each spontaneous group took charge of defining the indicators of one unit. Besides supporting the sense of a shared enterprise (Wenger, 1998) and the reorganisation of groups as suggested by the Jigsaw model, this activity was meant as an occasion for students to re-examine units that were unclear or not fully exploited. Therefore, students themselves were invited to select the indicators pertaining to the units in which they did not fully participate initially. About one week was allowed to finalise the grid and another week could be used to test the grid. The teacher prepared a list of the courses available to be observed. Each course was explored by one student from each group, so – having three groups - small clusters of three students were formed. A further, albeit adapted, application of the Jigsaw method was therefore achieved. Whilst observing the e-learning courses, the grid could be further edited and upgraded. Modifications proposed to the grid were plenary discussed in a forum dedicated to this activity. At the end of the course there was an empty grid and as many filled-in versions of it as the number of courses observed.

- *Self-evaluation.* For each unit students had to fill out a self-evaluation sheet composed of several questions. Students were required to describe qualitatively how much each activity - paper reviews, role-play, online and off line discussion, concept maps, and final grid–contributed to their learning, both in terms of content and skills. The aims of the self-evaluation sheet were to stimulate students' meta-cognitive processes and reflection on their own abilities and skills, as well as to support the development of critical self-evaluation. The structure of the sheet allowed the students to follow their learning and participation throughout the course; in fact, the sheet required repeating the assessment for each unit. All the students' sheets were collected in a specific folder that also hosted discussions of the activity. At the end of the course, the self-evaluation sheets contributed to the global evaluation of the students, together with the other individual and collective artefacts each of the students produced. To further support the self-evaluation process, students were invited to open personal folders where they could store their personal self-evaluation sheet, all the papers, reviews and maps they produced or made contributions to, and a selection of notes they considered as being most representative of their participation in the forums. This folder can be considered as a student's e-portfolio and as a meta-folder where students challenge themselves and the course.

Purpose and Method

The purpose of this chapter is to give an example of an architecture of a course able to put into practice principles coming from constructivism. The blended approach is used as a paradigm to mix face to face and online meetings to reciprocally enrich these two types of contexts. Furthermore, pedagogical models proven to be successful are mixed and reciprocally enriched. The ultimate aim of this chapter is to give teachers and educators indications of how to organize a blended course with little technology required – only the access to a platform – and where constructivism can be seen in practice.

To understand the efficacy of the course we used a method where the teacher's assessment was combined with the researcher's analysis and the students' feedback. The teacher assessed the students through two dimensions: the reviews students wrote and the online and off line discussions. Although the teacher read only two of the reviews for each group, a fairly good understanding of the general level of the task performance could be obtained. The socialisation via the commented reviews pushed all the students to re-read their own reviews and adjust their writing on the basis of the teacher's comments. Furthermore, the comparison between the initial reviews and the reviews produced at the end of the course, shows the changes in the students writing. In particular at the beginning of the course, reviews lacked the following elements: (a) academic style - colloquial style was more common; (b) ability to distinguish personal opinion from a scientific source; (c) the use of meta-text. During the course, students increasingly used these elements in their reviews and they became able to assess the material produced by their colleagues based on these elements.

The teacher assessed quantitatively the individual participation in the discussions according to how many notes were posted online and how much students talked during the face to face meetings. A qualitative assessment was also possible via the process of progressive inquiry and the quality of the students' end products. Furthermore, the quality of the final grid, and the way the grid was used to observe e-learning courses, gave many insights about students' learning and understanding. By combining all these elements the teacher could state whether each student and each group passed a unit at an appropriate level. The final grade assigned to each student was obtained by evaluating the participation and productivity, both online and off line, throughout the course. Actually, the whole architecture of this course was aimed at "forcing" students to be active, offering them many opportunities to go back to units or issues not well understood. Students' constructive participation was facilitated by the many products they had to submit to the teacher. In other words, failures were avoided by requiring students' constant commitment and by offering repeated routines.

From the research point of view, there was an enormous quantity of data collected. All the discussions, both online and off line (the latter of which were systematically video-recorded), and the material produced by the students (e.g. reviews, unit summaries, critical reports, maps, empty grid and its filled versions) could be analysed under different points of view. First, a longitudinal analysis is possible by comparing discussions and products at different points in time, in order to see the evolution of the course. Comparisons are also possible to track vis-à-vis the changes in the way the discussions were performed and how the various roles were played. Changes in the discussions can be studied by looking at many indicators. We have already examined – to some extent- the connections and reciprocal impacts between face to face and online discussions. For instance, in one study, the Social Network Analysis was performed to examine how participation evolved during the course, by comparing face to face and online discussions on the same topic and by analysing discussions occurring at different moments (at the starting of the course, half way, and at the end). Students were observed to present more democratic participation strategies when discussing online, whereas off line discussion strategies improved after performing online activities (Ligorio, Annese, Spadaro & Traetta, 2008).

A focus group discussion was also performed at the beginning of the course and another one at the end of it. During the first one, students were required to express their expectations; during the last one, students reported how they felt about the course and how they thought it could be improved. By qualitatively comparing the two focus groups, it was noticed that not only did students master collaborative strategies, but also their discourse strategies became more collaborative (Cucchiara, Spadaro & Ligorio, 2008). Students used more reciprocal reparations and the tendency to complete each other's sentences became more evident and frequent. Both these strategies are considered indicators of collective thinking and of a shared understanding of the topic under discussion (Schegloff, 1987).

Also positioning and participation styles changed over time. Students moved from an individual positioning toward a more social positioning. The movement from the periphery toward the centre of the activity was also observed, but it was not as linear and simple as that suggested by the community of practice model (Wenger, 1998). The trajectories we found were quite complex and they were sensitive to many dimensions, such as the role played, the topic under discussion and the general evolution of the small groups (Ligorio, Annese, Spadaro & Traetta, 2008).

Students offered feedback in many ways during this course, so we gathered a good understanding of how they perceived the course and of how they felt when participating in it. Furthermore, a final questionnaire was administrated through which they were required to assess the course and their own learning during the course. By examining the questionnaire, we found that students clearly sensed that this was a course where their active participation was needed. They considered this course quite demanding and hard, but they also recognized they learned crucial concepts about e-learning. They also learned how to work collaboratively in group, how to read critically and write academically, how to assess themselves, and, most of all, they learned to challenge themselves in new contexts and to be more self-confident.

CONCLUSION AND FUTURE TRENDS

In this chapter a course for higher education about e-learning was described. It is showed how concepts related to constructivism such as knowledge building, active and self-directed learning are put in practice in a blended course. The activities composing the course are all well rooted in pedagogical models such as collaborative learning, digital identities, communities, online role-play and e-portfolio.

A peculiarity of this course is that the blended approach was used as a strategy for putting into practice many of the suggestions of constructivism. In fact, the complex architecture described in the chapter allows a mixture of different educational models, an alternation of individual learning activities, and various types of collaborative learning and group activities. A sense of belonging to a community was also fostered through: (a) group and plenary discussions, (b) reflections on the roles students covered during the course, and (c) the construction and testing of a collective final artefact. All through the course, face to face meetings and online activities were carefully interwoven.

As already mentioned this chapter is meant to give an accurate overview of the architecture of the course and of its theoretical foundations. Therefore, we here only report briefly the general result concerning the validity of the course architecture and of the students' learning and performance. The feedback obtained from the students and their output in terms of participation and productivity, indicates that the structure described above is motivating and effective. Students gradually appreciate a course like this because it requires them to be active and overcome learning strategies based on studying by rote.

Many aspects of this course should be further analysed. We briefly list some of them, in particular those we think may have both a practical relevance – i.e. which could lead to an improvement of the course architecture, and a theoretical value, as they could foster a reconceptualization of some concepts concerning teaching and learning through e-learning and through the blended approach.

One of the aspects still to be analysed concerns the relationship between the teacher and the tutor.

Both of them were required to be very active and constructive. The teacher had to master the whole course architecture and understand the theoretical underpinning of each activity proposed. It was the teacher's responsibility to be attuned with the students' needs and to find the ideal balance between adapting the structure to the individuals' and groups' requirements while maintaining the constructivist and collaborative nature of the course. The tutor was required to be connected daily, prompting and monitoring the discussions, scaffolding the activities, and tailoring feedback. Therefore, strong interaction between the teacher and the tutor was needed throughout the course. This aspect has not yet been analysed.

Another interesting issue to be further explored is the triangulation between the technology, the teacher, and the students. We assume this relationship is very complex and a more in depth data analysis will be needed.

Among the aspects worthy of a finer understanding there are also the specific discussions around the role-play. One issue that is possible to be addressed by such an analysis is whether skills acquired by playing a certain role are still used when the role is no longer covered. Such retention could be considered an indicator of the role appropriation, and it could be included in the package of skills learned by the students.

Another activity that so far had not received adequate attention is the process of building conceptual maps. Questions remain to be asked, for example: What is the added value for students in collaboratively building a conceptual map after having written a text? Additional information could also be gathered by analysing discussions related to students' self-evaluation and the activities occurring in the personal folders.

As a consequence of the exposure to the platform and to the content of the course, we believe technological skills should increase. In particular, the educational vision of technology could be fostered. As stated in the introduction, often students know what the Internet is and how to use it, but they lack exposure to the learning potentialities of it. Participation in a course like this seems to offer this opportunity.

NOTE

The course presented in this chapter was designed during a National Project 2005-2007 funded by the Italian Ministery for Research and Instruction. We would like to thank all the researchers participating to this project. We also would like to thank Neil Schwartz for his valuable help in editing the chapter.

REFERENCES

Alvarez, S. (2005). Blended learning solutions. In B. Hoffman (Ed.), *Encyclopedia of Educational Technology.* Retrieved January 10, 2008, from http: //coe.sdsu.edu/eet/articles/blendedlearning/start.htm

Aronson, E., & Patnoe, S. (1997). *The jigsaw classroom: Building cooperation in the classroom (2nd ed.).* New York: Addison Wesley Longman.

Bakhtin, M. M. (1930). *The Dialogic Imagination: Four Essays.* C. Emerson & M. Holquist (Eds.) (Trans.), (1981). Austin, TX and London, UK: University of Texas Press.

Bell, M. (2001). A case study of an online role-play for academic staff. In G. Kennedy, M. Keppell, C. McNaught, & T. Petrovic (Eds.), *Meeting at the Crossroads. Proceedings of the 18th Annual Conference of the Australian Society for Computers in Learning in Tertiary Education* (pp. 63-72). Melbourne, VIC: Biomedical Multimedia Unit, The University of Melbourne.

Blatner, A. (2002). *Role Playing in Education. Adam Blatner's Web site.* Retrieved May 12, 2008, from http://www.blatner.com/adam/pdntbk/rlplayedu.htm

Blumenfeld, P. C., Marx, R. W., Soloway, E., & Krajcik, J. (1996). Learning with Peers: from small group cooperation to Collaborative Communities. *Educational Research, 25*(8), 37-40.

Bonk, C. J., & Graham, C. R. (Eds.), (2006). *The Handbook of Blended Learning: Global Perspectives, Local Designs*. San Francisco, CA: Pfeiffer Publishing.

Bonnet, C. (2000, May). The relevance of role playing in environmental education. *Proceedings of The International Union of Biological Sciences Commission for Biological Education (IUBS-CBE): International Symposium on Biological Education,* IUFM Versailles, Centre de Cergy, France.

Brown, A. L., & Campione, J. C. (1990). Communities of learning or a content by any other name. In D. Kuhn (ed.), *Contribution to human development* (pp. 108-126). New York: Oxford University Press.

Bruner, J. (1986). *Actual Minds, Possible Worlds*. Cambridge, MA: Harvard University Press.

Bruner, J. (1991). The Narrative Construction of Reality. *Critical Inquiry, 18*, 1-21.

Bruner, J. (1996). *The culture of education*. Cambridge, MA: Harvard University Press.

Clark, R. C., & Mayer, R. E. (2007). E-Learning and the Science of Instruction: Proven Guidelines for Consumers and Designers of Multimedia Learning, CA: Pfeiffer.

Cole, M. (1996). *Cultural psychology. A once and future discipline*. Cambridge, MA and London, UK: The Belknap Press of Harvard University Press.

Crook, C. K. (2002). The Social Character of Knowing and Learning: Implications of cultural psychology for educational technology. *Journal of Information Technology in Teacher Education, 10*, 19-36.

Cucchiara, S., Spadaro, P. F., & Ligorio, M. B. (2008). Identità e comunità in contesti collaborativi: un'esperienza blended universitaria [Identity and community in collaborative contexts: a blended university experience]. *Qwerty, 1*, 23-48.

Dillenbourg, P. (1999). What do you mean by collaborative learning? In P. Dillenbourg (Ed.), *Collaborative-learning: Cognitive and Computational Approaches* (pp. 1-19). Oxford, UK: Elsevier.

diSessa, A., & Minstrell, J. (1998). Cultivating conceptual change with benchmark lessons. In J. Green, & S. Goldman (Eds.), *Thinking Practices in Mathematics and Science Learning*. Mahwah, NJ: Lawrence Erlbaum Associates.

Dron, J. (2007). *Control and Constraint in E-learning: Choosing When to Choose*. New York: Idea Group Inc.

Engeström, Y., & Escalante, V. (1996). Mundane tool or object of affection? The rise and fall of the Postal Buddy. In B. A. Nardi (Ed.), *Context and consciousness: Activity theory and human-computer interaction* (pp. 119-161). Cambridge, MA: The MIT Press.

Fannon, K. (2005). *A Role-Play Simulation: Transformative Learning in Complex Dynamic Social Systems*. Retrieved May 4, 2007 from http://www.roleplaysim.org/papers/NeedleStick_web.pdf

Freeman, M., & Capper, J. (1999). Exploiting the web for education: An anonymous asynchronous role simulation. *Australian Journal of Educational Technology, 15*(1), 95-116. Retrieved May 11, 2008, from http://www.ascilite.org.au/ajet/ajet15/freeman.html

Gergen, K. J. (1994). *Realities and relationships: soundings in social construction*. London, UK: Harvard University Press.

Hakkarainen, K., Lipponen, L., & Järvelä, S. (2002). Epistemology of inquiry and computer-supported collaborative learning. In T. Koschmann, N. Miyake, & R. Hall (Eds.), *CSCL2: Carrying Forward the Conversation* (pp. 129-156). Mahwah, NJ: Erlbaum.

Hermans, H. (2004). Mediated identity in the emerging digital age: A dialogical perspective. *Identity: An international Journal of theory and research, 4*(4), 297-405.

Holmes, B., & Gardner, J. R. (2006). *E-Learning: Concepts and Practice.* London: Sage.

Horton, W. (2006). *E-Learning by Design.* CA: Pfeiffer.

Koschmann, T. (1999). Toward a dialogic theory of learning: Bakhtin's contribution to learning in settings of collaboration. In *Proceedings of Computer Supported Collaborative Learning (CSCL '99)*, Palo Alto, CA (pp.308-313). Available at: http://kn.cilt.org/cscl99/A38/A38.HTM

Ligorio, M. B., Annese, S., Spadaro, P. F., & Traetta, M. (2008). Building intersubjectivity and identity in on-line communities. In B. M. Varisco (Ed.), *Psychological, pedagogical and sociological models for learning and assessment in virtual communities of practice* (pp. 57-91). Milan: Polimetrica.

Linser, R., (2004). *Suppose you were someone else: The learning environment of a web-based role-play simulation.* Retrieved July 27, 2005, from http://www.simplay.net/papers/suppose.html

Novak, J. D. (1977). *A theory of education.* Ithaca, NY: Cornell University Press.

Novak, J. D., & Gowin, D. B. (1984). *Learning how to learn.* New York: Cambridge University Press.

Palincsar, A. S., & Brown, A. L. (1984). Reciprocal teaching of comprehension-fostering and comprehension-monitoring activities. *Cognition and Instruction, 2*, 117-175.

Salmon, G. (2002). *E-tivities: The Key to Active Online Learning.* London, UK: Kogan Page.

Scardamalia, M., & Bereiter, C. (1994). Computer Support for Knowledge-Building Communities. *The Journal of the Learning Sciences, 3*(3), 256-283.

Schegloff, E. A. (1987). Recycled Turn Beginnings: A Precise Repair Mechanism in Conversation's Turn-Taking Organization. In G. Button, L. Lee (Eds.), *Talk and Social Organization.* Clevelon, England: Multilingual Matters.

Van Ments, M. (1983). *The effective use of Role-Play.* London, UK: Kondon Page.

Vygotsky, L. (1978). *Mind in Society.* (Trans. M. Cole). Cambridge, MA: Harvard University Press.

Wartfoskij, M. (1973). *Models.* Dordrecht, NL: D. Reidel.

Wegerif, R. (2007). From Dialectic to Dialogic: A response to Wertsch and Kazak. In T. Koschmann (Ed.), *Theorizing Learning Practice.* Mahwah, NJ: Lawrence Erlbaum Associates.

Wenger, E. (1998). *Communities of Practice. Learning, Meaning and Identity.* Cambridge, UK: Cambridge University Press.

Chapter XV
Constructivism and Online Collaborative Group Learning in Higher Education:
A Case Study

Hwee Ling Lim
The Petroleum Institute, Abu Dhabi, UAE

Fay Sudweeks
Murdoch University, Australia

ABSTRACT

As educators utilize an increasingly wide range of technologies for facilitating interaction between distant learning parties, there are concerns over the ad hoc use of technology in online course design and activities that are not grounded in sound pedagogical frameworks. This chapter presents a case of a hybrid undergraduate course that is shaped by sociocultural constructivist principles. Survey findings on student experiences of online collaborative learning and group work processes in two constructivist-based learning activities are reported. Results reflecting sociocultural constructivist concepts of scaffolding and appropriation of shared knowledge are presented based on student learning experiences during online synchronous tutorials and collaborative team projects. The conclusion discusses the effectiveness of the two course activities in facilitating collaborative group learning and recommendations are offered to enhance overall student experiences of online collaborative-constructivist group learning processes.

INTRODUCTION

As educators access an increasingly wide range of technological tools for facilitating interaction between distant learning parties, there are concerns over the ad hoc applications of technology in online course design and activities that are not grounded in sound theoretical frameworks. This chapter reviews the philosophical assumptions of constructivism that influence interpretations of the form and function of educational interaction with specific focus on the branch of sociocultural constructivism. A qualitative case of a distance undergraduate course that is shaped by sociocultural constructivism is presented. The chapter also explains the influence of constructivist concepts on the instructional aims, virtual learning environment (VLE) design, the conduct and assessment of two main learning activities facilitated by computer-mediated communication (CMC) technologies, namely, online synchronous (chat) tutorial discussions and a collaborative team project.

Survey findings on student experiences of online collaborative learning and group work processes in the two learning activities are reported. Results reflecting sociocultural constructivist concepts of scaffolding and appropriation of shared knowledge are presented from student learning experiences during online synchronous tutorial discussions (Lim, 2006), and collaborative team projects (Sudweeks, 2003). The conclusion discusses the effectiveness of the two constructivist-based course activities in facilitating collaborative learning. Finally, several recommendations are offered to enhance the overall student experiences of online collaborative-constructivist group learning processes which may be of interest to researchers in the fields of CMC and educational technology, higher education faculty and other professionals involved in distance learning.

BACKGROUND

Although interaction is widely assumed to support learning, distance education literature offers various interpretations of its form and purpose, which are likely to stem from different theoretical beliefs of what constitutes knowledge and learning. This section reviews the philosophical basis of constructivism and discusses the sociocultural constructivist conceptualization of educational interactions in online learning and teaching contexts.

Objectivism and Constructivism

The objectivist view of the basic relationship between the individual and the environment is based on *realism*, which is "the doctrine that there is an independently existing world of objective reality that has a determinate nature that can be discovered" (Schwandt, 2001, p.176). Moreover, objectivist epistemology claims that knowledge, though produced by individual thought processes, is ultimately determined by real world structures and can be mapped on to learners (Jonassen, Davidson, Collins, Campbell, & Haag, 1995).

In contrast, constructivist philosophy is based on *subjectivism*, which holds that "all judgments ... are *nothing but* reports of an individual speaker's feelings, attitudes, and beliefs" (Schwandt, 2001, p.241-emphasis in original). Hence, constructivist epistemology claims that knowledge is a subjective interpretation imposed by the individual on the world. Also, since multiple individual interpretations would lead to multiple realities, no single interpretation is necessarily less valid than another (Jonassen, 1991a). Von Glasersfeld (1995) further argued that this stance is not extended to epistemological nihilism. Essentially, constructivism regards constructed knowledge as dynamic and subject to change when exposed to new perspectives during interaction. The two main schools of thought (radical and sociocultural constructivism) that emerged from this constructivist philosophical basis are explained later in this section.

Implications for Constructivist Learning and Teaching

Behaviourist and cognitivist learning approaches share most of the ontological and epistemological beliefs of objectivism. From both perspectives, learning is an attempt by the student to "mirror" the structure of the external world as interpreted by the instructor and gain this knowledge, which would be common to all learners. Teaching involves the transmission of information by the instructor that had been modelled on the "real" world. The ultimate goal of learning is to achieve complete and correct understanding of "reality" (Jonassen, 1991a).

In contrast, constructivism views learning as an active process involving individual interpretations of experiences, the sharing of multiple perspectives, and negotiation of meaning through interaction in authentic contexts. Teaching primarily involves establishing a facilitator relationship with learners for the purpose of scaffolding rather than strictly controlling learning processes. Hence, the aim of learning is not to reach complete understanding of some ultimate reality, but to gain self-awareness or reflexivity that would enable the learners to attain control and responsibility of their learning processes (Duffy & Cunningham, 1996).

In online education, given the belief that learners create multiple understandings, constructivist-based distance course objectives emphasize knowledge development processes rather than products (Ertmer & Newby, 1993) with course designs that reflect the conversational paradigm (Laurillard, 2002; Romiszowski & Mason, 1996). Hence, situated learning approaches (Brown, Collins, & Duguid, 1989) and strategies to reduce transactional distance (Moore & Kearsley, 1996) between distant learners are adopted which include student control of the learning path, accessibility to communication tools for supporting interaction, and goals-free evaluation through self-assessment or reflection on learning experiences (Jonassen, 1991b).

Sociocultural Constructivism and Online Educational Interaction

Within the constructivist paradigm, there are divergent perspectives on learning held by two schools of thought, termed radical and sociocultural constructivism. Both approaches acknowledge that learning involves interaction and individual cognitive activity but they differ in the emphasis placed on the primacy of each element's contribution to the learning process.

The epistemological basis for radical constructivism is widely attributed to von Glasersfeld who drew mainly from Piagetian theories on the nature of knowledge and cognitive development. The constructs of adaptation, equilibrium and interaction in Piaget's schema theory underlie radical constructivist view of learning as a product of self-organization (von Glasersfeld, 1989) involving an iterative process whereby interaction in an experiential world produces a state of mental dissonance in the individual, to be resolved by cognitive changes. Such mental changes entail the coordination of inner experiences with outer experiences, within the specific community, which would restore the individual to a state of equilibrium (von Glasersfeld, 1997). Essentially, radical constructivism regards cognitive self-organization as the primary process in learning with the implicit assumption of interactions in sociocultural practices serving as sources of perturbations.

However, sociocultural constructivism (Vygotsky, 1962), based on Vygotsky's work in developmental psychology, emphasizes the primacy of the social rather than individual processes in knowledge building. Hence, sociocultural theorists view learning as "enculturation into a community of practice" (Cobb, 1994, p.13) where guided social participation in shared knowledge construction, mediated by technical and/or psychological tools, provides learners with scaffolding enabling higher potentiality of cognitive growth. In particular, the construct of the zone of proximal development (ZPD) is used to explain the learner's potential capacity for intellectual growth when given guidance or scaffolding in the form of tutor/peer support through interaction. The sociocultural constructivist stance also assumes that from participation in social interaction, there is individual appropriation of the shared knowledge that s/he had a part in creating, which would in turn transform individual understandings (Rogoff, 1990). The social process is considered primary with the implicit assumption of the individual cognitive dimension. Therefore, social interactions are vital as they serve to provide opportunities for discourse and appropriation that influence intellectual development.

When guided by sociocultural constructivism, online course designs reflect the principle that group learning activities are opportunities for interaction during which there is sharing of multiple perspectives that is guided by tutor/peers, mediated by tools (such as print, web and/or CMC technologies, language), and lead to joint construction and individual appropriation of the shared knowledge. In particular, choices of CMC technologies are determined by their capabilities to facilitate interaction between distant learning parties. For instance, the asynchronous CMC mode enables delayed-time dialogue through applications such as e-mail, discussion forums and bulletin boards. Such interactions are usually text-based contributions that are composed, sent, accessed, without time and proximity constraints. However, the synchronous CMC mode requires communicating parties to be "present" at the same time for the dialogue to occur through services such as Voice over IP (VOIP) and Internet Relay Chat (IRC). Chat interactions are mainly text-based messages, composed and sent by parties simultaneously logged in chat rooms. Chat messages appear chronologically on-screen with preceding exchanges scrolling up and then off each party's screen at a speed corresponding to the conversational pace (Werry, 1996).

Even as technology facilitates engagement between learning parties, there have been concerns over the design of online courses that resemble electronic versions of traditional correspondence courses, offering limited interactional opportunities and providing content that "are no more than print-based text dumped online" (Cashion & Palmieri, 2002, p. 55). Additionally, Bonk and Cunningham (1998) cautioned that computer supported collaborative learning (CSCL) tools were being applied on an ad hoc basis by faculty who may not be informed by theoretical frameworks or conclusive studies on the capabilities of such tools in enhancing learning. Such a situation highlights the need to further current understanding of effective online course and learning activity design within a constructivist framework that facilitates knowledge construction processes and the transformations in individual understandings that occur when shared knowledge is "jointly produced and individually appropriated" (Rogoff, 1990, p.196).

The next section presents the case of an online undergraduate course and describes the influence of constructivist concepts on the instructional aims, conduct, virtual learning environment design, and assessment of two main learning activities.

THE CASE: ORGANISATIONAL INFORMATICS

Organisational Informatics (OI) is an undergraduate course available to third-year students from the School of Information Technology at Murdoch University (Perth, Western Australia). The evolution of the course to its current manifestation was reported in several publications (Lim, 2006; Lim, 2007; Lim & Sudweeks, 2008; Sudweeks,

2003; Sudweeks, 2004). This section presents an integrated description of the case from these main sources.

Course Aims and Assessment

The OI course, which focuses on computer-mediated work processes, aims to develop skills associated with the organizational aspects of information systems design and development; including the assessment and management of issues related to knowledge building organizations. The course topics include computer mediated communication, group dynamics, virtual organizations and communities, work and society in the information age, globalization, and computer-supported collaborative work. To foster active group learning, course assessment comprises mainly the deliverables associated with two learning activities: online synchronous (chat) tutorials and a team project involving the collaborative planning and presentation of a business proposal for a major function.

The WebCT Virtual Learning Environment

Organisational Informatics adopts a hybrid delivery design that offers face-to-face lectures and online tutorials to internal and external students who, respectively, undergo the course on-campus or via a distance learning mode. To reduce transactional distance between learning parties, there is significant use of Web Course Tools (WebCT) as a VLE where students have access to a rich variety of electronic resources for their educational needs and management of learning processes.

The VLE is organized into three main components: communication, unit materials, and administration (Figure 1). The communication component includes synchronous and asynchronous communication tools such as chat, bulletin boards, and e-mail. The administrative component supports course organizational services such as self-enrolment in tutorial groups through the Online Tutorial Signup System (OTSS), the dissemination of grades, access to the course syllabus (Learning Guide), and lecture/tutorial schedules. The unit materials component comprises three subcategories of learning materials: content materials, support resources, and assessment resources. Content materials and support resources provide access to main and secondary instructional materials such as iLecture (streamed audio links) and links to external sites. The assessment subcategory provides access to assignment resources such as project requirements and peer assessment forms.

WebCT offers a range of synchronous and asynchronous CMC technologies to mediate dialogic online interaction, which is a vital aspect of

Figure 1. Organisational Informatics virtual learning environment (adapted from Lim 2006, p.79)

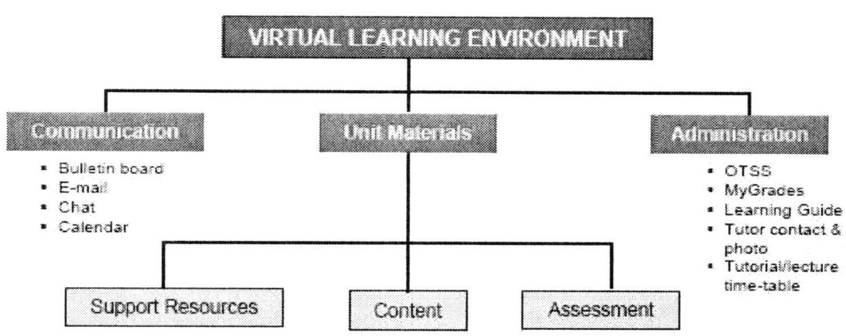

Figure 2. Organisational Informatics chat tutorial rooms (adapted from Lim 2006, p.83)

the two course instructional activities described below. Hence, the VLE design reflects use of information and communication technologies (ICTs) to facilitate engagement in chat tutorials and group work processes for the collaborative team project. Additionally, the ICTs support course administration and assessment; for instance, the electronic submission of coursework to the tutor via e-mail or posting of student critiques of readings (as journals) to bulletin boards.

Course Pedagogical Framework and Instructional Activities

As Organisational Informatics' pedagogical framework is based on sociocultural constructivism (Vygotsky, 1962), it regards learning as "a cycle of interpretation, evaluation and reflection of content evolving into individual and shared knowledge" (Sudweeks & Simoff, 2000, section 3). Hence, the two main learning activities described here, namely, the online synchronous tutorials and the collaborative team project, are designed to facilitate students' construction of knowledge through participation and reflection. Online interactions in these activities are supported by a range of WebCT CMC media including e-mail, bulletin boards, discussion forums and chat.

The Online Synchronous Tutorial

The online synchronous tutorials are held in WebCT chat rooms (Figure 2) and designed to introduce students to the theory and application of computer-mediated work processes which are directly related to the course topics. The tutorial activity aims to develop collaborative group knowledge building through active participation and sustain "students' continuous engagement in discovering and applying knowledge and skills in the context of authentic problem solving" (Sudweeks, 2004, p.92). The maximum number of students in each tutorial group is usually set at 16 and there is a tutor for each group. All groups undergo equivalent learning activities. Student preparation for the tutorial activity is supported by VLE resources such as the Reflective Journals which states the requirements for the critique; Guidelines for Tutorial Presenters which states the responsibilities of the presenter; and Ecoms Guidelines which highlights CMC conventions and netiquette.

As described in Lim (2006; 2007), the weekly chat tutorials are conducted in a seminar style, with a tutor-facilitator and one or two student presenters moderating the discussion. The presenter role is rotated among all the students in each tutorial group. In greater detail, presenters moderate a ½ hour discussion slot each in the one-hour tutorial session based on their critiques of the week's readings. During the tutorial, the presenter starts the discussion by highlighting the main issues in the selected readings, moderates and extends the discussion with questions, and encourages participation by all. Throughout the session, the tutor is present as a facilitator to evaluate the presenter's performance and the quality of

Figure 3. Tutorial logs (adapted from Lim 2006, p.85)

Group No.	Tutorial Day/Time	Week
Group 1	Monday 10:30	2 | 3 | 4 | 5 | 6 | 7 | 8 | 9 | 10 | 11 | 12 | 13
Group 4	Monday 19:30	2 | 3 | 4 | 5 | 6 | 7 | 8 | 9 | 10 | 11 | 12 | 13

contribution by other students. The other students are expected to participate actively in discussions, guided by peers' critiques as journals posted in bulletin boards prior to the tutorial session and evaluate the presenter with the aid of the archived discussion logs (Figure 3).

In congruence with the learning activity's pedagogical objectives, tutor/peer assessment of contribution emphasize participation quality, extent of collaborative effort, and sense of responsibility displayed. Essentially, the OI course's constructivist pedagogical framework is reflected in the tutorial activity design that involves critical review of readings, peer/tutor scaffolding of learning, dialogic exchange of multiple perspectives, and student reflection on learning.

The Collaborative Team Project

In conjunction with the weekly chat tutorial activity described earlier, the students are involved in a semester-long collaborative team project that requires the development of a business proposal for a major function by small teams in each tutorial group. Reflecting the course constructivist pedagogical framework that aims to sustain students' continuous engagement in discovering and applying knowledge and skills in authentic

Figure 4. Communication diary for a safari function in team project (Sudweeks, 2003, p.177, reprinted with permission)

	A	B	C	D
1	Tour the Wild	Group Communication Schemes		
2		Frequency	Time	
3	Category	(no of times)	(minutes/secs)	Overall Rating (out of 10) for Each Medium
4	Email	36	283.00	7.5
5	WebCT Private Mail	5	21.56	5
6	Bulletin Board	8	41.65	6.5
7	WebCT Chat	13	615.00	8
8	Internet Relay Chat (mIRC)	10	454.75	9
9	ICQ	13	330.89	10
10	Instant Messaging (Yahoo/MSN)	6	209.05	7
11	SMS	7	21.76	6.5
12	**Totals**	98	1977.66	
13				
14				
15				
16				
17				
18				
19				
20	***Summary and Findings*** : *There was more time spent on text based chat systems although emailing had*			
21	*the most number of frequencies. Overall ratings were based on* ***convenience***,			
22	***efficiency*** *and* ***effectiveness***			
23				

problem solving contexts, the project activity design fosters critical reading, thinking and argumentation skills. Preparation for the team project is supported by VLE resources such as online communication tools which enable interaction between team members and the delivery of the proposal presentation, as well as unit materials which include sample projects, main readings and peer assessment forms.

More specifically, the number of students in each project team is capped at four and four roles for team members are defined: Client, Consultant, Researcher, and Presenter. The Client is responsible for specifying the detailed requirements and budget for the function. The Consultant prepares the implementation steps and cost breakdown for the proposed function. The Researcher maintains a project diary and records the use of each communication medium (Figure 4). Specifically, the diary records the time, frequency, topic, and team members' reflections on the effectiveness of the communication medium in facilitating each type of communication topic (Figure 5). The Presenter integrates the project information and prepares an online presentation of the proposal to the entire tutorial group at the end of the semester. Typical functions that teams have selected include weddings, funerals, safaris, conferences, product launches, 21st birthday parties, concerts, and movie premiers. Throughout the project timeframe, a tutor and the course coordinator are available to scaffold the learning process through consultations via e-mail and the team discussion forums.

The unique aspect in the design of this learning activity lies in its restriction of communication media use among team members to only text-based electronic communication that encompasses e-mail (private and WebCT e-mail), bulletin boards, discussion forums, WebCT chat rooms, Internet Relay Chat (IRC), I Seek You (ICQ), Instant Messaging (IM) and Short Message Service (SMS). Such a restriction on text-based electronic communication media usage for supporting interaction between team members serves two main pedagogical purposes. First, this instructional strategy provides a level playing field for students studying in different modes (internal or external, part-time or full-time) and in different locations (intrastate, interstate and internationally). Second, it enables the practical application of theoretical issues covered in the course that are associated with the nature of computer-mediated communication, computer-supported collaborative work, and the dynamics of virtual work groups and organizations.

Moreover, with the definition of team member roles, the project has a clear structure and division of tasks while reinforcing both interdependency and accountability (McGookin, 2002) among

Figure 5. Reflections on effectiveness of IRC use in team project (Sudweeks, 2003, p.177, reprinted with permission)

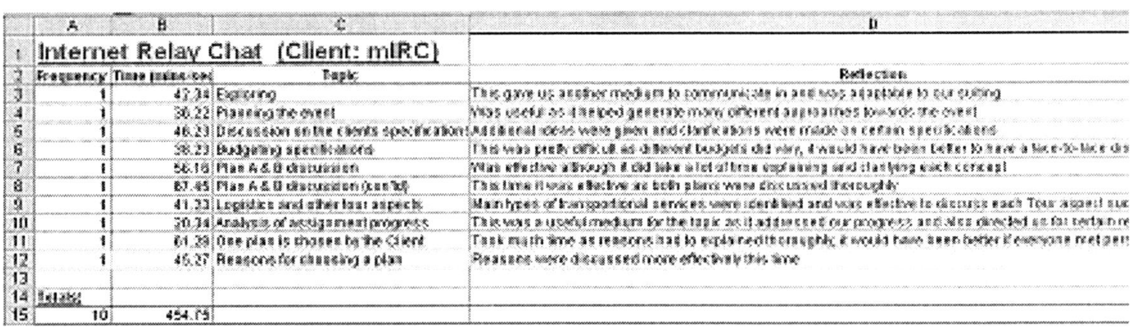

	A	B	C	D
1	Internet Relay Chat (Client: mIRC)			
2	Frequency	Time (mins.sec)	Topic	Reflection
3	1	47.34	Exploring	This gave us another medium to communicate in and was adaptable to our setting
4	1	36.22	Planning the event	Was useful as it helped generate many different approaches towards the event
5	1	46.23	Discussion on the clients specifications	Additional ideas were given and clarifications were made on certain specifications
6	1	38.23	Budgeting specifications	This was pretty difficult as different budgets did vary, it would have been better to have a face-to-face dis
7	1	56.16	Plan A & B discussion	Was effective although it did take a lot of time explaining and clarifying each concept
8	1	67.45	Plan A & B discussion (cont'd)	This time it was effective as both plans were discussed thoroughly
9	1	41.33	Logistics and other tour aspects	Main types of transportational services were identified and was effective to discuss each Tour aspect suc
10	1	30.34	Analysis of assignment progress	This was a useful medium for the topic as it addressed our progress and also directed us for certain ne
11	1	61.39	One plan is chosen by the Client	Took much time as reasons had to explained thoroughly, it would have been better if everyone met pers
12	1	45.27	Reasons for choosing a plan	Reasons were discussed more effectively this time
13				
14	Totals			
15	10	454.75		

members which are essential for the occurrence of peer scaffolding or learning support. The sense of accountability is further strengthened by tutor and peer assessment of contribution whereby the project (the online presentation and project documentation) is graded as a team effort by the tutor and each member initially receives a common grade, which may be adjusted after taking into account the students' evaluation of their members' contributions to the project.

THE RESULTS: SCAFFOLDING AND APPROPRIATION OF KNOWLEDGE

The constructivist pedagogical orientation of the OI course assumes that the design of an effective online course provides interactional opportunities in tandem with scaffolding that guides learning, leading to joint construction and appropriation of new knowledge. Hence, it is essential to evaluate the extent to which the course instructional activities are perceived to foster such active collaborative learning processes. This section reports survey findings on student experiences of the online synchronous tutorials and collaborative team project described earlier. The results form the basis for further discussion on the effectiveness of the two constructivist-based course activities in facilitating collaborative learning.

Results reflecting concepts of scaffolding and appropriation of shared knowledge presented here are integrated from two wider studies (Lim, 2006; Sudweeks, 2003). In both qualitative studies, the OI course is characterized as a particularly informative case (Stake, 2000; Yin, 1994), enabling an instrumental case study of "teaching and learning in one setting ... It is highly personalized because teaching and learning are highly personalized ... The system boundaries are not the skins of people, but are the boundaries around a particular experience" (Stake, 1988, p.257). Additionally, a common assumption is held that scaffolding and appropriation of knowledge in online collaborative learning processes are empirically observable through examining participant self-reflections on their learning experiences. Furthermore, the methodological approaches adopted in both studies are complementary in the following respects: (a) use of the survey method for gathering self-reports of learning experiences; (b) determination of the unit of analysis as the group at two levels, namely, the entire cohort of students enrolled in the OI course and the finer level of the tutorial group which enabled comparative group analysis; and (c) application of descriptive statistical analysis to the quantitative responses.

Scaffolding and Appropriation in Online Synchronous Tutorials

At the end of the teaching semester, a non-anonymous web survey was made available to 23 students from two chat tutorial groups resulting in 21 responses representing return rates of 93% (Group 1) and 89% (Group 4). The questionnaire included open-ended and closed questions on various aspects of the chat tutorial experience. Responses to closed questions were pre-coded by the survey software and descriptive statistical analysis was applied. Issues of content and construct validity were addressed at the pre-test and refinement stages of the survey development process, with modifications made that improved question clarity and instrument layout. A subset of results from Lim (2006) are presented here on student experiences of peer scaffolding defined as the extent of learning support perceived to be available from peers (other students) on clarifying content issues during tutorial discussions; and appropriation of shared knowledge defined as the attainment of learning from contributions to discussions.

Peer Scaffolding in Chat Tutorial Discussions

Peer scaffolding in chat tutorial discussions was measured by two questions on the availability of clarification and different ideas from other students in the tutorial group. The respondents indicated their extent of agreement on a scale from Strongly Agree (SA) to Strongly Disagree (SD), with an Unable to Judge (UJ) option for some questions. A between group comparison (Figure 6) found greater agreement (SA&A) among G4 respondents (100%) on the availability of clarification and different ideas from peers compared to G1 (84.6%).

Appropriation of Shared Knowledge from Chat Tutorial Discussions

The appropriation of shared knowledge was measured by two questions on the perception of attainment of learning from peer contributions to discussions. Figure 7 shows:

- all G4 respondents reported that *they learnt* from other students' contributions compared to 92.3% in G1.
- a small difference in the percentage of respondents from both groups who reported that other students learnt from their contributions to the discussion.
- approximately half the respondents in both groups *disagreed* and were *unable to judge* that peers learnt from their contributions.

Scaffolding and Appropriation in the Collaborative Team Project

A similar survey on the collaborative team project was administered at the end of the semester to all the teams resulting in 49 student respondents. The survey covered different facets of student experiences with team communication, group dynamics, their perceptions of the project and the overall online learning process. A selection of results from Sudweeks (2003) are presented here on student experiences of peer scaffolding defined as the extent of learning support perceived to be available from team members during collaborative activities within teams; and appropriation of shared knowledge defined as the attainment of learning from contributions to team activities.

Peer Scaffolding in Team Project

The extent of peer scaffolding perceived to be available from the online communication within teams was measured by two questions related to the exchange of views or information between team members. The respondents indicated their extent of agreement on a 5-point scale from *Strongly Agree* (SA) to *Strongly Disagree* (SD)

Figure 6. Extent of peer scaffolding in chat tutorial discussions

Questions	*Group*	*SA**	*A**	*D**	*SD**	*UJ**
The other students clarified issues on content that were raised during the discussion	**G1**	15.4%	69.2%	15.4%	0.0%	-**
	G4	50.0%	50.0%	0.0%	0.0%	-**
The other students contributed different ideas to the discussion	**G1**	15.4%	69.2%	15.4%	0.0%	0.0%
	G4	37.5%	62.5%	0.0%	0.0%	0.0%

**SA=strongly agree; A=agree; D=disagree; SD=strongly disagree; UJ=unable to judge.*

***The UJ option was not available for the first question.*

Figure 7. Appropriation of shared knowledge in chat tutorial discussions

Questions	*Group*	*SA**	*A**	*D**	*SD**	*UJ**
I learned from other students' contributions during the discussion	**G1**	30.8%	61.5%	7.7%	0.0%	0.0%
	G4	37.5%	62.5%	0.0%	0.0%	0.0%
I helped other students learn through my contributions during the discussion	**G1**	7.7%	46.2%	23.1%	0.0%	23.1%
	G4	12.5%	37.5%	25.0%	0.0%	25.0%

**SA=strongly agree; A=agree; D=disagree; SD=strongly disagree; UJ=unable to judge.*

Figure 8. Availability of peer scaffolding in team project

Questions	*SA**	*A**	*D**	*SD**	*N**
I am more likely to share my own opinion in an online group	14.3%	24.5%	18.4%	8.1%	34.7%
I am more likely to discover what other students think about a given topic in an online group	12.2%	34.7%	20.4%	6.2%	26.5%

**SA=strongly agree; A=agree; D=disagree; SD=strongly disagree; N=no opinion.*

Figure 9. Attainment of learning in team project.

Questions	*SA**	*A**	*D**	*SD**	*N**
I was able to understand many of the theoretical concepts taught in this unit by doing this project	14.3%	26.5%	10.2%	0.0%	49.0%
After this project, I have a better understanding of working in a virtual team	28.6%	51.0%	4.1%	0.0%	16.3%
I found the group project to be an effective learning experience	20.4%	57.2%	4.1%	2.0%	16.3%

**SA=strongly agree; A=agree; D=disagree; SD=strongly disagree; N=no opinion.*

and *No Opinion* (N). Figure 8 shows that most respondents agreed (*SA&A*) that they shared their views with other team members (38.8%) and were able to receive information or learn more about other members' opinions (46.9%).

Appropriation of Shared Knowledge from Team Project

The appropriation of shared knowledge was measured by three questions covering perceptions of attainment of learning from contributions to the project. Results in Figure 9 show that:

- more respondents (79.6%) agreed (*SA&A*) that by contributing to the project, they achieved better understanding of working in virtual teams than the theoretical concepts taught in the course (40.8%).
- most respondents (77.6%) agreed (*SA&A*) that engagement in team project was an effective learning experience

DISCUSSION AND RECOMMENDATIONS

The sociocultural constructivist learning perspective broadly assumes that knowledge building occurs during interaction which involves the sharing of multiple perspectives on experiences by members of the learning community. Such dialogic engagement, in formal instructional contexts, establishes a zone of proximal development where greater intellectual growth is enabled by the availability of scaffolding as guidance from peers and/or tutors, mediated by technology and language, leading to appropriation of the constructed meanings. To evaluate the extent to which the OI course fosters collaborative-constructivist learning processes, student experiences of two online instructional activities in a case study were examined.

Overall, for the chat tutorial activity, the findings from a comparative group analysis established that peer scaffolding was available to varying extents as efforts in clarification and provision of different ideas during discussions. In other words, even as there was peer support in both tutorial groups, there was also greater collaborative peer learning present in G4 compared to G1. Even though it is beyond the scope of this study to determine the exact form or extent of knowledge constructed, the attainment of learning perceived by respondents was examined to ascertain if there was appropriation of shared knowledge during the online learning process. Most of the respondents reported individual attainment of learning from other students' contributions during discussions. However, the results were equivocal regarding mutual attainment of learning since half the respondents in both groups disagreed or were unable to judge whether their peers had learnt from their contributions.

Results for the team project activity showed that peer scaffolding was also present as the exchange of information or opinions through online communication media within teams, albeit at a lesser degree compared to perceptions of learning support available in chat tutorial discussions. While engagement in the team project was largely reported to be an effective learning experience, more respondents indicated individual attainment of learning as achieving greater understanding in the actual practice of working in virtual teams than better comprehension of the abstract concepts taught in the OI course. This result is not unexpected and may be attributed to the purpose of the activity which requires the transfer of theoretical knowledge acquired, in course topics such as CMC processes and virtual work group dynamics, to application in the authentic context of developing a business proposal with virtual team members.

A comparison of the results from both online instructional activities revealed a greater extent of agreement on the presences of peer scaffolding and individual appropriation of knowledge in the chat tutorial activity compared to the team project activity. This finding may be explained by the nature of the learning activities and questionnaire design. In the team project, the sharing of multiple perspectives and negotiation of meaning occurred within teams that comprised four members while during chat tutorial discussions, participants had recourse to the shared knowledge of the entire tutorial group (9-14 students). Regarding the questionnaire design, the No Opinion option was available in most team project survey questions while the Unable to Judge option was offered in only one of 17 sets of questions in the chat tutorial activity survey. In survey perception analysis, it is acknowledged that the provision of neutral options can lead to errors of central tendency manifested as the consistent selection of mid points in rating scales due to the establishment of a comfortable pattern of responding reinforced by unvarying response format (Herzog & Bachman, 1981).

To account for the equivocal results regarding mutual attainment of learning from the chat tutorial activity (Figure 7), it could be argued that since the UJ option was limited to only one set of questions in the survey, the selection pattern for the question: I helped other students learn through my contributions during the discussion, validated the design assumption that respondents may not have sufficient information to provide an accurate answer pertaining to observed behaviour of other participants. Hence, it is likely that the equivocal results indicate that respondent difficulties in judging others' attainment of learning could be due to factors other than instrument design.

The results presented here on peer scaffolding and individual/mutual attainment of learning present certain possibilities for modifying the two course activities to enhance overall student experiences of online collaborative group learning processes. First, the reported inability by some respondents to provide an accurate answer on the behaviour of other participants during the chat tutorial discussions highlights an area for revising current assessment practices in the OI course. It is recommended that the peer assessment of contribution for both instructional activities, which is submitted by each student to the tutor at the end of the semester, be disseminated to other team or tutorial group members. The availability of peer feedback may raise individual student's awareness of the impact of one's own contributions on scaffolding the learning processes of others. Second, given that respondents in the team project survey reported better understanding of the actual practice of working in virtual teams than the abstract concepts taught in the OI course, it is recommended that regular online review meetings be held with each project team, during which the tutor can construct explicit links between the theory and practice of working in virtual groups.

FUTURE RESEARCH DIRECTIONS

The study presented in this chapter spans a number of fields such as education, information and communication technology, educational technology, and instructional design, which presents several potential areas for future research in these areas.

The single case design adopted by this study enabled an in-depth investigation of one particularly informative case. Although such cases are, by definition, not easily available, there is scope for further research with a methodological design that compares the OI course with other courses offering similar, though not necessarily identical learning experiences. In addition, the examination of interaction in the chat tutorial activity could be expanded to include those supported by the range of face-to-face, online asynchronous and synchronous instructional environments afforded by the OI course. Such research efforts could yield valuable insights on the appropriate integration of the various CMC technologies in the design of distance learning programmes.

Given that the equivocal results on mutual appropriation of knowledge during chat tutorials were not likely to stem from instrument design, they may be attributed to other factors such as the CMC medium. Future research could enable a clearer understanding of how the text-based and synchronous characteristics of the chat medium affect group learning processes. A final possible area for future investigation is to measure the relationship between student experiences of the online collaborative group learning processes and the quality of their coursework for both instructional activities in terms of the grades achieved.

CONCLUSION

This chapter presented a study examining the application of CMC technologies in a course framed by a constructivist pedagogical framework in the context of higher education. The chapter discussed the philosophical assumptions underlying behaviourist, cognitivist, and constructivist learning approaches that influenced interpretations of the form and function of educational interaction. Within the constructivist paradigm, the divergent perspectives on learning held by the radical and sociocultural constructivist schools of thought were contrasted. A qualitative case of an online undergraduate course shaped by sociocultural constructivism was presented, which included a detailed description of the course objectives, virtual learning environment design and two main online learning activities facilitated by synchronous and asynchronous CMC technologies. In particular, the influence of sociocultural constructivist principles on the aims, conduct and assessment of the online synchronous tutorials and the collaborative team project were described.

Survey results were presented and discussed on student experiences of collaborative learning and group work processes in the two online learning activities. Findings established that peer scaffolding was available during chat tutorials discussions and through online communication within the project teams. Similarly, there were perceptions of individual appropriation of knowledge from online interactions in the chat tutorial and team project activities. While there was greater extent of agreement on the presences of peer scaffolding and individual appropriation of knowledge in the chat tutorial activity compared to the team project activity, the findings were attributed to the questionnaire design and nature of the learning activities, which presented certain possibilities for modifying course activities to enhance overall student experiences of online collaborative-constructivist group learning processes.

Although the knowledge gained from this single case study is not intended for generalization to wider populations, the implications drawn from the findings may be extrapolated, in the form of recommendations, to similar contexts. Based the research presented, this chapter offered specific recommendations regarding the design of constructivist-based instructional activities to enhance overall student experiences of online group learning processes, which may be of interest to CMC researchers, higher education faculty, other professionals and staff involved in distance learning programmes.

REFERENCES

Bonk, C. J., & Cunningham, D. J. (1998). Searching for learner-centered, constructivist, and sociocultural components of collaborative educational learning tools. In C. J. Bonk & K. S. King (Eds.), *Electronic collaborators: Learner-centered technologies for literacy, apprenticeship, and discourse* (pp. 25-50). New Jersey: Lawrence Erlbaum Associates.

Brown, J., Collins, A., & Duguid, P. (1989). Situated cognition and the culture of learning. *Educational Researcher, 18*(1), 32-42.

Cashion, J., & Palmieri, P. (2002). *'The secret is the teacher': The learner's view of online learning.* SA: Australia: NCVER.

Cobb, P. (1994). Where is the mind? Constructivist and sociocultural perspectives on mathematical development. *Educational Researcher, 23*(7), 13-20.

Duffy, T., & Cunningham, D. J. (1996). Constructivism: Implications for the design and delivery of instruction. In D. Jonassen (Ed.), *Handbook of research for educational communications and technology* (pp. 170-198). New York: Simon & Schuster Macmillan.

Ertmer, P., & Newby, T. (1993). Behaviourism, cognitivism, constructivism. Comparing critical features. *Performance Improvement Quarterly, 6*(4), 50-70.

Herzog, A., & Bachman, J. (1981). Effects of questionnaire length on response quality. *The Public Opinion Quarterly, 45*(4), 549-559.

Jonassen, D. (1991a). Evaluating constructivistic learning. *Educational Technology, 31*(6), 28-33.

Jonassen, D. (1991b). Objectivism versus Constructivism: Do we need a new philosophical paradigm? *Educational Technology Research and Development, 39*(3), 5-14.

Jonassen, D., Davidson, M., Collins, M., Campbell, J., & Haag, B. (1995). Constructivism and computer-mediated communication in distance education. *The American Journal of Distance Education, 9*(2), 7-26.

Laurillard, D. (2002). *Rethinking university teaching: A conversational framework for the effective use of learning technologies* (2nd ed.). London: RoutledgeFalmer.

Lim, H. L. (2006). *Constructing learning conversations: A study of the discourse and learner experiences of online synchronous discussions.* Unpublished PhD Thesis, Murdoch University, Perth, Australia.

Lim, H. L. (2007). Community of inquiry in an online undergraduate information technology course. *Journal of Information Technology Education, 6*, 153-168.

Lim, H. L., & Sudweeks, F. (2008). Chatting to learn: A case study on student experiences of online moderated synchronous discussions in virtual tutorials. In S. Negash, M. Whitman, A. Woszcynski, K. Hoganson & H. Mattord (Eds.), *Handbook of distance learning for real-time and asynchronous information technology education* (pp. 170-191): Informing Science Reference, Idea Group.

McGookin, M. (2002). I can't let them down! Affiliative motivation and co-operative learning in higher education. In A. Goody, J. Herrington & M. Northcote (Eds.), *Proceedings of the 2002 Annual International Conference of the Higher Education Research and Development Society of Australia* (pp. 417-425). Perth, WA: HERDSA.

Moore, M., & Kearsley, G. (1996). *Distance education: A systems view.* California: Wadsworth Publishing Company.

Rogoff, B. (1990). *Apprenticeship in thinking.* New York: Oxford University Press.

Romiszowski, A., & Mason, R. (1996). Computer-mediated communication. In D. Jonassen (Ed.), *Handbook of research for educational communications and technology* (pp. 438-456). New York: Simon & Schuster Macmillan.

Schwandt, T. (2001). *Dictionary of qualitative inquiry* (2nd ed.). Thousand Oaks, California: Sage.

Stake, R. (1988). Case study methods in educational research: Seeking Sweet Water. In R. Jaeger (Ed.), *Contemporary methods for research in education* (pp. 253-265). Washington, D.C.: American Educational Research Association.

Stake, R. (2000). Case studies. In N. Denzin & Y. Lincoln (Eds.), *Handbook of qualitative research* (2nd ed., pp. 435-454). London: Sage Publications.

Sudweeks, F. (2003). Connecting students with group work. In C. Constantinou & Z. Zacharia (Eds.), *Computer-based learning in science* (Vol. 1, pp. 173-183). Nicosia, Cyprus: University of Cyprus.

Sudweeks, F. (2004). *Development and leadership in computer-mediated collaborative groups.* Unpublished PhD Thesis, Murdoch University, Perth, Australia.

Sudweeks, F., & Simoff, S. (2000). Participation and reflection in virtual workshops. In J. Burn (Ed.), *Proceedings of 3rd Western Australian Workshop on Information Systems Research.* Perth, Australia (CD-ROM).

von Glasersfeld, E. (1989). Cognition, construction of knowledge, and teaching. *SYNTHESE, 80*(1), 121-140.

von Glasersfeld, E. (1995). A constructivist approach to teaching. In L. Steffe & J. Gale (Eds.), *Constructivism in education* (pp. 3-15). Hillsdale, New Jersey: Lawrence Erlbaum Associates, Inc., Publishers.

von Glasersfeld, E. (1997, May). *Piaget's legacy: Cognition as adaptive activity.* Paper presented at the Presented at International Congress "Does Representation need Reality?" Vienna.

Vygotsky, L. (1962). *Thought and language.* Cambridge, MA: MIT Press. Revised and edited by A. Kozulin, 1986.

Werry, C. (1996). Linguistic and interactional features of Internet Relay Chat. In S. Herring (Ed.), *Computer-mediated communication* (pp. 47-64). Philadelphia, USA: John Benjamins Publishing Company.

Yin, R. (1994). *Case study research: Design and methods* (Vol. 5). Thousand Oaks, CA: Sage Publications.

Chapter XVI
Constructivist Strategies to Optimize Four Levels of Interaction in a Distributed Learning Environment:
A Case Study

Linda Lohr
University of Northern Colorado, USA

Nicholas Eastham
University of Northern Colorado, USA

David Kendrick
University of Northern Colorado, USA

ABSTRACT

This case study describes how a constructivist theory of learning guided the design of distributed learning environment for a three credit hour graduate level course on instructional design. Four types of interaction data were collected from 27 participants, one instructional designer/instructor, and two assistant designers. Overall, constructivist strategies appeared to contribute to a successful learning experience as measured by participant surveys, designer observations, and academic performance. A strong majority of students considered a number of constructivist strategies beneficial, such as the provision of a variety of reading and learning activity options, as well as participation in an authentic and relevant learning task. Academic quality of end of semester instructional products was high. A strong majority of participants received a rating of excellent, as determined by designer/instructor and mentor evaluations. Some constructivist strategies appeared to detract from the learning experience. Data related student-to-student, student-to-content, student-to-teacher, and student-to-interface interaction suggests the need to clarify expectations for small group discussions and participant blogs, rewriting or repositioning an instructional story as a case study, increasing design-document specific feedback, using a broad and shallow interface structure and moving selected course content to pre-packaged paper-based format to reduce cognitive demands related to reading while online.

INTRODUCTION

This case study identifies constructivist design principles behind the creation, implementation, and assessment of a semester-long graduate-level distributed education course. Distributed education involves multiple sources of information, occurring in or outside a classroom or campus, often mediated by technology, with time and location independence (Obinger, Barone, & Hawkins, 2001). The role of technology in facilitating constructivist instructional strategies is explored and analyzed in this chapter to direct future distributed learning case study design and assessment studies.

In its broadest sense, a constructivist learning experience is one in which the student is provided with opportunities to create knowledge and understanding as they reflect upon authentic learning activities. Students in this case study participated in a distributed education course titled "Instructional Design." Students, hereafter named participants, were required to create (construct) self-paced instructional modules for two audiences: a teaching mentor, and the mentor's students. Course content involved participant problem solving and creative thinking in the context of uniquely identified instructional problems. Participants were required to generate a number of documents and products including: analysis instruments, needs assessments, self-paced instruction, learning assessments, and weekly reflection blogs.

While the course was delivered online, participants interacted with a variety of learning resources characteristic of distributed learning environments. Participants in this student were involved in:

- one on one interaction with designers as questions arose
- online collaboration/interaction with other study participants, fellow graduate students who also were taking the class and also creating units of instruction
- interaction with a mentor based upon on instructions to "find someone whose teaching you admire and hope to emulate"
- interaction with the mentor's students

The unit of instruction developed by the participant involved:

- conducting a needs assessment and learner analysis
- developing instructional objectives, strategies and assessments
- implementing the instruction
- evaluating (formatively and summatively) the effectiveness of the instruction

The problem solving nature of the instructional design course, coupled with authentic design tasks and audiences, provided a strong opportunity to analyze and assess the implementation of constructivist learning strategies.

This chapter begins by describing key terms and concepts used in the context of this study. Constructivism, rich environments for active learning (REALs) and resource-based learning (RBL) are defined. The influence of cognitive load theory and interaction design in the implementation of constructivist learning strategies is discussed. Three overall research questions guide an assessment of the learning environments effectiveness. Trends in technology in light of the implications of the study conclude the chapter and provide the basis for continued research and application of instructional design.

BACKGROUND

A constructivist view of learning characterizes much of the research related to educational technology at this time (Molenda, 2008). While there are many definitions of constructivism, Driscoll (2005) finds most explain learning as an active process whereby the learner creates knowledge and understanding by making sense of learning experiences.

According to constructivist learning theory, individual and social experiences are the basis of knowledge experiences (Crotty, 1994; Jonassen, 1992; Savery& Duffy,1995; Valasidou, Sidiropoulos, & Makridou-Bousiou (2005). Learning is viewed as an active process of thinking, understanding, and reflecting that results in meaning making on an individual level.

Dolittle (1999), Driscoll (2005), and Grabinger (1996) have identified a variety of similar principles for design of constructivist learning environments. Doolittle (1999) identified eight constructivist principles important for a rich learning experience:

1. Learning should take place in authentic and real-world environments
2. Learning should involve social negotiation and mediation.
3. Content and skills should be made relevant to the learner
4. Content and skills should be understood within the framework of the learner's prior knowledge in which interfaces tend to become visible
5. Students should be assessed formatively, serving to inform future learning experiences
6. Students should be encouraged to become self-regulatory, self-mediated, and self-aware which often by nature excludes human intervention, including coercion by the teacher to complete assignments or learn the course material
7. Teachers serve primarily as guides and facilitators of learning, not as instructors.
8. Teachers should provide for and encourage multiple perspectives and representations of content

Driscoll (2005) identified five instructional strategies based upon constructivist theory:

1. Use of realistic, complex, and relevant learning contexts
2. Provision of opportunities for social negotiations
3. Presence of multiple perspectives and representations
4. Encouragement of learner responsibility
5. Nurturing learner awareness of knowledge construction

Grabinger (1996) presents constructivist-learning strategies using a Rich Environments for Active Learning (REAL) acronym. Five principles of a REAL include:

1. Providing an authentic learning context
2. Supporting cooperative learning between students and teachers
3. Providing authentic assessment
4. Encouraging student-initiated learning and responsibility
5. Use of generative learning strategies

Dolittle's eight characteristics derive from principles listed by both Driscoll (2005) and Grabinger (1996) and provide the design framework for this case study.

Resource-based learning (RBL) is a feature of many constructivist distributed education environments. RBL is defined as the application of available resources (assets) to support a variety of learning needs in a variety of learning contexts (Hannafin and Hill, 2008). Assets include resources such as people, media, tools, and ideas that support learning. Assets available within the case study context included mentors, peers, stories, text and media delivered instruction, to name a few.

Instructional design, the graduate level course described in this case study, describes the process behind creating instruction. While a number of models employ different steps and methods, five activities underlie all instructional design: analysis, design, development, implementation, evaluation (Dick and Carey, 2004); (Morrison, Ross, and Kemp, 2004); (Smith and Ragan, 2005).

Cognitive Load Theory

A critical issue in the design of constructivist strategies for the instructional design course relates to the cognitive load imposed by the content as well as the strategies designed to facilitate that load. Cognitive load is a term used to describe the mental energy needed to think about or process information. The research on cognitive load describes three categories of load: intrinsic load, extraneous load, and germane load (Paas, Renkle & Sweller, 2003).

Intrinsic load refers to the nature of the content and its level of complexity. Complexity can be defined in terms of element interactivity, or the extent to which a learner must understand instructional content that overlaps and interacts with other instructional content. The topic of instructional design is characterized by high content interactivity.

While instructional design is typically presented as a series of sequentially arranged design steps of analysis, design, development, implementation and evaluation, in reality the topic is a merging of these variables for every phase of the process. Analysis overlaps with design as the instructional designer attempts to create concise instructional messages through the sound development of instructional strategies. Implementation intersects with learner characteristics and needs as the instructional designer attempts to integrate learning products into the natural context of the learner. Assessment overlaps all phases of the design process. Not only is the learner assessed, but so are the objectives, goals, instructional strategies, and implementation plans. Learning each of the instructional design phases in isolation creates an artificial learning context. Reducing the cognitive load associated with instructional design may be in conflict with the need for an authentic learning experience.

While content complexity, or intrinsic load, cannot be altered, the designer can experiment with changes in extraneous as well as germane loads. Extraneous load can be thought of as the noise, or superfluous elements of communication, that act as barriers to learning due to the increased load placed on memory. Extraneous load in a distributed education environment can be found in a variety of sources. The nature of distributing information and experiences to a number of contexts may in itself pose an extraneous load as the learner adapts and readapts to changing resources inherent in RBLs. Extraneous load is realized when materials that attempt to entertain end up confusing or distracting the learner. Unless presented carefully, jokes, images, anecdotes, and the like, may inadvertently contribute to extraneous load.

The hypermedia format of distributed learning environments contributes to extraneous cognitive load as well. Additional hypertext layers in the form of links, folders, and links within folders, as well as different delivery formats (audio, video, text) may play a role in increasing extraneous load. Learners in distributed learning environments are susceptible to losing focus when attention is split while clicking or navigating through different presentation formats.

Germane load can be thought of as those things that a designer can do to facilitate optimal load. For example, textual techniques that reinforce the content, such as chunking content, sequencing it, and providing analogies to help people understand new information more quickly work to increase germane load. Use of relevant images, stories, music and other media may work to increase the learner's germane load. For example, instructors can provide graphic summaries of complex concepts or songs that act as mnemonics to facilitate optimal load.

Interaction Formats

Implicit in the Doolittle's eight principles (1999) is the concept of learner interactivity with the learning experience. Considered a sometimes overused, misunderstood, and potentially meaningless term (Reeves, 1999), interactivity is defined as a two-way message loop between a sender and receiver, a definition with roots in Shanon's communications theory. In one-way communication, the learner receives a teacher's message. To be considered interactive, the learner returns a mutually relevant message. The reverse is also true; the initial message might originate with the learner to be received by the teacher, other students, or even the computer system itself.

Moore and Anderson (2003) define four types of interactivity: (1) student-to-content, (2) student-to-student, (3) student-to-teacher, and (4) student-to-interface. Design to increase the effectiveness of each type of interaction is to some extent dependent upon the capacity of the interaction format to optimally engage the learner. In a distributed learning environment, student-to-content interaction is facilitated in part through use of a content space where instructional materials are posted. Student-to-student interactions are facilitated in discussion spaces, links to blogs, email, and private group spaces. Student-to-teacher interactions generally take place in discussion spaces, links to blogs, email and private group spaces. Student-to-interface interactions take place throughout a distance environment wherever a student is required to click or make a decision. In the past, a learner might have focused mostly on student-to-teacher interactions, the three additional interaction categories mentioned above now contribute to a potential information overload for the learner. According to Chang and Ley (2004), online learning environments impose a heavy cognitive load due to the number of choices available to the learner, from links to external sites to internal chat sessions and discussion threads, to separate email communications, readings embedded in the learning environment, and readings external to the learning environment.

In a traditional classroom, information is directed from one source. In distributed contexts, information originates from many sources, and that information is often a result of a learner-driven decision-making process. When knowledge originates not only from the instructor but also from content and other learners, and only through decisive interaction with an interface, the learner is more than a receiver of information. They must actively search for and select information. Given the array of choices available, the capacity of the learner to hold information is often challenged.

Section Summary

Three categories of information presented thus far influenced the design of the instructional design course described in this case study and include constructivist learning design principles, cognitive load theory, and interaction design. While constructivist principles ideally facilitate the open-ended and fuzzy problem-solving nature of instructional design, the content complexity and addition of different interactive formats for the distributed learning experience present a number of challenges.

The focus of the study centers on the integration of constructivist learning strategies into the design and development of four types of interactivity in a distributed-learning environment. Specific questions guiding this inquiry included:

1. Overall, was the course considered effective, as measured by student perceptions of learning gain and cognitive load as well as by instructional designer observations and academic ratings student of performance?
2. How successful were the implemented constructivist strategies in enhancing each of the four types of interaction, as measured by designer observations and student perceptions?
   a. How well did blogs enhance student-to-student interaction?
   b. How well did discussion groups enhance student-to-student interaction?
   c. How well did project sharing enhance student-to-student interaction?
   d. How well did content spaces that held instructional resources enhance student-to-content interaction?
   e. How well did question/answer spaces enhance student-to-teacher interaction?
   f. How well did e-mail enhance student-to-teacher interaction?
   g. How well did the three-step interface design improve student-to-interface interaction?
3. What design changes are needed to improve the instructional effectiveness of the four types of interaction?

DESIGN AND DEVELOPMENT OF A CONSTRUCTIVIST DISTANCE-LEARNING ENVIRONMENT

Dolittle's (1999) eight constructivist learning strategies were integrated into a semester-long graduate-level course teaching instructional design. Funds to support the design, development and assessment for the project were provided by the Center for the Enhancement of Teaching and Learning at the University of Northern Colorado.

Twenty-seven graduate students participated in one of two sections of the instructional design course. The designers of the environment included a lead designer (also the instructor), and two assistant designers. Design meeting notes, participant blogs, discussions, projects, and participant end of semester surveys provide the data needed to assess the effectiveness of the course

Course Description

The designer/instructor created an online Instructional Design course comprised of six two-week units: A Four Step Design Process, Cognitive Load Theory, Learner Analysis, Design of Objectives, Design of Strategies, and Design of Assessment. The overall objective of the course was to apply all readings, discussions, and reflections to the design, development, and assessment of a 20 - 30 minute self-paced instructional unit for a client or mentor.

Integration of Constructivist Principles

Each of Dolittle's (1999) constructivist principles were implemented as follows:

Principle 1: Provide an Authentic Real-World Environment

To address this principle while also addressing enhancement of germane load, the designer required each participant to create a unit of instruction for a mentor of their choice. Mentors were either a teacher or trainer in need of self-paced instruction for use in their classrooms/or distance education settings. Participants were evaluated by the designer/instructor as well as by the mentor. The Discussion Board and e-mail (student-to-teacher, student-to-student, and student-to-interface interaction formats), and PowerPoint and Flash technology presentations of examples and analogies (student-to-content interaction format) were used to facilitate this principle.

Principle 2: Provide an Environment that Allows Social Negotiation and Mediation

To address this principle and to also address strategies for working with intrinsic load, participants received weekly feedback from both their instructor and their mentor. Participants would first work with the mentor to create drafts of instruction. Participants would then work with the designer/instructor (using a detailed rubric that outlined required elements in the instructional piece) modifying work as necessary. A back and forth process between the participant, mentor and designer/instructor took place during each of the six units. E-Mail, the Discussion Board, and private discussion groups (student-to-teacher, student-to-student, and student-to-interface interaction formats), were used to facilitate this principle. A detailed project rubric accompanied each unit and served to guide communication related to feedback and revision between the designer/instructor, the participant, and the participant's mentor.

Principle 3: Provide Relevant Content

To address this principle as well as to employ germane strategies, students were asked to directly apply information from each of the six two-week units to the development of their mentor's lesson. This principle was facilitated through use of the participant content area titled "Go to class," e-Mail, the Discussion Board, and Private Group areas (student-to-teacher, student-to-student, and student-to-interface interaction formats). Small group discussions facilitated by both the assignment and the assignment rubrics helped focus participant work.

Principle 4: Recognize Prior Knowledge

To address this principle, participants worked in team environments to discuss readings and experiences in light of their own experiences. Participants were also asked to keep a weekly blog reflecting on what they learned and how they might reconcile that information with previous experiences and knowledge. Private Group Areas, Internet blog sites, and blog space set up in Discussion areas were used to facilitate this principle (student-to-student, and student-to-interface formats).

Principle 5: Formatively Assess Students

To address this principle, participants created units of instruction with bi-monthly formative feedback from the designer/instructor of the course. This feedback was provided using a detailed rubric that specified how unit content must be implemented into the project. E-mail and Discussion board interactions were use to facilitate this principle (student-to-teacher interaction).

Principle 6: Provide Opportunities for Reflection and Self-Mediation

To address the principle, the designer /instructor asked participants to describe their thinking and learning in a weekly blog. Participants were also encouraged to share changes in their thinking in the small group discussions as well. Participants selected from a number of blog sites (such as Edublogs) or used a series of indented threads within the Discussion space. Each unit provided many reading and activity options, which allowed participants to mediate their own learning. This principle reflects student-to-content interaction and potentially student-to-student interaction.

Pinciple 7: Promote Teacher as Facilitator

To address this principle, the designer/instructor asked participants to learn from their interactions with their mentor, as well as interactions with other participants in discussion areas. The designer/instructor provided a number of readings, many optional, for student selection, as well. To guide participants, the designer/instructor blogged her experiences with content for each unit, and provided worked examples for student reference. Edublogs and Blackboard content, small group, and Discussion areas helped promote the designer/instructor as a facilitator of the learning experience (student-to-content, student-to-interface, student-teacher interaction formats).

Principle 8: Provide Multiple Representations of Content

To address the principle, the designer/instructor asked participants to work with mentors and other participants. Additionally, a story, an analogy, teacher modeling through development of a sample lesson, and a variety of reading materials were used to teach critical content. The Blackboard content area, Internet links, and Edublogs facilitated access to different types of instructional content (student-to-content, student-to-student, student-to-teacher, and student-to-interface interaction formats).

Four Interaction Formats

The Blackboard learning management system was used to deliver instruction to students. Student-to-interface interactions originated with a main menu with three options. Participants were introduced to the objectives of each unit in a "Start here" space that directed participants to a " Go to class" link designed for student-to-content interaction. This space was chunked into five categories: Unit Overview, Unit Readings, Tools for Richer Learning, Activities, and Self-check.

Student-to-student interaction was facilitated in private group spaces for teams of four to five participants as well as general discussion board in a "Talk to others and post work" space. Participant blogs were accessed through embedded links posted in this space.

Student-to-teacher interactions took place in this team space as well. The top thread in this space was designated for daily questions and clarifications. Participants had e-mail, phone, and blog access to the designer/instructor as well.

Enhancement of Cognitive Load

To increase germane load and enhance overall cognitive load, the designer/instructor chunked each unit into five categories: 1) Unit Overview and Highlights, 2) Unit Readings, 3) Tools for Richer Learning (where participants could access additional readings, and related web-sites, and an ongoing story about a fictional designer's experiences creating self-directed learning, as well as examples of the evolving instruction, 4) Activities where participants applied unit readings to short assignments designed to allow participant practice, and 5) a Self-check where participants took a 4 - 5 item quiz over general unit content.

METHODS AND FINDINGS

Three sources of data were collected: (1) Notes from the lead-designer and teaching assistant/designer discussions were kept in the form of daily e-Mail correspondence, three project journals of weekly meetings categorized according to perceived effectiveness of group discussions and blogs, as well as in the designer/instructor blog; (2) Qualitative and quantitative data from end-of-semester participant surveys; and (3) Designer/instructor and Mentor project grades.

Designer Journals

The designer/instructor and assistant designers monitored weekly participant blogs and group discussions in order to identify problems with course organization and general participant concerns. Entries were marked according to topic depth (rich entry, moderate entry, superficial entry), as well as questions raised. Of the seven participant groups (approximately four participants each), problems centered in one discussion group; these participants rarely took part in discussions. Three of the five participants in this group were doing above average "B - level" work with their mentors and blogs, but aside from the weekly posting requirement, were absent from group discussions.

Designer/instructor notes during this time focused on interactions with one of the assistant designers who shared the need for clarification or instructor attention to specific participants based upon a review of blog and discussion entries. Another problem was noted when the other assistant designer expressed dissatisfaction with the puppeteer metaphor and story. This assistant designer was an international student who was struggling with the English language and admitted to experiencing significant problems with cognitive overload during the course. The lead designer noted in her journal "if <name withdrawn> is experiencing difficulty, the story may be contributing to cognitive load for international students rather than reducing it."

Of the comments shared by the assistant designers, the design issues that impacted a majority of students focused on problems with the interface, including technical issues caused by the learning management system. During one unit, the Blackboard learning management system was upgraded, causing browser problems for some participants as well as lost attachments (including rubrics and instructional projects).

Another problem noted occurred when some students disregarded content that they thought was optional reading. A story that described a designer working through the instructional design process was provided in the "tools for richer learning" area of the environment. Reading the story in its entirety was optional, but lesson examples and artifacts (designed to provide an example that participants could use to direct their own development of instruction for their mentors,) extracted from the story were required content. The interface, titled "Tools for richer learning" implied that everything related to the story was optional, perhaps due to an implication that "richer learning" meant "optional learning."

Participant Grades

The average grade for student projects (development of a self-paced unit of instruction) was high. The designer/instructor and mentor grades had an average score of 90%, an A-. Participant grades were calculated by the designer/instructor, the mentor, and the participant's self-assessment of the needs statement, learner analysis, learning objectives, instructional strategies, and learner assessment as well as formative assessment.

Participant Survey Results

Student reactions to the experience were collected using a 22 item 5-point Likert survey consisting of: 3 questions focused on the overall effectiveness of the course, 8 questions focused on student-to-content interaction, 5 questions focused on student-to-teacher interaction, 2 questions focused on student-to-student interactions, and 4 questions focused on student-to-interface interaction. Three open-ended survey responses asking students to share what they liked most about the course, least about the course, as well as any additional comments concluded the survey.

Overall Effectiveness of the Course

A strong majority of participants responded Agree or Strongly Agree to the statement "I learned a lot from this course." For this survey item 28.57%, 57.14% and 14.29% indicated Strongly Agree, Agree, and Neutral respectively. Likewise, but with a heavier emphasis on Strongly Agree, a strong majority of participants felt challenged by the class, with 57.14%, 28.57%, and 14.29% respectively. All students responded Agree" (57.14%) or Strongly Agree (42.86%) to the statement "Overall, I liked this class".

Student-to-Content Interaction

Participant responses to the amount of information covered and the workload with respect to the credit hours earned were consistent. A majority (71.43%) of participants indicated that the class was the right amount of work for the number of credit hours, 57.14% of participants were neutral to the statement that there was too much information

Instructional strategies designed to improve an understanding of the content had mixed results with 42.86% of participants responding Neutral to the statement "The story helped me understand the content of the class." An equal percentage of participants agreed (28.57%) and disagreed or strongly disagreed (14.29% and 14.29%, respectively) to this statement as well. A majority (42.86% Strongly Agree, 28.57% Agree) of participants agreed that the puppeteer analogy was helpful. Participant use of the story and the examples extracted from the story in the lesson was less effective, with a majority of students equally responding Agree, Neutral, and Disagree (28.57% each) to the statement "I used the story lesson and design document" as a model for my lesson and design document." A strong majority (85.71%) of participants agreed or strongly agreed that the end of unit self-assessment items helped them gain confidence in their understanding of unit information.

A strong majority of participants favored the variety of options available in the course. 85.71% of participants strongly agreed to the statement "I liked the option to choose alternative levels of activities (beginner, intermediate, advanced) and reading assignments (chapter highlights, full chapter text). The remaining 14.29% of participants agreed to this statement as well. 85.72% of participants equally agreed or strongly agreed to the statement "I learned more because of the option to choose alternative levels of activities and reading assignments."

One consistent response from students related to the perceived effectiveness of participant blogs. A strong majority of participants (71.43%) found blogging to be unhelpful with the remaining 28.57% neutral to the instructional benefits of this strategy.

Student-to-Student Interaction

The effectiveness of participant interaction with other participants was minimal. A strong majority of participants was neutral or disagreed with the effectiveness of student discussions (71.53%). Looking at fellow participant projects, however, was considered helpful by a strong majority of participants (85.72%).

Student-to-Teacher Interaction

Participants considered almost all student-to-teacher interactions helpful. Participants unanimously agreed or strongly agreed that the designer/instructor provided timely and helpful feedback. A majority (71.43%) of participants were satisfied with the amount of feedback provided by the designer/instructor regarding their participation in unit activities. Participants were less positive about feedback specific to the design document with 42.86% indicating that they needed more help writing the document. Most participants (71.43%) were neutral to the benefit of viewing teacher feedback given to other students.

Student-to-Interface Interaction

Participant responses to the effectiveness of the interface were mixed. All of the participants agreed or strongly agreed that the course would be easier to use and more effective if it were prepackaged as a book and available before class started. Follow up interviews with a few of the participants indicated that students disliked online reading. While chunking content for online viewing worked online, it was more of a hassle when printing and reassembling materials after they were printed. With that said, however, 71.43% indicated the course was structured in a way that made it easy to find things. However, the other 28.57% strongly disagreed with that assessment. A strong majority of participants responded favorably to the use of the unit map (71.43%) and the top "start here" thread (71.43%) as helpful navigational elements.

Open-Ended Responses

When responding to the question about what they liked most about the class, a majority of the comments related to the many options available to students. Participants appreciated being able to select from a variety of unit activities and readings. Likewise, they appreciated the ability to work with self-selected mentors or clients. Two participants mentioned the presence of many examples extracted from the designer story as beneficial to their learning.

When responding to the question about what they liked least, participant comments regarding the difficulty of the course layout and interface were more prominent than what might be expected from survey responses alone. Blogging, and its effectiveness, on the other hand, was equally criticized by participants in the open-ended responses and survey items.

General participant comments were overall very positive, indicating they felt the course content was important and well taught. A few comments underscored the importance of providing an improved interface to future participants. One participant remarked "it took me 4 - 5 weeks to get the hang of the layout ... way too long to be effective."

SUMMARY OF DESIGN ISSUES

Based on the data, a number of design changes are planned. To increase student-to-content interaction, and potentially cognitive load, extracted examples from the story will be located at a prominent level of the interface and will be labeled "required".

Although the story was intended as a strategy to increase germane load, the response of some students, and from one of the assistant designer suggests that the story may have inadvertently increased extraneous load. Given the international status of the assistant designer, problems related to cultural misunderstanding are a likely cause of misunderstood information. Development of a case study as an alternative option to the story will be created and available to students needing an example of instructional design more relevant to the needs of a face-to-face instructor. Though the course was offered in a distributed-learning environment, participant of the course often teach in face-to-face settings and might be more interested in supplemental reading related to a more relevant example.

Improvements in student-to-student interaction will be based on a combination of implications from the study. Both participant blogs and participant discussions were not considered by a strong majority of participant as beneficial to learning. While constructivist strategies promote the importance of socially constructed knowledge, blogs as well as group discussions were not embraced as effective. Creating detailed instructions on how to blog and how to participate in group discussion might improve perceptions of the effectiveness of these learning strategies. Given the already heavy load of course content, however, additional activities are potential opportunities for cognitive overload. Borrowing from some of the positive results of the study, perhaps the best strategy for learning based on social construction of knowledge might be to make blogs optional.

Improvements to student-to-teacher interactions are likely to focus on providing additional feedback related to the design documents. Many participants indicated wanting more feedback from the designer/instructor related to their design document. Although a detailed rubric had been used to provide the feedback, participants indicated a need for more help with the document. To increase interaction, the designer/instructor might create an additional level of feedback by requesting participant feedback of feedback. For example, the designer/instructor might ask for each participant to respond to feedback by asking for additional clarification or indicating an understanding of what was requested.

Improvements to student-to-interface interaction will be the largest change to the learning environment design. While the designer/instructor created a simple interface consisting of three numbered links, the depth of one of the links created an extraneous load for many. For each unit, a student visiting this space would be presented with six or more additional links. Each link, in turn, would branch to additional links, making the space on the surface simple, but once inside very complex. An improved design will focus on creating a broad and shallow structure for the entire menu The interface design field considers two types of navigation: the broad and shallow option or the narrow and deep option. While debated, most designers conclude that it was better to have many menu options at the highest level (a broad and shallow design) to reduce the number of clicks needed to access information. That participants requested a book compiling all of the contents in the problem space attests to the challenges of the narrow deep structure. A potentially better option than paper would be the design of a Web space using a broad and shallow interface requiring only one or two clicks to reach content rather than the four or five clicks now part of the interface.

TRENDS

Participants in this study, on average, had a Bachelor's degree with six to 10 years of teaching experience. While these students did not embrace blogs and discussion groups, it is unlikely that future students will share this lack of interest. Recent educational technology-related forums (WCET, 2008) center on the "Gen-Net" or "millennial" population only a few years away from higher education. These students possess advanced skills in social networking and Web 2.0 technologies. Future students are likely to have skill (and bad habits as well) in the construction and use of wikis, blogs, and avatar-based communication. Second Life, a virtual network of millions of users, considers education its next frontier. Large corporations already host virtual marketing and training related to products and service. Business schools, art departments, and communication programs are increasingly present to tap into instructional opportunities provided by these virtual worlds.

Donald Phillip's message during the 15th Annual Education Technology Conference, Campus Technology 2008, stressed that innovation will drive the economy and that we are in the grasp of the knowledge economy, but that only a third of our current students are among the "Net Generation" (Philips, 2008). This begs the question of how many instructors are ready to use the technology for this new generation of learners.

SUMMARY

Constructivist epistemology was the lens through which the designers of this study framed their work. The open-ended learning experiences described in this case study employed a student-centered pedagogical approach that was contextually centered and aligned with constructivism. Participants in the study interacted with an instructional environment characterized by a high content interactivity and potential cognitive overload.

This design approach differs from direct instruction with roots in objectivist epistemology and behavioral psychology, which is characterized by hierarchically arranged information structures accompanied by specific behavioral objectives aligned with equally specific outcomes. Rather than experiencing a hierarchical ordering of instructional content, these participants took part in an iterative and interactive instructional design process consisting of analysis, design, development, implementation, and evaluation phases.

By positioning instruction within an authentic and relevant context, participants worked with real instructors and students as they created and implemented instruction using the theoretical instructional design model. Participants socially negotiated complex design decisions using instructional rubrics interpreted by both their designer/instructor and mentor. Social networks such as discussion spaces, whole-class project display spaces, reflective blogs, and e-mail correspondences provided a mixture of instructional perspectives.

Focusing design of the course around four types of interaction assisted the designers in accommodating this variety of instructional perspectives. Student-to-student interaction was focused on design of blogs, discussion groups, and project sharing spaces. Student-to-content interaction was focused on content spaces that held instructional resources in the form of articles, information presentations, stories and metaphors. Student-to-teacher interaction was facilitated through use of question/answer spaces, bi-monthly use of an instructional design performance rubric, and e-mail. Student-to-interface interaction was facilitated using a three-step interface design, with expanded depth in the second level of the interface.

Overall, data collected from the designers, 27 student participants, and their mentors suggests a positive and effective learning experience for a strong majority of the students. A strong majority of students earned a rating of "excellent" from both the course instructor and their mentor/client on the quality of lessons they designed while taking part in the case study. Likewise, a strong majority of students indicated "liking the class" and "learning a lot", but at the same time indicating that the amount of content covered for the credit hours received was not wholly congruent.

When looking specifically at the effectiveness of constructivist strategies within different interaction formats, a number of design issues arose. Several constructivist strategies implemented to enhance student-to-content interaction were not always effective. The effectiveness of the story and the worked examples from the story received a mixed, but mostly neutral response from students. Much stronger student support was noted for the provision of reading options and practice activities. Students clearly liked being able to choose among basic, intermediate, and advanced reading and practice options. The benefit of blogging as a means of self-reflection was considered by most students to be minimal. Some students noted in the open-ended response "blogging was a waste of time."

Student-to-interface interaction data conflicted with respect to data source. Participant surveys were overall positive about the course layout and organization, however many participants found the depth of layers in the content presentation space to be a barrier to ease of navigation within the learning environment.

CONCLUSION

This case study provides insight into the design of a distributed learning environment employing constructivist learning strategies. High content interactivity characterizes the topic of instructional design, making it a logical choice for the implementation of constructivist learning strategies to reduce cognitive overload and to increase relevant and authentic interaction. Participants worked with individually selected mentors and their students to design, develop, implement, and assess self-paced instructional units. A rubric generated and socially negotiated formative assessment between the participant, the designer/instructor and the mentor was employed as instruction was developed. Participant responsibility for learning and reflection was encouraged by the availability of blogs and discussion forums as well as the provision multiple perspectives and representations of content. Attention to four types of learner interaction facilitated the implementation of a number of constructivist strategies important for the presentation of multiples perspectives and content representations.

REFERENCES

Chang, S. L., & Ley, K. (2003). *The Effect of Using Printed Online Materials on Learning Achievement in Web-based and Hybrid Courses.* Paper presented at the national conference of the American Association of Research in Education, San Diego, CA.

Crotty, T. (1994). Integrating distance-learning activities to enhance teacher education toward the constructivist paradigm of teaching and learning. *In Distance Learning Research Conference Proceedings,* (pp. 31-37). College Station, TX.

Dick, W., & Carey, L. (2004). *The systematic design of instruction.* New York: Harper Collins.

Doolittle, P. (1999). *Constructivism and online education.* Retrieved May 3, 2008 from: http://edpsychserver.ed.vt.edu/workshops/tohe1999/text/doo2.pdf.

Driscoll, M. P. (2005). *Psychology of Learning for Instruction*, 3rd ed. Bostono Allyn & Bacon.

Grabinger, S. R. (1996). Rich environments for active learning. In D. H. Jonassen (Ed.), *Handbook of Research for Educational Communications and Technology*, (pp. 665-692), New York: Simon & Schuster Macmillan.

Hammond, M. (2005). A Review of Recent Papers on Online Discussion in Teaching and Learning in Higher Education. *Journal of Asynchronous Learning Networks, 9*(3).

Hannifin, M. J., & Hill, J. R. (2008). Resource-Based Learning. In J. Spector, D. Merrill, J. Van Merrienboer, & M. Driscoll (Eds.), *Handbook of Research on Educational Communications and Technology, 3rd ed.* (pp. 525-526). New York: Lawrence Erlbaum Accociates.

Innovate.(2008). *Innovate 2008: The real and the ideal.* Retrieved April 22, 2008 from: http://www.education-2008.org/.

Jonassen, D. (1992) Objectivism versus constructivism: Do we need a new philosophical paradigm? *Educational Technology Research & Development*, Vol. 39, No.3, pp.5-14.

Moore, M. G., & Anderson, W. G. (2003). *Handbook of Distance Education.* NJ : L. Erlbaum Associates.

Molenda, M. (2008). Historical Foundations. In J. Spector, D. Merrill, J. Van Merrienboer, & M. Driscoll (Eds.), *Handbook of Research on Educational Communications and Technology, 3rd ed.* (pp. 525-526). New York: Lawrence Erlbaum Accociates.

Morrison, G., Ross, S., & Kemp, J. (2004). *Designing effective instruction.* New York: John Wiley & Sons.

Oblinger, D., Barone, C. A., & Hawkins, B. L. (2001). *Distributed education and its challenges: An overview.* American Council on Education

(ACE). Retrieved June, 2003: http://www.acenet.edu/bookstore/pdf/distributed-learning/distributed-learning-01.pdf

Paas, F., Renkl, A., & Sweller, J. (2003). Cognitive load theory and instructional design: Recent developments. *Educational Psychologist, 38*(1), 1-4.

Philip, D. (2008). *Economics, knowledge building, and the Net Generation: A model of learning for Net Generation students*. Retrieved April 30, 2008 from: http://education-2008.org/speakers/don_philip/don_philip_abstract.cfm.

Reeves, T. (1999). *"Keynote Address" Presented at ED-MEDIA 1999 World Conference on Educational Multimedia, Hypermedia & Telecommunications*. Seattle, Washington.

Savery, J., & Duffy, T.(1995). Problem based learning: An instructional model and its constructivist framework. *Educational Technology, 35*(5), 31-38.

Smith, P. L., & Ragan, T. J. (2005*). Instructional design*. Upper Saddle River, NJ: Merrill/Prentice Hall.

Valasidou, A., Sidiropoulos, D., & Makridou-Bousiou, D. (2005). The constructivist perspective in distance learning environments. In P. Kommers & G. Richards (Eds.), *Proceedings of World Conference on Educational Multimedia, Hypermedia and Telecommunications 2005* (pp. 1932-1935). Chesapeake, VA.

WCET, (2008). *WCET Annual Conference*. Retrieved January 15, 2008, from:http://www.wcet.info/2.0/.

Chapter XVII
Facilitating Student-Driven Learning of Computer Programming with Technology

Alessio Gaspar
University of South Florida Polytechnic, USA

Sarah Langevin
University of South Florida Polytechnic, USA

Naomi Boyer
University of South Florida Polytechnic, USA

ABSTRACT

This chapter discusses a case study of the application of technology to facilitate undergraduate students' learning of computer programming in an Information Technology department. The authors review the evolution of the didactic of introductory programming courses along with the learning barriers traditionally encountered by novice programmers. The growing interest of the computing education research community in a transition from instructivist to constructivist strategies is then illustrated by several recent approaches. The authors discuss how these have been enabled through the use of appropriate technologies in introductory and intermediate programming courses, delivered both online and face to face. They conclude by discussing how the integration of technology, and the switch to online environments, has the potential to enable authentic student-driven programming pedagogies as well as facilitate formal computing education research or action research in this field.

INTRODUCTION

The question of whether directed teaching methodologies are more efficient than self-directed or constructivist ones is still being debated in both global and discipline-based educational communities (Kirschner, Sweller & Clark, 2006; Sweller, Kirschner & Clark, 2007). As far as computing education research is concerned, applying constructivism in programming courses is a recurrent theme as illustrated by general studies (Kolling, Quig, Patterson, Rosenberg, 2003; Wulf, 2005). More specific reference will be considered as we review specific constructivist contributions in the remainder of this chapter. In this context, our work has been focused on studying the impact of constructivism in a specific context (i.e. university-level programming courses), rather than in "education at large", while also further developing our understanding of how technology can actually support it. The introduction provides the context of the case study presented in this chapter (e.g. courses taught, characteristics of student population) as well as an overview of the main learning barriers encountered by novice programmers. The background section discusses how these learning barriers can be addressed by a transition from instructivist programming pedagogies to constructivist ones. We then describe and discuss the impact of specific pedagogies which have been designed for and evaluated on Information Technology (IT) undergraduate students. The next section then introduces the various technologies which were used to support our constructivist approach in both face to face and online offerings. We share the lessons learned so far and underline how the transition from face to face to online delivery resonated with these technologies. The future trend section discusses how the above-mentioned technologies might end up re-shaping the nature of computing education research. We then conclude with a summary of opinions and findings.

Context of this Case Study

Our case study focuses on the application of technology to improve the teaching of computer programming courses in the Information Technology department (IT) at the University of South Florida Polytechnic (USFP). This department offers small size classes (often in the evening) for mostly non-traditional undergraduate students (older age groups, already in the workforce). The pedagogical interventions described hereafter were deployed in two programming courses. COP2510 Programming Concepts is designed as a first programming course introducing fundamental concepts to students. During the first 5 weeks, the Raptor visual flowchart interpreter (Carlisle, Wilson, Humphries, Hadfield, 2005) is used to introduce programming building blocks and concepts while developing design skills without burdening students with the syntactical complexities of a mainstream programming language. The following five weeks are devoted to reviewing the concepts already covered with Raptor and transferring them to the Java programming language. The last five weeks explore more advanced Java concepts. This pedagogy of contents has been independently explored by others (Adams, 2007) and is based on a two-steps, syntax-light then syntax-heavy, redundant, introduction to programming. Unlike other studies, we use a fundamental-first approach (Liang, 2006), which turned out to be more welcoming to our students than the original objects-first approach we evaluated for this course (Kolling et al., 2003). Enrollment was respectively 13 and 35 students for the face to face (fall 2007) and the online offerings (spring 2008).

COP3515 IT Program Design is an intermediate programming course which requires COP2510 as pre-requisite. It strengthens students' programming skills by (1) further developing their design and troubleshooting skills and (2) teaching a new programming language which will lead to a better understanding of "under the hood" mechanics to facilitate the learning of other languages. The C programming language is used throughout the semester. The first five weeks are devoted to review fundamental programming concepts in our new language. The next five weeks focus on technical aspects (i.e. pointers, memory allocation) and the implementation of standard language constructs (e.g. strings, arrays). This course is also used to prepare students for upper-level undergraduate system-oriented courses. The last five weeks focus on recursion and data structures implementations. This further develops design skills while preparing students to their next programming course. We use the Deitel text (Deitel, 2007). Enrollment included 15 students in fall 2007 and 13 students in spring 2008.

What is Programming?

Programming can be seen as a multi-faceted activity. The process entails transforming specifications, written in plain English (and often ambiguous), into a working program implemented in a given programming language. This requires students to analyze requirements, set objectives to reach, design and implement an algorithmic solution, and design and implement a set of test cases. The latter, if successfully applied to their program, will validate it. This activity is summarized by the following diagram (Fig 1) used as an overview of the programming activity during the first week of class.

This view of programming has evolved with the advent of new methodologies and technologies (e.g. testing frameworks, test-driven development methodologies (Beck, 2002)). Today, it is clear that teaching programming is not only about teaching facts and static knowledge such as syntactical rules, domain jargon, etc. The programming thought process itself must be taught to students through hands-on practice; one learns programming by *programming*, instead of *talking about* programming. We are affecting how students think by teaching them a skill rather than facts they can absorb and later regurgitate. It is therefore critical to leverage the students' cognitive models of the programming activity in this transformative step. This essentially constructivist perspective is pervasive in the various learning outcomes of our programming syllabus.

Programming Learning Outcomes

Program design and implementation skills must be acquired in a way which reflects the fact that programming is a creative endeavor aimed at solving algorithmic problems. The various studies leveraging Bloom's taxonomy of the cognitive domain (Bloom, 1956) to better understand

Figure 1. A overview of the two programming processes taught in introductory courses

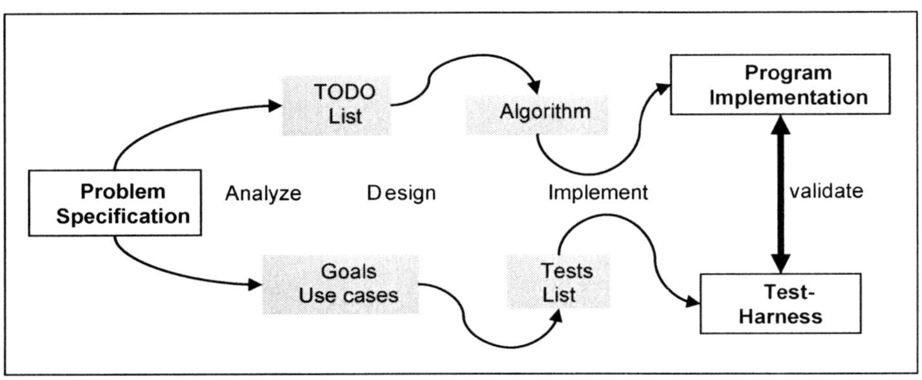

programming pedagogies, converge on observing that programming can't be taught as a simple regurgitation of facts. Bloom's work has been used to to evaluate programs' comprehension (Buckley and Exton 2003), debugging (Shaochun, Rajlich, 2004), testing activities (Scott, 2003), student's performance (Lister and Leaney, 2003) and characterize learning barriers (Oliver, Dobele, Greber and Roberts, 2004). However, many still fail to see that it is equally unacceptable to lead students to only mimic the programming thought processes without internalizing the necessary skills to develop original solutions to problems never encountered before. This observation led us to identify "programming with intention" as a difficult skill to learn for students who are used to absorb-regurgitate knowledge. These often attempt to learn programming by rote memorization of a set of problems-solutions pairs (Gaspar & Langevin, 2007). When facing a new problem, these students try to match its specification with that of a previously studied problem. They then either re-use the existing solution "as is" or attempt to modify it. Because of their lack of understanding of how this solution was designed, they often resort to almost purely random modifications until the adapted solution "behaves" as requested by the assignment.

Program testing skills are also critical to develop in the novice programmer. A previous study already indicated that novice programmers have difficulties with the concept of program validity and therefore do not know when their solutions are fully satisfying the requirements (Kolikant, 2005). We believe that teaching students how to test their programs is a promising way to develop their understanding of the notion of valid program and provides a platform that helps the student to address the level of ambiguity found in some programming activities (Mingus & Grassi, 1998). There is no clear evidence of whether test-early (or test driven development – TDD) or test-late approaches are more beneficial. Research on assessing the individual benefits is still relatively recent and hasn't yet compared alternatives to one another. In the former, students approach a problem by first developing its test harness (goal), then the program itself. In the latter, they design a solution first, and then assess its validity by developing a test-harness, i.e., a series of tests the program must successfully pass. In both cases, they develop a wider perspective on the programming activity which leads them to build solutions guided by their understanding of the objectives to be met.

The final skill used in programming is troubleshooting. This term encompasses the student's capability to identify erroneous behaviors in their solutions and trace back their origins in the program. Such activities are tightly integrated with testing; students need to generate test cases, which will help validate or falsify hypotheses made about the origin of the bug. This thought process requires students to read programs rigorously (Buckley and Exton, 2003), without assumptions or shortcuts, and interpret them as the machine would (Shaochun and Rajlich, 2004). Students need to develop a solid discipline and a solid understanding of how their programs are interpreted so they will not think only in terms of the author's intentions when reading them, but rather in terms of what the computer system will do when executing them. Ultimately, and just as in testing skills (Scott, 2003), these reflexes end up being internalized in the design and implementation steps (Kolikant, 2005). They also help the students "program defensively" and validate their statements as they write them with a wider perspective on their impact on the entire program being developed.

From the didactic perspective, programming requires students to acquire both knowledge (definitions, concepts, syntax) and skills (designing, implementing, troubleshooting programs). This diversity of skills, and therefore learning outcomes, spawns the entire Bloom's taxonomy thus making programming education extremely challenging. The next section will discuss the evolution of programming pedagogies toward constructivist approaches.

BACKGROUND

This section reviews the impact of constructivism in programming education, stresses its benefits and discusses obstacles to its implementation. We then introduce various instructional strategies and discuss their preliminary evaluation before considering, in the next section, how technology has helped overcome some of their limits.

From Instructivist Programming Education...

While lectures can be used to introduce elementary programming concepts, ideas and syntax, it is unclear that they efficiently convey programming skills to students. Such skills are by nature dynamic and require students to develop a new way of thinking. This is true at many levels, e.g., designing solutions, implementing them, testing their validity, and debugging errors.

From the perspective of the design and implementation skills, common practice still consists in (1) presenting students with a problem to solve, (2) giving them a certain amount of time to work on it individually or in groups and (3) presenting them with the detailed solution (Deitel, 2006). This approach does not focus on the programming process, but rather on the programming process's outcome: a working program. It offers students a rather static picture of the programming process; the thought process used to develop the solution is omitted and only mentioned to retrospectively justify it. This is akin to teaching someone how to build a house, not by apprenticeship (having them work with you) but by inviting them to visit a finished house and discussing how it was built.

This approach also severely limits the number of alternative solutions being presented. A single solution, which embodies the standards we want to teach, is learned as the direct (only?) answer to the problem. To remedy this, instructors often develop (or lead students to develop collaboratively) "live" solutions which illustrate specific error patterns and the misconceptions which lead to them. However, even when doing so, it is not obvious that the instructor's choice of "bad examples" will be relevant to the cognitive models of the students. In most cases, such choices are based on personal experience (e.g. "my students always fall for this") and on the literature (e.g. "this error has been established as common in novice programmer in a study at University of Elsewhere"). However, even if statistically significant, such results are rarely able to tell the instructor what will work best with a specific student population. Significantly different student populations (e.g. selective universities vs. less selective ones, commuter-only vs. residential students, regular vs. evening classes, full time students vs. full time employees) exhibit unrelated responses to similar pedagogical interventions. In order to affect their own students, instructors need to base pedagogies on their very particularities, using various feedback-enabling instruments. Therefore, the optimal source of knowledge for guiding the presentation of alternatives is the relevant experiences drawn from the student population being taught. Gholson and Craig (2006) support "Learners experience new phenomena, interpret experiences in terms of what they already know, reason about new experiences, reflect on the experiences, and reflect on the reasoning process itself" (p. 122).

This argument ties in with the need to take into consideration, in the teaching process, the students' preconceptions and misconceptions. With regards to programming education, previous studies have established the richness of novice programmers' grasp of algorithmic problems before any formal training (Lewandowski, Bouvier, McCartney, Sanders & Simon, 2007; Chen, Lewandowski, McCartney, Sanders & Simon, 2007). It is important that students exhibit algorithmic thinking in dealing with disciplines that include algorithms, in which they think about the steps that they are implementing to solve the problem and create a program in a way that further internalizes the learning experience (Mingus & Grassi, 1998). Other disciplines, such as physics, have long embraced the idea of leveraging the knowledge of novices to inform their teaching practice (Halloun and Hestenes, 1985).

Finally, presenting students with problems' descriptions on one slide, and their solution on the next, only reinforces their natural tendencies to build a "big book of all programming problems' solutions" made of problem-solution pairs that they will later reuse when facing a new problem (Gaspar & Langevin, 2007). This then limits the possibility for the student construction of solutions to new problems.

To Cognitive Apprenticeship...

These issues have been addressed so far by developing pedagogies based on the idea of apprenticeship. Many crafts are taught through apprenticeship; however, in the case of intellectual activities it is difficult for the learner to observe the expert's thought process. This difference led to the development of cognitive apprenticeship strategies (Collins, Brown, & Newman, 1987) which have motivated many different approaches to teaching programming; e.g. (Kolling et al., 2003).

Such approaches address the lack of focus of instructivist strategies on the programming process rather than its outcomes. In this respect, they considerably improve on the previously discussed methods. We implemented such approaches in our programming courses through so-called "instructor-led live coding" activities. During these, the instructor designs, tests, implements, and debugs a program "live". Whether in face-to-face or synchronous online settings (using conferencing tools, e.g., Elluminate, Dim Dim), these enable the teaching of the programming process by example and narration of the thought process. Students also learn more about programming by witnessing hesitations, errors, the non-linear development of the program itself and how erroneous decisions are mitigated.

Despite these advantages, this approach is still instructionist. The "sage on the stage" demonstrates skills rather than lecturing about them, but students are still passively observing. Instead of memorizing entire solutions, they mimic the processes without necessarily internalizing them and therefore without having to learn the craft of programming.

... To Constructivist Programming Education

This limit illustrates the need for a constructivist framework in programming education. In order to minimize the ways students can be successful (and sometimes pass) without acquiring the skills we are trying to teach, it is important to not only focus our teaching efforts on the right skills but also to involve students in the learning process. This can be achieved by structuring the active development of skills rather than passive observance and mimicry.

Social constructivist approaches' main tenets are the recognition of (1) the importance of taking into consideration the students' preconceptions and misconceptions in the teaching process (Gholson & Craig, 2006), (2) the benefits of building a community of learners to scaffold the students' learning process and (3) the recognition of learning as being an active process. The second point is particularly important as it distinguishes between cognitive constructivism and social constructivism which focuses on the importance of peers and instructors to promote learning. In other words, the learner utilizes the culturally bound activities to make sense of the presented material (Cobb, 2005). While most constructivists would agree that activity and engagement with material can lead to deeper understandings, the internal thought process versus external culturally organized practices separate the philosophies (Fosnot & Perry, 2005; Cobb, 2005). For the purposes of this work, instructional strategies have been designed to accommodate both approaches incorporating both cognitive algorithmic problem solving and socially situated activities.

It is safe to assume that, in a typical programming class, (1) students have different levels of expertise and (2) student-to-student expertise differences are smaller than student-to-instructor ones. Under these assumptions, peer learning activities in which students challenge one another are more likely to foster effective learning, as defined by the concept of zone of proximal development (ZPD) (Vygotsky, 1978), compared to instructor-student interactions. Of course, an instructor's skill often lies in the ability to develop activities which will be located in the students' ZPD. However, as previously mentioned, most instructors base their pedagogies on experiences or studies which might not apply to their students. This is especially true in an emergent field such as programming which continuously evolves technologically and thus introduces new educational challenges regularly. Therefore, taking your own student population characteristics into consideration is the only way to align teaching practices with their needs.

These observations indicate that peer learning is beneficial to constructivist programming education and enables an "auto-tuning" of the learning challenges occurring in the class so that they remain, without need of tuning from the instructor, in the students' ZPD.

Impact of this Transition

Both constructivism and peer learning converge toward the idea of a student-driven programming pedagogy. Our ongoing research attempts to enable such dynamics in face to face and online programming courses. This endeavor relies heavily on instructional technologies whose benefits and limits are discussed in the next section.

TECHNOLOGIES FOR STUDENT-DRIVEN PROGRAMMING EDUCATION

This section provides the reader with concrete examples of constructivist programming activities which we introduced in introductory and intermediate programming courses. We then review the various technologies which have been used to support them in face-to-face and online synchronous courses. We underline how they allowed us to overcome previously discussed shortcomings and identify remaining limitations before to formulate recommendations.

Examples of Constructivist Activities

The idea of constructivist apprenticeship (Gaspar, Langevin, Boyer, 2007b) encompasses a multitude of learning activities suitable for introductory and intermediate programming courses and based on social constructivism, peer and competitive learning. The following describes the general pedagogical principles involved in these approaches as they have been implemented during fall 2007 in face to face offerings of both Programming Concepts (COP2510) and IT Program Design (COP3515). The next section will review technological support as well as the technological development which was necessary to transition to online synchronous offerings.

The first learning activity, instructor-led live coding, was used initially during class meetings for the instructor to correct practice exercises. These instructor-led programming demonstrations were then made available as videos using Camtasia to record screencasts. These were progressively augmented by presenting alternative solutions, showing students how to debug programs, test them or by discussing design options.

This transition from "live" to videos allowed us to refocus our class meeting time on students' practice. We introduced student-led live coding sessions (Gaspar & Langevin, 2007a) during which one selected student works on an exercise using a wireless keyboard connected to the data projector PC. The solution was developed "live" in front of other students and the instructor. The latter facilitates discussions of possible alternatives and train students to critically analyze the program being written (e.g., debug it, optimize it, and justify its validity). This allowed instructors to evaluate the students' proficiency level and respond to their misconceptions while exploring a particular problem. Two variants of this type of activities were used. In the first one, a single student was selected for the duration of an exercise. Others were put in a role very close to the "observer role" as defined in pair programming studies (McDowell, Hanks, Werner, 2003). Through the questions and comments of the instructor, this variant would involve the entire class working collaboratively on solving a given exercise. The second variant also consisted in picking a single student for the duration of an exercise, but it also invited all students to work individually on their own solution. Students unsure about how to start would benefit from being able to peek at the lead student's work on the podium PC. After a set time, all students would take a look at the podium PC solution and the instructor would lead the class in discussions comparing it to the alternative others came up with. This strategy has been particularly useful when working on problems which have a wide range of possible solutions such as problems accepting both recursive and iterative solutions (Gaspar & Langevin, 2007a).

In fall 2007, both programming classes used Peer Learning Forums (discussion boards) intensively to encourage students to be more engaged in their reading assignments and to come prepared to weekly meetings. After each session, a chapter from the text was assigned as reading assignment. Students had two days to read it and post, on a weekly forum, questions about anything unclear to them. They then had two days to revisit their readings and address their peers' questions. These activities were graded with points awarded for posting questions and responses. To count, these posts had to be on-topic, within allowed deadlines and not boil down to banter. Our grading focused on rewarding meaningful and regular participation while not penalizing students for bringing up wrong hypothesis. While we did not expect students to solve each other problems, this structure motivated a more in-depth reading of the material and forced them to revisit it in search of responses to specific questions. In intermediate programming courses, but also in introductory ones after about 5-6 weeks, the capability of students to solve each other's problems improved over time as evidenced by the increased pertinence of the responses. This is consistent with our previous hypothesis that, given a basic understanding of the foundations, student-to-student interactions would naturally align themselves within the ZPD.

In addition to these peer learning interactions, these forums also helped the instructor to develop weekly lectures which focused on the very learning barriers encountered by students. This allowed using class meeting time better by lecturing about relevant aspects of the material rather than about what students already understood. This "just in time lecturing" approach introduced explanations after students felt a need for them. During these lectures, the Ubiquitous Presenter software (Wilkerson, Griswold, Simon, 2005) was used to foster active learning.

These activities required technological support in one form or the other; Blackboard forums for peer learning forums, a wireless keyboard for student-led live coding sessions, Ubiquitous Presenter (Wilkerson, Griswold, & Simon, 2005) for active learning, Camtasia for instructor-led live coding screencasts, etc. The next section will review the strengths and weaknesses of this technological support while also underlining how the switch from a face to face offering to a synchronous online one affected both programming courses.

Technological Support and its Evolution During Online Offerings

This section discusses the limits of the technologies used to support the above-mentioned pedagogies as well as how they had to evolve once we made the transition to online offerings.

As explained in the previous sub-section, Camtasia videos have been used to extend the principle of instructor-led live coding outside the classroom. In addition to face to face and online synchronous teaching sessions, this approach has also been useful to provide students with corrections of practice exercises. It is easier to list an exercise and provide students with its complete solution. Once again, this approach tends to reinforce the tendency to assimilate programming with a memorization of problem-solutions pairs. By recording screencasts of the instructor developing a solution from scratch and narrating his or her thought process while doing so, one can develop videos which extend the instructor-led live coding approach to out-of-class activities. Students are invited to work individually on exercises, use the videos for further explanation and pause them to resume trying on their own and, once they are done, watch the entire video to confirm the validity of their solution. Students particularly appreciated the ability to revisit difficult explanations (by rewinding) but also to work on an exercise and play just enough video to enlighten them, pause it, work some more on their exercise and be able to resume playing the video as needed. From the instructor's perspective, such videos are a great opportunity to introduce a problem-based learning (PBL) component to the instructional toolset. After the transition to an online offering, Camtasia turned out to be efficient to develop small-duration lectures to illustrate specific topics complementing the textbook coverage or to present material for which the textbook is not a suitable medium or which it does not include.

From the perspective of the weekly designed mini-lectures, based on the problems encountered by students while working on the reading assignments and posted on the peer learning forums, we plan on using Camtasia videos in the near future as a way to prepare problem-driven tutorials. Our long-term objective is to constitute a collection of videos responding to the most common student questions on the reading assignments and make them available to the community at large. This would allow the move to a completely asynchronous online model where, each week, appropriate mini-lectures would be disseminated in response to the peer learning forums' posts. This could also lead to improved use of synchronous interaction time that focuses on group practice sessions (e.g., student-led live coding).

The Peer Learning Forums have been implemented during the face to face and online synchronous classrooms through Blackboard's discussion boards. This tool allowed us to implement this activity with some contortions but also revealed that a custom-designed tool might facilitate better interactions. A first missing feature is the possibility to allow the instructor to track down a post's authors (for grading purposes) while making the posts look anonymous. This would make students more comfortable, early in the semester, when posting questions and reduce the potential for instructor's responses to discourage further participations. In our experience, some students confuse forums with chat venues early in the semester. While this misusage needs to be rectified, doing so often makes students feel singled out. Anonymous posts can also be useful when student participation is particularly low, allowing instructors to inject questions into the forums to bring up traditional issues with the reading material. In addition to this anonymous post feature, Blackboard also lacks facilities to encourage peer evaluation. Ideally, students should be able to rate their level of interest in the questions posted (e.g. "trivial" to "we definitively need to discuss this during the next class meeting") but also their level of satisfactions with the answers posted by their peers and the instructor. Given the multiplicity of ratings possibilities, the grading process could be automated to a great extent while providing the instructor with additional information about the overall class interest in the weekly topics. These issues have been exacerbated with the move to online classes as it became more difficult to clarify postings and posting evaluations during class meetings. Blackboard does not support this type of "advanced" forum usage and we are planning to develop an in-house solution to be tested in our courses.

Ubiquitous Presenter (Wilkerson, Griswold, & Simon, 2005) provided an excellent environment for face to face active learning during the "just in time lecturing" activities. However, latency issues (the delays for instructors' actions to be communicated to students' workstations) prevented us from using it in an online context. Other synchronous and asynchronous approaches to foster active learning, such as Message Grid (Pargas, 2006) might be more suitable in an online environment due to the fact that they (1) provide better response times during synchronous interactions and (2) allow for asynchronous interactions. However, to date, such systems are rather centralized (e.g. hosted on a single machine at the author's institution) and their availability is therefore subject to the continuing interest of the faculty involved in its development. This makes it particularly risky to base pedagogies on the availability of a given tool which might become unavailable at any time.

Elluminate was used to host online synchronous sessions. It enabled voice and text communication in a virtual classroom. Overall, the students reacted positively to the possibility of interrupting the instructor by typing a question in the chat window. This motivated far more interaction than in a face to face setting where some might be shy about talking in front of everyone if not explicitly prompted to do so by the instructor. In addition, Elluminate also allowed students to share their desktop, thus providing others with a view of what they are working on. This feature has been extensively used for both instructor-led and student-led live coding, replacing advantageously the data projector / wireless keyboard scheme. Further, Elluminate enables the shared desktop to be temporarily controlled by anyone in the group, thus enabling anyone to easily "jump in" to provide help. Last but not least, Elluminate also allows dispatching participants into isolated virtual rooms. This feature was critical to enable student-led live coding in higher enrollment courses. However, this is still a feature we are working on learning to use efficiently. By creating multiple groups of up to 5 students, the instructor reduces his or her ability to be present and to facilitate the student-led live coding for all groups simultaneously. This has proved so far to be an obstacle to fully realizing the potential of this type of learning activities as it was observed in the face to face, small classes setting.

What Did We Learn From Using These Approaches?

The above discussion summarizes the motivation behind the selection of specific instructional technologies but also the additional constraints and opportunities brought up by the switch to an online synchronous setting. Overall, these technologies enabled the constructivist peer learning approaches at the core of our programming pedagogy.

From the student perspective, a survey from the face to face offering revealed that this instructional method was received positively; students identified the potential for it to change their overall attitude toward learning. This motivated a study of the impact of these pedagogies and their technological implementation on the self-direction capabilities of our students (Boyer, Langevin & Gaspar, 2008). This study will be complemented by results collected in spring 2008 and allow us to evaluate the differences between face to face and online offerings based on the supporting technologies since the course contents and pedagogy have been kept identical.

So far, the enrollment increase in the introductory online course has been linked with an attrition increase. We conjecture that this phenomenon is linked to the levels of self-direction required by our pedagogy (e.g., come prepared to lecture, actively work on reading assignments rather than following passively a lecture or reading, etc.).

FUTURE TRENDS

Both the constructivist activities discussed in this chapter and the supporting technologies which enable them have, in our opinion, a fruitful future in terms of computing education research. This section reviews how the transition to online delivery methods might constitute a boon to computing education research.

Nationwide in the US, computing curricula are embracing the challenge of online courses, whether using synchronous or asynchronous delivery methods. As we presented, these new courses require us to rethink and adapt our teaching pedagogies and thus present new challenges both educational and technological as we strive to develop new technologies to either support further our pedagogical strategies or simply remedy the teaching aspects lost in the transition from face to face to online delivery methods.

Most educators are aware of the potential of online courses to foster the development of new tools or pedagogies, along with the difficulties associated with such a transition. Students' opinions form based on practical considerations (e.g., schedule and location flexibility) or preferences relative to the student's learning style; self-starter learners tend to love flexibility while other self-directed learners prefer the constraints associated with weekly meetings. It is also common to observe that students enrolled in online classes tend to come with the expectation that the instructional process will be different. Because of that, they are more likely to accept a focus on self-direction, peer learning activities and, more generally, an increased personal responsibility about their learning than when they meet weekly in a traditional face to face setting.

To the best knowledge of the authors, no study has yet reported this transition to online delivery modes as an opportunity to facilitate computing education research. By adopting the technologies discussed in the previous sections; instructors are not only improving their teaching practice and implementing well-established pedagogical strategies. They are also putting themselves, and their students, in a situation where what happens during class meetings is automatically archived, documented and therefore easier to study than ever before. Let us examine this idea in light of the previously mentioned technologies.

Discussion forums are a transcribed form of the type of interaction which would usually take place in class (assuming a non solely instructivist setting). They facilitate the observation of how the learning barriers encountered by students evolve over the course of the semester. Once correlated to the schedule of topics covered, they allow the instructor to accumulate feedback on where to expect the most difficulties. This feedback is constantly updated and can be used to establish long term trends in separate student populations.

Similarly, student-led live coding sessions can be recorded via Elluminate or screencast software. Such archives are invaluable windows into the students' thought processes and allow the instructor to engage in qualitative and quantitative studies allowing the computing education research community to identify trends and the impact of pedagogical approaches under a constantly updated student-driven feedback.

The impact of these technological tools on the research agenda of any computing education researcher can therefore not be understated. Our team is working on disseminating a virtual appliance meant to help instructors easily adopt a complete teaching environment which will allow them to collect and analyze easily such information. We believe that such projects have the potential to not only generate interesting results from the team developing the tools but, most importantly, motivate and enable a critical mass of instructors to collect their own feedback and either use it to guide their own teaching practice or report it back to the research community and therefore, contribute to improving our knowledge of students' cognitive processes in programming activities.

CONCLUSION

This chapter presented a case study of the application of technology to facilitate undergraduate students' learning of computer programming in an Information Technology department. The importance of constructivist peer learning pedagogical strategies to teaching programming was illustrated with respect to the epistemological and didactic foundations of this approach as well as through its impact on machine learning techniques. Our own experience in two programming courses taught both online and face to face during fall 2007 and spring 2008 confirmed the potential benefits of such approaches from the pedagogical point of view. These results also enabled us to reflect on the role played by technology to support these efforts. We concluded with the unsuspected impact of these technological tools on the computing education research agenda by allowing the automatic collection of abundant data on students' participation (in forums, wikis, synchronous sessions...) which is usually harder to collect in face-to-face offerings.

REFERENCES

Bloom, B. S. (1956). Taxonomy of educational objectives: The classification of educational goals. *Handbook I, cognitive domain.* New York: David McKay Co, Inc.

Beck, K. (2002). *Test-driven development by example.* Boston, MA: Addison-Wesley.

Adams, J. (2007). *Alice in action with Java.* Boston, MA: Thomson Course Technology.

Boyer, N., Langevin, S., & Gaspar, A. (2008, Jan.-Feb.). *Computer programming meets education: The intersection between information technology and self-direction.* Paper Presented at the 22nd Annual International Self-Directed Learning Symposium, Cocoa Beach, Florida.

Buckley, J., & Exton C. (2003). Blooms' Taxonomy: A Framework for Assessing Programmers' Knowledge of Software Systems. In Proceedings of the *11th IEEE International Workshop on Program Comprehension* (pp. 165). Washington, DC: IWPC. IEEE Computer Society.

Carlisle, M. C., Wilson, T. A., Humphries, J. W., & Hadfield, S. N. (2005). Raptor: A visual programming environment for teaching algorithmic problem solving. *Proceedings of the 36th SIGCSE technical symposium on Computer science education* (pp. 176-180). St. Louis, MO.

Chen, T., Lewandowski, G., McCartney, R., Sanders, K., & Simon, B. (2007). Commonsense computing: using student sorting abilities to improve instruction. In *Proceedings of the 38th SIGCSE Technical Symposium on Computer Science Education* (pp.276-280). New York: Association for Computing Machinery. Retrieved October 3, 2008, from http://doi.acm.org/10.1145/1227310.1227408.

Cobb, P. (2005). Where is the mind? A coordination of sociocultural and cognitive constructivist perspectives. In C. T. Fosnot (Ed.), *Constructivism: Theory, perspectives, and practice* (pp. 39-61). New York: Teachers College Press.

Collins, A., Brown, J. S., & Newman, S. E. (1987). *Cognitive apprenticeship: Teaching the craft of reading, writing and mathematics* (Technical

Report No. 403). Cambridge, MA: Centre for the Study of Reading, University of Illinois. (ERIC Document Reproduction Service No. ED 284 181).

Deitel, H., & Deitel, P. (2006). *C How to program* (5 ed). Upper Saddle River, NJ: Prentice Hall.

Fosnot, C. T., & Perry, R. S. (2005). Contructivism: A psychological theory of learning. In C. T. Fosnot (Ed.), *Constructivism: Theory, perspectives, and practice* (pp. 8-38). New York: Teachers College Press.

Gaspar, A., Langevin, S., & Boyer, N. (2008). Constructivist apprenticeship through antagonistic programming activities. *Encyclopedia of Information Science and Technology* (2nd ed.). Hershey, PA: Information Science Reference.

Gaspar, A., & Langevin, S. (2007a, July). *Active learning in introductory programming courses through student-led "live coding" and test-driven pair programming.* Presented at the International Conference on Education and Information Systems, Technologies and Applications, Orlando, FL.

Gaspar, A., & Langevin, S. (2007b). Restoring "coding with intention" in introductory programming courses. *Proceedings of the 8th ACM SIGITE conference on Information Technology Education* (pp. 91-98). New York: Association for Computing Machinery. Retrieved October 3, 2008, from http://portal.acm.org/citation.cfm?id=1324302.1324323&coll=GUIDE&dl=ACM&CFID=5033070&CFTOKEN=27681204

Gholson, B., & Craig, S. D. (2006). Promoting constructivist activities that support vicarious learning during computer-based instruction. *Education Psychology Review, 18*(2), 119-139.

Halloun, I., & Hestenes, D. (1985). The Initial Knowledge State of College Physics Students, *American Journal of Physics, 53*, 1043-1055.

Kirschner, P. A., Sweller, J., & Clark, R. E. (2006). Why minimal guidance during instruction does not work: An analysis of the failure of constructivist, discovery, problem-based experiential and inquiry-based teaching. *Educational Psychologist, 41*(2), 75-86.

Kolikant, Y. B. (2005). Students' alternative standards for correctness. *In Proceedings of the International Workshop on Computing Education Research* (pp. 37-43). New York: Association of Computing Machinery, 37-43. Retrieved October 3, 2008, from http://doi.acm.org/10.1145/1089786.1089790

Kolling, M., Quig, B., Patterson, A., & Rosenberg, J. (2003). The BlueJ system and its pedagogy. *Journal of Computer Science Education, special issue on learning and teaching object technology, 13(*4), 249-268.

Lewandowski, G., Bouvier, D. J., McCartney, R., Sanders, K., & Simon, B. (2007). Commonsense computing (episode 3): concurrency and concert tickets. *In Proceedings of the Third international Workshop on Computing Education Research* (pp. 133-144). New York: Association of Computing Machinery. Retrieved October 3, 2008, from http://doi.acm.org/10.1145/1288580.1288598

Liang, Y. D. (2006). *Fundamentals first introduction to Java programming* (6 ed.). Upper Saddle River, NJ: Prentice Hall.

Lister, R. & Leaney, J. (2003). Introductory programming, criterion-referencing, and bloom. In *Proceedings of the 34th SIGCSE Technical Symposium on Computer Science Education* (pp. 143-147). New York: Association of Computing Machinery. Retrieved October 3, 2008, from http://doi.acm.org/10.1145/611892.611954

McDowell, C., Hanks, B., & Werner, L. (2003). Experimenting with pair programming in the classroom. *ACM SIGCSE Bulletin Proceedings of the 8th annual conference on Innovation and technology in computer science education, 35*(3), (pp 60-64). ACM New York: Association for Computing Machinery. Retrieved October 3, 2008, from http://portal.acm.org/citation.cfm?id=961290.961531&coll=GUIDE&dl=GUIDE&CFID=5033070&CFTOKEN=27681204

Mingus, T. T. Y., & Grassi, R. M. (1998). Algorithmic and recursive thinking current beliefs and their implications for the future. In L. J. Morrow & M. J. Kenney (Eds.), *The teaching and learning of algorithms in school mathematics* (pp. 32-42). Reston, VA: National council of Teachers of Mathematics.

Oliver, D., Dobele, T., Greber, M., & Roberts, T. (2004). This course has a Bloom Rating of 3.9. In R. Lister & A. Young (Eds.), *Proceedings of the Sixth Conference on Australasian Computing Education – Vol. 3.* (pp. 227-231). Darlinghurst, Australia: Australian Computer Society, Inc. Retrieved October 3, 2008, from http://portal.acm.org/citation.cfm?id=979968.979998&coll=GUIDE&dl=ACM&CFID=5033070&CFTOKEN=27681204

Pargas, R. P. (2006). Reducing lecture and increasing student activity in large computer science courses. *Proceedings of the 11th annual SIGCSE conference on Innovation and technology in computer science education, 38*(3), (pp. 3-7). New York: Association for Computing Machinery. Retrieved October 3, 2008, from http://portal.acm.org/citation.cfm?id=1140124.1140129&coll=GUIDE&dl=ACM&CFID=5033070&CFTOKEN=27681204

Scott, T. (2003). Bloom's taxonomy applied to testing in computer science classes. *Journal for Computing in Small Colleges, 19*(1), 267-274.

Shaochun X., & Rajlich, V. (2004). Cognitive process during program debugging. *Proceedings of the Third IEEE International Conference on Cognitive Informatics, (pp.* 176-182). Retrieved October 3, 2008, from http://www2.computer.org/portal/web/csdl/doi/10.1109/COGINF.2004.1327473

Sweller, J., Kirschner, P. A., & Clark, R. E. (2007). Why minimally guided teaching techniques do not work: A reply to commentaries. *Educational Psychologist, 42*(2), 115-121.

Vygotsky, L.S. (1978). *Mind and society: The development of higher mental processes.* Cambridge, MA: Harvard University Press.

Wilkerson, M., Griswold, W.G., & Simon, B. (2005). *Ubiquitous Presenter: Increasing student access and control in a digital lecturing environment* (pp 116-120). St. Louis, Missouri: Association for Computing Machinery. Retrieved October 3, 2008, from http://portal.acm.org/citation.cfm?id=1047344.1047394&coll=GUIDE&dl=ACM&CFID=5033070&CFTOKEN=27681204

Wulf, T. (2005). Constructivist approaches for teaching computer programming. In *Proceedings of the 6th Conference on information Technology Education,* (pp. 245-248). New York, Association for Computing Machinery. Retrieved October 3, 2008, from http://www2.computer.org/portal/web/csdl/abs/proceedings/icci/2004/2190/00/21900176abs.htm

Chapter XVIII
Designing Asynchronous Discussions to Teach Critical Thinking

John Miller
National University, USA

ABSTRACT

A central component of constructivist pedagogy at the college level is the modeling and practicing of critical thinking, and since Socrates, discussion has been the basic vehicle for accomplishing this. Advocates of online teaching have argued that the written and asynchronous nature of online discussions enhance its ability to teach critical thinking. Unless online discussions are properly designed, however, these apparent advantages may in fact have the opposite effect. This chapter sets forth six basic principles for designing online discussions that model and shape critical thinking experiences for students, and illustrates them with examples from two different online college literature classes.

INTRODUCTION

It is a basic premise of a social constructivist pedagogy that learning arises from interaction with others, and from the comparison and evaluation of different voices and viewpoints. Such interaction is also central to what we mean by, and value in, the activity of "critical thinking." While this term has been defined in many ways, critical thinking revolves fundamentally around the testing of ideas and opinions—especially our own—against alternative ideas and opinions. Through this process, we probe our ideas for weaknesses of coherence, validity, and completeness, and this in turn allows us to synthesize new and more sophisticated ideas. Thus, the activity of critical thinking is an essential component of the construction of knowledge at the higher levels of cognition we expect from students in higher education.

Discussion has been regarded as the fundamental method of enabling such learning since Socrates, who (though he would not have accepted a constructivist definition of knowledge as an ultimately subjective possession) believed that dialog was the best tool for evaluating our ideas and thus for approaching Truth. He complained against the nascent pedagogical technology of his day—writing—that it would make it harder for us to think clearly because a written text couldn't talk back to us or answer our questions. And despite the advent of writing, and of more recent instructional technologies, face-to-face oral interaction in a classroom has remained to this day the basic paradigm of education from kindergarten through graduate school for essentially the same reason: "[t]the overarching purpose of discussion is to help learners to explore their experience so that they become more critical thinkers" (Brookfield, 1991, p. 192).

ONLINE DISCUSSION AND CRITICAL THINKING

Many early advocates of online instruction argued that one of its defining features, asynchronous discussion, had the potential to combine the learning benefits of discussion with the advantages of a written medium. Because students have time—often days—to formulate their responses to discussion prompts, their contributions to discussions tend to be more thoughtful and coherent than those of students asked to respond on the spur of the moment, and under a certain degree of social pressure, in a classroom. Students can avail themselves of the precision and structural devices of writing to compose their thoughts completely and precisely, with clarity, complexity, and nuance that would not be possible in a classroom discussion. Kaye (1989), for example, argued that student contributions to asynchronous discussions "are often more thoughtfully composed because of the text-based nature of the medium"; he found the medium therefore particularly appropriate for teaching in disciplines which try to encourage students to learn through "discussion and debate" (p. 11). Harasim (1990) claimed that the "need to verbalize all aspects of interaction within the text-based environment can enhance such metacognitive skills as self-reflection and revision in learning" (p. 49), and Cooper and Selfe (1990) made similar arguments in their pioneering study of the use of computer conferencing in literature classes.

Some research attempted to identify differences between the structure and content of the dialog, and thus the cognitive activity, that takes place in face-to-face settings and that which occurs in asynchronous online discussions. Harasim (1989, pp. 55-57), for instance, cited studies suggesting higher rates of interaction between students in online discussions than in face-to-face situations. Newman et al. (1996) tried to analyze the content of online discussions for "depth of critical thinking" in five distinct categories or stages of the critical thinking process. Their findings, though qualified by a small sample of student work studied, were nevertheless intriguingly mixed. Generally, they concluded that "the computer conference discussions showed a significantly deeper overall critical thinking ratio than the face-to-face seminars." However, for certain cognitive categories, asynchronous discussion seemed to offer fewer advantages: specifically, in "the stage where most creativity is required, including the generation of new ideas" within the group, oral face-to-face discussion in some cases appeared to be more productive than asynchronous, written discussion. Interestingly, the authors speculated that this phenomenon might have something to do with the written medium in which the online discussions took place, the result of "the self-censorship of new ideas before committing finger to keyboard, or less spontaneity at the slower pace of asynchronous computer conferencing" ("Conclusions").

As this speculation suggests, the very qualities praised by these early advocates of online education may in fact be impediments to encouraging and modeling critical thinking through class discussion. More recent studies, informed by longer and wider experience with online teaching, have tended to qualify the optimistic claims of some of the early adopters. They have observed that online discussions, while possibly "more thoughtful and substantive," tend to be "less spontaneous and sustained" (Koory, 2003); that "online reflection . . . is more of a solitary activity in which an individual explores his or her own understanding than an activity that focuses on learner co-exploration of concepts" (Dennen & Wieland, 2007); that synchronous discussants may exhibit "higher levels of critical reflection" (Levin et al, 2006); and that the formal structures of online discussions may encourage conformity rather than individuality (Gulati, 2008).

Harasim (1990) argued that participation in online discussions might offer a student "the opportunity to make explicit to oneself the aspects of an activity that are usually tacit—for example, expressing the thinking processes by which a decision or conclusion is reached" (p. 49). This indeed is a fundamental goal of classroom discussion: to model and practice the processes of critical thinking, of ideas being formed and tested. However, it is these very processes which the written medium of asynchronous discussion tends to obscure. A well-conducted face-to-face classroom discussion will move from initial responses to a prompt towards a more reasoned, balanced, nuanced and sophisticated understanding of the issue being discussed. Initial responses in face-to-face class discussions tend to be brief and tentative; they are more likely to express an emotional reaction, an unexamined prejudice, or relate a personal experience. As are members of oral cultures, students in oral discussions are more likely to begin thinking in concrete terms that arise from their "lifeworld" (Ong, 1982), rather than with an abstract generalization or conclusion.

Such responses, however, grounded in emotions, prejudices, and concrete experience, are the proper raw material of critical thinking, the place where such thinking starts and the stuff with which it works. In fact, they are the habits of mind which critical thinking addresses as its object and which it tries to refine and correct. Such unedited, unconsidered reactions are valuable because analyzing them can reveal the emotions, prejudices, ignorance, and simplifications that underlie unexamined opinions. When students are encouraged to consider concrete examples of the issue at hand and to try to apply theoretical explanations to those examples, the complexities of the issue, and perhaps the insufficiencies of generalized theories applied to understanding it, can often be fruitfully revealed. The ensuing discussion is a process by which those raw materials are refined, through dialog, challenge, comparison with alternative responses, and revision. Within a constructivist pedagogy, the goals of discussion should be to model this process of critical thinking and give students opportunities to practice it.

With these goals in mind, the very thoughtfulness of online contributions becomes a liability. Formal, essay-like asynchronous discussion posts are likely to be longer, more coherent, and more organized than the typical oral comment. They tend to offer reasoned dissertations in defense of a generalization. They tend to represent the finished product of a thought process, not the processes itself. Students who make such initial posts are likely to be less open to revising those thoughts, which they have spent so much care composing on their own. And the finished appearance of such postings may tend to discourage other students from challenging them.

DESIGNING DISCUSSIONS TO PROMOTE CRITICAL THINKING

If online discussions are going to model and engage students in processes of critical thinking, they need to be designed to replicate as far as possible the qualities of face-to-face discussion that allow this to occur in the classroom. Students need to be given genuinely challenging questions to deal with, not "pseudo-questions" designed to lead them to a predetermined result, but questions that can actually be debated, and to which evidence from the course, and perhaps from beyond the course, can be brought to bear (Jonassen, 1999). Brookfield and Peskill (2005) note that "Discussion works best when students who are somewhat knowledgeable about a topic are prepared to apply that knowledge to untried areas or to explore questions that take them beyond the simple information given" (p. 237); as Palloff and Pratt (1999) write, "Questions posed in the online environment need to be the jumping-off point of a discussion promoting deep exploration of a topic and the development of critical thinking skills" (p. 119). Though some researchers have noted possible drawbacks to requiring student participation in online discussions (Gulati, 2008; Miller, 1999), if discussions are properly structured to attain these goals, they may spread their benefits more "democratically" to all students in the class by making it feasible to require, or at least facilitate, all students' participation. In several years of researching and experimenting with online instructional design of university-level literature courses, both undergraduate and graduate, I have concluded that while the efficiency of face-to-face discussion may never be matched in writing-based online discussions, certain steps can be taken to make online discussion serve the purposes of discussion within a constructivist pedagogy that aims at developing students' critical thinking skills as outlined above.

Make Discussion Central to the Class

If we accept the premise that discussion is one of the fundamental activities of a constructivist pedagogy, then we need to find ways to put it at the center of our instruction, in whatever medium. In the classroom, where we have a defined number of hours to spend with students, it is easy to measure this time commitment. To whatever degree is feasible, we would like to have students spend that limited resource interacting with each other and with the instructor in activities that allow them to experiment with ideas, test ideas, challenge others' ideas, and begin to work towards synthesizing those ideas. If we are teaching in a classroom, therefore, we are likely to try to devote as many of the three hours a week we typically have with students to such activities. By moving the class online, we are really moving that discussion online. Students, therefore, should regard the online discussion as the online equivalent of the class meetings; other course work is "homework," time spent outside that foundational activity of collective and collaborative learning (though such "homework" might certainly involve further collaborative activities).

Online it is harder to measure the amount of time or effort a student will spend on any particular activity. In one sense, all work in an online course is homework: work done at home or out of the classroom. In designing online courses, teachers typically think in terms not so much of time but of components. The online course is likely to have some sort of "content": the digital equivalent of lectures or other materials delivered to students for consumption and reaction. It is likely to have other sorts of assignments that require students to read or respond, what would be "homework" in a traditional format. The third leg of this stool is likely to be discussion, whether conducted asynchronously or synchronously, or other forms of "interactive" work. In the classroom, the instructor can exert some control over how much time students spend on this third component of the class, but online this becomes more difficult. The ability to schedule such investment oneself is indeed one of the attractions, for students, of online education.

One way to encourage students to invest effort in a particular component of the online class is to give that component a certain weight in the grade. If you want students to spend a certain amount of effort in discussions, you can encourage them to do so by giving discussion work a larger weight in determining the final course grade. In an on-ground course, even one heavily emphasizing discussion, such an incentive isn't really necessary. Typically, "participation" on ground is likely to count, if at all, for no more than 10% or so of a course grade. The real incentive for a student to come to class and participate is that that is where the learning will take place. Online courses should strive to provide the same incentive.

However, I have seen many online classes in which the discussion boards are weighted much more heavily than "participation" would normally be in an on-ground class, sometimes exceeding 25% of the grade. Partly this is due to—or justified by—the fact that discussion postings require students not just to offer ideas but to write small essays. As I will discuss below, I believe that this too is a mistake. But such weighting also reflects the teacher's desire to emphasize the discussions' importance and to make sure students take it seriously. Because the student interfaces with the discussions in the same way that he or she interfaces with most of the rest of the course—i.e., by sitting in front of the computer and reading or typing—students tend to think of it as just another assignment, and they often expect that its weighting in their course grade will be proportional to the amount of time they spend doing it.

A better way to make the discussions seem important to students is to make them important in fact, as the site of learning in the class. To do so, discussion boards should, first of all, ask students lots of questions. In a 50 minute discussion class a teacher might raise a dozen or so questions, followed by additional questions prompted by student responses. Too often, in online classes, students are asked only one or two questions in a whole unit, too few to capture the complexity of any subject worthy of college level study. Such limited questioning may suggest to students either that the subject is not that complex, that the discussion is not that important to actually understanding the subject, or that the question serves more as a means of assessment than a prompt for learning.

Of course, the online medium can make it difficult to ask too many questions. Students, after all, need to write their answers, and the asynchronicity of discussion boards means that it may take them a day or two, or more, to respond even to an initial question, let alone follow-up postings. Nevertheless, each discussion board prompt can offer multiple questions for students to choose from. This practice will reduce redundancy in student responses, since students will not all be answering the same question. Redundancy may be further reduced by organizing students in groups, and instructing them to respond to questions to which others in their group have not already made an initial posting. Thus, a number of discussions, each on different questions, can be conducted simultaneously on a single discussion board. Once students have made their initial response to one question, they can read and participate in discussions of other questions.

In my own classes, I typically require students to participate in 4-8 discussion boards per week, and most of those boards contain 4-8 distinct questions. Thus, I may be asking 60 or more questions each week. Students need to make initial responses in each of the boards (i.e., 4 to 8 initial responses), and I generally require them to make a minimum of five responses to other participants in the discussions over the course of the week. Because I grade discussion board participation holistically at the end of the class, students are encouraged to continue discussions from earlier weeks up to the end of the class.

In my experience, students who have taken other online courses often enter my course with different expectations about the role and form of online discussions. They often don't expect to spend as much time on them as I ask them to, and they often expect to get more credit for that time than I usually grant them. Thus I have found it helpful to explain to students in the syllabus not only what role I see the discussions playing in the class, but how much time I expect them to spend on them. I point out that in an on-ground class they would be spending 3 hours a week in class and that (in my classes at least) most of those three hours would be spent in discussion. Thus, they should expect to spend at least that much time, and possibly more (if only due to the fact of having to type responses rather than just blurting them out), in the class discussion boards each week. The relatively low weighting of the discussion encourages students to see the discussions as the site of learning, rather than as another graded assignment. It is, rather, the place where they will learn how to do well on the more heavily weighted assignments.

Make Discussions Informal

One way to make it more feasible to ask students more questions is to keep their responses brief and informal. This allows students to participate more frequently and to discuss more questions. "Informal" has a specific meaning here; it means that I will not be grading their responses as I would grade a formal piece of writing. I will not be concerned with their spelling, grammar, or organization. I emphasize to students that I don't want them to spend too much time on any one response; again, I would rather have them make quicker, less considered responses to several different questions, and to several different posts by other students, than to write one or two short formal essays (I may have them do that elsewhere, but not in the discussion boards). In some cases, where the question warrants it, I may impose a minimum length requirement, usually no more than 100 words. But generally, the only requirement I make is that they cite a specific detail or passage from our reading in their response.

Their follow-up responses to one another should be similarly informal, and there I never impose a minimum length. I do require that their response add something to the discussion, move it forward in some way, not just agreeing with what has already been said ("cheerleading"). Some of the best responses, I tell them, are questions for further discussion, and I try to model this practice in my own responses.

Beyond the practical advantage of allowing more questions to be discussed, asking for informal responses brings online discussion closer to the experimental, exploratory, adventurous sort of activity that characterizes the most successful face-to-face discussions. I want students to make their initial responses quickly, with as little self-editing as possible. Students should not think of their contributions to the discussion boards as conclusive statements. Too often writing implies such conclusiveness, especially in school, where it is seen as a "product," and usually one to be graded. As Bender (2003) points out in her book on online discussion, "student postings that are too long and sound more like essays . . . [are] harder to respond to than in a face-to-face discussion in which interruptions or rebuttals are more common, stimulating, and expected"; she recommends instead what she calls an "informed conversational style" in discussion board posts (70). Making discussion postings informal and not grading individual posts reinforce their role not as a product but as part of a process. I want students to feel safe to experiment with ideas in the discussion boards, as they would in a classroom discussion (with the additional safety offered by physical absence), and to offer their ideas with the understanding that they are always only "drafts."

Ladder Discussion Questions to Scaffold Inductive Thinking Processes

A constructivist approach to learning incorporates both inductive and deductive thinking. The metaphor seems primarily inductive: if we imagine knowledge as "constructed," we might think of a building being put together piece by piece to create a larger coherent whole. The ongoing nature of constructivist learning, though, reveals the deductive element in the process: the testing of those constructions against new evidence, ideas, or insights. This completes the recursive circle of the critical thinking process: invention or analysis (the equivalent of "drafting" in the writing process), revision, and synthesis (which becomes material for the further revision, and so on). In learning, the building is never finished. Rather—to push the metaphor—it is frequently upgraded, refitted, even torn down and rebuilt.

Discussion boards can be structured to provide scaffolding to help students develop these inductive and deductive processes of constructivist learning. In my online Shakespeare course, we generally read one play a week, and students participate in eight threads each week. These threads are usually divided into two groups: initial postings to the first group are due earlier in the week, initial postings to the second group later in the week, and responses by the end of the week (or weekend). Because each group of discussion boards has three or four different threads, I can ask different sorts of questions in each. The sequencing of questions can shape an inductive process. Thus, the early questions are likely to ask students to comment on specific passages of the texts we're studying. These questions force them to deal with textual details, either ones I've selected as important, or ones that strike them as significant or interesting. These initial questions are followed by threads that ask them more general questions. If well-designed, the early, more focused questions will have provided them with ideas from which they can develop their responses to the more general questions.

Thus, after students have read the first half of the play, I ask them to respond to the following sequence of questions:

DB 1: Respond to one of the "Questions for Readers" from the Reading Guides for the first half of the play. [From 1 to 10 of these questions appear after the discussion of each scene of the play in the lecture material; thus students may have as many as 40 or 50 such questions to choose from for this DB.]
DB 2: Respond to another of the "Questions for Readers" or to one of the "Questions for Actors and Directors" from the Reading Guides for the first half of the play. [The "Questions for Actors and Directors" also appear at the end of the lecture discussions of each scene; thus, students have an even larger field of questions to choose from here.]
DB 3: Pick a brief passage (2-10 lines) from the first half of the play and explain what makes it interesting or significant, and what poetic or literary devices are used to help it create its meanings or effects.
DB 4: Respond to one of the following questions in a paragraph of at least 100 words, making specific reference to passages from the text. [Here would follow 4-10 questions of a more general nature, asking about themes, characters, or other significant issues raised in the first half of the play.]

The first three discussion boards ask students to attend to specific details of the text. The questions are open-ended, asking for their interpretations of these details (Shakespeare's plays provide rich fields for such questions). Students are encouraged not to respond to questions or passages already discussed by their classmates, so these three boards end up covering lots of different aspects of the first half of the play. The third question requires students to practice some traditional "close reading" and to consider how stylistic details might lead certain passages to stand out from others; the students themselves identify passages stand out to them. The fourth discussion board asks questions that might be answered by collecting evidence from various places in the play that they and their classmates have been looking at in the first three.

The second set of discussion boards for each play is due later in the week, after students have finished reading the play. DBs 5 and 6 are identical to DBs 1 and 2 above, but dealing with the second half of the play. DB 7 will be similar to DB 4, but now dealing with the play as a whole. Some questions in DB 7 may return to questions raised after reading the first half of the play. For instance, one of the questions in DB 4 might ask students to comment on a character as he or she is presented in the first half of the play, and a similar question in DB 7 might ask students to reassess that character based on the second half of the play. In DB 8, the focus widens; there I ask students questions that have them compare the play we are reading that week with the other plays read up to that point in the class.

This laddered structure of discussion boards, then, puts students through the process of inductive thinking, beginning with a focus on particular details, then using those details to build larger structures of understanding or interpretation. At every step, there is opportunity for students to discuss, compare, and revise their interpretations of details and of larger issues.

Ladder Discussion Questions to Move from Subjective to More Objective Responses

Also embedded in this laddering of questions is a movement from subjective to objective responses. As noted above, subjective responses are important to the process of constructing knowledge. The interpretation of a literary character, for instance, always begins with "first impressions," our personal response to him or her. Do we like this character? Why or why not? These are the first questions on the path to constructing the sort of complex, objective judgment of a character that characterizes what we would call literary interpretation, and that would be of interest to other readers.

In my online Introduction to Literature class, I always start discussions with just such basic, subjective questions: What did you like or not like about the text? Why did you like it or not like it? For example, in a lesson on the role of sound (rhyme, meter, alliteration and assonance, etc.) in poetry, the initial discussion board prompt asks students to "Pick one of the poems read for this part of Unit Three that you liked or didn't like, and try to explain why you liked or didn't like it. No minimum length required." This is following by three discussion boards, each focusing on a different element of sound; typical questions in those boards ask students to identify examples of the sonic device in one of several poems and to try to find connections between the use of that device and their responses to the poem.

The goals of instruction in such a unit include learning to correctly identify and label different poetic devices, but they also include, more importantly, being able to use those terms to help explain how poems work or mean to a reader. The reader's subjective response to the poems is thus part of the subject matter of the lesson; it is something to investigate and try to explain; it is, in other words, part of the content of the course. From a constructivist point of view, this would be the case in any class; the material to be worked with, from which knowledge is built, is first of all what the student brings to the class in the form of prior knowledge, opinions, and experiences. Subjective questions therefore represent an appropriate place to begin discussion.

Ask Lots of Questions, and Allow Students to Choose their Questions

As I've already suggested, there are some practical reasons to do this. For one thing, it keeps online discussions from becoming redundant. In a classroom, you might (if you're lucky) get two or three different responses to any one question asked. Online, everyone can be required to participate, which is certainly one of the benefits of the medium. But getting twenty-five responses to the same question is inefficient, as many of them will inevitably be redundant to varying degrees. By offering 4-8 questions in each discussion board, or designing questions, such as the one just discussed, which allow students to pick an issue or a part of the text to comment on, you can set a variety of different discussion threads going within each discussion board. Each student will only make an initial response to one of these questions, and might be required to make a certain number of responses to other students' or the instructor's initial posts, but all students (if they want to) can peruse all of the different discussion questions.

This offers the potential to expand discussions well beyond the constraints of three 50 minute class sessions a week. In any course, there are always many more interesting questions to discuss than one has time to address, especially within the time constraints of a synchronous class meeting. Online one can ask a lot of questions, and even if they don't all get fully discussed, students can see the variety of questions that might be asked, and can pursue those that interest or intrigue them.

Each discussion board will have some common thread. In the Shakespeare course, each board asks several questions about the same particular part of the play. In the Intro to Lit course, each one focuses on a particular course topic (e.g., the use of sound in poetry). As a result, students may be prompted to see connections between different characters or parts of the same scene, or between different poems that use similar poetic devices.

Encourage Students to Ask their Own Questions

This may seem obvious, but it is important to explicitly encourage students to do this. As Brookfield (1991) declares, "[t]he cardinal principle underlying discussion is that of flexibility" (p. 197), and a truly constructivist pedagogy will encourage students' own authentic questions to be the starting point of learning The online environment, however, can often seem inflexible and overly structured to students. In the classroom, we solicit student questions almost without thinking about it: "Any questions?" This needs to be done regularly and explicitly online. I always have an open-topic discussion board in each Unit that invites students to make a comment or ask a question about the material not covered by the other discussion boards in the Unit. In some classes participation in this discussion board is optional; in others I require it. Invariably, students seem to respond to their classmates' questions and comments with greater enthusiasm than they do to mine.

It also helps to give students multiple avenues for asking questions. I usually tell them that they should post any questions about the material in the open topic discussion boards in each Unit. For questions about the mechanics of the course (e.g., deadlines, assignments, unclear instructions), I create a separate discussion board; this one usually gets a lot of action at the beginning of the course, as students are figuring things out. Finally, I tell them that for questions whose answer would only be of interest to them, they should e-mail me. I conclude every Announcement to the class and every e-mail to a student with a boilerplate phrase, "Let me know if you have any questions."

THE TEACHER'S ROLE

The recommendations above concern the design of threaded discussions and of their role in an online course in which critical thinking is a goal—as I imagine it would be in any post-secondary course. This leaves undiscussed the equally important question of how to conduct and manage these courses, that is, of the instructor's role in the discussions once they begin. There is a considerable body of literature on this subject, dealing with both traditional and online classroom discussions. While the subject is beyond the scope of this chapter, a few general suggestions might be made in light of the goals of the design strategies outlined above. In a course in which the goal is to not only to allow students the opportunity to practice the skills of critical thinking, but to model those skills for students, the best role model is likely to be the instructor herself. The design of the course is one way to model those skills, as I have tried to show. But the instructor can further model those processes in some of the following ways.

As in the classroom setting, the teacher has some obligation to keep the discussion on track, to prevent it from wandering into fruitless or unending debates on small points, friendly but fruitless mutual congratulation, or personal chat. When this happens the instructor needs to intervene to redirect the students back to the question at hand. If the goal of the critical thought process is to move towards a point of synthesis, in which the competition of points of view produces, at least provisionally, a more complex, sophisticated understanding of the issues being discussed, the instructor may have to provide, or at least hint at, such syntheses when students don't quite arrive at them themselves. An instructor, for instance, might point out how two points raised in a discussion reinforce or contradict one another, or how students pursuing different threads of a discussion have come across related ideas.

Some online instructors like to stand back from the discussion until the students seem to have exhausted the topic, then step in to sum up. While seeming to allow students more room, this approach ultimately reinforces the notion that the instructor has the "last word" (Muilenberg & Berge, 2006). Better, the instructor should involve herself as a collegial participant in the discussion, offering, as the student participants do, brief, informal contributions that carry the discussion forward rather than bringing it to a stop.

Finally, the instructor can model the process of such synthesis, rather than doing the work herself, by asking questions. Questioning, after all, is the initial prompt for critical thinking. To teach students to think critically, they need to be taught to ask questions. Questions start the discussion, and at crucial junctures questions can lead it forward towards more complex constructions.

I have observed in my own classes that my postings do serve as models for students. One (possibly somewhat disturbing) indication of this is that I notice students picking up some of my own stylistic quirks, the use of words or phrases I tend to favor in my own posts. More usefully, though, after several weeks of discussion I begin to see some students ending their responses with questions to their classmates, a practice I model in almost every one of my own posts. I state in my initial instructions for the discussion boards that a good question is an especially good form of response, but it usually takes a couple weeks of my modeling such questioning responses before students begin adapting the practice themselves.

CONCLUSION

However instructors design or conduct their online discussions, I believe a good instructional design touchstone is the old-fashioned classroom discussion, which has successfully proved its usefulness as a way of teaching thinking since Socrates defended it against the revolutionary distance technology of his day: the written word. While Socrates' objections, fortunately, did not prevent writing from assuming its crucial role in the way we learn and think, neither did the advent of writing—nor of television or computers—displace dialog from its central place in our pedagogy. One of the important consequences of the debates over online education is that they have led us to re-examine what we have been doing all these centuries in our classrooms. Such re-examination is always healthy, and it should make us more effective in using discussion to help students learn how to think critically in any medium.

REFERENCES

Bender, T. (2003). *Discussion-based online teaching to enhance student learning.* Stirling, VA: Stylus Publishing.

Brookfield, S. D. (1991). Discussion. In M. W. Galbraith (Ed.), *Adult learning methods: A guide for effective instruction* (pp. 187-204). Malabar, FL: Krieger Publishing Company.

Brookfield, S. D. & Preskill, S. (2005). *Discussion as a way of teaching: Tools and techniques for democratic classrooms.* San Francisco, CA: Jossey-Bass.

Cooper, M. M., & Selfe, C. L. (1990). Computer conferences and learning: Authority, resistance, and internally persuasive discourse. *College English, 52,* 847-69.

Dennen, V. P., & Wieland, K. (2007). From interaction to intersubjectivity: facilitating online group discourse processes. *Distance Education, 28*(3), 281-97.

Gulati, S. (2008). Compulsory participation in online discussions: Is this constructivism or normalisation of learning? *Innovations in Education and Teaching International, 45*(2), 183-92.

Harasim, L. M. (1989). Online education: A new domain. In R. Mason & A. R. Kaye (Eds.), *Mindweave: Communication, computers, and distance education* (pp. 50-62). New York: Pergamon Press.

Harasim, L. M. (1990). Online education: An environment for collaboration and intellectual amplification. In L. M. Harasim (Ed.), *Online education: Perspectives on a new environment* (pp. 39-64). New York: Praeger.

Jonassen, D. (1999). Designing constructivist learning environments. In C. M. Reigeluth (Ed.), *Instructional-design theories and models* (Vol. 2): *A new paradigm of instructional theory* (pp. 215-39). Mahwah, NJ: Lawrence Erlbaum Associates.

Kaye, A. R. (1989). Computer-mediated communication and distance education. In R. Mason & A. R. Kaye (Eds.), *Mindweave: Communication, computers and distance education* (pp. 3-21). New York: Pergamon Press.

Koory, M. A. (2003). Differences in learning outcomes for the online and F2F versions of "An Introduction to Shakespeare. *Journal of Asynchronous Learning Networks, 7*(2). Retrieved August 28, 2008 from http://www.sloan-c.org/publications/jaln/v7n2/v7n2_koory.asp.

Levin, B. L., He, Ye, & Robbins, H. H. (2006). Comparative analysis of preservice teachers' reflective thinking in synchronous versus asynchronous online case discussions. *Journal of Technology and Teacher Education, 14*(3), 439-60.

Miller, J. (1999). Critical thinking and asynchronous discussion. *Inquiry: Critical Thinking Across the Disciplines, 19*(1), 18-27.

Muilenburg, L., & Berge, Z. L. (2006). A framework for designing questions for online learning. *Welcome to emoderators.* Retrieved May 28, 2008, from http://www.emoderators.com/moderators/muilenburg.html

Newman, D. R., Johnson, C., Cochrane, C., & Webb, B. (1996). *An experiment in group learning technology: Evaluating critical thinking in face-to-face and computer-supported seminars.* Retrieved May 28, 2008, from http://www.helsinki.fi/science/optek/1996/n1/newman.htm/contents.html

Ong, W. J. (1982). *Orality and literacy: The technologizing of the word.* London: Routledge.

Palloff, R. M., & Pratt, K. (1999). *Building learning communities in cyberspace: Strategies for the online classroom.* San Francisco, CA: Jossey-Bass.

Chapter XIX
"To Be in Occasional Touch": Goddard College's Progressive Principles and Distributed Learning

Mark Schulman
Goddard College, USA

ABSTRACT

The challenges for Goddard College posed by 21st Century information technologies are their incorporation into, and reflection of, the foundational principles of the College's low residency delivery model. This learning model is complex; the many aspects of this complexity forces the College (and other progressive institutions) to insist on a values base that helps determine attitudes to and uses of technology. If administrative staff and faculty have a deep commitment to those values, when done well, information technology can enhance the progressive model.

INTRODUCTION

This chapter offers a brief case study of a small, rural college with a long-term commitment to progressive principles: Goddard College in Plainfield, Vermont. As it happens, the story of Goddard's seven-decade commitment to apply progressive principles to post-secondary education intersects with new modes of educational delivery and emerging information technologies. The study of what has happened in Plainfield, Vermont (and globally in Goddard's spread-out student and faculty population) may provide lessons for how technology can aid in distributing learning across distances.

The current age of information technology has revolutionized the way institutions of higher education operate. Contrary to the notions of some,

Goddard College does not engage in "distance learning," in the sense of offering online courses which have been structured to facilitate use of the machinery through which they are disseminated. Goddard strives to remain true to progressive education as an ideal while availing itself of the vast resources as made available by technological advances. Hence, Goddard College is driven by ideas, not technology. The difference is vast.

Our case here is to examine, in an exploratory and illustrative manner, how that vast difference is manifest in Goddard's use of information technology. Before we examine that, a brief review of progressive principles is useful.

WHAT ARE PROGRESSIVE PRINCIPLES?

There is a huge literature delineating progressivism as a philosophy and its application to education; a full review of that literature is beyond the scope of this brief case study. Goddard consciously adopted these progressive principles in its 1938 inception, and continues to espouse them in its community standards and educational practices.

Like many other paradigmatic societal shifts, the progressive education movement experienced its share of fits and starts from the mid-1800s through the mid-1900s. Proponents of progressive education were rethinking American notions of public education in an era dominated by a state-sponsored philosophy designed to promote an unquestioning conformity by the masses rather than to stimulate critical thinking skills. Dewey and other progressive education theorists sought to broaden education's reach into the social and political fabric of society. Progressivists sought to encourage wider political, cultural, and social participation in governance and gave primacy to the individual's lived experiences as necessary enrichments to the educational process (Bode, 1938; Dewey, 1997a; 1997b; 2006; Westbrook, 1991).

Various perspectives on progressive education were developed by theorists, including Booker T. Washington's methods employed at Tuskegee Institute some thirty years earlier than Dewey's writings. Although Washington's experiment at the Tuskegee Normal and Industrial Institute goes largely uncited by historians as one of the earliest examples of progressive education in action, his curriculum combined academic theory classes – geometry, mathematics, science – with carpentry and electrical skills in a hands-on building construction project (Generals, 2000). The Tuskegee experiment illustrates how the tenets common to virtually all Progressivists included experiential learning, the development of problem solving skills, an eclectic approach to curriculum development, the development of critical thinking skills, the promotion of democratic participation, and social responsibility, among others.

Dewey's specific approach, often referred to as instrumentalism, had its origins in the pragmatist school of educational theory. Instrumentalism posits that humans use truth as an instrument to solve problems. This viewpoint renders culture, experience, and vocation, among other human phenomena, as the "truths" of each individual. In fact, Dewey argues that both progressive and traditional education are inadequate unless they respect experience, regardless of the source (Dewey, 1997ab).

Much denigrated during the McCarthy era, progressive education ideals enjoyed a resurgence in the late 1950s and early 1960s. John Dewey's principles of curriculum development became the basis for numerous iterations of student-centered private and public schools. Goddard is a college started on Dewey and other progressive theorists' educational principles. This makes Goddard unusual in the landscape of North American education: progressive principles applied to K-12 are much more common, and more highly articulated, than those in post-secondary schools.

SITUATING GODDARD

Goddard became a college (actually, a junior college, becoming a four-year institution a bit later) in 1938. With its change from its former incarnation as Goddard Seminary in the city (albeit a tiny one of 12,000 or so inhabitants) of Barre, Vermont, to a rural college on a former sheep farm in Plainfield, Vermont, Goddard built on clearly voiced principles of progressive education in the context of residential liberal arts colleges.

For twenty-five years, most of the action at Goddard was on the couple of hundred acres in central Vermont. (The College always had an externally focused element, encompassing field study, local outreach, and college-community collaboration.)

In the boom time of the Sixties, Goddard grew tremendously, as did many other liberal arts colleges, setting up ventures far from the original campus, in Goddard's case in such locales as Cambridge, MA, and California. Experiments abounded (this was the time when the Union for Research and Experimentation in Higher Education – later the Union for Experimenting Colleges and Universities – was founded at Goddard). Budgets were (nearly) adequate to the perceived needs. Optimism about progressive values, both in education and politics, held sway.

In that environment Goddard began to rethink its commitment to a solely rural, residential, traditional-age, liberal arts college. Questions were asked with vigor and purpose: about how learning could take place in other places, how people who were not free-wheeling 18-22 year olds could get a degree, and what it would mean to reconceive education outside the context of fifteen-week semesters with students living in dorms. Such inquiry became foundational in the quest to discover what could and should be done to break down the walls and open up the schools.

And thus was born the Adult Degree Program (ADP). This program was conceived as the first full-fledged liberal education baccalaureate program for men and women (but mostly women) who were "college-leavers." How could Goddard, the originators pondered, make it possible for older people to complete their college degrees while maintaining their family and work responsibilities? In retrospect from our vantage point in 2008, the question does not seem so puzzling.

But this was the early 1960s, before the Internet as facilitator and the recognition that learning styles might vary with multiple delivery systems. In the stuffy educational environment that defined what a real education was, the notion that older people could return to campus for a short period and then work independently at home and receive a legitimate BA was revolutionary.

ADP, the brainchild primarily of Evalyn Bates when she was founding Goddard president Tim Pitkin's assistant, has gone through many iterations in the 45 years it has been in operation. But there has always been, and continues to be, a deep commitment to the values of progressive learning: the framework differed dramatically from that of the residential college, but the values did not change.

The story of the "situated (and contextualized) Goddard" begins with a section on the history of ADP and its successors. But this is not a narrative on one segment of the College: "ADP" is Goddard now: the College is one of a few freestanding, solely adult-education, solely learning-at-a-distance-cum-short-residency institutions of higher education in the United States. The history of ADP illustrates the issues pertinent to this book for a unique progressive institution like Goddard. The focus of this discussion is the relationship of this model of education developed at Goddard to information technology. The questions we seek to address include:

- What roles have information technologies played over the course of ADP and its successors?

- Were there criteria used in selecting technology to apply to learning? Are there now?
- Does the curriculum as it has evolved show a reliance upon (or resistance to) the technologies under consideration?
- How is information technology used to foster active learning? Collaborative learning?
- What can be said about assessment of outcomes in Goddard's experiment?

THE HISTORICAL ADP AND WHAT GODDARD HAS BECOME

The first Adult Degree Program (ADP) residency enrolled 17 students in 1963 and the program grew rapidly, increasing to over three hundred in five years and to over five hundred in ten years. The program continued to get bigger and to mature for its first two decades. But its history as a Goddard program becomes discontinuous: the "original" program was sold to Norwich University in 1981. As part of Norwich's Vermont College and then of Union Institute and University, it has flourished there as well.

For the last 25 years, Goddard has continued to refine the concept originally articulated and implemented in 1963. Through name changes and tinkering with such aspects as length of residencies throughout the history of the program(s), the essence of ADP remains the DNA of Goddard's current manifestation.

In 1997, Richard Hathaway, a founding faculty member of ADP, wrote a pamphlet entitled "The Friendly Pioneer: An Informal History of the Adult Degree Program at Goddard and Vermont College." He asked: "What are the key reasons the ADP has proven to be a resilient educational model?" and he answered:

1. The ADP originated in a supportive progressive culture, anointed with status equal to the prevailing undergraduate, residential, campus-based model – allowing it time to develop its own critical mass and tested practices.
2. The ADP succeeded in enlisting and keeping the discriminating loyalties of key faculty and staff; these provided continuity and a historic sense of the program as it grew and developed....
3. On a practical level, the ADP allowed its learners to complete fifteen credits for each four and one-half (weekend model) or six month long (original cycle model) semester – an attractive pattern for an alternative college experience.
4. Its low-residency format allowed students to pursue college degrees while still maintaining jobs and obligations within their families and home communities. Indeed, the alternation between practical experiences (praxis) and academic theory enriched students' educational ventures substantially.
5. Its role as incubator of new initiatives at both Goddard and Norwich suggest the model's vitality and adaptability. We can also cite the myriads of analogous programs (some 200 in number) developed using aspects of the ADP model throughout the American educational establishment over the last thirty-four years.
6. Perhaps most importantly, the ADP offers membership in an invented, mutually supportive educational community appearing only rarely in contemporary society. As veteran faculty member John R. Turner observes: "We live in a socially autistic age which manifests itself through the incessant busyness which each person directs towards his own set of affairs." ADP counters this isolation, providing "students the opportunity for genuine dialogue which, if it is married to knowledge ... can lead to educational and intellectual achievement," (Hathaway, 1997, pp.12-13).

As a press release in 1969 boasted when "the first of 300" adult students arrived for their semi-annual residency,

> *The Adult Degree Program, a first-of-its-kind innovation in continuing education, was created ... to offer studies in the liberal arts to adults whose formal education had been interrupted.... The program is unique in that Goddard College is the only institution conferring regular Bachelor of Arts degrees on students successfully completing the program."* (Mahoney, 1969, pp. 1-2)

As we know from Hathaway's 1997 statistics, that is no longer true: at that time the "analogous programs" were something like 200 (compared to only a handful at the ADP program's official start in 1963); in 2008, there are between 1,500 and 2,000 programs that have borrowed significantly from the original ADP model.

What is crucial to understand in differentiating this program from others, including correspondence courses before and on-line learning after, is that it was and is (as the only mode of delivery at contemporary Goddard) a liberal arts curriculum. In a pamphlet published in 1967, "Grown-Up Education: The Goddard College Adult Degree Program," the strongly-worded advice for prospective students states:

> *It is not a vocational program, and those primarily interested in vocational training or in meeting the requirements of a narrowly defined profession should not apply.* (Author Unknown, 1967, p.3)

The short/intensive/low residency aspect of ADP and its successors has a flavor and power that defines a unique educational experience with robust progressive principles throughout. It is primarily in the "independent study" segment, the work at home for fifteen weeks or so between residencies, that we see demonstrated the interface between technology and progressive pedagogy.

The examination of information technology in its relationship to what might be called progressive independent study is something to which we shall return. But a motif that emerges and runs through our narrative – that values determine the use of technology and not vice versa – can be seen in the first statement of the first group's schedule – in 1963:

> *The student will be in occasional touch with his faculty advisor through the exchange of tape recordings, telephone conferences, regional gatherings of participants, traveling faculty members, correspondence.* (Author Unknown, 1962, p. 4)

To be in occasional touch is the key aspect of technology's role in Goddard's version of progressive education in the distance phase. As Evalyn Bates, who invented the program, said,

> *Well, it made sense to me.... When you get to the person who wants to learn after normal college age, you have to design a program for them.... You have to have a program that meets their needs and allows them to have the time and space for learning.* (Collar, 2007, p. 7)

Reflecting the spirit of the origins of the Adult Degree Program, our exploration aims to show how we shall understand the time and space for learning for adults as the nexus of progressive educational principles and information technology.

The question might be asked: What about the residential students – where are they in this picture? The simple answer is: they are not in it. In 2002, the Board of Goddard closed the residential program and the ADP model is the only pedagogical model now.

Goddard's nearly constant financial struggles had taken it to the precipice of closure several times since its inception. However, those struggles, coupled with intra-organizational disputes, culminated in the most recent near-death experience in 2002, and the decision to save the college by closing the residential program.

For more than five years, Goddard has pursued a goal of low-residency, adult education. Goddard no longer has an "on campus" program but consists of nine intensive/low residency programs that serve students from their twenties to their seventies. Remaining true to the progressive lineage means enabling a modus operandi that adheres to a mission of being driven by ideas and enabling the ability to take advantage of technology to advance those ideas.

Many colleges and universities tout their low residency programs today, and that notion covers a wide range of programs. The rubric, with a proper use, should be clear: a contrast between models which demand "high" residency (a quarter- or semester- long sojourn on campus) versus the low residential option, which may involve a day or a week or more studying on location coupled with a longer period of independent study at home.

TECHNOLOGIES IN LOW-RESIDENCY EDUCATION

For the purposes of this chapter, a useful distinction, again often cited in the literature, is between technique and technology. Technique refers to particular methods for singular phenomena; technology refers to systemic treatments. Thus technology is the collective, systematic application of techniques. Our question, then, would be what systemic (as in the aforementioned "delivery systems") technological applications have been employed in Goddard's low residency model?

In today's parlance, the subject matter with which this book, and this chapter, concerns itself is information technology. "Information," of course, has a long history of varied uses, but has come to be identified with the machinery (actual and conceptual) of "the Digital Age," or roughly the last 30-50 years. In recent times, Information technology, or IT, has become tightly identified with the uses of computers, telephony and other communications, and fundamentally all things digital. "Digital" itself, beyond its technical meanings, has become a metaphor for Modern Times – the ubiquitous, always-in-touch, mobile phone/Wi-Fi/IM/laptop new world order, rather than the Chaplinesque, old-fashioned version.

However, for our historical trace of Goddard's program, we are not yet ready to examine the Digital Age. Information technology, as applied to Goddard, appropriately includes the educational context of everything implemented at the College and should not be limited to computers and like devices. In other words, a distinction between administrative and academic uses of information technology, and between computerized and non-computerized applications, while having situational accuracy, would not aid an understanding of technology's role in the Goddard information technology model.

In the Goddard College model of low residency, terms such as "distance education," "on-line (or Web-based) courses," and "virtual campus" are not applicable. The official description of the way all Goddard programs operate is that each semester begins with an 8-9 day residency in Vermont or Washington, followed by a fifteen-week semester of independent study mentored by a faculty member.

Throughout the transformations of the Adult Degree Program of the 20th century through the exclusively low-residency, adult education campus of 21st century Goddard, this pattern – though varying in length of days and weeks and presented with different names – has remained consistent. Other institutions of higher education have experimented with variants (including attempting to replicate the residency experience on-line and interspersing the independent study with more or less frequent residencies such as once a month or once a year). What occurs on campus during the residency depends on technology, in a broad sense, for the creation of an environment in which a learning community can flourish. What occurs after the learners return home depends on technology to facilitate continued communication with the faculty mentor (and, in some cases, with peers). In both contexts, the Library and Information technology department plays a key role in providing academic support. As with all institutions, the technologies to support services such as student records, tutoring, admissions, alumni relations also play crucial roles.

The history of information technology in Goddard's low residency programs over the last four decades parallels in many ways such histories at other small, independent, liberal arts colleges. Differences exist in the calculus of institutional wealth, location (rural versus urban), vision of technological support (by senior administrators), and stakeholder buy-in (especially faculty). Goddard can by no means claim leadership in the application of such technologies: it is not on the leading edge but on the trailing edge, partly by conscious choice, partly for financial reasons.

Thus to understand technology in the context of low residency education at Goddard one starts from the premise that whatever technologies are available, possible, and affordable will be appropriate if they enhance the values of the college in the context of improving the co-learning of faculty member and student. The Goddard community has identified those values as: an ethical commitment to inclusiveness, social responsibility, and sustainability; an educational project that is individual, experiential, and participatory; a community that is rigorous, caring and accessible; an organizational ethos of dignity, collaboration and enthusiasm.

On campus there is an array of audio, video and communication technologies for presentation and discussion. There is not a highly sophisticated array of high-tech on either the Vermont or the Washington campus, but the capacity to employ the technologies of display, interaction, and staging – appropriate to what relatively sophisticated faculty and students require – exists. It is facilitated by portable equipment deployed around campus, two computer labs, a couple of well-equipped rooms (one, the Media Room, a "smart classroom" stocked with projection, interactive and presentation equipment), and Wi-Fi available in most buildings (and in some areas outside).

Off campus the ability to communicate over distances has depended on mail, parcel services, telephones, and more recently computer connectivity. It would be fair to say that information technology, both in residencies and independent studies, has been added to the mix in creative and experimental ways. It would also be fair to say that no one at Goddard is committed to technology qua technology (except, of course, in the sense that transportation, shelter, and long-distance communication are necessary to keep the machine running).

Technology supports progressive teaching and learning at Goddard by:

- Making possible the submission of materials during the independent study/at home segment;
- Allowing a quicker response from the advisors to those submissions;
- In the one program with courses (Psychology and Counseling), encouraging on-line work as part of the semester's coursework;
- Providing the machinery to enhance presentations during the residencies;
- Creating electronic and telephonic spaces for groups to gather and converse;
- Permitting faculty to use Internet research resources with each other and their students;
- Facilitating community meetings with dispersed participants to increase the possibilities of democratic governance in at-a-distance networks.

Think of the venerable adage (which originally involved Mark Hopkins in a log cabin on one end of a bench) as an old yet new refrain: The best school is a log with a teacher on one end and a student at the other. Technology has played and continues to play the role at Goddard of making each end of the log comfortable enough to sit on, making sure many logs (like strewn pick-up sticks) are conjoined for a communal experience, and extending the log hundreds or thousands of miles for the one-on-one study segment.

ACTIVE, COLLABORATIVE LEARNING

Knowing now that we would be unwise to feature technology per se to explain the success of Goddard's collaborative learning model, what explains Goddard's outcomes? And what is the role of technology in achieving those outcomes?

The success of the learning paradigm lies in the roots of the College's philosophy. Goddard, unlike other colleges that espoused a Deweyan progressive ideal at their inception but have moved on, clings tenaciously to its history of and commitment to that conceptual framework. The current tagline for the College – Progressive Education for Creative Minds – exemplifies that deep value stance.

Consciously experimental, academically rigorous, radical, and grounded in experience ("thoughtful action"), Goddard hires faculty, carefully screened, who embody these values. Through a process of self-selection and a complex admissions process, but with no reliance on traditional measures such as test scores, students ready to benefit from this kind of intellectual guidance and active learning-by-doing eagerly matriculate.

The key element of the collaborative milieu is the residency. The intense and intimate learning community which faculty and students form as co-learners, albeit for only a week or so, leads to an atmosphere of a co-construction of knowledge. As each student must, during that time, put together and have vetted (not only by faculty but by peers) a learning plan for the upcoming independent study period, each departs with a sense of collective wisdom, well-prepared to work individually during the following phase.

Through a combination of learning opportunities, from 8 a.m. in the morning through late at night each day, a Goddard student picks and chooses elements of the residency smorgasbord relevant to his or her desired outcomes. Among the mediated presentations, small group meetings, lectures, informal and formal one-to-one interactions, computerized instruction, research skills training, and more, are the elements that students assemble into a foundation for the individual learning plan.

The dialectic of collaborative collectivity and individualized scholarship creates a dynamism that infuses each program with not only an active but also an activist stance. Because of the deep commitment to social justice and the belief that learning by doing is essential, the student demonstrates both a zeal to combine reflection with action and (on campus and off) an eye to applying principles of equity and inclusiveness to the world around him or her. And, after returning home to their "regular" lives, the learners know that they will be expected, and indeed have already contemplated how, to build their studies around their lives: to blend reflection with an involvement that will be shared and analyzed in their next residency. They learn that the examined life is worth living – and sharing with others.

Goddard employs IT to aid this learning model. Few choices are imposed by dictum: until recently, it was not even required to own and use a computer to be a student (from a social justice and inclusion perspective, it remains undesirable to demand computer ownership). But there is no "penalty" imposed for technological literacy: in the spirit of collaborative learning, the technologically literate are expected to help those less literate, and they do. Both in the use of IT and in core values Goddard eschews competition.

Possibly the most collaborative aspect of IT in pursuit of these values is the creation of online conferences, formal and informal cyberspace gathering places.

Using the collaborative server software environment FirstClass, anyone may set up a chat room and invite anyone else. In addition, there are lists for communicating both academic ("Residency Schedule Posted") and social ("Anyone Want to Buy Winter Tires?") information. Both curricular and extra-curricular subjects pop up on the "unified digital campus," as it is called. Depending on the topic and the times, vigorous discussion occurs, sometimes attracting dozens of participants.

For example, students (and some faculty and staff) initiated and have continued for months ongoing conversations on "Sustainable Goddard," "Funding for the Arts," and anti-oppression and diversity discussions in the "Campus at Large" chat area of FirstClass. Faculty have long-term discussions of the evaluation process and professional development funding in their area. Staff members have Human Resources colloquia and event planning, plus student evaluations of the residencies, online.

As stated previously, there is no attempt to impose or require a technological application in most instances. The desire and usage arises organically, usually by mutual consent, by as few as two or as many as 200 people. Collaborative, active learning is central to the Goddard experience. Information technology becomes important when it contributes to the collaborative and active presence. It is not surprising, though perhaps counterintuitive, that, within the technology-enhanced learning spaces, qualitative methods continue to predominate: lengthy narrative evaluations enumerate the collaborative and active products and processes for each student and for the group (however large it may be).

Within each program, faculty have a high degree of autonomy in determining how each will construct the residency learning opportunities as well as in what modes each student will study independently. And, consequently, the ranges of uses of IT academically are wide and deep.

- The Internet is used for research, for the individual or a group: a group of students choose a topic and combine the results of their research in a collaborative learning space.
- Digital media presentations are offered, both for faculty show-and-tell and student project completion: on any day of a residency, scheduled events include computer-generated art or scholarship presented for varied audiences.
- Films and photographs are projected or digitally distributed for interpretive exercises: the notion of visual literacy pervades the assignments for those students on programs where that judgment capacity is relevant.
- Students collaborate with each other with and without their advisers on projects over great distances: they connect electronically and produce a group project or aid each other on individual efforts.
- Courses or learning modules are presented during the independent study phase: students in some programs may call up lectures or other materials that were offered during past residency presentations.

It is not the case that these academic uses of IT are extraordinary. It is the case that the nuanced understanding of their "costs and benefits" as curricular aids creates a different intellectual atmosphere than the usual one. In academic as in administrative uses of technology, from the earliest uses of materials like books and chalkboards to the latest uses of computers, Goddard holds to the notion that values are primary, machines secondary, and the learning outcome as a co-

construction of knowledge trumps the delivery of content.

CURRICULAR ISSUES

Goddard offers undergraduate and graduate programs in diverse curricular areas. These range from the free flow of individualized programs at the BA and MA level through more-regulated (through licensure) offerings in Education and Psychology to professional degrees: the MFA in Writing and the MFA in Interdisciplinary Arts.

With the exception of the Psychology program, no program offers courses. Instead, each program's curriculum reflects a synthesis of the desires of a student and the expertise of a faculty member. Within a rigorous intellectual framework for the entire College and each program, the curriculum starts from a question to the student: What do you want to learn? Building on a broad but rigorous set of required competencies for graduation and molded by the expertise of the mentor's skillful probing and questioning, each student defines their large (the total program) and small (the upcoming semester) learning plan at every step of the process. It should be noted that this approach is evident at both the introductory and more advanced curricular levels. The faculty offer guidance and assume a growing level of sophistication that will encourage ever-increasing freedom to self-define the student's learning. The student assumes that the passion and curiosity he or she brings in combination with the respect and caring of the faculty mentor energizes a process that produces spectacular results. Indeed, there is a saying at Goddard, acknowledging how messy and complex such an approach can be: Trust the Process.

It does work and this is one area where information technology has played an especially salutary role.

"Process" at Goddard has a unique and, perhaps contradictorily, precise meaning. Since the learning goals may be diffuse and certainly are student-generated, faculty use IT to track what has occurred and influence what will occur. Thus, a process that can be messy and inchoate benefits from mechanisms of control and monitoring that, while employed frequently in the past, has become, with IT support, a far more profound and empowered aspect of the rigorous yet unfettered learning paradigm. It becomes trustworthy.

As one might imagine, management and record keeping for the Goddard model is complicated, even onerous. There is the individual student tracking, semester by semester, and from matriculation to graduation. There are the requirements for constructing and implementing many residencies each year, each with a different start and end date. There is the ongoing and semester-end evaluation, compounded by the fact that Goddard does not use grades/GPAs and credits, but instead narrative evaluations. Not to mention the tracking of faculty fulfillment of obligations and their remuneration, governed by a 100+ page collective bargaining agreement.

Administrative computing improvements over the last twenty years have produced advances in the College's capacity to manage this demanding mélange of records and evaluations. As the College has become more sophisticated in its set up and use of IT for administrative purposes that touch upon the unique curricular model, administrators have been able to improve quality and build capability. By diminishing the need to spend time attending to the more menial aspects of the recordkeeping, though not diminishing the importance of good records, there can be a concomitant increase of time for more significant intellectual tasks, such as carefully constructed narrative evaluations. IT has helped here. While all colleges do their administrative work more efficiently with more sophisticated computing services, at Goddard the digitizing of records also directly impacts the curriculum, since the more sophisticated the entries in the student records can be, the more differentiated and comprehensive the individual analyses become.

The curriculum continues to evolve, changing within programs, and in the addition of new programs (the two latest: a Master's of Arts in Socially Responsible Business and Sustainable Communities and a Bachelor of Fine Arts in Creative Writing). Evolution of curricular offerings refers to not only the ability to examine what will be a marketable area of study but also to the ability of staff, faculty and students to stay up-to-date during the development of new areas or refinement of current areas. The necessity for academic administration, for which IT provides aid, is to assure quality and parity with the external curriculum possibilities while maintaining the radically massive degree of choice within those overall parameters. The evolutionary principle involved, like the trusting of process, is to heighten the confidence with which stakeholders can assert that a Goddard education, though "bizarre," has credibility in the coin-of-the-realm sense.

ASSESSMENT

Though progressive education is sometimes excoriated for its lack of structure and rigor, the way Goddard manages assessment in its academic programs is anything but structureless and slack. Learning in this environment is typically difficult to assess by traditional means and modes. But, throughout its history, Goddard has relied on qualitative and narrative techniques, epitomized by the transcript: no grades, no GPA, and many pages of prose by both student and faculty members.

At the end of every semester, the student prepares a self-evaluation and the faculty member prepares a narrative evaluation. These are lengthy and thorough and they are also an integral part of the collaborative learning paradigm: the final product is the result of a discussion that might even become, in the process (Trust It!), a difficult dialogue and/or negotiation. And the student's transcript, a rather thick document, consists of all the semester evaluations plus summative commentary for the entire program.

As the semester unfolds, it is not bereft of evaluation, however – far from it.

Programs structure the work of the semester around the submission and evaluation of "packets." A packet is often a written piece, equivalent to a research paper. It may be creative writing or journaling, or an experiment with the written word. Or it may not be verbal; students submit aural or visual projects, computer games or programs, two-dimensional or three-dimensional art, documentation of performances.

There is variance program by program; understandably, the Interdisciplinary Arts MFA has more visual and aural project submissions than Psychology and Counseling, with its more-rigid licensure requirements. But across programs, with the infusion of IT as a readily available option, there has been a loosening of the written word as the primary medium of exchange.

Because the digital means of communication encourage mixed media, a student desirous of expanding beyond writing can now do so more readily, whereas in the past the need to package a bulky art project (as opposed to a PowerPoint of the project) or send a videocassette (as opposed to a DVD) reduced the use of these newer media. In a sense, the option of IT as students can choose to employ it has created a liberating sense of what's possible and what's acceptable. This is by no means a retreat to a lack of rigor, but a progressive redefinition of what intellectual challenge and meticulous presentation of ideas can stand for to allow different ways of knowing to flourish. This phenomenon is in its infancy, but it is possible to see the contours, as the technological support becomes more sophisticated, of exciting movements to diversify what it means to be scholarly.

Any combination is possible. The way the packets will be completed is specified in the learning plan developed at the residency – the requirement, generally held as non-negotiable though there are exceptions (as with everything at Goddard), is that the student submit a packet every three weeks, totaling five during the fifteen week semester. In addition, a senior project or other culminating experience, and the equivalent thesis or project at the graduate level, are required to complete the degree.

Goddard students, remarkable people working with equally remarkable mentors, produce remarkable results. The work they do has benefitted from their own applications of technology.

The assessment process has benefitted as well. In the old days, studenst submitted their packets by mail or UPS or FedEx. Faculty responded by non-electronic means: a letter or occasionally a phone call. With the digital age, students in these times now often submit electronically via email and faculty respond also via email. Keep in mind that what's usually involved is 20-30 pages of single-spaced document in the packet and 5-6 pages in the evaluative response.

The benefit of electronic communication is that the feedback, and ensuing dialogue, can take place with much less down time: the interval between submissions and responses is hours or days instead of a week or more. In addition, it is now possible to submit multi-media work, including moving images, digitally altered art, and PowerPoint presentations, recalibrating the bias toward the written word. As required, the College's IT department aids this process; their intervention is usually not necessary due to the convergence and ready accessibility of individually-oriented computer-based (e.g., sending and receiving PowerPoint files) and web-based modalities (e.g., social networks such as Face Book and Second Life).

INFORMATION TECHNOLOGY IN GODDARD'S FUTURE

Leadership at Goddard firmly believes that values should drive the use of technology and it cannot be the other way around. In the New Goddard, discussions are now taking place that examine what forms the deep-structured academic revitalization of the College must take in the next dozen years or so. Underway this year is a participatory and wide-ranging process to design a campus plan fully integrated with the current academic nature of the institution and, in addition, sustainable and friendly to the surrounding community. The processes that encourage an examination of curriculum and policies centered on the learner are becoming infused throughout the institution. IT is essential to this effort since the dispersed academic community (students and faculty living all over the world in the fifty weeks they are not on campus) must have sophisticated communication and collaboration forms to engage.

The sense of participation, involvement, and democratic functioning is, of course, part of the philosophy of progressive education, and, at Goddard, the hope is to extend it to every stakeholder: students and faculty by definition, but also staff, neighbors, alumni, and trustees. Goddard has a long history of such participatory decision-making and collective wisdom. From its inception as a college in 1938, a community meeting (modeled on the New England town meeting) has been a constant and essential element of campus life. Recent experiments, recognizing the dilemma attributable to the significant segment of Goddard not in proximity for a physical town meeting, bring electronic and telephonic communication milieus into the mix. For example, though the governance councils (Student, Faculty, Staff, Program Directors, Alumni/ae) meet occasionally face-to-face, most of their meetings are scheduled as a computer or telephone conference. On the IT agenda for the near future is making generally available videoconferencing technology. How else could there be community meetings in the New Goddard without digital modes of connection?

A question that comes up with increasing frequency these days is whether Goddard's progressive model can and should include a solely online learning model. Can Goddard provide the education it values completely on-line?

This question, as asked, can unpack deeper aspects of what it means to be progressive and electronic. While it is useful from a marketing perspective, for instance, for Goddard's name to show up in searches for "distance learning" or "on-line education," the prevalent attitude at Goddard is to eschew a correlation with such competitors.

The distinction is between a values-based commitment to asynchronous learning and a market-driven approach to off-campus convenience. Goddard epitomizes an institution committed to asynchronous learning because over 75% of the interaction a student has in a semester occurs in the at-home, independent study setting. But the synchronous aspects of the residency, as well as technologically enhanced synchronous experiences, such as chats, allow a blend that would not be as robust in a solely on-line model. Put fundamentally, Goddard has not discovered a rationale for changing what it does well to a different, perhaps more marketable, you-can-get-a-degree-without-leaving-home stance.

Cyberspace experiments occur frequently: the psychology department offers online courses; the undergraduate program sets up a social justice on-line study group; faculty continue meetings with their advising groups after the residencies electronically or telephonically to supplement independent study. But there is a mostly unspoken assumption that such uses of IT are not the essence of what is done. It needs to be asserted that this is not because Goddard folks are backward-thinking Luddites: on the contrary, Goddard folks have not been swept up into the "gee-whiz" technocratic whirlwind so pervasive in 21st century culture.

The New Goddard must continue and deepen a commitment both to progressive learning and to experimentation, innovation and transformation. Information technology will be a component of the experiments. And so will potlucks, worker-student dialogues, community workdays, locally based volunteerism, trips to help rebuild a hurricane-ravaged Gulf Coast – all of which have happened in the distant and recent past. Goddard believes its engagement will be part of its curriculum. It will be radical as well, in the etymological meaning of that word: "to the roots."

CONCLUSION

It is not mere rhetoric to say Goddard invented an adult college degree comparable to the traditional age student experience. It is not mere rhetoric to assert that Goddard invented a low residency format for education that allowed those who could not spend weeks away from home to become a member of a learning community. But we have to stress the inconclusiveness of assessing whether Goddard (and, for that matter, any of the colleges which call themselves progressive) can find potency in IT applications for its core values.

The question, then, is what can we learn from the nearly five decades of experience that Goddard brings to this project? Several observations may be helpful in illuminating the dual foci of this book, information technology and progressive/constructivist learning, as the case of Goddard guides our agenda for future considerations:

- The learning model that Goddard uses is complex and unwieldy in its management. However, as evidenced by the fact that Goddard managed such a model successfully many years before the digital age afforded easy access to information technology, all that is really required is a student and a teacher on either end of the log. While that will omit the social and scholarly collaboration of the group, it is enough for much of what the model offers; the well-constructed intensive residency can add the collective dimension to the individualized model.
- Goddard lacks the capability to rigorously validate propositions it might advance about the use of technology. This is a flaw that the administrative leadership perceives

requires short-term correcting. Lacking the institutional research capacity, it is difficult to assess whether the structures (email, Web sites, collaborative cyberspaces) and the applications (presentation software, groupware, course management software) are appropriately represented as assets or liabilities in developing College policies.

- An irreplaceable component of the model is the right kind of faculty members. The "right stuff" for faculty in this model is neither affinity toward nor experience with online programs though clearly there is a benefit if that capacity is there. It is the attitudinal stuff that counts at Goddard. The primary attitude needed is to believe fully in the progressive tenet that learning comes from the learner's interests, that learning how to learn is more important than content, and that all learning is based in and reflects back upon experience. No habitué of the educational world in which the banking model of teaching is at the center can survive at Goddard. What counts is less a particular proficiency (e.g., with computers) and more a general stance vis-à-vis how adults learn. Methods and machinery follow.
- Discussions are just starting with other progressive colleges about IT issues and the utility of digital solutions. In fact, Goddard is taking the lead in bringing together progressive higher education institutions for discussions of and comparisons regarding an array of areas of mutual concern, including information technology. The hope is that sharing of information, and ultimately collaboration in calculating and implementing best (progressive) practices, will grow over the next decade.
- Though it is the case that IT is not an essential component, it is not the case that IT is irrelevant. What Goddard does is easier to do in 2008 than it was in 1968 because of the digital revolution. Not only easier but also more successful, which will be the case by a greater order of magnitude in 2048. Indeed, mapping the future demonstrates that Goddard will require more planning in its approach to IT and its model: the aforementioned insouciance of Goddard leaders and faculty toward digital applications will need to be replaced by a seriousness about thoroughly integrating technology into the model heretofore lacking.
- Goddard is fortunate that the audience – adult students – and the delivery system – short residencies – are market-favorable elements at this time. There is no evidence that the appeal of this model will diminish. There are fewer distractions and complications, like tuition discounting and demographic challenges, that plague traditional-age, residential institutions. That is not to say that Goddard doesn't face challenges. To mention two: When Goddard began this model, there were less than 10 competitors – now there are nearly 2,000. And how will Goddard contend with the dislocations of skyrocketing transportation costs: the programs are probably responsible for over 1,500 airborne round trips a year within a community of less than 1,000. There is an economic contradiction and an ethical quandary in the after-peak-oil, quadruple-current-airfare possible (likely?) world of the not-too-distant future.
- As is the case with many, probably most, colleges, Goddard will be inhibited in its work in this area by resource scarcity. Not just for Goddard, but also for all except the wealthiest liberal arts colleges, ways need to be discovered to maximize how technology can be used while minimizing the shocking expense of keeping up to date.
- As the 21st century matures, more students will have been brought up with an affinity for technology. When they arrive as adults at Goddard – or as traditional age students

at progressive residential colleges – the need will increase to integrate their lived experience, with deep technological connectivity, to their intellectual reflection. It needs to be figured out how both curricula and modes of delivery will acknowledge and ultimately advance this emergent precept.

- Can applications of information technology put joy back in the educational enterprise? Will progressive educators bring play into the work of learning with IT? How can the charge that progressive education is unstructured and sloppy be countered with a joyful, playful, rigorous, intense assemblage of technology-enhanced learning communities?
- Will this activity and line of investigation contribute to a more just and more inclusive academy? Does IT have a specific role to play, or (on the other hand) a bias against, signifying and striving for justice and inclusivity as goals for progressive tenets?

There should be an affinity of new means of communication and presentation, which transcend distances and boundaries, with the ideals of progressive colleges such as Goddard. To the extent that that is not the case, the priority must be to make it so. To the extent that it is the case, the priority is to enhance the efficacy of the technological interventions.

In either instance, the goal is to redefine progressive education for the 21st century and to claim a place, but not the only place, for information technology in that redefined objective. Moving toward that goal will both humanize technology and pump up education. The stakes for Goddard and other progressive colleges are high, but, as was the case in the past, the case in the future can be the most exhilarating and provocative educational experience for all students, including adults who cannot participate in the traditional modes.

REFERENCES

Bates, E. (1962). *Adult Degree Program, mimeo ephemera* (September 20, 1962). Plainfield, Vermont: Goddard College.

Hamlin, W. (1967, August). *Grown-Up education: The Goddard College Adult Degree Program.* Plainfield, Vermont: Goddard College Publications.

Bode, B. (1938). *Progressive education at the crossroads.* New York, New York: Newson and Co.

Collar, K. (2007). Getting to the roots of adult education. *Clockworks* (Winter/Spring 2007). Plainfield, Vermont: Goddard College Publications,

Dewey, J. (1997a). *Democracy and education.* New York, New York: Free Press (Reprint edition).

Dewey, J. (1997b). *Experience and education.* New York, New York: Free Press (Reprint edition).

Dewey, J. (2006). *The school and society.* Whitefish, Montana: Kessinger Publishing (Reprint edition).

Generals, D. (2000). *The architect of progressive education: John Dewey or Booker T. Washington?* In National Association of African American Studies and Hispanic and Latino Studies 2000 Literature Monograph Series: Proceedings (Education Section). Houston, Texas.

Hathaway, R. (1997). *The friendly pioneer: An informal history of the Adult Degree* Program at Goddard College and Vermont College. Montpelier, Vermont: Word and Image.

Westbrook, R. B. (1991). *John Dewey and American Democracy.* Ithaca, New York: Cornell University Press.

Chapter XX
Using Blogs to Foster Inquiry, Collaboration, and Feedback in Pre-Service Teacher Education

Carol R. Rinke
Gettysburg College, USA

Divonna M. Stebick
Gettysburg College, USA

Lauren Schaefer
Carroll County Public Schools, USA

M. Evan Gaffney
Gettysburg College, USA

ABSTRACT

This chapter presents a critical case study on the use of information technology in a pre-service teacher education program. The authors integrated Weblogs (blogs) into two constructivist-oriented teacher preparation courses with the goal of helping students learn to think like a teacher through enhanced inquiry, collaboration, and feedback. The authors found that, through the use of blogs, pre-service teaching candidates grew in their abilities to reflect on their own teaching and to provide constructive comments to peers. The authors' experience also indicated that while instructor and peer feedback via blogs was valuable, it functioned best when paired with face-to-face meetings between the instructors and students. They discussed design principles for combining online and face-to-face environments and offer possibilities for the expanded use of blogs in pre-service teacher education.

INTRODUCTION

Constructivist theory suggests the centrality of active participation and social interaction to the learning process (Phillips, 1995). Although forms of constructivism in education range broadly from Dewey's (1938) notion of experience to Vygotsky's (1978) zone of proximal development, constructivism in education maintains a focus on student participation and interaction as central to learning. Information technology is uniquely situated to facilitate dynamic and collaborative learning environments supportive of constructivist principles. In higher education in general, information technology can provide structures to support learners in both making sense of their own experiences and communicating these understandings to a larger community (Garrison & Anderson, 2003). In teacher education in particular, it can bring authentic classroom experiences to the college curriculum through videos and artifacts, strengthening the theory-practice connection (Gomez, Sherin, Griesdorn, & Finn, 2008).

Working together as teacher educators and pre-service teachers, we used information technology in the form of weblogs (blogs) to enhance student inquiry, collaboration, and feedback in two teacher preparation courses at a highly-selective liberal arts college. In this chapter, we illustrate the partnership between constructivism and information technology through a critical case study of the use of blogs in our courses. In particular, we examine the ways in which blogging helped students learn to think like a teacher through inquiry, collaboration, and feedback. We then introduce data from the blogs and peer comments, analyzing it in relation to constructivist principles. We conclude with recommendations for future use of blogs in pre-service teacher education.

THEORETICAL FRAMEWORK

As teacher educators, we support a foundation of constructivist principles in our practice. Our classes, advanced undergraduate courses for pre-service secondary teachers, work toward the ultimate goal of teaching how to think like a teacher (Crowe & Berry, 2007). We believe that beginning teachers should have not only the knowledge and skills necessary for teaching, but also the ability to think critically about instances of classroom practice from a variety of perspectives. Five key principles for thinking like a teacher include:

- **Principle One:** Thinking like a teacher involves learning to see teaching from the viewpoint of the learner. Experiencing the role of learner is an important means of developing an understanding of the learner's viewpoint
- **Principle Two:** Prospective teachers need opportunities to "see into" the thinking like a teacher of experienced others
- **Principle Three:** Prospective teachers need opportunities to try out thinking like a teacher in order to develop their thinking as a teacher
- **Principle Four:** Prospective teachers need scaffolding (guidelines, questions, structures) to support them in the process as they begin thinking like a teacher
- **Principle Five:** Developing responsive relationships is at the heart of learning to think like a teacher and at the heart of supporting our students (Crowe & Berry, 2007, p. 33)

We believe that these principles create a core set of abilities necessary for pre-service teachers to reflect and think deeply about their practice and learn from experience. Exposure to these higher-level abilities during the undergraduate years can prepare future teachers to continue learning from their practice and interacting with communities of learners throughout their professional lives.

The Study

We supported these broad principles of teacher learning through the use of inquiry, collaboration, and feedback in our classes. We structured assignments to foster systematic inquiry into student learning and reflection on practice, integrating these principles into our practice in the following ways:

1. To apply principles one and three, we structured assignments to foster systematic inquiry into student learning and reflection on practice. Thus, our students learned to analyze learners' viewpoints and to practice thinking like a teacher.
2. To support principle five, we included online and face-to-face discussions of specific instances of practice to encourage meaningful collaboration which supported the development of responsive relationships. The face-to-face meetings included both the instructors and the student teachers.
3. To address principles two and four, we encouraged instructor and peer feedback to provide reflective, constructive suggestions as well as support to see into the thinking of others and provide scaffolding to support the process.

We wanted our pre-service teachers to develop a general understanding that inquiry is embedded in teaching. We drew upon the work of Cochran-Smith and Lytle (1999), who proposed the notion of inquiry-as-stance among educators. This idea, in combination with critical reflection (Gore & Zeichner, 1991; Richert, 1991; Smyth, 1989), encouraged pre-service teachers to continually examine their teaching in a systematic manner in order to construct knowledge collaboratively and ultimately enact change. In our classes, we asked pre-service teachers to select a focus for their inquiry, look carefully at student learning and experience, reflect on their instructional decisions, and revise their lessons and assessments to better meet the needs of learners. This systematic investigation into teaching and close attention to student learning was designed to foster a stance of inquiry among future teachers to prepare them for their future classrooms.

Next, we wanted our pre-service teachers to operate within a collaborative community of professionals reflecting upon their own teaching. Here we based our efforts on the notion of professional learning communities, which provided opportunities for intellectual growth, knowledge development, and leadership for teachers (Grossman, Wineburg, & Woolworth, 2001; Lieberman & Grolnick, 1996; Little, 2003). We encouraged peer collaboration in our courses through blog comments and face-to-face discussions. We promoted instructor-student collaboration through formal and informal interaction on the blog, over email, in the college course, and in the secondary classroom. We designed multiple avenues for peer and instructor interaction with the idea that strong professional relationships are a critical context for ongoing professional growth among both beginning and veteran teachers.

Finally, we believed ongoing and interactive feedback is critical for developing effective practitioners (Hattie & Timperley, 2007). We used the strengths of instructional technology to encourage feedback about concrete instances of practice. Based on the recommendations of Ball and Cohen (1999), Lampert and Ball (1998), and others who advocate practice-based teacher education, we used blogs to illustrate instances of practice otherwise invisible in the college classroom. Pre-service teachers used their blogs to display edited videos of their lessons, scanned student work, lesson plans, and rubrics, demonstrating exactly what was taking place in the secondary classrooms where they were student teaching. We created these structures to foster concrete feedback on particular classroom situations, connecting theory with practice and providing more targeted and relevant feedback and support for pre-service teachers.

BLOGS

Blogs are basic websites that allow users to display text, pictures, videos, and links on the Internet. Regularly maintained and updated, blogs display entries in reverse chronological order so that viewers see the most recent entry first. Blogs distinguish themselves from traditional websites because they function as a social interface. Anyone with access to a blog may post comments, facilitating a form of online conversation. Bloggers can permit the public to view or restrict their blog to a select group of users. In order to maintain the privacy and confidentiality of both teachers and students, we restricted access to our blogs, including only those individuals involved in each pre-service teacher education course.

We selected blogs rather than various other forms of information technology for two key reason. The first was practical: blogs are freely available and quite simple to use. As instructors, neither of us had any significant prior experience with blogs. We wanted a form of information technology which would support our purposes without burdening students to learn new and challenging technical skills. We selected Blogger (http://www.blogspot.com), a widely-used and freely-available weblog system, because it was free, simple, and more visually attractive than the blogging feature available on our campus' course management system.

The second was more conceptual. Blogs provided unique tools to support the implementation of constructivist principles in the classroom. In particular, blogs contained a combination of features to support inquiry, collaboration, and feedback:

- *Inquiry* We wanted our students to inquire into classroom contexts through ongoing reflection. Blogs provided a centralized environment for that reflection and also archived posts in reverse chronological order, allowing students to look back at the end of the semester and observe their own growth over time. This provided an opportunity for metacognition, or thinking about one's thinking (Bransford, Brown, & Cocking, 1999), in which students could read over their posts and literally reflect on their reflection. We hoped this feature would support the development of reflection skills and an inquiry stance in pre-service teachers.
- *Collaboration* Blogs also supported our goal of collaboration through their social interface. The pre-service teachers read and commented on their classmates' reflections on a regular basis. This feature allowed them to operate in an online professional community, giving and receiving feedback from both peers and instructors. This feedback often came very quickly, permitting the pre-service teachers to use feedback almost immediately in the classroom, a critical feature for supporting theory-into-practice. Finally, the individual nature of each blog allowed the pre-service teachers to direct the conversation to their own teaching.
- *Feedback* Blogs also offered tools to support our goal of providing feedback, in which pre-service teachers reflected and commented upon specific, practice-based instances from the classroom. Blogs allowed students to display artifacts from the classroom, grounding reflection and feedback in particular instances of practice. However, unlike traditional websites, blogs also encourage comments on those artifacts, fostering more sophisticated practice-based reflection on teaching.

Overall, blogs served our purposes by providing a free and easy-to-use online environment with specific tools for encouraging constructivist goals among pre-service teachers.

LITERATURE BASE

Earlier work examining the use of blogs in teacher education has primarily focused on the logistical challenges of creating a social network via blogs. This is particularly true in courses with large numbers of students, where participation becomes an issue. For example, Makri and Kynigos (2007) used blogs in a pre-service mathematics methods course and found participation varying from blog enthusiast to frequent visitor to blog skeptic. Work by West, Wright, Gabbitas, and Graham (2006) in a pre-service instructional technology course identified additional implementation challenges. They suggested that successful use of blogs depends upon advance preparation, technical training, and comfort with the particular online writing style. Lock (2006) postulated that the effective design and development of an online community, along with its reliability and scalability, directly affect the quality of online professional development. Together, these studies suggest that student skepticism and lack of familiarity with technology pose considerable challenges to blog implementation in pre-service teacher education courses. Increasing student comfort with various skills necessary for blogging is an important prerequisite to implementation. Our study aimed to move beyond such implementation issues in order to examine the conceptual dimensions inherent in blog use.

Some studies examined issues of reflection and social interaction among pre-service teachers in blog contexts. Barab, MaKinister, Moore, and Cunningham (2001) noted that collaboration among individual community members in an online environment fostered the notion of a collective whole working toward the joint goals of the community. Stiler and Philleo (2003) found that students were generally satisfied with the use of blogs in pre-service multicultural education and technology courses. Additionally, student reflections appeared to be at higher levels than in previous courses, although they were not systematically evaluated using a rubric. Kuzu (2007), using blogs in an online information technology in education course, found that students generally held positive views about the social interaction fostered through blogs. Finally, Dickey (2004) found that blogging prevented feeling of isolation and alienation among distance learners. Although these studies took a critical first step toward examining the ways in which blogs can support teacher growth, they relied heavily upon reports of self-reports of teacher satisfaction.

Our case study aimed to build upon existing work by providing direct evidence of inquiry, collaboration, and feedback through rich examples of student reflections, comments, and face-to-face interactions. We operated in an idealized environment, with small groups of students and a distinct lack of implementation problems. Because our context was unique, it allowed us to push our analysis past a focus on implementation and reports of pre-service teacher satisfaction and toward an identification of the types of reflection, collaboration, and feedback which took place in the blogging context. Although the small scale of our study limited it in certain ways, this same small scale allowed us to look deeply into what blogs afforded our pre-service teachers and the ways in which blogs helped them to think like a teacher.

METHODS

We studied the implementation of blogs in two pre-service teacher preparation courses using a qualitative case study methodology, selected for its ability to look closely at bounded situations of interest (Stake, 1995). Our idealized instructional context certainly created a situation of interest because it allowed us to look past issues of implementation yet began to provide an opportunity for us to take an initial look at the growth of pre-service teachers through blogging. We specifically focused on understanding the ways in which blog use in pre-service teacher education courses facilitated inquiry, collaboration, and feedback. We collected and analyzed data in a variety of forms, including pre- and post-course surveys, weekly blog entries, student comments, emails, and field notes on face-to-face communications.

Context

This case study focused on two courses with extremely limited enrollment. Once again, this ideal teaching situation allowed us to develop close instructor-student relationships and avoid some of the implementation issues faced by other teacher educators. It also provided us with the ability to collaboratively analyze our practice, continuing the work of reflection that we began during our semester together. Because of the constraints inherent in this, and any methodology, findings from this study are not intended to be generalized to the larger population. However, we hope they can provide insight transferable to educators of all forms looking to integrate instructional technology into higher education classrooms (Donmoyer, 1990). We focus on the relationship between course goals, in the form of constructivist principles, and their outcomes as mediated by one particular technology, blogging. Lessons from our critical case study can be translated to a variety of teaching and learning contexts using information technology as a pedagogical tool.

Participants

We worked with five pre-service teacher candidates enrolled in two different upper-level education methods courses during the fall semester of 2007 (Table 1).

All students were seniors, three females and two males. Data from all five participants were used for analysis, although absence from consistent peer-to-peer blogging comments led to some difficulty during analysis. The pre-study survey indicated that while all of the students were familiar with a variety of forms of information technology, including email, internet, and social networking, some were not familiar with blogs. Three of the five pre-service teachers had previously used a blog for other purposes and two of the participants had no or very limited previous blog experiences. Despite this range of interest and experience levels regarding blogs, students generally found the blogs easy to use and reported no substantial technical difficulties.

One course focused on secondary English education and the other, concurrent with the student teacher semester, on secondary science and math education. Expectations in both courses included using theory to develop and refine effective instructional practices within their respective fields of study. Teacher candidates in both courses maintained ongoing blogs in which they systematically inquired into and reflected upon

Table 1. Characteristics of students using blogs in teacher education courses and study participants

Student	Gender	Prior Blogging Experience	Year	Certification Area
Evan	Male	Yes	Senior	Science
Lauren	Female	No	Senior	Science
Joyce	Female	No	Senior	Mathematics
Jarred	Male	Yes	Senior	English
Marilyn	Female	Yes	Senior	English

issues of practice using narrative and classroom artifacts. These frequent entries included reflections on student learning, classroom management, instructional strategies, assessment practices, as well as the politics of teaching in a public education setting. As instructors, we also maintained our own blogs with ongoing entries focused on the process of integrating instructional technology into pre-service teacher education. These blogs served to enhance our own reflective practice and as a model for pre-service teachers.

Procedure

As instructors, we incorporated blogging into our course syllabi. We informed the pre-service teachers that the course requirements included the completion of non-traditional reflective journals using Blogger. We explained that this blogging opportunity would provide a method to express their concerns, celebrations, and confusions and better apply instructional theory to practice. In both courses, teacher candidates interacted collaboratively with the instructor to provide both online and face-to-face feedback on instructional practice. In order to develop collegial relationships with peers and the instructors, the courses began with face to face sessions in a traditional classroom context. These sessions included course overviews, blogging expectations, and questions for discussion regarding lesson planning and archiving student artifacts online. Throughout the semester, the English education students held weekly and then bi-weekly face to face meetings with the instructor. The secondary science and math students met bi-weekly throughout the entire semester. Instructors and peers also commented on each blog entry with questions, connections, and suggestions in order to encourage the blogger to move to a higher level of reflective thinking. These face-to-face and online sessions worked together to support students' pedagogical needs to learn in a variety of instructional contexts.

The course expectations about blogging differed based on the structure of the course. For the English education course, which took place prior to the student teacher semester, teacher candidates were expected to post to their blogs after every teaching session, usually twice per week. As the teaching load increased later in the semester, their blogging expectation changed to one post per week. The science and math education course was taken concurrently with the student teaching semester; thus the pre-service teachers were in the secondary classroom full time. They were expected to conduct a systematic case study through their blogs, posting topic selection, lesson plans, lesson revisions, classroom artifacts, and reflections throughout the semester. Although the particular structure of blog use differed slightly between these two courses, the overall goal to foster thinking like a teacher remained constant.

Instruments

We analyzed the data using the three categories of inquiry, collaboration, and feedback. In order to better understand the development of inquiry abilities over time, we coded all blog entries using a rubric based on Arter and McTighe's (2001) work (Table 2).

Students used this rubric throughout the education program to evaluate their levels of reflection. The rubric also provided clear standards to allow for consistency when evaluating blog comments. In coding the data, we noticed that three levels did not entirely capture the complexity of reflection present, and thus we added a fourth category, medium-high reflection, which included some but not all features of high-level reflection. This revised rubric captured individual levels of reflection as well as change in reflection over time.

We scored the reflections between one and four, with one indicating low self-reflection and four indicating high self-reflection. A level one reflection consisted primarily of description:

Table 2. Departmental rubric used to evaluate blog posts and comments based on degree of self-reflection

High Self Reflection	Medium Self Reflection	Low Self Reflection
Detailed, with specific examples provided	Identifies ideas but does not provide a full discussion	Vague
Many points covered	Begins to take ownership	Simple restatements
Self-revelation	Describes performance but leaves rationale out	Descriptions only
Student is motivated	Focuses on only a few ideas	Focuses mainly on surface features
Sets goals for the future	Considers content, context or process, but not all three	Obvious content reported with no new insights
Considers more content, context, and process	Is sometimes inaccurate with theory applications	No examples
Takes thinking risks	Doesn't seem completely honest	I like it/I don't like it
Growth is supported with examples	Lacks sincerity	One dimensional
Organized	There are descriptions of some key events affecting thinking, but no comparisons over time	Superficial
Sincere & honest	I have to make inferences about what you mean	Low to no awareness of need to set goals or make progress
Gives comparisons of personal thinking over time		Inaccurate
Shares feelings, but includes theory anchors		No personal ownership of ideas
Examines both strengths & weaknesses		Disorganized-difficult to follow your meaning
Analyzes in depth looking at various perspectives		Hard to read
Provides good reasons & explanations		
Reveals thoughts		
Develops personal voice and ownership of ideas		
Easy to read & neat		
Examines skill improvements		
Gives personal reaction		
Sets goals & looks ahead		
Interesting		
Considers ethics		

Mrs. Patrick forgot that it was junior picture day during my class period...I was surprised that when the students returned to the classroom, they immediately began reading or studying their recitation without any prompting from Mrs. Patrick or I (Marilyn).

A level two reflection identified key ideas embedded in practice but did not provide a full discussion of those ideas:

This lesson definitely had its flaws. But in some ways it's better not to succeed at your first go at something. I have definitely learned more from this lesson than I have from other lessons that have gone smoothly (Lauren).

A level three reflection used explanations and examples, including personal reactions to the situation:

We've done a lot of activities and labs, but this is the first real world application of physics I've exposed the students to. They were engaged the whole time and did exactly what I expected of them. Kevin was able to participate in the entire experience, with the aid of his teammates and his graphic organizers (Evan).

The highest level of reflection, level four, included a discussion of feeling and also anchored those feelings in theory. In this example, the student linked her experience in the classroom to the educational notion of scaffolding, or support for student learning:

So as I continued scaffolding my students to make the connections between the meanings of word elements and word definitions, I never actually showed them how to do this, or what to do when they come across unknown words in a text (Marilyn).

Using these benchmarks, we categorized initial blog reflections and their change over time.

We also analyzed student-to-student comments on blog posts in order to better understand the nature of the collaboration taking place. Initially, we coded these comments superficial versus constructive, although during the process we refined our categories to include constructive without theory and constructive with theory. This analytical process helped us to better understand the content as well as the structure of peer collaboration in each of the courses. Superficial feedback was largely descriptive in nature, praising the students' ideas without offering concrete suggestions for improvement or making connections to theory:

I really like your lesson. I like how you begin it with video clips—something the kids will really relate to and enjoy. Then, you have the kids explore the material in different ways. I love the assignment where they have to come up with their own examples. I think your plan sounds awesome, and I can't wait to see how it turns out! (Joyce)

On the other hand, constructive feedback analyzed the initial post and offered questions or suggestions to enhance teaching skills:

I thought the lesson was very creative, however, I do have one question for you. Do you think your guardians will be able to follow this type of instruction? As a guardian who loves structure and formatted notes... what will you do for these learners? (Lauren)

Some of these comments even included theory and methods drawn from the education curriculum. The following comment connected to the idea of assessment in education:

About informal assessment, do you call on the kids that you worry about? I try to call on everyone at least once every-other day. Also, I don't know how your class is set up, but we do a ton of group work, so while they're working, I go around and monitor and probe for explanations. Finally, do you ever have students come up to teach the material? For me, I've found it works really well because, even if the student got the answer wrong, it helps everyone learn... especially because there are almost always other people who have done the same error (Joyce).

Finally, we analyzed student responses on the pre- and post-course surveys and field notes from face-to-face meetings to integrate student feedback into our analysis and examine student growth in inquiry, collaboration, and feedback. These data provided students' perceptions of the process as they experienced it and supported our desire to integrate student feedback into our analysis.

In investigating the use of blogs in these pre-service teacher education courses, we worked together as teacher educators and pre-service teachers. We saw this study as a co-construction of knowledge, with blurred lines between investigators and participants. Although the teacher educators were primarily responsible for designing and implementing the courses and the pre-service teachers for reflecting collaboratively on their practice, we worked together to analyze our experiences. We saw this collaboration as an extension of, rather than a departure from the constructivist approach we took in the course themselves. We also saw this collaborative self-study as enhancing, rather than detracting from the validity of the research (Lather, 2001).

RESULTS

The data from this study supported the use of blogs in constructivist-oriented pre-service teacher education courses. The data showed that pre-service teachers improved their reflection abilities on blogs over the course of the semester and most reached a high level of self-reflection. Our data also indicate that the feedback among peers was supportive and helpful. We found the feedback as overwhelmingly constructive and traced the development of some lessons directly to peer feedback. Data from surveys however, suggested that blogging should be only one element in a comprehensive teacher education course.

Inquiry

Scored student reflections (Figure 1) showed that reflective abilities, as demonstrated on the blog, consistently improved over time.

Even though students started out on different levels, all students continually advanced in their reflective abilities. In Lauren's case, she initially reflected with more description. She identified problems and discussed them, but her posts failed to show her personal thoughts on the important issues embedded in practice:

The school in which I am student teaching does not believe in separating students according to their abilities. There are advanced (honors) classes, but all the other classes are a melting pot of alternative education students, ESL students, special education students, autistic students, and everyone else. I believed this method of inter-mixing the students would not only be difficult for me to teach but also be unfair to the children without disabilities (Lauren).

Figure 1. Graph showing student reflection level as measured by the rubric over the course of the semester

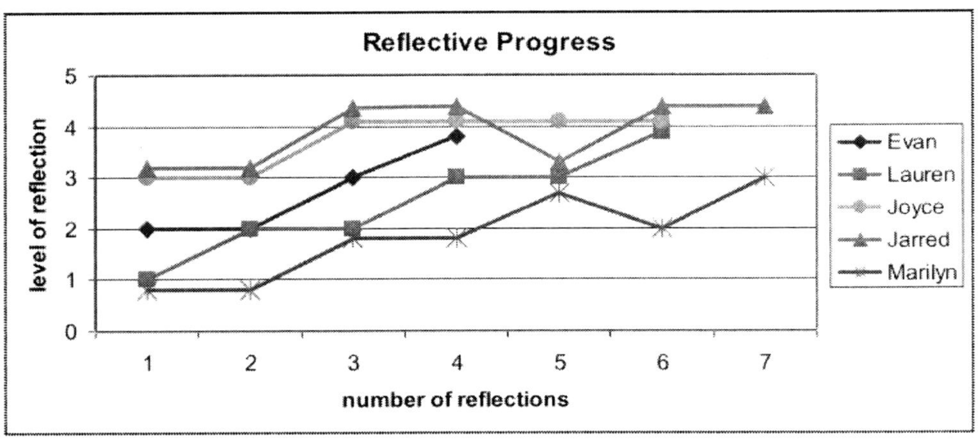

A later post illustrates her growth. She discussed the same topic but with greater emphasis on specifics from her classroom:

I want all the students in this class to receive the same education and do not want to slow down the pace to suit one student if it means holding back for students with higher abilities. The solution I pose to this question is differentiated instruction. This is modeled after "fair is not always equal." I intend on using tools such as graphic organizers, fill-in notes, and other means of organization to heighten the educational experience for these students with varying abilities. I plan on measuring the success through observing informal assessment/class participation. I observed that the majority of students who volunteered to answer a question during the class period were those students who did not suffer from a learning disability. My goal is to have equal voluntary participation in the class (Lauren).

In this reflection, Lauren discussed the recurring problem in greater depth and outlined a solution using strategies learned from courses within the education program. Over the course of the semester, she demonstrated an improved her ability to think critically about her classroom and connect theory to practice.

Marilyn grew over the semester as evident in her blog posts. Marilyn began reflecting on a very basic level, simply describing the classroom environment or the day's events. From there she improved slightly as she started to incorporate personal conflicts that arose such as:

My biggest fear throughout this entire lesson was that they would ask a question or insert a comma somewhere and I would not know if they were correct or how to answer their question. Would they discover I was a fraud and my weakness was grammar? There were a few questions and I am not sure I answered them clearly or correctly but overall, the students seemed to understand. One of the students actually found an error in my paragraph that I didn't catch! (Marilyn)

Toward the end of the semester, Marilyn's reflections began to do more than paint a picture of the classroom. In the following excerpt from Marilyn's reflection, she identified her concern and also posed a potential solution, grounded in the assessment strategy of rubrics:

I still doubt myself. My biggest fear is having a student demand to know why they got a certain grade and I would not know what to say. Having a rubric makes it easier for the students to understand their grade and easier for me to determine what grade they should get. I know that I always hate when I spend the time to write a paper or write an essay exam and I get it back with just a grade on it, no comments or indication of why I got the grade. Rubrics make it easier on everyone! (Marilyn)

These examples from Lauren and Marilyn illustrated improvement in reflective abilities and a stronger grounding of practice in theory over the course of the semester.

Collaboration and Feedback

Data also indicated that substantial student collaboration was achieved in the blogging context. We analyzed peer comments on student blogs and found that the comments were overwhelmingly constructive, as opposed to superficial in nature. The majority of blog comments encouraged growth and development in teaching. Students posed questions, offered advice, made connections to their own teaching, and referred to theory or in-class discussions. One such constructive comment offered:

The easiest way to fill in the gaps, especially in a lesson like this, has to be informal assessment. One of the things that I like to do is anticipate potential hang ups for the students and pre-formulate some questions that address them. Then when you're walking around during group activities, you can pull them out to dipstick the students you know may be struggling. Sometimes (rarely) I even think of specific questions for specific students based on misconceptions identified in previous assessments. Try this and watch your students light up when you individually question them and seemingly pluck their confusion out of mid air. Your exit tickets can be a great tool for this type of practice (Evan).

This comment gave advice and suggested methods, motivating improvements.

Eighty-three percent of all comments were, as with this example, constructive in nature. Moreover, 35% of the constructive comments went an additional step and connected to various aspects of educational theory. This was a surprising result for us. We developed a term, constructive-with-theory, to refer to these events. The following example of a constructive-with-theory comment suggested engaging students by activating their prior knowledge, a technique discussed in numerous education courses:

I think it is very important to keep the students engaged which you could have done a number of ways, but utilizing your knowledge of their CSI interest was above and beyond. Good job :) (Lauren)

This sophisticated level of feedback, constructive-with-theory, connected prior experience in the education program with specific instances of practice within the context of a professional community of educators.

In some cases, improvements in particular lessons could be traced back to constructive feedback given by peers. In one example, Lauren posted:

The high school I have been placed at does not use tracking. The classes are a mixture of students at different ability levels. In one such classroom I have six learning disabled students, one ESL student, and an autistic student. This class has brought me to question what the best way of teaching to these students would be. I want all the students in this class to receive the same education and do not want to slow down the pace to suite one student if it means holding back for students with higher abilities (Lauren).

She received the following comments:

Comment 1: I find it's also important to prevent ability divisions within your classroom. Don't allow a "smart group" of students to form. Encourage collaboration between all of your students. You said yourself you need more teachers, so recruit some students! Don't be afraid to use that "special helper" tactic to get your bored gifted students helping your challenged students" (Joyce).
Comment 2: The same thing is happening with a class at our school. I think that one of the hardest things may even be not completely bashing all the confidence of the lower-level students. Have you all tried peer scaffolding groups? Maybe make groups according to academic ability, say 1-7, 1 being the most advanced. Then 1 scaffolds 2; 3, 4; etc (Jarred).

Lauren took the advice of her peers and developed student groups in advance of her next small-group activity, taking into consideration student ability. Through direct peer collaboration on blogs and constructive feedback, pre-service teachers connected to theory and even improved their practice.

Student Response

Students responded with overwhelmingly supportive comments to using blogs in their pre-service teacher education courses. In a post-course survey, participating students rated and answered questions about their experience with blogging throughout their course. All students responded that blogs were helpful to very helpful in allowing them to practice thinking like a teacher. They also reported that the blogging process was helpful to very helpful in allowing them to receive support for the process of learning to teach. Seventy-five percent of the pre-service teachers found that maintaining blogs allowed them to understand how experienced teachers think. All of the students found that their colleagues' comments on blogs were somewhat helpful to very helpful.

Although the data supported the use of blogs to enhance reflection and collaboration, student feedback suggested that blogs should only be one element of a comprehensive teacher education course. The final question on the post-course survey asked students to rank the importance of other means of communication used throughout the course. Participants ranked formal in-person meetings with instructor and students, email communication with instructor, out-of-class communication with classmates, and blog comments from 1, most helpful, to 4, least helpful (Table 3).

The results showed that students found formal in-person meetings to be the most productive source of feedback in the course. Marilyn commented:

I liked the feedback I could receive on each lesson/day. Because we didn't meet often during the semester, I could ask questions and get advice that I could implement in my next lesson. It was also helpful to see how other students were doing and how their problems/issues were similar or different from my own.

Her comment supported blog use but suggested that they were most helpful when face-to-face meetings were not possible.

Just as blogs provide certain affordances for learning, face-to-face discussions also have particular advantages. During a face-to-face meeting, classmates and instructors can build on one another's ideas in a more immediate fashion to reach a general consensus. One student developed a lesson and received minor comments on it via the blog. Later, during a subsequent face-to-face conversation, students and the instructor worked together to substantially revise the instructional approach of the lesson. This type of collaborative knowledge-sharing would not have been possible in the turn-taking approach inherent in blog comments. Face-to-face meetings also provided the opportunity to examine numerous student work samples at the same time. Although it is possible to look at student work on a blog, clicking on large numbers of scanned classroom artifacts can be time-consuming and laborious. Using the low-tech alternative of hard copies is far simpler and more effective. While technology continues to

Table 3. Student responses to the post-course survey question, "Which of the following did you find to be the most productive source of feedback in the course? Please rank (1= most helpful; 4 = least helpful)"

	In-person	Email	Out of class	Blog comments
Evan	1	3	4	2
Lauren	3	1	2	4
Jarred	1	4	3	2
Marilyn	2	4	3	1
Group Average	1.75	3	3	2.25

**Joyce did not complete the post-course survey*

evolve, it is important for us to not lose sight of the comfort and advantages that can be provided by face-to-face discussions. Also, since there is a learning curve involved when initiating a blogging forum, educators must consider the advantages and disadvantages of this form of communication, based upon the student teachers' strengths and needs. Blogging should not replace in-person communication. Rather, it should be one element of many in the further development of collaborative communities.

CONCLUSION

We used blogs in the context of two pre-service teacher education courses designed around constructivist principles. In particular, we wanted pre-service teachers to learn to think like a teacher (Crowe & Berry, 2007) and designed a variety of online and face-to-face experiences in support of that goal. Our in-depth case study of student experience in these two courses investigated the ways in which blogging fostered inquiry, collaboration, and feedback among teacher educators and pre-service teachers. We found that student reflection on blogs increased in sophistication over the course of the semester, from mainly descriptive to more theory-based. We also found that student collaboration on blog comments was overwhelmingly constructive and at times spontaneously linked theory to practice. However, one important pedagogical drawback to blogging did emerge. The pre-service teachers felt that the blogging experience was a critical element of their preparation, but could not operate in isolation. Rather, it worked best in conjunction with face-to-face meetings which allowed for group analysis and synthesis of ideas.

Based on this case study, we concluded that blogging can be considered an effective strategy for fostering high levels of reflection, peer-to-peer collaboration, and concrete feedback based on theory and practice. In doing so, blogging supported general constructivist principles of student participation and interaction in learning. This case study in teacher education strengthened earlier work by Gomez et al. (2008), indicating a role for information technology in building social relationships and encouraging reflective teaching. It also exemplified the framework suggested by Garrison and Anderson (2003), in which information technology can be used in higher education for sense-making and community building.

Our case study, together with earlier research (Kuzu, 2007; Stiler & Phileo, 2003), indicated that blogging can be used successfully in teacher education. Our experiences indicated however, that the key to effective use in the college classroom is not only thoughtful implementation (e.g., Makri & Kynigos, 2007; West et al., 2006) but also an appropriate course design combining blogging with more conventional instructional methods. For example, both of our courses included face-to-face meetings and artifact-based collaborative conversations which yielded results unique to that particular teaching context. Moreover, one particular student valued one-on-one email interaction with the instructor above collaborative blog comments. This reflected findings by Dickey (2004) indicating that while blogs are successful on the whole, they may pose serious challenges for particular students. We suggest that future teacher education courses using blogs carefully consider the most effective combination of methods in order to achieve the best possible educational experience for students.

FUTURE TRENDS

We also envision future possibilities in which blog use can be extended beyond the higher education context to bring in valued professionals from the field. This integration of authentic, real-world connections into the college classroom is supportive of the constructivist ideal of learning from experience (Dewey, 1938). In the context of teacher education, these blogs could be expanded to incorporate individuals from the K-12 school setting, such as student teaching supervisors or cooperating teachers. In an even more elaborate form, blogs could connect pre-service teachers with other beginning or veteran educators nationally or even internationally, providing a far-reaching professional network. Although issues of student confidentiality, pre-service teacher comfort, and commitment to the online environment would have to be addressed in order for these models to succeed, we see broad possibilities for using and even expanding this model of information technology in teacher education. Technological refinements such as integrated university-school networks, increased sophistication of video streaming, and enhanced online artifact display would facilitate the achievement of these greater constructivist goals.

FUTURE RESEARCH DIRECTIONS

Our case study had a number of limitations, including its small scale, idealized instructional environment, and absence of implementation challenges. However, the in-depth examples of student reflection, collaboration, and feedback offered a rich context for understanding the types of student learning which took place in an online environment. Further research on this topic should expand upon our findings to investigate an ideal combination of instructional methods for pre-service teacher education. More broadly, it should also consider the integration of inquiry, collaboration, and feedback through blogs with fields outside of teacher education. Finally, it should investigate the possibilities inherent in a more inclusive blog context, one which brings together individuals from higher education, K-12 schools, and beyond. Information technology has been used successfully to promote constructivist principles in teacher education; future research should move toward a more nuanced understanding of its use in supporting pre-service teachers and educators in general.

REFERENCES

Arter, J., & McTighe, J. (2001). *Scoring rubrics in the classroom.* Thousand Oaks, CA: Corwin Press.

Ball, D. L., & Cohen, D. K. (1999). Developing practice, developing practitioners: Toward a practice-based theory of professional education. In L. Darling-Hammond & G. Sykes (Eds.), *Teaching as the learning profession: Handbook of policy and practice* (pp. 3-32). San Francisco: Jossey-Bass.

Barab, S., MaKinister, J., Moore, J., & Cunningham, D. (2001) Designing and building an online community: The struggle to support sociability in the inquiry learning forum. *Educational Technology Research and Development, 49*(4), 71-96.

Bransford, J. D., Brown, A. L., & Cocking, R. R. (1999) *How People Learn: Brain, Mind, Experience, and School.* Washington D.C.: National Academy Press.

Cochran-Smith, M., & Lytle, S. (1999). Relationships of knowledge and practice: Teacher learning in communities. In A. Iran-Nejad & P. D. Pearson (Eds.), *Review of research in education, 24,* 249-305. Washington DC: American Educational Research Association.

Crowe, A. R., & Berry, A. (2007). Teaching prospective teachers about learning to think like a teacher: Articulating *our* principles of practice. In T. Russell & J. Loughran (Eds.), *Enacting a pedagogy of teacher education: Values, relationships, and practices.* London: Routledge.

Dewey, J. (1938). *Experience and education.* New York: Simon & Schuster.

Dickey, M. D. (2004). The impact of web-logs (blogs) on student perceptions of isolation and alienation in a web-based distance-learning environment. *Open Learning, 19*(3), 279-291.

Donmoyer, R. (1990). Generalizability and the single-case study. In E. Eisner & A. Peshkin (Eds.), *Qualitative inquiry in education: The continuing debate* (pp. 175-200). New York: Teachers College Press.

Garrison, D. R., & Anderson, T. (2003). E-learning in the 21st century: A framework for research and practice. London: Routledge Falmer.

Gomez, L. M., Sherin, M. G., Griesdorn, J., & Finn, L.-E. (2008). Creating social relationships: The role of technology in preservice teacher preparation. *Journal of Teacher Education, 59*(2), 117-131.

Gore, J. M., & Zeichner, K. (1991). Action research and reflective teaching in preservice teacher education: A case study from the United States. *Teaching and Teacher Education, 7*(2), 119-136.

Grossman, P., Wineburg, S., & Woolworth, S. (2001). Toward a theory of teacher community. *Teachers College Record, 103*(6), 942-1012.

Kuzu, A. (2007). Views of pre-service teachers on blog use for instruction and social interaction *Turkish Online Journal of Education 8*(3), 34-51.

Lampert, M., & Ball, D. L. (1998). *Teaching, multimedia, and mathematics.* New York: Teachers College Press.

Lather, P. (2001). Validity as an incitement to discourse: Qualitative research and the crisis of legitimation. In V. Richardson (Ed.), *Handbook of research on teaching* (4th ed., pp. 241-250). Washington DC: American Educational Research Association.

Lieberman, A., & Grolnick, M. (1996). Networks and reform in American education. *Teachers College Record, 98*(1), 7-45.

Little, J. W. (2003). Inside teacher community: Representations of classroom practice. *Teachers College Record, 105*(6), 913-944.

Lock, J. V. (2006). A new image: Online communities to facilitate teacher professional development. *Journal of Technology and Teacher Education, 14*(4), 663-678.

Makri, K., & Kynigos, C. (2007). The role of blogs in studying the discourse and social practices of mathematics teachers. *Educational Technology and Society, 10*(1), 73-84.

Phillips, D. C. (1995). The good, the bad, and the ugly: The many faces of constructivism. *Educational Researcher, 24*(7), 5-12.

Richert, A. E. (1991). Case methods and teacher education: Using cases to teach teacher reflection. In B. R. Tabachnich & K. Zeichner (Eds.), *Issues and practices in inquiry-oriented teacher education* (pp. 130-150). London: Falmer Press.

Smyth, J. (1989). Developing and sustaining critical reflection in teacher education. *Journal of Teacher Education*, (pp. 2-9).

Stake, R. E. (1995). *The art of case study research.* Thousand Oaks, CA: Sage Publications.

Stiler, G., & Phileo, T. (2003). Blogging and blogspots: An alternative format for encouraging reflective practices among preservice teachers. *Education, 123*(4), 789-797.

Vygotsky, L. S. (1978). *Mind in society.* Cambridge, MA: Harvard University Press

West, R. E., Wright, G., Gabbitas, B., & Graham, C. R. (2006). Reflections from the introduction of blogs and RSS feeds into a preservice instructional technology course. *TechTrends, 50*(4), 54-60.

Chapter XXI
The Intersection of Theory, Tools and Tasks in a Postgraduate Learning Environment

Michal Zellermayer
Levinsky College of Education, Israel

Nili Mor
Levinsky College of Education, Israel

Ida Heilweil
Levinsky College of Education, Israel

ABSTRACT

This chapter describes the learning environment that the authors created for veteran teachers, graduate students in Teaching and Learning who are interested in developing professionally as designers and moderators of Information Communication Technology (ICT) learning environments. It is based on the assumption that learners construct knowledge and understanding when they are an integral part of the learning environment, when they are members of local and global communities where learning is conceived as acquiring the necessary skills for participation through participation. The program is structured as three concentric circles in terms of ICT use. In the first, the teachers learn about constructivism and socio-cultural theory while they collaboratively investigate their own school cultures with the support of ICT tools. In the second, they expand their acquaintance with ICT tools and learning environments and further develop their skills for learning and teaching in such environments. In the third, they design and moderate E-Learning environments, document them and reflect on the process. This chapter describes the activities and tasks in each of these circles with illustrations of how the students respond to these tasks both as learners who interpret and evaluate information and theoretical concepts and as teachers of relevant virtual communities who reflect on their practice.

INTRODUCTION

In *Understanding Media, The Extensions of Man*, Marshall McLuhan (1964) argued that communication technologies influence the learning process and products of their users. Based on the principle that "the medium is the message" (McLuhan, 1964), one can further argue that a learning environment is an extension of a technology. Generally invisible, the learning environment consists of the cognitive and social adjustments that its participants, as well as their society, undergo when they adopt a new form of technology. It is a "message" about teaching and learning sent by the new technology. In other words, new communication technology activates a cognitive and social transformation as well as the implementation and dissemination of more progressive ideas about learning and teaching.

This has indeed been the story of constructivism. Constructivism has been known as an epistemology and a theoretical stance on knowledge, teaching and learning since Dewey, Piaget and Vygotsky in the first half of the twentieth century. However, this theoretical stance began to transform learning environments only with the spread of computer-mediated learning. Today the majority of computer-mediated learning environments are grounded in constructivist principles of learning. These learning environments skillfully utilize the strengths of various media attributes in order to develop powerful learning strategies such as problem solving, inquiry and critical reflection and to engage learners in meaningful and motivating learning tasks. Thus, the computer is both a cognitive tool and a medium for reflection.

In the last twenty years, as communication has changed, the social processes and structures of teaching and learning have changed as well. The widespread adoption of ICT in society at large caused the network society to continually redefine the meaning of knowledge, and learning environments changed in ways that expressed the designers' assumptions about learning and teaching. This process affected school systems as well as higher education. In many traditional institutes of higher education, college administrators and teachers are realizing that computer-mediated education requires developing a new contemporary vision of learning. As a result, constructivist learning environments have been increasingly apparent in higher education in general and in teacher education and professional development in particular (Robinson & Latchem, 2003). Unmistakably, constructivism in its diverse forms is gaining considerable influence in the integration of information technology into educational contexts.

Theorizing ICT Use in Graduate Programs for Teacher Professional Development

Although ICT learning environments for teacher professional development are relatively new, there have already been three important phases in theorizing such environments (Anderson & Henderson, 2004; Pulkkinen, 2004) that, according to Koschman (1996), are guided by different models of learning. Each model replies differently to the question of the role of internet-technology in education. The first, concentrating more on the individual learner's perspective, focuses mainly on the learner's interaction with the learning environment. The purpose of learning is understanding and understanding is constructed in our interactions with the environment (Savery & Duffy, 2001). The learning environment is often conceived as a physical place where learning takes place (Wilson, 1996). The expectation is that the interaction with ICT in mindful ways will enable knowledge construction and reflection and will lead to cognitive change. Jonassen (2000), a major representative of this phase, developed the ideas of *mindtools* - computer based tools (i.e. databases, semantic networks, expert systems and cognitive simulations for student modeling and multimedia/hypermedia) and learning environments that have been adapted or developed to function as intellectual partners with the learner in order to engage and facilitate critical thinking and higher-order learning (e.g., Zellermayer, Salomon, Givon & Globerson, 1991). According to Jonassen, the roles of mindtools are to extend the learner's cognitive functioning during the learning process and to engage the learner in higher order thinking on a level that s/he would not have been able to accomplish otherwise. Jonassen, Peck and Wilson (1999, p.16), who demonstrate how technology can be utilized to support constructivist learning, argue that higher order thinking can develop when technologies engage learners in:

- Knowledge construction, not reproduction;
- Conversation, not reception;
- Articulation, not repetition;
- Collaboration, not competition; and
- Reflection, not prescription.

They further emphasize that students cannot learn from teachers or technologies but rather from critical thinking: "thinking about what they are doing or what they did, thinking about what they believe, thinking about what others have done and believe, thinking about the thinking processes they use... technologies can foster and support learning... if they are used as tools and intellectual partners that help learners to think" (p. 2).

In this first phase, the main rationale for training teachers in ICT-supported pedagogies is that they must become acquainted with constructivist pedagogies and master the level of computer literacy and cognitive skills that their students will need in order to succeed in the social and economic system of the 21st century.

The second phase is an extension of the first, with a focus on social interaction among learners. The work of Vrasidas, Zembylas and Legaspi (2004, p.83) is a typical representative of this phase:

In addition to making technology an integral part of teacher preparation, the integration of technology in schools will be facilitated by transformation of pedagogical practices from teacher-centered to student-centered approaches. For a constructivist teacher, learning is a result of construction, collaboration, reflection, and negotiation within a rich context in which learning is situated (Brown, Collins, & Duguid, 1989; Vrasidas, 2000). Technology has the potential to support constructivist learning and be used for active, authentic, reflective, and collaborative activities (Jonassen, Peck, & Wilson, 1999). Pedagogically sound uses of technology can provide an environment within which learners can take control of their learning and become active builders of knowledge while collaborating with others on solving real-world problems.

In this second phase, the focus is on training teachers in constructivist pedagogies so that they can support autonomous learners in their inquiry projects. There is a growing understanding that teachers must learn to take advantage of the social interactionist affordances of ICT learning environments, since ICT connects autonomous learners to each other. The main responsibility of the teacher is to create opportunities for collaborative learning so that they can take advantage of the technology to function as a community of inquiry.

The main theoretical model for this phase is Garrison, Anderson and Archer's (2000) analysis of the optimal ICT learning environment for higher education. In this environment, deep and meaningful learning takes place: the students develop both as independent autonomous learners and interdependent members of a community of inquiry composed of instructors and learners, the key participants in the educational process. In a community of inquiry, learning occurs through the interaction of three core components: cognitive presence, teaching presence, and social presence:

1. **Cognitive presence** is the extent to which the participants in any particular configuration of a community of inquiry are able to construct meaning through sustained communication.
2. **Social presence** is the ability of learners to project their personal characteristics into the community of inquiry, thereby presenting themselves as 'real people'.
3. **Teaching presence** is defined as the design, facilitation, and direction of cognitive and social processes for the purpose of realizing personally meaningful and educationally worthwhile learning outcomes.

There have been numerous attempts to theorize teacher learning and professional development on the basis of the Community of Inquiry model. For example, for Wells, Chang, & Maher, 1990 as cited in Matusov, Hayes and Pluta (2005, p. 1), an academic community of teacher education learners is an academic community of inquiry when "it grounds inquiries and dilemmas emerging in their practice in an academic discourse based on considering alternatives and providing argumentation and evidence for their claims." Zhou, Varnhagen, Sears, Kasprzak and Shervey's (2007) study of an online professional development program in ICT suggests that further online ICT professional development should incorporate face-to-face sessions and enroll more than one teacher from the same school to enable the participants to develop a social presence. Kreijns and Kirschner (2001) claim that teachers should become familiarized with the social affordances of an environment to trigger social interaction and to sustain engagement as a part of their own learning and as a part of their attempts to design ICT learning environments. Other designers of online communities for teachers draw on Gilly Salmon's (2000) model of CMC e-moderating. Salmon argues that web-based discussions can facilitate reflection-in-action, collaborative reflection that takes place "in the midst of practice" (2002, p. 380). In fact, in one survey of preservice teachers inspired by Salmon's claims, over 90% reported that participation in an online web discussion provided important support for them during their practicum (Khine & Lourdusamy, 2003).

Salmon's model has five stages. Each stage requires that participants acquire certain technical skills as well as e-moderating skills. The ability of participants to use CMC are essential prerequisites for conference. In stage one, the participants acquire individual access. Stage two involves individual participants' establishing their online identities and then finding others with whom to interact. In stage three, participants cooperate by providing information relevant to the course to each other. In stage four, course-related group discussions occur and the interaction becomes more collaborative. The communication depends on the establishment of common understandings. In stage five, participants look for more benefits from the system to help them achieve personal goals, explore how to integrate CMC into other forms of learning and reflect on the learning processes.

In the third phase of ICT learning environments, there is an expansion of some of the ideas emphasized in the two former phases, while the focus moves toward developing communities of practice, and studying the processes in which knowledge is produced and disseminated within these communities. This third phase draws on Vygotsky's (1978) socio-cultural theory of learning and development. Wenger (1998, p. 3), one of its leading contemporary representatives, explains that his theory comes to contradict the assumption of most educational institutions that "learning is an individual process, that it has a beginning and an end, that it is best separated from the rest of our activities, and that it is the result of teaching" Within the context of socio-cultural theory, the idea of learning in this sense is displaced. Learning becomes, fundamentally, a social phenomenon and is placed in the context of our lived experience and participation in the world. According to Wenger, the basic assumptions of socio-cultural theory are the following:

- Humans are social beings. The social is a central aspect of learning
- Knowledge is a matter of competence with respect of valued enterprises
- Knowing is a matter of participating
- Meaning is ultimately what learning is to produce

Within this theoretical framework, "the 'environment for learning' is a learning community rather than a place or determinant. Learning is an activity that takes place in interaction and negotiation with one's social environment or in enculturation or adoption of a group's ways of seeing, acting and using tools and artifacts" (Pulkkinen, 2004, p. 67). When the main metaphor for learning is participation, practical applications of ICT integration could mean digital tools and resources to interact within collaborative working environments. Accordingly, the important factors that support teacher professional development of web-based learning in the community are: the possibility to learn from colleagues and from everyday working practices in an emotionally safe atmosphere that creates possibilities for learners to reach their potential.

Wenger (2006, p. 5) explains that the perspective of communities of practice affects educational practices along three dimensions:

- *Internally*: How to organize educational experiences that ground school learning in practice through participation in communities around subject matters or a theoretical framework
- *Externally*: How to connect the experience of students to actual practice through peripheral forms of participation in broader communities beyond the walls of the school or the university
- *Over the lifetime of students*: How to serve the lifelong learning needs of students by organizing communities of practice focused on topics of continuing interest to students beyond the initial schooling period

This third phase in theorizing learning environments (Barab & Duffy, 2000) and teacher professional development is in concert with two important developments in teacher professional learning. In the first, studies of teachers' learning of ICT pedagogy adopt the model of the adult learner whose main responsibility is to prepare pupils for collaborative, life-long, work-based learning (see review by McKenzie, 2001). These studies show that veteran teachers are more satisfied with professional development that they themselves have planned and developed, that actively engages them in their learning, that is clearly and effectively organized, allows time for reflection and discussion, and has a balance between theory and practice.

In parallel, there have been efforts (Lieberman & Miller, 1999) to construct networks of school-university partnerships that support school-based inquiry-oriented professional development for teachers. The success of these partnerships prompted many schools of education to shape their graduate programs in the form of communities of practice in which teachers collaboratively investigate their own school cultures with the support of ICT learning environments.

In what follows, we will describe and analyze the structure of the graduate program for teachers that we developed in terms of ICT use, the activities and tasks developed for the graduate students in the program, and will present analyzed illustrations of how the students respond to these tasks both as learners who interpret and evaluate information and theoretical concepts and as teachers of relevant virtual communities who reflect on their practice.

A GRADUATE PROGRAM IN TEACHING AND LEARNING

This section focuses on a Master of Education program in Teaching and Learning, emphasizing the design and analysis of school-based learning environments. The online courses in this MEd program targeted education practitioners with three years and more teaching experience and are operated in a complex, e-learning, technology-rich space designed specifically for each course. The program websites include both information and communication tools that provide learners with the necessary support for active participation. The information tools include a detailed rationale written by the instructor, the course outline, a list of students' responsibilities and the assessment scheme. In different sections, there is a list of learning topics and clear and detailed instructions for performing the tasks designed for each of them. In addition, each website provides a limited list of online resources, links to articles and Internet sites relevant to the learning topics (to be expanded by students). The communication tools include discussion forums in which the students pose questions, post their tasks and peer responses, and in which the instructor makes general comments that relate to the whole class. The "Participants" section includes the students' and instructor's email addresses, for one-to-one or group communication.

The use of ICT in the program is enacted in two modes: ICT Supplemented (*The face-to-face component is the main component*) or ICT Dependent (*participation is a compulsory requirement although some face-to-face activity is retained*). In the ICT dependent courses, synchronous meetings (1-2 per course) are held in a special space (through InterWise) and are used mainly when a guest lecturer (either from the country or abroad) gives a lecture. Learners are asked to keep a personal journal (weblog) on their impressions of their learning process and their reflections on the relevance of the tasks. Wiki technology provides the learners with a space for collaborative writing and assessment through constant dialogue and reflection. Wiki is both an information and a communication tool. E-mail correspondence enables communication

Figure 1. Course website

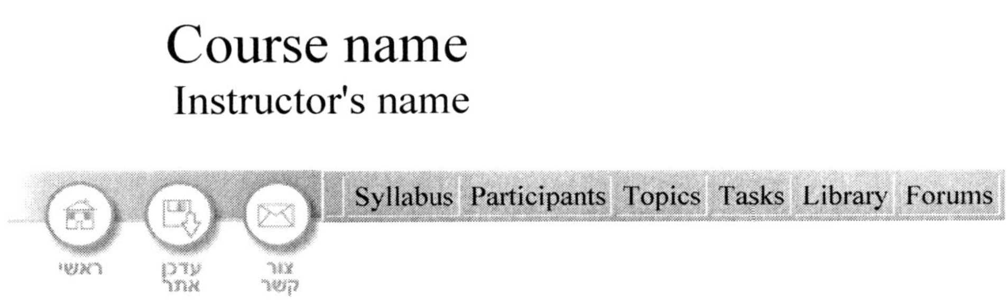

according to the user's needs (personal and/or group communication).

In both modes, the following communication principles are upheld:

1. Student interaction with faculty and other students is essential and is facilitated through a variety of means,
2. The focus is on the learning process rather than products,
3. Students are encouraged to engage in web-based research and learn to assess the validity of web-based resources.

The program is organized in three concentric circles:

The External Circle: The Constructivist Theoretical Framework

This includes courses and seminars on Teaching and Learning that provide the theoretical platform for the analysis of learning environments. Many of these courses (i.e., Constructivism, Vygotsky and Education, Literacy as a Dialogue, Learning in Communities) rely on ICT as a complementary technology for participation. The web-supported courses aim to build a theoretical understanding of the basic premise of the program, namely that knowledge and higher order thinking develop in the context of social activities involving interaction with tools. In these courses, PowerPoint presentations, e-mail, search engines, listservs and/or threaded discussion groups assume a central role in the ongoing collaborative learning processes that take place among the students and between the students and their teachers. The students are asked to provide school-based evidence for the theoretical concepts and models that they are studying and to negotiate the meaning of this evidence using these tools. These academic tasks actually render the contribution of the tools to higher order thinking visible to them.

The choice of Socio-cultural Activity Theory as the theoretical platform for the program is based on its power to explain children's de-

Figure 2. Theory, technology and tasks in the program

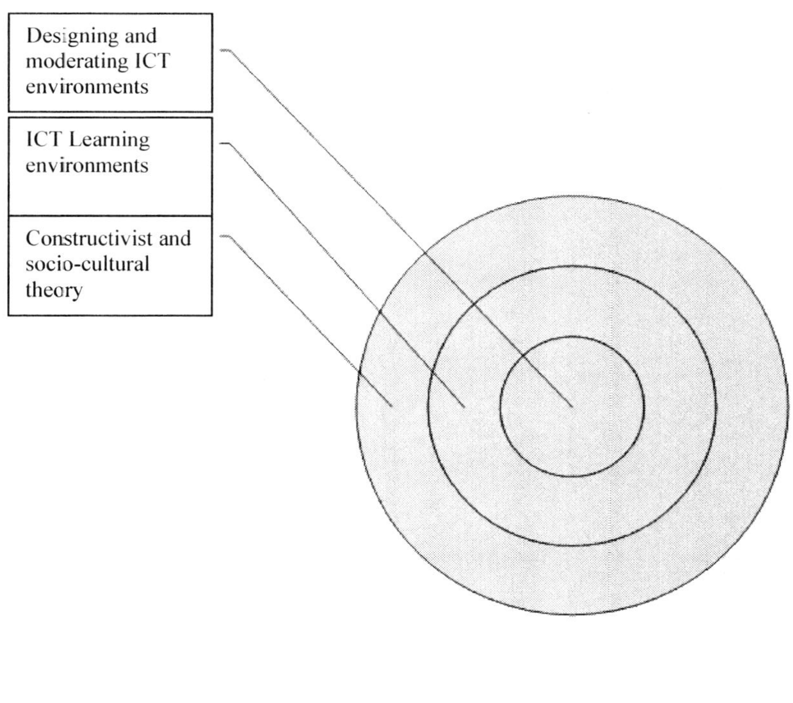

velopment and its implications for teaching as well as the importance of interaction with tools for development and the centrality of language as the most important psychological tool. The studies that they read on children's development become the students' lens for inquiring into their own teaching as well as their own professional learning. Specifically, this theory provides them with the language and ideas that enable them to investigate these issues. For example, they think about their own professional development within a zone of proximal development, which they can often identify, and the significance of tools in mediating their own learning process. Finally, the students' growing understanding of socio-cultural theory opens opportunities for them for peripheral participation in the broader scientific community that does research and writes on learning environments beyond the boundaries of our program (Bonk & Cunningham, 1998; Barab & Duffy, 2000). Through their virtual interactions with texts produced by the members of this community, they seek ways to name their own experience in teaching and learning.

From the socio-cultural point of view, the technical environments in the program can be seen as part of the social environment: "ICT applications are social artifacts that have a specific function and meaning in education processes, and are usually expressions of the designers' assumption about teaching and learning" (Pulkkinen, 2004, p. 69). The technologically supported theoretical platform provides a common professional language that gives the students membership in a task-centered professional learning community in which they perform as adult learners. They choose the topics for their assignments and the context and focus of their inquiry as well as the persons who become their critical friends. As critical friends, they participate in the reciprocal evaluation of the assignment drafts by suggesting ways to improve each other's work. This is done in the electronic forum on each course website. While working, they benefit from the guidance of their instructors, the advice of their peers as well as free access to the latest research-based knowledge about managing school-based online learning communities. Through their engagement in the tasks, in their relationships and in theory, they appreciate the importance of activity-based learning in community that is scaffolded by technology.

The Middle Circle: Focusing on ICT Learning Environments

This is a cluster of on-line courses, in which technology is perceived by the students as a goal as well as a means for professional development, and practice itself becomes the curriculum. This is particularly evident in *ICT environments for school teaching and learning,* a five-unit course designed to enhance the students' skills in using ICT technology for learning and for teaching that leads to teacher learning about on-line learning. Each unit focuses on a specific learning activity and reflection on that activity, introduces students to a particular example of such environments (e.g., Computer Supported Collaborative Learning) and a theoretical concept related to designing technology-rich learning environments (e.g., synchronous and asynchronous communication tools). The students are invited to critique these environments or discuss the relevant theoretical concepts via the forum. They are then asked to reflect on what they can take from these environments for their own classrooms.

In this course, the students interact with a wide range of digital tools. They experience learning on an e-learning platform adapted for the course (Highlearn by Britannica), engage in individual writing using Office applications (PowerPoint and Word) as well as collaborative writing on the Wiki open source, and participate in synchronous and asynchronous communication through e-mail, discussion groups, chat rooms, group work, polls and WebQuests. The digital tools are gradually introduced in a functional mode: the forum and the e-mail serve as authentic communication tools among the students and between them and the instructor; the personal blog enhances reflective thinking and writing, and peer assessment with the teacher; the online resource library of exemplars of rich technological learning environments is constantly updated and available on the net; the technology of the online course platform (Highlearn by Britannica) enables sharing teaching materials with colleagues as well as communicating with students; while the search engines enable browsing among various sources of information and finding the most suitable ones.

The above activities are accompanied by functional reading and writing. For example, the students read the taxonomy for the evaluation educational web-sites (Nachmias, Mioduser, Oren & Lahav, 1999) for the purpose of evaluating educational sites that they discover in the online resource library that are relevant to the discipline that they teach and the age of their pupils. They collaborate in writing summaries of the course readings using the Wiki platform.

The course site is constantly updated according to students' needs and interests. For example, forums are opened (and closed) whenever needed, either for small groups of students, for the whole class, or for their pupils. Students are invited to reflect upon what was new, what was already known, what they liked (or did not like) in this learning mode, what was difficult and how the difficulties were overcome. While reflecting upon the possibilities opened through the course for individual and group learning, and while evaluating the use of the digital tools and the relevance and authenticity of the materials presented, knowledge constructed and thinking skills practiced, the ICT learning environment becomes more visible to the students and they begin to consider it as a possible model for their own classroom teaching.

The Inner Circle: Designing and Moderating School-Based E-Forums

Many students choose to appropriate the pedagogical model that they encounter in the program for their school-based teaching and their tutoring of other teachers. They design and/or moderate school-based forums that support their work in the schools. In parallel, they also participate in Managing Virtual Learning Communities, an advanced theory-driven and practice-based course conducted in a blended online and face-to-face environment that supports this activity.

The primary goal of the course is to enable students to construct their theoretical and practical knowledge about virtual learning communities from the point of view of participants as well as designers and moderators. A secondary goal is to provide the students with a model for e-moderating using a variety of digital tools (Wiki, blog, forum, e-mail, the course platform) in an authentic and relevant way. The students master these web-based tools for communication and collaboration as they inquire into various theoretical concepts that serve as the rationale for their learning activities, such as *Mindtools, Mediation, Distributed Learning, Communities of Learning and Communities of Practice.*

The course is designed according to Salmon's (2000) five stages of e-moderating. Each unit corresponds to a learning stage:

Stage 1: Access and Awareness

In the first phase of the course (two weeks), the students gain access to the system and use it to get to get to know each other as future tutors of online environments. Their first task helps them to do so: they use the Wiki platform to make a collective list of metaphors for online instructors that fit their experiences, and to discuss their choices. The students tend to compare instructors in online environments to navigators and discoverers (i.e., a pilot, an air controller, a ship captain, a diving or parachuting master, an astronaut, a construction engineer, a tour guide) or to navigating tools (a compass, GPS, a street light, a steering wheel). Some emphasize the caring qualities needed for teaching in an online environment and compare online instructors to lions nurturing their cubs or a tree with many branches, or even a spider and his web. They are subsequently asked to compare their own list with Bonk's (2001) Metaphors of Online Instruction that includes many of their own as well as an additional section of help providers, coaches and negotiators. They are then invited to deliberate on the differences between their own list and Bonk's. Via this activity, students learn about their peers in the course and their personal and professional preferences. They use Wiki, and develop the necessary skills for learning through this technology. When the teachers investigate their own figurative speech, they become aware of their own conception of knowledge, learning and instruction, and of the role of ICT in these processes. Finally, as critical users of ICT, they become more aware of different ways that moderators control new learning environments, and of signs of the abuse of that control.

Stage 2: Online Socialization

Stage 2 (two weeks): Students visit open online communities on the web and join as temporary participants. They document, analyze and describe their culture in the course forum. In order to do so, they inquire collaboratively into some related theoretical concepts, such as "communities of inquiry" and "communities of practice." Occasionally, they request and receive advice online.

Stage 3: Information Exchange

In Stage 3 (one week): Participants analyze an article describing the relationship between the moderator and participants of forums, which deals with the question: "Who controls online communities?" (Barzilai-Nahon, 2004). They present examples from the forums they visited that support or refute the claims of the article. This helps them envision their forthcoming task: to develop and moderate authentic learning environments for their own pupils.

Stage 4: Knowledge Construction

This is a more extensive (seven-week) double-experience period of simultaneously moderating a learning environment for schoolchildren or colleagues (usually a forum or a Wiki environment), and participating as students in the course forum. In this phase of the course, the participants begin to interact with each other in more participative ways: sharing the difficulties that they encounter as moderators, and offering each other help. For example, one student (R.), who had recently transferred to a new school where she was also a technology consultant for the teachers, felt that building a teachers' online community of practice was a good idea, but wondered if it was not too soon for them as beginners and for her in her new role:

R.: I am considering moderating a forum with my colleagues; we have a lot in common, we spend a lot of time together preparing professional tasks; we can save time, be more at home and be more efficient if we work online. But I'm a new teacher in school and a new tutor of k-3 teachers. Isn't it too innovative? Isn't it too dangerous to begin a new job in a new school and lead the team to a new learning environment?
S.: I'm also a tutor for K-3 teachers and I plan to introduce the forum and work with my team there. Although I'm not new to my team, your situation seems simpler than mine. You start with online communication from the onset.... Let's work as a team and plan our moderating together.

This example shows how risky moderating an on-line environment can be for our students and how important the peer support is.

Stage 5: Development

In Stage 5 (three final weeks), "participants become responsible for their own learning and need little additional support.... Participants start to challenge the system and the conference.... Participants and moderator are using a constructivist approach to learning" (Salmon, 2000, p. 35-36). This is a period of individual and mutual reflection, enhanced by the final task of the course. The students are required to describe the learning environments that they manage, focusing on the process from the point of view of the moderator, emphasizing difficulties, decision checkpoints, alternatives and roads not taken, feelings and thoughts, practical and principal lessons learned.

Their reflective writing revealed that some important lessons were learned, either on a personal or on a more general level. Some participants focused on personal flaws in their own conduct as e-moderators, such as responding too rapidly. This happened to G., who wrote:

The turning point occurred when I realized that my immediate responses to each pupil in the forum cause harm; not only was it very hard and unnecessary work for me, but also a destructive factor in the forum; students didn't have the chance to respond to their peers and so no dialogues developed.... But my "on the job" learning caused a change in the conduct of the forum....

In his reflection, L. discovered that he imposed his agenda on the participants. He wrote:

After a painful experience, I learned that interaction in a community of practice (of adults) should be constructed with the participants according to their needs and authentic fields of interest, and not imposed by the moderator.

Two other participants compared their experience as participants in the course forum and their experiences as moderators. M. wrote:

I learned a lot from your teaching presence.... Especially from the opportunity we were given to work differently, in an environment that meets our authentic special needs ...

D. said:

The parallel work as a participant in the learning community of the course and as a moderator of my pupils' forum gave me insights. Experiencing the mild interactions in our community in the beginning (in contrast to the heavy traffic in the forum in the advanced stages) enabled me to diagnose, understand, accept, and deal with my pupils' minimal interactions in the beginning. The real-time parallel processes enhanced my understanding of the importance of time and task in online learning.... This local experience had a permanent influence on my school, my colleagues, the school principal and the school community: virtual learning and communities of practice became an integral part of learning, although still in a minimal phase.

Organizing the course according to Salmon's model enables students to develop toward autonomous e-moderating. The gradual change they undergo while facing challenging tasks accompanied by individual and collective reflection and peer and instructor support, is reflected in their knowledge construction, technical skills and their better understanding of constructivist pedagogy and constructivist learning environments.

DISCUSSION

This chapter describes the learning environment that we created for veteran teachers, graduate students in Teaching and Learning, who are interested in developing professionally as designers and moderators of ICT learning environments. The courses in the program are organized in three concentric circles in terms of the students' online learning about online learning (Figure 2). The three circles are based on the assumption that learners construct knowledge and understanding when they are an integral part of the learning environment, when they are members of local and global communities where learning is conceived as acquiring the necessary skills for participation through participation.

An additional important principle of our program is the dynamics between theory, technology and tasks. The tasks enable the constant shifting between theory and praxis. They require the use of theory to interrogate praxis and the use of praxis to understand theory. Although the dynamics are different in each of the concentric circles, the interaction between theory, tools and tasks prompts higher order thinking. In the external circle, the students begin with theory. They deliberate on the theoretical concepts and discover the relevance of these concepts to their school-based practice. They construct a case from their practice that can provide evidence for the theory. In the inner circle, they begin with action and then reflect and deliberate on the problematic situations that they encounter with the aid of certain theoretical concepts. On the whole, the practical work-based tasks demand a varying capacity in the use of technological tools that suit these teachers' professional activities and are always contextualized in the wider constructivist or socio-cultural theoretical framework of human development as well as in the more local theory of learning in ICT environments. Students are particularly enriched by socio-cultural theory that views technology as a set of cultural tools, which have evolved from a long history of human activity and social interaction, in which language is and will always be the most important tool. Throughout the program, they engage in a large number of activities that demonstrate time and again that the main function of tools is to enable a reflective process that affords the development of higher order thinking that is in turn reified in language and in the ability to design and create more complex tools. In fact, all of the courses described above set aside time for deliberation and reflection, as well as observation and discussion of alternative practice and feedback. All the tasks demand inquiry and the collection of evidence from the teachers' classrooms, from websites and from the students' online learning communities that illustrate how technology functions as a set of social/cultural tools.

This is only the beginning of our work on conceptualizing learning environments for teacher professional development. Our understanding is that over the last two decades, the significant developments in theorizing ICT learning environments moved from behaviorism to constructivism, to social constructivism and subsequently to socio-cultural communities of practice. These developments mainly emphasized the cognitive, communicative and collegial potentials of ICT learning environments. As textbooks become obsolete and the emphasis on acquiring factual information is reduced, online resources gain greater importance. In our program, reading lists are only partly supplied by the instructor; most are discovered by students and are relevant to the specific contexts of their professional lives. We believe that the next phase will highlight the creative potential of digital technology. Technological developments in sound and video will enable students to use a range of senses for representing their experiences while they access resources that they discover and participate in new networks that offer learning opportunities far beyond those offered by the Web in its current form.

In line with Weiland's (2008, p. 1219) conceptualization of "the new liberal education," we anticipate that in the future, our students will increase their use of online resources and will encourage their pupils to visit virtual learning communities outside their classrooms or course websites. Our hope is that when our students venture into a borderless learning environment, they will become more familiar with what Weiland calls "the different meanings of civic life" and learn about their responsibilities as teachers to help civic life continue and thrive. Weiland (2008, p. 1219) explains that

In the same way that liberal education has always included attention to the meanings of civic life, of living and working in groups, knowledgeable and thoughtful use of ICT requires understanding of how the phrase "virtual community" signifies new forms of electronic communication and of how people live together.

Our understanding is that Weiland's ideal of liberal education is an expansion of the concept of Communities of Practice. It signifies a shift in the focus of the design of learning environments to the creation of new forms of participation, new forms of engagement, and new forms of knowledge. This ideal takes into account the potential of digital technologies to enable new connections among resources and various funds of knowledge that open up possibilities for creating new representations of our lived experience.

REFERENCES

Anderson, N., & Henderson, M. (2004). E-PD: Blended models of sustaining teacher professional development in digital literacies. *E–Learning, 1, 3*. Retrieved May 29, 2008 from http://www.wwwords.co.uk/pdf/viewpdf.asp?j=elea&vol=1&issue=3&year=2004&article=4_Anderson_ELEA_1_3_web&id=89.139.44.175

Barab, S., & Duffy, T. M. (2000). From practice field to communities of practice. In D. Jonassen & S. M. Land (Eds.), *Theoretical foundations of learning environments* (pp. 25-56). Mahweh, NJ: LEA.

Barzilai-Nahon, K. (2004). *Gatekeepers and gatekeeping mechanisms in networks*. Unpublished dissertation, Management of Technology and Information Systems Program, Tel-Aviv University. Retrieved August 15, 2008 from http://www.itu.org.il/Index.asp?ArticleID=2170&CategoryID=579& Page=2

Bonk, C. J. (2001). *The Vision Thing: 40 metaphors of online instruction*. Retrieved April 29, 2008 from http://www.trainingshare.com/download/metaphors.doc

Bonk, C. J., & Cunningham, D. J. (1998). Chapter 2: Searching for learner-centered, constructivist, and socio-cultural components of collaborative educational learning tools. In C. J. Bonk & K. S. King (Eds.), *Electronic collaborators: Learner-centered technologies for literacy, apprenticeship, and discourse* (pp. 25-50). Mahwah, NJ: Erlbaum.

Delfino, M., & Stefania, M. (2007). The expression of social presence through the use of figurative language in a web-based learning environment, *Computers in Human Behavior, 23*, 5, 2190-2211. Retrieved May 29, 2008 from http://portal.acm.org/citation.cfm?id=1234416. 1234613

Garrison, R., Anderson, T., & Archer, W. (2000). Critical inquiry in a text-based environment: Computer conferencing in higher education. *The Internet and Higher Education, 2*(2-3), 87-105.

Jonassen, D. H., Peck, K. L., & Wilson, B. G. (1999). *Learning with technology: A constructivist perspective*. Upper Saddle River, NJ: Prentice-Hall.

Jonassen, D. H. (2000). *Computers as mindtools for schools: Engaging critical thinking* (2nd ed.). Upper Saddle River, NJ: Prentice-Hall.

Khine, M. S., & Lourdusamy, A. (2003). Blended learning approach in teacher education: combining face-to-face instruction, multimedia viewing and online discussion. *British Journal of Educational Technology, 34*(5), 671-675.

Koschmann, T. (1996) Paradigm shifts and instructional technology: An introduction. In T. Koschmann (Ed.) *CSCL: Theory and practice of an emerging paradigm* (pp. 1-22). Mahwah; NJ: Lawrence Erlbaum Associates, Publishers.

Kreijns, K., & Kirschner, P. A. (2002). The social affordances of computer-supported collaborative learning environments. In *Proceedings of the 31th ASEE/IEEE Frontiers in Education Conference, 1*(T1F, pp. 12-17) October 10-13, 2002, Reno, NV, USA.

Lieberman, A., & Miller, L. (1999). *Teachers – Transforming their world and their work*. New York: Teachers College Press.

Matusov, E., Hayes, R., & Pluta, M. J. (2005). Using discussion webs to develop an academic community of learners. *Educational Technology & Society, 8*(2), 16-39. Retrieved May 29, 2008 from http://www.ifets.info/journals/8_2/ets_8_2.pdf#page=21

McKenzie, J. (2001). How teachers learn technology best. *The Educational Technology Journal, 10***(6)** [http://www.fno.org/mar01/howlearn.html]

McLuhan, M. (1964). *Understanding media: The extensions of man*. Cambridge, MA: MIT Press.

Nachmias, R., Mioduser, D., Oren, A., & Lahav, A. (1999). Taxonomy of educational websites - A tool for supporting research, development and implementation of web-based learning. *International Journal of Educational Telecommunications, 5*(3), 193-210. Retrieved on 29.3.08 from http://muse.tau.ac.il/publications/taxonomy.html

Pulkkinen, J. (2004). The paradigms of e-Education: An analysis of the communication structures in the research on information and communication technology integration in education in the years 2000–2001. Unpublished doctoral dissertation, Faculty of Education, University of Oulu, Finland. Retrieved May 29, 2008 from http://herkules.oulu.fi/isbn9514272463/

Robinson, B., & Latchem, C. (2003). Teacher education: challenges and change. In B. Robinson, & C. Latchem (Eds.), *Teacher education through open and distance learning* (pp. 1-27). London: RoutledgeFalmer.

Salmon, G. (2000 & 2004). *E-moderating: The key to teaching and learning online*. London: Taylor and Francis.

Salmon, G. (2002). Mirror, mirror, on my screen.. Exploring online reflections. *British Journal of Educational Technology, 33*(4), 379-391.

Savery, J. R. & Duffy, T. M. (2001). *Problem based learning: An instructional model and its constructivist framework*. Indiana University: CRLT Technical Report No. 16-01. Retrived May 29, 2008 from http://cee.indiana.edu/publications/journals/TR16-01.pdf

Vrasidas, C., Zembylas, M. and Legaspi, J. M. B. (2004). EmpowerICT: eLearning for in-service teacher education and support, *International Journal of the Computer, the Internet and Management, 12,* 2, 81-86. Retrieved May 26.2008 from http://www.ijcim.th.org/v12n2/pdf/p81-86-vrasidas-EmpowerICT.pdf

Vygotsky, L. S. (1978). *Mind in society: The development of higher psychological processes.* Cambridge MA: Harvard University Press.

Weiland, S. (2008). Teacher education toward liberal education. In M. Cochran-Smith, S. Feiman-Nemser & D. McIntyre (Eds.), *Handbook of research on teacher education: Enduring questions and changing contexts* (3rd ed.) (pp. 1204-1227). New York: Routledge.

Wells, G., Chang, G. L. M., & Maher, A. (1990). Creating classroom communities of literate thinkers. In S. Sharan (Ed.), *Cooperative learning: Theory and research.* (pp. 95-121), New York, NY: Praeger Publishers.

Wenger, E. (1998). *Communities of practice, learning, meaning, and identity*. Cambridge, MA: Cambridge University Press.

Wenger, E. (2006). Communities of practice, a brief introduction. Retrieved October 2, 2008 from http://www.ewenger.com/theory/communities_of_practice_intro_WRD.doc

Wilson, B. G. (1996). What is a constructivist learning environment? In B. G. Wilson (Ed.), *Constructivist learning environments: Case studies in instructional design* (pp. 3–8). Englewood Cliffs NJ: Educational Technology Publications

Zellermayer, M., Salomon, G., Globerson, T. & Givon, H. (1991). Enhancing writing-related metacognitions through a computerized writing partner. *American Educational Research Journal, 28*(2), 372-391.

Zhou, G., Varnhagen, S., Sears, M. R. Kasprzak, S., & Shervey, G. (2007). Online professional development for inservice teachers in Information and Communication Technology: Potentials and challenges, *Canadian Journal of Learning and Technology, 33*(2). Retrieved August 15, 2008 from http://www.cjlt.ca/content/vol33.2/zhou.html

Compilation of References

Abrahamson, D., diSessa, A. A., Blikstein, P., Wilensky, U., Uttal, D. H., Amaya, M. M., Mariulis, L. M., & Collins, A. M. (n.d.). *What's a situation in situated cognition? – A constructivist critique of authentic inquiry* (pp. 1015-1021).

Acevedo, D., Vote, E., Laidlaw, D. H., & Joukowsky, M. S. (2001, Oct.). *IEEE Visualization 2001* (p. 493).

Ackermann, E. (1990). *Piaget's constructivism, Papert's Constructionism: What's the difference?* Retrieved April 15, 2008, from: http://learning.media.mit.edu/

Adams, J. (2007). *Alice in action with Java*. Boston, MA: Thomson Course Technology.

Addison, C. (2006). A new chapter in web development? *Development in Practice, 16*(6), 623-627.

Akkerman, S., Van den Bossche, P., Admiraal, W., Gijselaers, W., Segers, M., Simons, R. J., & Kirschner, P. (in press). considering group cognition: From conceptual confusion to a bouRendary area between cognitive and socio-cultural perspectives? *Educational Research Review*.

Akyol, Z., & Garrison, D. R. (in press). The development of a community of inquiry over time in an online course: Understanding the progression and integration of social, cognitive and teaching presence. *Journal of Asynchronous Learning Networks*.

Alamargot, D., & Andriessen, J. (2002). The "power" of text production activity in collaborative modelling: Nine recommendations to make a computer supported situation work. In M. Baker, P. Brna, K. Stenning, & A. Tiberghien (Eds.), *The role of communication in learning to model* (pp. 275-302). Mahwah, NJ: Lawrence Erlbaum Associates.

Albon, R. (2003). *Assessment drives the learning: Raising the bar*. Paper presented at the Evaluation and Assessment Conference, Adelaide, Australia, Nov. 2003.

Alexander, B. (2006). A new wave of innovation for teaching and learning. *Educause* (March/April) 33-44. Retrieved April 5, 2008 from http://www.educause.edu/ir/library/pdf/ERM0621.pdf

Alexander, B. (2008). Web 2.0 and emergent multiliteracies. *Theory Into Practice, 47*(2), 150-160.

Alexander, B. (March/April 2006). , Web 2.0: A new wave of innovation for teaching and learning? http://net.educause.edu/ir/library/pdf/ERM0621.pdf. *EDUCAUSE Review, 41*(2), 32–44.

Al-Khalifa, H. S., & Davis, H.C. (2007). Replacing the monolithic LOM: A folksonomic approach. *Seventh IEEE International Conference on Advanced Learning Technologies (ICALT 2007).*

Allen, I. E., & Seaman, J. (2007). *Online Nation: Five Years of Growth in Online Learning*. The Sloan Consortium. Retrieved June 8, 2008 from http://www.sloan-c.org/publications/survey/index.asp.

Allesi, S., & Trollip, S. (2001). *Multimedia for Learning: Methods and Development*. Massachusetts: Ally & Bacon.

Ally, M. (2004). The Theory and Practice of Online Learning. In T. Anderson & F. Elloumi (Eds.), *Foun-*

dations of Educational Theory. Athabasca, Canada: Athabasca University.

Alvarez, S. (2005). Blended learning solutions. In B. Hoffman (Ed.), *Encyclopedia of Educational Technology*. Retrieved January 10, 2008, from http: //coe.sdsu.edu/eet/articles/blendedlearning/start.htm

Anderson, N., & Henderson, M. (2004). E-PD: Blended models of sustaining teacher professional development in digital literacies. *E–Learning, 1, 3*. Retrieved May 29, 2008 from http://www.wwwords.co.uk/pdf/viewpdf.asp?j=elea&vol=1&issue=3&year=2004&article=4_Anderson_ELEA_1_3_web&id=89.139.44.175

Anderson, T. (2004). Toward a theory of online learning. In T. Anderson & F. Elloumi (Eds.), *Theory and practice of online learning* (pp. 33-60). Athabasca, AB: Athabasca University.

Anderson, T. (2008). Towards a Theory of Online Learning. In T. Anderson (Ed.), *The Theory and Practice of Online Learning* (2nd Ed., ch. 2). Edmonton, Alberta: AU Press. Retrieved June 9, 2008 from http://www.aupress.ca/books/Terry_Anderson/TerryAndersonEntireBook.pdf .

Anderson, T. (2008). Social Software to Support Distance Education Learners. In T. Anderson (Ed.), *The Theory and Practice of Online Learning* (2nd Ed., ch. 9). Edmonton, Alberta: AU Press. Retrieved June 9, 2008 from http://www.aupress.ca/books/Terry_Anderson/TerryAndersonEntireBook.pdf .

Anderson, T. (Ed) (2008). *The Theory and Practice of Online Learning, Second Edition*. Edmonton, Alberta: AU Press. Retrieved June 9, 2008 from http://www.aupress.ca/books/Terry_Anderson/TerryAndersonEntireBook.pdf.

Anderson, T., Rourke, L., Garrison, D. R., & Archer, W. (2001). Assessing teaching presence in a computer conferencing context. *Journal of Asynchronous Learning Networks, 5*(2), 1-17.

Andersson, P., & Harris, J. (2006). *Retheorising the recognition of prior learning*. Leicester, UK: NIACE.

Andriessen, J. (2006). Arguing to learn. In K. Sawyer (Ed.), *Handbook of the learning sciences* (pp. 443-459). Cambridge, MA: Cambridge University Press.

Andriessen, J., & Sandberg, J. (1999). Where is education heading and how about AI. *International Journal of Artificial Intelligence in Education, 10,* 130-150.

Andriessen, J., Baker, M., & Suthers, D. (2003). Argumentation, computer support, and the educational context of confronting cognitions. In J. Andriessen, M. Baker, & D. Suthers (Eds.), *Arguing to learn: Confronting cognitions in Computer-Supported Collaborative Learning environments* (pp. 1-25). Dordrecht: Kluwer Academic Publishers.

Ansell, B. W. (2008). Traders, Teachers and Tyrants; Democracy, Globalization and Public Investment in Education. *International Organization, 62*(2), 289–322.

Applefield, J., Huber, R., & Moallem, M. (2001). Constructivism in Theory and Practice: Towards a Better Understanding. *High School Journal, 84*, 35-53.

Aragon, C. R., & Hearst, M. A. (2005, Apr.). Improving aviation safety with information visualization: A flight simulation study. *CHI 2005* (pp. 441-450). ACM.

Arbaugh, J. B. (2007). An empirical verification of the community of inquiry framework. *Journal of Asynchronous Learning Networks, 11*(1). Retrieved July 27, 2008, from Sloan-C-Publications database http://www.sloan-c.org/publications/jaln/index.asp

Arbaugh, J. B., & Hwang, A. (2006). Does "teaching presence" exist in online MBA courses? *The Internet and Higher Education, 9*(1), 9-21.

Argyris, C. (1993). *Knowledge for action: A guide to overcoming barriers to organizational change.* San Francisco: Jossey-Bass Publishers. 20.

Arnett, J. J. (2000). Emerging adulthood: A theory of development for the late teens through the twenties. *American Psychologist, 55*(5), 469-480.

Arnold, N., & Ducate, L. (2006). Future foreign language teachers' social and cognitive collaboration in an online environment. *Language, Learning & Technology, 10*(1), 42–66. Retrieved January 14, 2006 from http://llt.msu.edu/vol10num1/pdf/arnoldducate.pdf

Aronowitz, S. (2000). *The knowledge factory.* Boston: Beacon Press.

Aronson, E., & Patnoe, S. (1997). *The jigsaw classroom: Building cooperation in the classroom (2nd ed.).* New York: Addison Wesley Longman.

Arter, J., & McTighe, J. (2001). *Scoring rubrics in the classroom.* Thousand Oaks, CA: Corwin Press.

Ashburn, E. (2008, May 2). After the deluge, the drought? How admissions offices are preparing for a shrinking pool of traditional applicants. *Chronicle of Higher Education* (pp. B5-6).

Audi, R. (2002). *Epistemology: A Contemporary Introduction to the Theory of Knowledge.* Routledge.

Bach, S., Haynes, P., & Smith, J. L. (2007). *Online Learning and Teaching in Higher Education.* Maidenhead: Open University Press.

Bain, K. (2004). *What the Best College Teachers Do.* Cambridge, MA: Harvard University Press.

Bain, K. (2004). *What the best college teachers do.* Cambridge, MA: Harvard University Press.

Baker, M., Hansen, T., Joiner, R., & Traum, D. (1999). The role of grounding in collaborative learning tasks. In P. Dillenbourgh (Ed.), *Collaborative learning: Cognitive and computational approaches* (pp. 31-63). Amsterdam: Pergamon/Elsevier Sience.

Bakhtin, M. M. (1930). *The Dialogic Imagination: Four Essays.* C. Emerson & M. Holquist (Eds.) (Trans.), (1981). Austin, TX and London, UK: University of Texas Press.

Ball, D. L., & Cohen, D. K. (1999). Developing practice, developing practitioners: Toward a practice-based theory of professional education. In L. Darling-Hammond & G. Sykes (Eds.), *Teaching as the learning profession: Handbook of policy and practice* (pp. 3-32). San Francisco: Jossey-Bass.

Ball, E. & Lai, A. (2004). What's special about online education. Retrieved Sept. 20, 2008, from: http://empire2.esc.edu/facultyweb/ericball.nsf/3c6bed5f67edce6185256889006cd8e8/c6faea6fc298b883852573b400548a55?OpenDocument

Barab, S. A., & Duffy, T. (2000). From practice fields to communities of practice. In D. Jonassen & S. M. Land (Eds.), *Theoretical Foundations of Learning Environments* (pp. 25-56). Berkley, CA: McCutchan Publishing.

Barab, S., MaKinister, J., Moore, J., & Cunningham, D. (2001) Designing and building an online community: The struggle to support sociability in the inquiry learning forum. *Educational Technology Research and Development, 49*(4), 71-96.

Barr, R. B., & Tagg, J. (1995). From teaching to learning: A new paradigm for undergraduate education. *Change,* 27(6), 12-25.

Barrett, H. C. (2006). *Authentic Assessment with Electronic Portfolios using Common Software and Web 2.0 Tools.* Retrieved January 31, 2008 from http://electronicportfolios.org/web20.html.

Barrett, H. C. (2008). *The REFLECT Initiative: Researching Electronic portFolios: Learning, Engagement and Collaboration through Technology with a focus on high school students in the Arizona Education Professions program.* Paper presented at the American Educational Research Association Annual Meeting, New York, NY.

Barrett, H.C. (2008). *Using Technology to Support Alternative Assessment and Electronic Portfolios.* Retrieved June 9, 2008 from http://electronicportfolios.org/portfolios.html

Barrett, T. (2008). *Students Talk About Problem-based Learning in PBL Tutorials: Illuminative Concepts.* Unpublished doctoral dissertation, Coventry University, UK.

Barzilai-Nahon, K. (2004). *Gatekeepers and gatekeeping mechanisms in networks.* Unpublished dissertation, Management of Technology and Information Systems

Program, Tel-Aviv University. Retrieved August 15, 2008 from http://www.itu.org.il/Index.asp?ArticleID=2170&CategoryID=579& Page=2

Bates, E. (1962). *Adult Degree Program, mimeo ephemera* (September 20, 1962). Plainfield, Vermont: Goddard College.

Batson, T. (2008). Is Web 2.0 "Designed for Education?" *Campus Technology*, March 5, 2008. Retrieved May 28, 2008, from http://www.campustechnology.com/article.aspx?aid=59341

Bauerlein, M. (2008). *The dumbest generation.* NY: Tarcher/Penguin.

Beck, K. (2002). *Test-driven development by example.* Boston, MA: Addison-Wesley.

Becta (2008). *Web 2.0 technologies for learning at Key Stages 3 and 4.* Retrieved August 13, 2008 from: http://becta.org.uk

Beedle, J. B., & Wright, V. H. (2007). Perspectives from Multiplayer Game Players. In D. Gibson, C. Aldrich & M Prensky (Eds.), *Games and simulations in online learning: Research and development frameworks* (pp. 153-155). Hershey: Idea Group.

Beers, P. J. (2005). *Negotiating common ground: Tools for multidisciplinary teams.* Heerlen, The Netherlands: Open University of the Netherlands.

Bell, M. (2001). A case study of an online role-play for academic staff. In G. Kennedy, M. Keppell, C. McNaught, & T. Petrovic (Eds.), *Meeting at the Crossroads. Proceedings of the 18th Annual Conference of the Australian Society for Computers in Learning in Tertiary Education* (pp. 63-72). Melbourne, VIC: Biomedical Multimedia Unit, The University of Melbourne.

Bellamy, R. (1997). *Support for learning conversations.* Abstract retrieved November 11, 2002, from http://kmi.open.ac.uk/seminars/abstract.cfm?id=10.

Ben-Ari, M. (1998). Constructivism in computer science education. *SIGSCE 98* (pp. 257-261).

Bender, T. (2003). *Discussion-based online teaching to enhance student learning: Theory, practice and assessment.* Sterling, VA: Stylus.

Bereiter, C. (2001). *Education and mind in the knowledge age.* Mahwah, NJ: Lawrence Erlbaum.

Bereiter, C. (2002). *Education and mind in the knowledge age.* Mahwah, NJ: Lawrence Erlbaum Associates.

Bereiter, C., & Scardamalia, M. (2000). Commentary on Part I: Process and Product in Problem-based Learning (PBL) Research. In D. H. Evensen & C. E. Hmelo (Eds.), *Problem-based Learning. A Research Perspective on Learning Interactions* (pp. 185-195). Mahwah, NJ: Lawrence Erlbaum Associates.

Berge, Z. L. (May-June 2008). Multi-User Virtual Environments for Education and Training? A Critical Review of Second Life. *Educational Technology* (pp. 27-31).

Bernheim Brush, A. J., Bargeron, D., Grudin, J., Borning, A., & Gupta, A. (2002). Supporting interaction outside of class: Anchored discussion vs. discussion boards. In G. Stahl (Ed.), *Computer support for collaborative learning: Foundations for a CSCL community. Proceedings of CSCL 2002* (pp. 425–434). Hillsdale, NY: Lawrence Erlbaum Associates.

Biggs, J., & Tang, C. (2007). *Teaching for quality learning at university* (3rd ed.). Maidenhead, Berkshire: Society for Research into Higher Education & Open University Press.

Billig, M. (1988). *Ideological dillema's; a social psychology of everyday thinking.* London: Sage Publications.

Blatner, A. (2002). *Role Playing in Education. Adam Blatner's Web site.* Retrieved May 12, 2008, from http://www.blatner.com/adam/pdntbk/rlplayedu.htm

Bloom, B. S. (1956). Taxonomy of educational objectives: The classification of educational goals. *Handbook I, cognitive domain.* New York: David McKay Co, Inc.

Bloom, M. (1996). Transformative moments in education. *Holistic Educational Review*, 9(4), 44-47.

Blumenfeld, P. C., Marx, R. W., Soloway, E., & Krajcik, J. (1996). Learning with Peers: from small group cooperation to Collaborative Communities. *Educational Research, 25*(8), 37-40.

Blumenstyk, G. (2008, May 9). In turbulent times, two small colleges brace for the worst: They find strategies to keep enrollments up. *Chronicle of Higher Education*, (pp. A1, A12-13).

Bode, B. (1938). *Progressive education at the crossroads*. New York, New York: Newson and Co.

Boettcher, J. (2007). Ten core principles for designing effective learning environments: Insights from brain research and pedagogical theory. *Innovate,* 3(3). Retrieved November 5, 2007, from http://www.innovateonline.info/index.php?view=article&id=54

Boghossian, P. (2006). *Behaviorism, Constructivism, and Socratic Pedagogy, 38*, 713-722.

Bok, D. (2006). *Our underachieving colleges: A candid look at how much students learn and why they should be learning more*. Princeton, NJ: Princeton University Press.

Bonk, C. J. (2001). *The Vision Thing: 40 metaphors of online instruction*. Retrieved April 29, 2008 from http://www.trainingshare.com/download/metaphors.doc

Bonk, C. J., & Cunningham, D. J. (1998). Chapter 2: Searching for learner-centered, constructivist, and socio-cultural components of collaborative educational learning tools. In C. J. Bonk & K. S. King (Eds.), *Electronic collaborators: Learner-centered technologies for literacy, apprenticeship, and discourse* (pp. 25-50). Mahwah, NJ: Erlbaum.

Bonk, C. J., & Cunningham, D. J. (1998). Searching for learner-centered, constructivist, and sociocultural components of collaborative educational learning tools. In C. J. Bonk & K. S. King (Eds.), *Electronic Collaborators: Learner-Centered Technologies for Literacy, Apprenticeship, and Discourse* (pp. 25-50). Mahwah, NJ: Lawrence Erlbaum. Retrieved on August 25, 2008 from http://www.publicationshare.com/docs/Bon02.pdf.

Bonk, C. J., & Graham, C. R. (Eds.), (2006). *The Handbook of Blended Learning: Global Perspectives, Local Designs*. San Francisco, CA: Pfeiffer Publishing.

Bonk, C. J., & Kim, K. A. (1998). Extending sociocultural theory to adult learning. In C. M. Smith & T. Pourchot (Eds.), *Adult learning and development: Perspectives from educational psychology* (pp. 67-88). Mahwah, NJ: Lawrence Erlbaum.

Bonnet, C. (2000, May). The relevance of role playing in environmental education. *Proceedings of The International Union of Biological Sciences Commission for Biological Education (IUBS-CBE): International Symposium on Biological Education,* IUFM Versailles, Centre de Cergy, France.

Booth, W. (2006). Insuring the nation's destiny: reducsing the digital divide. *Journal of Asynchronous Learning Networks, 10*(3). Retrieved July 27, 2008, from Sloan-C-Publications database http://www.sloan-c.org/publications/jaln/index.asp

Bottino, R. M., & Robotti, E. (2007). Transforming classroom teaching & learning through technology: Analysis of a case study. *Educational Technology & Society, 10*(4), 174-186.

Boudourides, M. A. (1998, July). *Constructivism and Education: A Shopper's guide*. Paper presented at the International Conference on the Teaching on Mathematics, Samos, Greece. Retrieved May 2, 2008, from: http://math.upatras.gr/~mboudour/articles/constr.html

Bowers, C. A. (2000). *Let them eat data*. Athens, GA: University of Georgia Press.

Boyer, N., Langevin, S., & Gaspar, A. (2008, Jan.-Feb.). *Computer programming meets education: The intersection between information technology and self-direction*. Paper Presented at the 22nd Annual International Self-Directed Learning Symposium, Cocoa Beach, Florida.

Boyes, E. (2008). The Sims Passes 100-Million Mark. *Gamespot News*, April 16, 2008. Retrieved June 9, 2008 from http://www.gamespot.com/news/6189330.html.

Boys, J. (2001). Learning creativity via the web. In R. Shaw & McKay (Ed.), *Digital creativity: crossing the border. Proceedings of CADE 2001: The 4th Computers in Art & Design Education Conference* (pp. 61-64). Glasgow: Glasgow School of Art.

Brandt, D. S. (1997, Oct.) Constructivism: Teaching for understanding of the Internet. *Communications of the ACM*, *40*(10), 112-117.

Bransford, J. D., Brown, A. L., & Cocking, R. R. (1999) *How People Learn: Brain, Mind, Experience, and School.* Washington D.C.: National Academy Press.

Bransford, J. D., Franks, J. J., Vye, N. J., & Sherwood, R. D. (1989). New approaches to instruction: Because wisdom can't be told. In S. Vosniadou & A. Ortony (Eds.), *Similarity and analogical reasoning* (pp. 470-497). NY: Cambridge University Press.

Bransford, J. et. al. (2000). *How people learn.* Washington, DC: National Academy Press.

Bransford, J., Brown, A., & Cocking, R. (1999). *How people learn: Brain, mind experience and school.* Washington, DC: National Research Council. Retrieved June 9, 2008 from http://www.nap.edu/html/howpeople1/

Brookfield, S. (1986). *Understanding and facilitating adult learning.* San Francisco: Jossey Bass.

Brookfield, S. (1987). *Developing critical thinkers: Challenging adults to explore alternative ways of thinking and acting.* San Francisco: Jossey-Bass.

Brookfield, S. (Ed.) (1987). *Learning democracy: Eduard Lindeman on adult education and social change.* London: Croom Helm.

Brookfield, S. D. (1991). Discussion. In M. W. Galbraith (Ed.), *Adult learning methods: A guide for effective instruction* (pp. 187-204). Malabar, FL: Krieger Publishing Company.

Brookfield, S. D. & Preskill, S. (2005). *Discussion as a way of teaching: Tools and techniques for democratic classrooms.* San Francisco, CA: Jossey-Bass.

Brookfield, S., & Preskill, S. (1999). *Discussion as a way of teaching.* San Francisco: Jossey-Bass.

Brophy, J. (Ed). (2002). *Social Constructivist Teaching: Affordance and Constraints.* Elsevier: London.

Brophy, J., & Alleman, J. (1991). Activities as instructional tools: A framework for analysis and evaluation. *Educational Researcher, 20*(4), 9-23.

Brown, A. L., & Campione, J. C. (1990). Communities of learning or a content by any other name. In D. Kuhn (ed.), *Contribution to human development* (pp. 108-126). New York: Oxford University Press.

Brown, J. S., & Adler, R. P. (2008). Minds on fire: Open education, the long tail, and learning 2.0. *Educause Review, January/February,* (pp. 17-32).

Brown, J. S., & Adler, R. P. (2008, Jan. – Feb.). Minds on fire: Open education, the long tail, and learning 2.0. *Educause Review. 20 & 30.*

Brown, J., Collins, A., & Duguid, P. (1989). Situated cognition and the culture of learning. *Educational Researcher, 18*(1), 32-42.

Bruce & Goodall (2001). I have some pages up! In D. Murphy, R. Walker, & G. Webb (Eds.), *Online learning and teaching with technology: Case studies, experience and practice* (pp. 107-113). London: Kogan Page.

Bruner, J. (1990). *Acts of meaning.* Cambridge, MA: Harvard University Press.

Bruner, J. (1986). *Actual Minds, Possible Worlds.* Cambridge, MA: Harvard University Press.

Bruner, J. (1991). The Narrative Construction of Reality. *Critical Inquiry, 18*, 1-21.

Bruner, J. (1996). *The culture of education.* Cambridge, MA: Harvard University Press.

Bruning, R. H., Schraw, G. J., & Ronning, R. R. (1995). *Cognitive psychology and instruction*, 2nd ed. Englewood Cliffs, NJ: Prentice Hall.

Buckley, J., & Exton C. (2003). Blooms' Taxonomy: A Framework for Assessing Programmers' Knowledge of Software Systems. In Proceedings of the *11th IEEE International Workshop on Program Comprehension* (pp. 165). Washington, DC: IWPC. IEEE Computer Society.

Burbules, N. C. (2000). Does the internet constitute a global educational community? In N. C. Burbules & C. A. Torres (Eds.), *Globalization and education: Critical perspectives* (pp 323-355). New York: Routledge.

Burbules, N. C., & Callister, Jr., T. A. (2000). *Watch IT: The risks and promises of information technologies.* Boulder, CO: Westview/Perseus.

Bureau of Labor Statistics, U.S. Department of Labor, *Occupational Outlook Handbook, 2008-09 Edition, Tomorrow's Jobs*. Retrieved *August 29, 2008 from* http://www.bls.gov/oco/ocos114.htm

Burgan, M. (2006). *Whatever happened to the faculty? Drift and decision in higher education.* Baltimore: John Hopkins University Press.

Buzan, T., & Buzan, B. (1996). *The Mind Map Book: How to Use Radiant Thinking to Maximise your Brain's Untapped Potential* (reprint ed.). New York: Plume Books.

Campbell, J., Oblinger, D., & Colleagues (2007). Creating a culture of evidence tops the list of important issues as the academic technology profession moves to an "Instruction 2.0" world. *Educause* Quarterly, 30(3). Retrieved March, 11, 2008 from http://connect.educause.edu/library/abstract/TopTenTeachingandLea/44831

Cannella, G. (1998). Early Childhood Education: a call for the construction of revolutionary images. In W. Pinar (Ed.), *Curriculum: Towards new identities* (pp. 157-184). New York: Garland.

Carlisle, M. C., Wilson, T. A., Humphries, J. W., & Hadfield, S. N. (2005). Raptor: A visual programming environment for teaching algorithmic problem solving. *Proceedings of the 36th SIGCSE technical symposium on Computer science education* (pp. 176-180). St. Louis, MO.

Carrier, S. I., & Moulds, L. D. (2003). Pedagogy, andragogy, and cybergogy: Exploring best-practice paradigm for online teaching and learning. *Sloan-C 9th International Conference on Asynchronous Learning Networks (ALN)*, Orlando, USA [PowerPoint slides]. Retrieved May 15, 2004, from http://www.sloanc.org/conference/proceedings/2003/ppt/1471.ppt /

Carter, Bryan. (2006). Virtual Harlem in the Beginning: Retrospective Reflections. In J. J. Sosnoski, P. Harkin, & B. Carter (Eds.), *Configuring History: Teaching the Harlem Renaissance through Virtual Reality Cityscapes* (pp. 47-57). New York: Peter Lang.

Cashion, J., & Palmieri, P. (2002). *'The secret is the teacher': The learner's view of online learning.* SA: Australia: NCVER.

Cavanaugh, J. K. (2005). Are online courses cannibalizing students from existing courses? *Journal of Asynchronous Learning Networks, 9*(3). Retrieved July 27, 2008, from Sloan-C-Publications database http://www.sloan-c.org/publications/jaln/index.asp

Chang, S. L., & Ley, K. (2003). *The Effect of Using Printed Online Materials on Learning Achievement in Web-based and Hybrid Courses.* Paper presented at the national conference of the American Association of Research in Education, San Diego, CA.

Chao, J. (2007). Student project collaboration using wikis. *20th Conference on Software Engineering Education & Training.*

Chen, S. (2007). Instructional Design Strategies for Intensive Online Courses: An Objectivist-Constructivist Blended Approach. *Journal of Interactive Online Learning, 6*(1).

Chen, S., & Macredie, R. (2002). Cognitive Style and Hypermedia Navigation: Development of Learning Model. *Journal of the American Society for Information Science and Technology.* Retrieved April 16, 2008 from http://www.brunel.ac.uk/~csstsyc/sherry_paper.pdf

Chen, T., Lewandowski, G., McCartney, R., Sanders, K., & Simon, B. (2007). Commonsense computing: using student sorting abilities to improve instruction. In *Proceedings of the 38th SIGCSE Technical Symposium on Computer Science Education* (pp.276-280). New York: Association for Computing Machinery. Retrieved October 3, 2008, from http://doi.acm.org/10.1145/1227310.1227408.

Cherubini, M., Van der Pol, J., & Dillenbourg, P. (2005, July). *Grounding is not shared understanding: Distinguishing grounding at an utterance and knowledge level.* Paper presented at CONTEXT'05, the Fifth International and Interdisciplinary Conference on Modelling and Using Context, Paris.

Chew, S. C., & Beaumont, C. (2004, June). *Evaluating the effectiveness of ICT to support globally distributed PBL teams.* Paper presented at the ITiCSE Conference, ACM/SIGCSE, (pp. 47-51).

Chicoine, D. (2004). Ignoring the obvious: A constructivist critique of a traditional teacher education program. *Educational Studies, 36*(3), 245-263.

Chieu, V. M., Milgrom, E., & Frenay, M. (2004, Sept.). Constructivist learning: Operational criteria for cognitive flexibility. *Proceedings of the IEEE International Conference on Advanced Learning Technologies (ICALT '04)* (pp. 221-225).

Clark, H. H. (1992). *Arenas of language use*. Chicago: University of Chicago Press.

Clark, H. H., & Marshall, C. R. (1981). Definite reference and mutual knowledge. In A. K. Joshi, B. L. Webber, & I. A. Sag (Eds.), *Elements of discourse understanding* (pp. 10-63). New York: Cambridge University Press.

Clark, H., & Brennan, S. (1991). Grounding in communication. In L. Resnick, J. Levine & S. Teasley (Eds.), *Perspectives on socially shared cognition* (pp. 127–149). Washington, DC: American Psychological Association.

Clark, H., & Wilkes-Gibbs, D. (1986). Referring as a collaborative process. *Cognition, 22,* 1-39.

Clark, R. C., & Mayer, R. E. (2007). E-Learning and the Science of Instruction: Proven Guidelines for Consumers and Designers of Multimedia Learning, CA: Pfeiffer.

Clark, R. C., & Mayer, R. E. (2008). *E-learning and the science of instruction.* San Francisco: Pfeiffer.

Cleveland-Innes, M., & Campbell, P. (2006). *Understanding emotional presence in an online community of inquiry.* Paper presented at the 12th Annual SLOAN-C ALN Conference, Orlando, Florida.

Cobb, P. (1994). Where is the mind? Constructivist and sociocultural perspectives on mathematical development. *Educational Researcher, 23*(7), 13-20.

Cobb, P. (2005). Where is the mind? A coordination of sociocultural and cognitive constructivist perspectives. In C. T. Fosnot (Ed.), *Constructivism: Theory, perspectives, and practice* (pp. 39-61). New York: Teachers College Press.

Cochran-Smith, M., & Lytle, S. (1999). Relationships of knowledge and practice: Teacher learning in communities. In A. Iran-Nejad & P. D. Pearson (Eds.), *Review of research in education, 24,* 249-305. Washington DC: American Educational Research Association.

Coghlan, M. (2001). *Emoderation: Managing a new language?* Paper presented at the Net*Working 2001 Conference - from Virtual to Reality. Brisbane. Retrieved January, 12, 2008 from http://www.chariot.net.au/~michaelc/nw2001/emod_newlang.htm

Cole, M. (1996). *Cultural psychology. A once and future discipline.* Cambridge, MA and London, UK: The Belknap Press of Harvard University Press.

Collar, K. (2007). Getting to the roots of adult education. *Clockworks* (Winter/Spring 2007). Plainfield, Vermont: Goddard College Publications,

Collins, A., Brown, J. S., & Newman, S. E. (1987). *Cognitive apprenticeship: Teaching the craft of reading, writing and mathematics* (Technical Report No. 403). Cambridge, MA: Centre for the Study of Reading,

University of Illinois. (ERIC Document Reproduction Service No. ED 284 181).

Collins, M. (1998). The use of email and electronic bulletin boards in college-level biology. *Journal of Computers in Mathematics and Science Teaching*, *17*(1), 75-94.

Collis, B., & Moonen, J. (2008). Web 2.0 tools and processes in higher education: quality perspectives. *Educational Media International*, *45*(2), 93. doi: 10.1080/09523980802107179.

Collison, G., Haavind, S., & Tinker, R. (2000). *Facilitating online learning: Effective strategies for moderators*. Madison, WI: Atwood.

Compos, M. (2004). A constructivist method for the analysis of networked cognitive communication and the assessment of collaborative learning and knowledge-building. *Journal of Asynchronous Learning Network*, 8(2), 1-29.

Computerworld. (2008). *IT will "take head out of sand" on Web 2.0 in 2008*, *42*(6), 5.

Conference of European Ministers of Education (1999). *Bologna Declaration. 19 June 1999*. Retrieved May 28, 2008, from http://www.eua.be/eua/jsp/en/upload/OFF-DOC_BP_bologna_declaration.1068714825768.pdf

Connor, E. (2007). Using Wiki technology to build a faculty publications database. *Journal of Electronic Resources in Medical Libraries, 4*(4), 11-25.

Conole, G., & Oliver, M. (Eds.) (2007). *Contemporary Perspectives in e-Learning Research: themes, methods, and impact on practice*. Abingdon: Routledge.

Conrad, R. M., & Donaldson, J. A. (2004). *Engaging the online learner: Activities and resources for creative instruction*. San Francisco: Jossey-Bass.

Cooper, M. M., & Selfe, C. L. (1990). Computer conferences and learning: Authority, resistance, and internally persuasive discourse. *College English, 52,* 847-69.

Corcoran, M. (2000). Interview with James Elkins. *CIRCA,* 94. 26-29. Retrieved January, 10, 2008 from http://www.recirca.com/backissues/c94/jelkins.shtml

Cormier, D. (2008). Rhizomatic education: Community as curriculum. *Innovate,* 4 (5). Retrieved August, 15, 2008 from http://www.innovateonline.info/index.php?view=article&id=550

Coulter, X. (2006). What do we know about student learning? A case study. *All about mentoring*, Issue 31, 35-39.

Coulter, X., & Herman, L. (1994). Individualization in distance education. *Proceedings: International Conference on Distance Education* (pp 190-192). Moscow, Russia: Association for International Education.

Coulter, X., with Oaks, S., Gadbow, N., & Gerardi, J. (2000). Technology and education. *All About Mentoring*, Issue 18, 3-10.

Council for Adult and Experiential Learning (CAEL) (2005). *Principles in practice: Assessing adult learning focused institutions – Case studies*. Chicago: CAEL.

Cowie, H. (1992). Peer commentaries on Keith Topping's overview of cooperative learning and peer tutoring. *The Psychologist: Bulletin of British Psychology Society, 5*, 158-9.

Craig, E. M. (2007). Changing paradigms: managed learning environments and Web 2.0. *Campus-Wide Information Systems*, *24*(3), 152 - 161. doi: 10.1108/10650740710762185.

Crapo, A. W., Waisel, L. B., Wallace, W. A., & Willemain, T. R. (2000). *Visualization and the process of modeling: A cognitive-theoretic view* (p. 219). Association for Computing Machinery, Inc.

Creswell, J. (1998). *Qualitative Inquiry and Research Design: Choosing Among Five Traditions*. Thousand Oaks CA: Sage Publications.

Croft, W., & Cruse, D.A. (2004). *Cognitive Linguistics*. Cambridge: Cambridge University Press.

Cronin, J. (2005, July). *Joining the digital dots: Promoting resource integration and* autodidactic learning in teaching art history to adult learners (Towards a global understanding of lifelong learning: making a difference). Paper presented to the Forum for the Advancement of

Continuing Education (FACE) Conference, University Cork.

Cronin, J. (2008). Scaffolding the learner: Teaching for Understanding (TfU) and the adult learning experience, University College Cork. [Poster session]. In N. Ryan (Ed.), *International perspectives on teaching and learning in higher education*. National Academy for Integrating Research & Teaching & Learning (NAIRTL) Conference Proceedings, November 2007. Cork: NAIRTL.

Crook, C. K. (2002). The Social Character of Knowing and Learning: Implications of cultural psychology for educational technology. *Journal of Information Technology in Teacher Education, 10*, 19-36.

Crosier, D., Purser, L., & Smidt, H. (2007). *Trends V: Universities Shaping the European Higher Education Area.* Brussels: European Universities Association.

Cross, K. P. (1992). *Adults as learners: Increasing participation and facilitating learning.* San Francisco: Jossey-Bass.

Crotty, T. (1994). Integrating distance-learning activities to enhance teacher education toward the constructivist paradigm of teaching and learning. *In Distance Learning Research Conference Proceedings,* (pp. 31-37). College Station, TX.

Crowe, A. R., & Berry, A. (2007). Teaching prospective teachers about learning to think like a teacher: Articulating *our* principles of practice. In T. Russell & J. Loughran (Eds.), *Enacting a pedagogy of teacher education: Values, relationships, and practices.* London: Routledge.

Cucchiara, S., Spadaro, P. F., & Ligorio, M. B. (2008). Identità e comunità in contesti collaborativi: un'esperienza blended universitaria [Identity and community in collaborative contexts: a blended university experience]. *Qwerty, 1*, 23-48.

D'Amico, A., & Kocka, M. (2005, Oct.) *Information assurance visualizations for specific stages of situational awareness and intended uses: Lessons learned* (p. 110). Workshop on Visualization for Computer Security. IEEE.

Dalgarno, B. (2001). Interpretations of constructivism and consequences for computer assisted learning. *British Journal of Educational Technology, 32*(2), 183-194.

Daloz, L. A. (1999). *Mentor.* San Francisco: Jossey Bass.

Danchak, M. M., Walther, J. B., & Swan, K. (2001). *Presence in mediated instruction: bandwidth, behavior, and expectancy violations.* Orlando, FL: Paper presented at the Seventh Annual Sloan-C International Conference on Online Learning.

Dark, M. J., & Winstead, J. (2005, Sept.). Using educational theory and moral psychology to inform the teaching of ethics in computing. *Information Security Curriculum Development Conference '05* (p. 29). Association for Computing Machinery, Inc.

Davies, A., Ramsay, J., & Linfield, H. (2005). Building learning communities: Foundations for good practice. *British Journal of Educational Technology, 36*(4), 615-628.

de la Harpe, B., & Radloff, A. (2008). Developing graduate attributes for lifelong learning – how far have we got? In D. Orr, P.A. Danaher, G. Danaher, & R. E. Harreveld (Eds.), *Lifelong Learning: Reflecting on successes & Framing Futures* (pp. 136-142). Refereed proceedings of the Lifelong Learning Conference, Yeppoon, Queensland: Central Queensland University.

De Rossi, L. C. (Ed.) (2007, November 9). Online Social Networking and Education: Study Reports on New Generations Social and Creative Interconnected Lifestyles. *Robin Good's MasterNewMedia*, (pp. 1-11). Retrieved May 30, 2008 from http://www.masternewmedia.org/.

Dearstyne, B. W. (2007). Blogs, mashups, & wikis. T*he Information Management Journal, 41*(4), 24-33.

Dede, C. (2005). Planning for Neomillennial Learning Styles. *Educause Quarterly, 28*(1), 7-12.

Dede, C. (1996). Distance learning: Distributed learning: Making the transformation. *Learning and Leading with Technology, 23*, 25-30.

Deitel, H., & Deitel, P. (2006). *C How to program* (5 ed). Upper Saddle River, NJ: Prentice Hall.

Delanty, G. (2001). *Challenging knowledge: The university in the knowledge society*. Buckingham: Open University Press.

Delanty, G. (2003). Ideologies of the knowledge society and the cultural contradictions of higher education. *Policy Futures in Education*, 1(1), 71-82. Retrieved December, 10, 2007 from http://www.wwwords.co.uk/pdf/viewpdf.asp?j=pfie&vol=1&issue=1&year=2003&article=4_Delanty_PFIE_1_1&id=86.41.151.59.

Deleuze, G., & Guattari, F. (1987). *A thousand plateaus: Capitalism and schizophrenia*, 2. (B. Massumi, Trans.). Minneapolis: University of Minnesota Press (Original work published 1980).

Delfino, M., & Stefania, M. (2007). The expression of social presence through the use of figurative language in a web-based learning environment, *Computers in Human Behavior, 23*, 5, 2190-2211. Retrieved May 29, 2008 from http://portal.acm.org/citation.cfm?id=1234416.1234613

Dennen, V. P., & Wieland, K. (2007). From interaction to intersubjectivity: facilitating online group discourse processes. *Distance Education, 28*(3), 281-97.

Derry, S. J. (1996). Cognitive schema theory in the constructivist debate. *Educational Psychologist, 31*(3/4), 163-174.

Dewey, J. (1897) My Pedagogic Creed. *The School Journal, LIV*(3), 77-80. Retrieved April 23, 2008 from http://www.infed.org/archives/e-texts/e-dew-pc.htm

Dewey, J. (1910/1991). *How we think*. Amherst. New York: Prometheus.

Dewey, J. (1916/1966). *Democracy and education*. New York: Free Press.

Dewey, J. (1929/1979). *The Quest for Certainty*. New York: Paragon.

Dewey, J. (1933). *How we think* (rev ed.). Boston: D. C. Heath.

Dewey, J. (1938). *Education and experience*. New York: Touchtone Books.

Dewey, J. (1959). My pedagogic creed. In J. Dewey (Ed.), *Dewey on education* (pp. 19-32). New York: Teachers College, Columbia University. (Original work published 1897).

Dewey, J. (1997). *Democracy and education*. New York, New York: Free Press (Reprint edition).

Dewey, J. (1997). *Experience and education*. New York, New York: Free Press (Reprint edition).

Dewey, J. (2006). *The school and society*. Whitefish, Montana: Kessinger Publishing (Reprint edition).

Dick, W., & Carey, L. (2004). *The systematic design of instruction*. New York: Harper Collins.

Dickey, M. D. (2004). The impact of web-logs (blogs) on student perceptions of isolation and alienation in a web-based distance-learning environment. *Open Learning, 19*(3), 279-291.

Dickey, M. D. (2005). Three-dimensional virtual worlds and distance learning: Two case studies of active worlds as a medium for distance education. *British Journal of Educational Technology. 36*(3), 439-451.

Dijkstra, S. (2005). Cognition and Instructional Design for Problem-Based Learning. In Spector, M., Ohrazda, C., Van Schaack, A., Wiley, D. (Eds.), *Innovations in Instructional Technology*, (pp. 187-206). New Jersey: Lawrence Erlbaum Associates.

Dillenbourg, P. (1999). What do you mean by collaborative learning? In P. Dillenbourg (Ed.), *Collaborative-learning: Cognitive and computational approaches* (pp.1–19). Oxford, UK: Elsevier.

Dillenbourg, P. (1999). What do you mean by collaborative learning? In P. Dillenbourg (Ed.), *Collaborative-learning: Cognitive and Computational Approaches* (pp. 1-19). Oxford, UK: Elsevier.

Dillenbourg, P. (2002). Over-scripting CSCL: The risks of blending collaborative learning with instructional design. In P. A. Kirschner (Ed.), *Three worlds of CSCL. Can we support CSCL?* (pp. 61-91). Heerlen, The Netherlands: Open University of the Netherlands.

Dillenbourg, P., & Bétrancourt, M. (2006). Collaboration load. In J. Elen, & R. E. Clark (Eds.), *Handling complexity in learning environments: Research and theory.* Amsterdam: Elsevier.

Dillenbourg, P., Traum, D., & Schneider, D. (1996) Grounding in multi-modal task-oriented collaboration. In P. Brna, A. Paiva, & J. Self (Eds.), *Proceedings of the European Conference on Artificial Intelligence in Education* (pp. 401-407). Lisbon, Portugal: Fundacao Calouste Gulbenkian.

Dinter, F. (1998). Constructivism in Instructional Design Theory. *Journal of Structural Learning & Intelligent Systems, 13*(2). Retrieved January 19, 2007 from Academic Search Premier.

diSessa, A., & Minstrell, J. (1998). Cultivating conceptual change with benchmark lessons. In J. Green, & S. Goldman (Eds.), *Thinking Practices in Mathematics and Science Learning.* Mahwah, NJ: Lawrence Erlbaum Associates.

Dolence, M., & Norris, D. (1995). *Transforming higher education: A vision for learning in the 21st Century.* Ann Arbor, Michigan: Society for College and University Planning.

Doll, W. (1993). *A post-modern perspective on curriculum.* New York: Teachers College Press.

Donmoyer, R. (1990). Generalizability and the single-case study. In E. Eisner & A. Peshkin (Eds.), *Qualitative inquiry in education: The continuing debate* (pp. 175-200). New York: Teachers College Press.

Donnelly, R. (2004). The effectiveness of teaching 'Online Learning' in a problem-based learning classroom environment. In M. Savin-Baden & K. Wilkie (Eds.), *Challenging Research into Problem-based learning.* Buckingham: Open University Press.

Donnelly, R. (2006). Blended problem-based learning for teacher education: lessons learnt. *Journal of Learning, Media and Technology, 31*(2), 93-116.

Donnelly, R. (2006). The Academic Developer as Tutor in PBLonline in Higher Education. In M. Savin-Baden & K. Wilkie (Eds.), *Problem-based Learning Online* (pp. 79-97). Maidenhead: Open University Press.

Donnelly, R. (2008). Virtual Problem-based Learning Communities of Practice for Teachers and Teacher Educators: An Irish Higher Education Perspective. In C. Kimble & P. Hildreth (Eds.), *Communities of Practice: Creating Learning Environments for Educators.* USA: Information Age Publishing.

Donnelly, R., & O'Farrell, C. (2006). Blended E-Learning for Continuous Professional Development of Academic Staff. In J. O'Donoghue (Ed.), *Technology Supported Learning and Teaching: A Staff Perspective.* Hershey PA: Information Science Publishing.

Donnelly, R., & O'Rourke, K. (2007). What now? A critical evaluation of E-Learning CPD practice in Irish third level education. *Journal of Further & Higher Education.*

Doolittle, P. (1999). *Constructivism and online education.* Retrieved May 3, 2008 from: http://edpsychserver.ed.vt.edu/workshops/tohe1999/text/doo2.pdf.

Dougiamas, M. (1998). *A Journey into Constructivism.* Retrieved August 25, 2008, from http://dougiamas.com/writing/constructivism.html.

Dougiamas, M. (2003). *Moodle. Using Learning Communities to Create an Open Source Course Management System.* Refereed paper, presented at EDMEDIA 2003. Retrieved May 2, 2008, from http://dougiamas.com/writing/edmedia2003/

Dougiamas, M. (2008). *Pedagogy.* From MoodleDocs. Retrieved May 1,2008, from http://docs.moodel.org/en/Pedagogy

Downes, S. (2006). E-learning 2.0. *ELearn Magazine*

Downing, K. F. & Holtz, J. K. (2008). *Online science learning: best practices and technologies.* Hershey, PA: Information Science Publishing.

Draper, S. W., & Anderson, A. (1991). The significance of dialogue in learning and observing learning. *Computers & Education, 17*(1), 93-107.

Driscoll, K. (2007, May – June). Collaboration in today's classrooms: New web tools change the game. *Multimedia & Internet @ Schools* (pp. 9-12).

Driscoll, M. P. (2005). *Psychology of Learning for Instruction*, 3rd ed. Bostono Allyn & Bacon.

Dron, J. (2007). *Control and constraint in e-learning: Choosing when to choose,* 33. Hershey: Idea Group Publishing.

Du Mont, R. (2002). *Distance Learning – A Systems View: An Assessment and Review of Literature*. Retrieved February 12, 2008 from http://www.rcet.org/research/ATT-OLN/Dumont-finalreport.pdf

Du, H. S., & Wagner, C. (2005). Learning with weblogs: An empirical investigation (pp. 1-9). IEEE.

Duffy, T., & Cunningham, D. J. (1996). Constructivism: Implications for the design and delivery of instruction. In D. Jonassen (Ed.), *Handbook of research for educational communications and technology* (pp. 170-198). New York: Simon & Schuster Macmillan.

Duit, R. (1993). The Constructivist View: A Fashionable and Fruitful Paradigm for Science Education and Practice. In L. Steffe (Ed.), *Epistemological Foundations of Mathematical Experience.* New York: Springer-Verlag.

Ebner, M. (2007). E-Learning 2.0 – e-Learning 1.0 + Web 2.0? *Second International Conference on Availability, Reliability and Security (ARES'07)* (pp. 1-5).

Eijkman, H. (2008). Web 2.0 as a non-foundational network-centric learning space. *Campus-Wide Information Systems, 25*(20), 93-104.

Eisner, E. (2003). Objectivity in Educational Research. In M. Hammersley (Ed.), *Educational Research. Current issues* (pp. 57-72) London: Paul Chapman Publishing.

Ellen, E., & Seaman, J. (2007). *Online Nation: Five Years of Growth in Online Learning.* Sloan Foundation. Retrieved May 22, 2008 from http://www.sloan-c.org/publications/survey/pdf/online_nation.pdf

Empire State College Office of Institutional Research (2007). 2006-2007 fact book. Retrieved May 18, 2008, from http://www.esc.edu/ESConline/Across_ESC/InstResearch.nsf/db1a77fb2f6bcf2085256bfa005466b0/

Engeström, Y., & Escalante, V. (1996). Mundane tool or object of affection? The rise and fall of the Postal Buddy. In B. A. Nardi (Ed.), *Context and consciousness: Activity theory and human-computer interaction* (pp. 119-161). Cambridge, MA: The MIT Press.

Entwistle, N., Skinner, D., Entwistle, D., & Orr, S. (2000). Conceptions and beliefs about "good teaching": An integration of contrasting research areas. *Higher Education Research & Development, 19*(1), 5-26.

Erikson, E. H. (1959). *Identity and the life cycle.* Madison, CT: International Universities Press, Inc.

Ernest, P. (1995). The One and the Many. In P. Steffe & J. Gale (Eds.), *Constructivism in Education* (pp. 355-366). Hillsdale/New Jersey: Lawrence Erlbaum Associates.

Ertmer, P., & Newby, T. (1993). Behaviorism, cognitivism, constructivism: Comparing critical features from an instructional design perspective. *Performance Improvement Quarterly, 6*(4), 50-72.

Ertmer, P., & Newby, T. (1993). Behaviourism, cognitivism, constructivism. Comparing critical features. *Performance Improvement Quarterly, 6*(4), 50-70.

Fannon, K. (2005). *A Role-Play Simulation: Transformative Learning in Complex Dynamic Social Systems.* Retrieved May 4, 2007 from http://www.roleplaysim.org/papers/NeedleStick_web.pdf

Fardanesh, H. (2006). *A Classification of Constructivist of Instructional Design Models based on Learning and Teaching Approaches.* Retrieved March 27, 2008 from ERIC database.

Farmer, J. (2004). *Communication dynamics: Discussion boards, weblogs and the development of communities of inquiry in online learning environments.* Retrieved

June 9, 2008 from http://incsub.org/blog/2004/communication-dynamics-discussion-boards-weblogs-and-the-development-of-communities-of-inquiry-in-online-learning-environments.

Fay N., Garrod, S. C., & Carletta J. (2000). Group discussion as interactive dialogue or as serial monologue: The influence of group size. *Psychological Science, 11*(6), 481–486.

Finder, A. (2007, Nov. 20). Adjuncts outnumber tenured professors on U.S. campuses. *International Herald Tribune*. Retrieved May 14, 2008, from http://www.iht.com/ articles/2007/11/20/america/college.php

Finger, M., & Asun, J. M. (2001). *Adult education at the crossroads: Learning our way out.* London: Zed Books.

Fink, L. D. (2003). *Creating significant learning experiences.* San Francisco, CA: Jossey-Bass.

Finke, W. F. (2000). *Lifelong learning in the information age.* Bueren, Germany: Fachbibliothek Verlag.

Flower, L. (1990). Negotiating academic discourse. In L. Flower (Ed.), *Reading to write: Exploring a cognitive and social process.* Oxford, UK: Oxford University Press.

Floyd, I. R., Jones, M. C., Rathi, D., & Twidale, M. B. (2007). *Web mash-ups and patchwork prototyping: User-driven technological innovation with Web 2.0 and open source software* (p. 2). IEEE.

Fosnot, C. T., & Perry, R. S. (2005). Contructivism: A psychological theory of learning. In C. T. Fosnot (Ed.), *Constructivism: Theory, perspectives, and practice* (pp. 8-38). New York: Teachers College Press.

Fountain, R. (2007). *Conceptual enhancement via textual plurality: A pedagogical wiki bow towards collaborative structuration* (pp. 43-46). Association for Computing Machinery, Inc.

Franklin, G. (2000). Augmenting print-based distance learning with other technologies can enhance its effectiveness for adult learners. *Journal of Educational Technology Systems, 28*(4), 327-334.

Franklin,T., & van Harmelen, M. (2007). *Web 2.0 for Learning and Teaching in Higher Education.* London/ UK: The Observatory on borderless Higher Education.

Freeman, M., & Capper, J. (1999). Exploiting the web for education: An anonymous asynchronous role simulation. *Australian Journal of Educational Technology, 15*(1), 95-116. Retrieved May 11, 2008, from http://www.ascilite.org.au/ajet/ajet15/freeman.html

Freire, P. (1972). *Pedagogy of the oppressed.* New York: Herder and Herder.

Friedman, T. (2000). *The Lexus and the olive tree.* New York: Anchor Books.

Gagne, R. M., Wager, W. W., Golas, K. C., & Keller, J. M. (2005). *Principles of instructional design.* (5th ed) Belmont, CA: Wadsworth Publishing.

Gardner, H. (2006). *Changing minds: The art and science of changing our own and other people's minds.* Boston, MA: Harvard Business School Press.

Garrison, D. R. (2007). Online community of inquiry review: social, cognitive, and teaching presence issues. *Journal of Asynchronous Learning Network,s 11*(1). Retrieved July 27, 2008, from Sloan-C-Publications database http://www.sloan-c.org/publications/jaln/index.asp

Garrison, D. R., & Anderson, T. (2003). *E-Learning in the 21st century: A framework for research and practice.* London: Routledge/Falmer.

Garrison, D. R., & Arbaugh, J. B. (2007). Researching the community of inquiry framework: Review, issues, and future directions. *Internet and Higher Education, 10*(3), 157-172.

Garrison, D. R., & Archer, W. (2000). *A transactional perspective on teaching-learning: A framework for adult and higher education.* Oxford, UK: Pergamon.

Garrison, D. R., & Cleveland-Innes, M. (2005). Facilitating cognitive presence in online learning: Interaction is not enough. *American Journal of Distance Education, 19*(3), 133–148.

Garrison, D. R., & Kanuka, H. (2004). Blended learning: uncovering its transformative potential in higher education. *The Internet and Higher Education, 7*(2), 95-105.

Garrison, D. R., & Vaughan, N. D. (2008). *Blended learning in higher education: Framework, principles, and guidelines*. San Francisco, CA: Jossey-Bass.

Garrison, D. R., Anderson, T., & Archer, W. (2000). Critical inquiry in a text-based environment: Computer conferencing in higher education. *The Internet and Higher Education,* 2, 87–105.

Garrison, D. R., Anderson, T., & Archer, W. (2001). Critical thinking, cognitive presence and computer conferencing in distance education. *American Journal of Distance Education, 15*(1), 7-23.

Garrison, D. R., Anderson, T., & Archer, W. (2003). A theory of critical inquiry on online distance education. In M. G. Moore & W. G. Anderson (Eds.), *Handbook of Distance Education* (pp. 113-127). New Jersey, London: Lawrence Erlbaum Associates, Publishers.

Garrison, D. R., Cleveland-Innes, M., & Fung, T. (2004). Student role adjustment in online communities of inquiry: Model and instrument validation. *Journal of Asynchronous Learning Networks, 8*(2), 61-74. Retrieved August 13, 2005, from http://www.sloan-c.org/publications/jaln/v8n2/pdf/v8n2_garrison.pdf

Garrison, R., Anderson, T., & Archer, W. (2000). Critical inquiry in a text-based environment: Computer conferencing in higher education. *The Internet and Higher Education, 2*(2-3), 87-105.

Gaspar, A., & Langevin, S. (2007, July). *Active learning in introductory programming courses through student-led "live coding" and test-driven pair programming.* Presented at the International Conference on Education and Information Systems, Technologies and Applications, Orlando, FL.

Gaspar, A., & Langevin, S. (2007). Restoring "coding with intention" in introductory programming courses. *Proceedings of the 8th ACM SIGITE conference on Information Technology Education* (pp. 91-98). New York: Association for Computing Machinery. Retrieved October 3, 2008, from http://portal.acm.org/citation.cfm?id=1324302.1324323&coll=GUIDE&dl=ACM&CFID=5033070&CFTOKEN=27681204

Gaspar, A., Langevin, S., & Boyer, N. (2008). Constructivist apprenticeship through antagonistic programming activities. *Encyclopedia of Information Science and Technology* (2nd ed.). Hershey, PA: Information Science Reference.

Gay, G., Sturgill, A., Martin, W., & Huttenlocher, D (1999). Document-centered peer collaborations: An exploration of the educational uses of networked communication technologies. *Journal of Computer-Mediated Communication, 4*(3).

Gaytan, J., & McEwen, B. C. (1997). Effective online instructional and assessment strategies. *The American Journal of Distance Education, 21*(3).

Ge, S., & Tok, M. (2003). Enhancing online education using collaboration solutions. *Journal of Educational Technology Systems, 31*(4), 361-380.

Gee, J. P. (2004). *What Video Games Have to Teach Us About Learning and Literacy.* New York: Palgrave Macmillan.

Generals, D. (2000). *The architect of progressive education: John Dewey or Booker T. Washington?* In National Association of African American Studies and Hispanic and Latino Studies 2000 Literature Monograph Series: Proceedings (Education Section). Houston, Texas.

George, S., & Labas, H. (in press). E-learning standards as a basis for contextual forums design. *Computers in Human Behavior.*

Gergen, K. J. (1994). *Realities and relationships: soundings in social construction.* London, UK: Harvard University Press.

Gergen, K. J. (1995). Social Construction and the Educational Process. In P. Steffe & J. Gale (Eds.), *Constructivism in Education* (pp. 355-366). Hillsdale/New Jersey: Lawrence Erlbaum Associates.

Gholson, B., & Craig, S. D. (2006). Promoting constructivist activities that support vicarious learning during computer-based instruction. *Education Psychology Review, 18*(2), 119-139.

Giddens, A. (2000). *Runaway world: How globalization is reshaping our lives.* New York: Routledge.

Gilder, G. (2000). *TELECOSM: How infinite bandwidth will revolutionize our world.* New York: The Free Press, Simon and Schuster.

Giles, J. (2005). Special Report: Internet encyclopaedias go head to head. *Nature, 438*, 900-901 (15 December 2005).

Giroux, H. A. (1988). *Teachers as Intellectuals. Towards a Critical Pedagogy of Learning.* Critical Studies in Education Series. New York: Bergin & Garvey.

Giroux, H. A. (2007). *The university in chains.* Boulder, CO: Paradigm.

Glasersfeld, E. von (1989). Constructivism in Education. In T. Husen, & T. N. Postlethwaite, (Eds.), *The International Encyclopedia of Education,* Supplement, 1, 162-163. Oxford/New York: Pergamon Press.

Glogoff, S. (2005). Instructional blogging: Promoting e-learning, student-centered learning, and peer input. *Innovate, 1* (5). http://www.innovateonline.info/index.php?view=article&id=126

Gold, S. (2001). A constructivist approach to online teaching for online teachers. *Journal of Asynchronous Networks, 5*(1). Retrieved April 8, 2007 from http://www.sloan-c.org/publications/jaln/

Gomez, L. M., Sherin, M. G., Griesdorn, J., & Finn, L.-E. (2008). Creating social relationships: The role of technology in preservice teacher preparation. *Journal of Teacher Education, 59*(2), 117-131.

Goody, A. E., & Malone, V. M. (1999). Cybergogy and adult learning: a case research study used as a decision making tool in higher education. In H. E. Klein (Ed.), *Interactive teaching and the multi media revolution: case method & other techniques.* World Association for Case Method Research and Application (WACRA), Needham: MA.

Gore, J. M., & Zeichner, K. (1991). Action research and reflective teaching in preservice teacher education: A case study from the United States. *Teaching and Teacher Education, 7*(2), 119-136.

Götschi, T., Sanders, I., & Galpin, V. (2003, Feb.). Mental models of recursion. *SIGCSE '03* (pp. 346-350).

Grabinger, S. R. (1996). Rich environments for active learning. In D. H. Jonassen (Ed.), *Handbook of Research for Educational Communications and Technology*, (pp. 665-692), New York: Simon & Schuster Macmillan.

Graham, C. (2006). Blended Learning Systems: Definition, Current Trends, and Future Directions. In C. J. Bonk & C. R. Graham (Eds.), *The Handbook of Blended Learning. Global Perspectives, Local Designs* (pp. 3-21). San Francisco: Pfeiffer.

Gregorian, V. (2005). Six challenges to the American university. In R. H. Hersh & J. Merrow (Eds.), *Declining by degrees: Higher education at risk* (pp. 77-96). New York: Palgrave Macmillan.

Grossman, P., Wineburg, S., & Woolworth, S. (2001). Toward a theory of teacher community. *Teachers College Record, 103*(6), 942-1012.

Guarino, N. (1998). Some ontological principles for designing upper level lexical resources. In *Proceedings of the First International Conference on Lexical Resources and Evaluation.* pp. 1-8. Granada, Spain, May 28-30.

Gulati, S. (2008). Compulsory participation in online discussions: Is this constructivism or normalization of learning? *Innovations in Learning and Teaching International, 45*(2), 183-192.

Gulati, S. (2008). Compulsory participation in online discussions: Is this constructivism or normalisation of learning? *Innovations in Education and Teaching International, 45*(2), 183-92.

Gunawardena, C. (1995). Social presence theory and implications for interaction and collaborative learning in computer conferencing. *International Journal of Educational Telecommunications, 1*(2-3), 147 – 166.

Gunawardena, C., & Zittle, F. (1997). Social presence as a predictor of satisfaction within a computer mediated conferencing environment. *American Journal of Distance Education, 11*(3), 8-26.

Gurstein, M. (2003). Effective use: A community informatics strategy beyond the digital divide. *First Monday*, 8(12) Retrieved April, 2, 2007 from http://firstmonday.org/issues/issue8_12/gurstein/index.html

Guye-VuiliEme, A., Capin, T. K., Pandzic, S., Magnenat Thalmann, N., & Thalmann, D. (1999). Nonverbal Communication Interface for Collaborative Virtual Environments. *Virtual Reality*, 4, 49-59. Retrieved June 9, 2008 from http://www.springerlink.com.ezproxy2.lib.depaul.edu/content/q2767j5xw7216555/fulltext.pdf.

Guzdial, M., & Turns, J. (2000). Effective discussion through a computer-mediated anchored forum. *Journal of the Learning Sciences, 9*(4), 437–469.

Hakkarainen, K., Lipponen, L., & Järvelä, S. (2002). Epistemology of inquiry and computer-supported collaborative learning. In T. Koschmann, N. Miyake, & R. Hall (Eds.), *CSCL2: Carrying Forward the Conversation* (pp. 129-156). Mahwah, NJ: Erlbaum.

Halloun, I., & Hestenes, D. (1985). The Initial Knowledge State of College Physics Students, *American Journal of Physics, 53*, 1043-1055.

Halpern, D. F. (2002). Cognitive science and the work of reform. *New Directions for Higher Education, 119*(Fall 2002), 41-43.

Ham, V., & Davey, R. (2005). Our first time: Two higher education tutors reflect on becoming a 'virtual teacher'. *Innovations in Education and Teaching International, 42*(3), 257-264.

Hamlin, W. (1967, August). *Grown-Up education: The Goddard College Adult Degree Program*. Plainfield, Vermont: Goddard College Publications.

Hammond, M. (2005). A Review of Recent Papers on Online Discussion in Teaching and Learning in Higher Education. *Journal of Asynchronous Learning Networks, 9*(3).

Hannifin, M. J., & Hill, J. R. (2008). Resource-Based Learning. In J. Spector, D. Merrill, J. Van Merrienboer, & M. Driscoll (Eds.), *Handbook of Research on Educational Communications and Technology, 3rd ed.* (pp. 525-526). New York: Lawrence Erlbaum Accociates.

Hannifin, M. J., Hannifin, K. M., Land, S. M., & Oliver, K. (1997). Grounded practice and the design of constructivist learning environments. *Educational Technology Research & Development, 45*(3), 101-117.

Hanson, K., & Clem, F. (2006). To Blend or Not to Blend: A Look at Community Development Via Blended Learning Strategies. In C.J. Bonk & C.R. Graham (Eds.), *The Handbook of Blended Learning. Global Perspectives, Local Designs* (pp. 136-149). San Francisco: Pfeiffer.

Harasim, L. (1990). *On-line Education: Perspectives on a New Environment*. New York: Praeger.

Harasim, L. M. (1990). Online education: An environment for collaboration and intellectual amplification. In L. M. Harasim (Ed.), *Online education: Perspectives on a new environment* (pp. 39-64). New York: Praeger.

Harasim, L. M. (1989). Online education: A new domain. In R. Mason & A. R. Kaye (Eds.), *Mindweave: Communication, computers, and distance education* (pp. 50-62). New York: Pergamon Press.

Harvard University (2008). *Instructional Technology Survey, January–February 2008*. Department of Romance Languages and Literatures. Retrieved October 15, 2008 from http://blogs.law.harvard.edu/digitalnatives/files/2008/03/instructionaltechnologysurvey.pdf

Hathaway, R. (1997). *The friendly pioneer: An informal history of the Adult Degree* Program at Goddard College and Vermont College. Montpelier, Vermont: Word and Image.

Hendler, J. (2007). The dark side of the semantic web. A letter from the editor. *IEEE Intelligent Systems* (pp. 2-4). IEEE Computer Society..

Herman, L., & Mandell, A. (2004). *From teaching to mentoring*. New York: Routledge.

Hermans, H. (2004). Mediated identity in the emerging digital age: A dialogical perspective. *Identity: An international Journal of theory and research, 4*(4), 297-405.

Herring, S. (1999, June). Interactional coherence in CMC. *Journal of Computer-Mediated Communication 4*(4). Retrieved March 4, 2004, from http://jcmc.indiana.edu/vol4/issue4/herring.html

Herrmann, T., & Kienle, A. (2003, January). *Integration of communication, coordination and learning material A guide for the functionality of collaborative learning environments.* Paper presented at the 36th Annual Hawaii International Conference on System Sciences, Hilton Waikoloa, Hawaii.

Hersh, R., & Merrow, J. (Eds.) (2005). *Declining by degrees.* New York: Palgrave/Macmillan.

Herzog, A., & Bachman, J. (1981). Effects of questionnaire length on response quality. *The Public Opinion Quarterly, 45*(4), 549-559.

Hew, K. (2004). Past Technologies, Practice and Applications: A Discussion on How the Major Developments in Instructional Technology in the 20th Century. *Association for Educational Communications and Technology.* Retrieved March 31, 2008 from ERIC database.

Heylighen, F. (1997). Epistemological Constructivism. *Principia Cybernetica Web.* Retrieved May 5, 2008 from http://pespmcl.vub.ac.be/CONSTRUC.hyml

Hildebrand, D. L. (2008). *John Dewey: A Beginner's Guide.* Oxford, UK: Oneworld Press.

Hiltz, S. R., Shea, P., & Kim, E. (2007). Using focus groups to study ALN faculty motivation. *Journal of Asynchronous Learning Networks, 11*(1). Retrieved July 27, 2008, from Sloan-C-Publications database http://www.sloan-c.org/publications/jaln/index.asp

Hine, C. (2000). *Virtual Ethnography.* London: Sage.

Hirumi, A. (2002). Student-Centered, Technology-Rich Learning Environments (SCenTRLE): Operationalizing Constructivist Approaches to Teaching and Learning. *Journal of Technology and Teacher Education, 10*(4), 497-537. Retrieved on August 25, 2008 from http://www.ryerson.ca/activelearning/files/student-centered.pdf.

Ho, C., & Swan, K. (2007). Evaluating online conversation in an asynchronous learning environment. *Internet and Higher Education, 10*(1), 3-14.

Hoare, S. (2006). *The Guardian.* Tuesday June 20 2006. Retrieved September 24, 2008, from http://www.guardian.co.uk/education/2006/jun/20/elearning.technology15

Hodgkinson-Williams, C., Slay, H., & Sieborger, I. (2008). Developing communities of practice within and outside higher educational institutions. *British Journal of Educational Technology, 39*(3), 433-442.

Hofmann, J. (2006). Why Blended Learning Hasn't (Yet) Fulfilled Its Promises: Answers To Those Questions That Keep You Up at Night. In C. J. Bonk & C. R. Graham (Eds.), *The Handbook of Blended Learning. Global Perspectives, Local Designs* (pp. 27-40). San Francisco: Pfeiffer.

Holec, H. (1981). *Autonomy and Foreign Language Learning.* Oxford: Pergamon

Holmes, B., & Gardner, J. R. (2006). *E-Learning: Concepts and Practice.* London: Sage.

Horton, M. (1990). *The long haul: an autography by Myles Horton.* New York: Anchor.

Horton, W. (2006). *E-learning by design.* San Francisco: Pfeiffer.

Houle, C. O. (1961/1993). *The inquiring mind: A study of the adult who continues to learn.* Norman, OK: Oklahoma Research Center.

Howe, K. R., & Berv, J. (2000). Constructing Constructivism, Epistemological and Pedagogical. In D. C. Phillips (Ed.), *Constructivism in Education: Opinions and Second Opinions on Controversial Issues.* Ninety-ninth Yearbook of the National Society for the Study of Education. Part I. Chicago, Illinois, The University of Chicago Press.

Huber, M. T., & Hutchings, P. (2005). *The advancement of learning: Building the teaching commons*. San Francisco, CA: Jossey-Bass.

Huitt, Caroline. (2003). Constructivism. *Educational Psychology Interactive*. Valdosta, GA: Valdosta State University. http://chiron.valdosta.edu/whuitt/col/cogsys/construct.html.

Hung, D. (2001). Design principles for web-based learning: Implications for Vygotskian thought. *Educational Technology, 41*(3), 33-40.

Hung, D. (2001, May – June). Design principles for Web-based learning: Implications from Vygotskian thought. *Educational Technology* (pp. 33-41).

Hung, W., Bailey, J. H., Jonassen, D. H. (2003). Exploring the tensions of problem-based learning: Insights from research. In D. S. Knowlton & D. C. Sharp (Eds.), *Problem-based learning in the information age* (pp 13-23). New Directions for Teaching and Learning, 95. San Francisco: Jossey-Bass.

Hunt, R. A. (1998). Electronic discussions in learning and teaching: Why they don't work, and how they might. *Connexions: The Newsletter of the International Society for the Exploration of Teaching Alternatives, 10*(2), 1–7.

Hwang, J.-H., Lee, K. W., & Kwon, D.-S. (2005). The role of mental model and shared grounds in human-robot interaction. *2005 IEEE International Workshop on Robots and Human Interactive Communication* (pp. 623-628).

Ice, P. (2008). *The impact of asynchronous audio feedback on teaching, social and cognitive presence*. Paper presented at the 2008 Conference of the Canadian Network for Innovation in Education, Banff, Alberta, Canada.

Ice, P., Curtis, R., Philips, P., & Wells, J. (2007). Using asynchronous audio feedback to enhance teaching presence and students' sense of community. *Journal of Asynchronous Learning Networks*, 11(2). Retrieved May 6, 2008, from http://www.sloan-c.org/publications/JALN/v11n2/v11n2_ice.asp

Illich, I. (1971). *Deschooling society*. New York: Harper and Row.

Innes, R. B. (2004). *Reconstructing undergraduate education: Using learning science to design effective courses*. Mahway, NJ: Lawrence Erlbaum Associates.

Innovate.(2008). *Innovate 2008: The real and the ideal*. Retrieved April 22, 2008 from: http://www.education-2008.org/.

Interview: Media Grid to Take Education Across the Virtual World--and the XO. January 29, 2008. http://www.virtualworldsnews.com/2008/01/interview-media.html.

It's a Web 2.0 world: Are you ready for it? (2003, April). *Campus Technology* (pp. 43-50).

Ivins, T. Z. (2008). Microfranchising microlearning centers: a sustainable model for expanding the right to education in developing countries? *Journal of Asynchronous Learning Networks, 12*(1). Retrieved July 27, 2008, from Sloan-C-Publications database http://www.sloan-c.org/publications/jaln/index.asp

Jackson, R. L., & Fagan, E. (2000). Collaboration and learning within immersive virtual reality. CVE 2000 (pp. 83-92). Association of Computing Machinery, Inc.

Jamieson, I., & Naidoo, R. (2004). How the market economy is undermining HE performance. *Management in Education, 18*(2004), 13–16.

Järvelä, S., & Häkkinen, P. (1999, March). *Web based cases in teaching and learning - reciprocal understanding and perspective taking in conversation*. Paper presented at the conference for Computer Assisted Learning, London.

Jarvela, S., & Hakkinen, P. (2002). Web-based cases in teaching and learning: The quality of discussions and a stage of perspective-taking in asynchronous communication. *Interactive learning environments, 10*(1), 1-22.

Jennings, D. (2006). PBLonline: A Framework for Collaborative E-Learning. In M. Savin-Baden & K. Wilkie (Eds.), *Problem-based Learning Online* (pp. 105-123). Maidenhead: Open University Press.

Jiang, M., & Ting, E. (2000). A study of factors influencing students' perceived learning in a web-based course environment. *International Journal of Educational Telecommunications*, 6(4), 317-338.

JISC (Joint Information Services Committee, UK), (2008). *Exploring Tangible Benefits of eLearning: Does investment yield interest?* Northumbria: JISC infoNet publication. Retrieved October 15, 2008, from http://www.jiscinfonet.ac.uk/publications/publications/info/tangible-benefits-publication

Johnson, J. A. (1995, June). *A comparison of user interfaces for panning on a touch-controlled display.* Paper presented at the Conference on Human Factors in Computing Systems, Denver, CO.

Jonassen, D. (1991). Evaluating constructivistic learning. *Educational Technology, 31*(6), 28-33.

Jonassen, D. (1991). Objectivism versus Constructivism: Do we need a new philosophical paradigm? *Educational Technology Research and Development, 39*(3), 5-14.

Jonassen, D. (1992) Objectivism versus constructivism: Do we need a new philosophical paradigm? *Educational Technology Research & Development*, Vol. 39, No.3, pp.5-14.

Jonassen, D. (1994). Thinking technology. *Educational Technology*, (pp. 34-37).

Jonassen, D. (1999). Designing constructivist learning environments. In C. M. Reigeluth (Ed.), *Instructional-design theories and models* (Vol. 2): *A new paradigm of instructional theory* (pp. 215-39). Mahwah, NJ: Lawrence Erlbaum Associates.

Jonassen, D. (2005). Problem Solving: The Enterprise. In Spector, M., Ohrazda, C., Van Schaack, A., Wiley, D (Eds), *Innovations in Instructional Technology*, (pp. 91-110). New Jersey: Lawrence Erlbaum Associates.

Jonassen, D. H. (2000). *Computers as mindtools for schools: Engaging critical thinking* (2nd ed.). Upper Saddle River, NJ: Prentice-Hall.

Jonassen, D. H. (2000). Revisiting activity learning theory for designing student centered learning environments. In D. H Jonassen & S. M. Land (Eds.), *Theoretical foundations of learning environments* (pp. 89-122). Mahwah, NJ: Lawrence Erlbaum Associates.

Jonassen, D. H., & Land, S. M. (2000). Preface. In D. H. Jonassen & S. M. Land (Eds.), *Theoretical foundations of learning environments* (pp. iii-ix). Mahwah, NJ: Lawrence Erlbaum Associates

Jonassen, D. H., Peck, K. L., & Wilson, B. G. (1999). *Learning with technology: A constructivist perspective.* Upper Saddle River, NJ: Prentice-Hall.

Jonassen, D., & Land, S. (2000). (Eds.) *Theoretical Foundations of Learning Environments.* Mahwah, NJ: L. Erlbaum Associates.

Jonassen, D., Davidson, M., Collins, M., Campbell, J., & Haag, B. B., (1995). Constructivism and computer-mediated communication in distance education. *American Journal of Distance Education, 9*(2), 7-26.

Jones, D. and Kelly, P. (2007). *Mounting pressures facing the U.S. workforce and the increasing need for adult education and literacy.* National Commission on Adult Literacy. Retrieved October 10, 2008 from http://www.nationalcommissiononadultliteracy.org/content/nchemspresentation.pdf

Jones, P. H., & Popken, D. A. (n.d.). *Interactive prototyping methodology for a domain dependent user interface* (p. 814). US Government.

Jones, R. A., & Spiro, R. J. (1995). Contextualization, cognitive flexibility, and hypertext: The convergence of interpretive theory, cognitive psychology, and advanced information technologies. In S. L. Star (Ed.), *The cultures of computing* (pp. 146-157). Oxford, UK; Blackwell Publishers.

Julyan, C., & Duckworth, E., (1996). A constructivist perspective on teaching and learning science. In C. Twomey-Fosnot (Ed.), *Constructivism: Theory, perspectives, and practice.* (pp. 55-72). Columbia University. Teachers College Press.

Juwah, C. (2002). *Using communication and information technologies to support problem-based learning.* HEA

Academy. Retrieved February 28, 2007, from http://www.heacademy.ac.uk

Kant, I. (1784). *An Answer to the Question: "What is Enlightenment?"* Konigsberg in Prussia, 30th September, 1784. Retrieved May 27, 2008, from http://philosophy.eserver.org/kant/what-is-enlightenment.txt

Kanuka, H., & T. Anderson. (1998). Online social interchange, discord, and knowledge construction. *Journal of Distance Education 13 (1),* 57–75.

Karagiorgi, Y., & Symeou, L. (2005). Translating Constructivism into Instructional Design: Potential and Limitations. *Educational Technology & Society, 8*(1), 17-27.

Karpinski, J. L. (2008). Disconnected in a connected world: Knowledge and understanding of Web 2.0 tools at the University of Pennsylvania Medical Center. *Medical Reference Services Quarterly, 27*(1), 53-72.

Kaye, A. R. (1989). Computer-mediated communication and distance education. In R. Mason & A. R. Kaye (Eds.), *Mindweave: Communication, computers and distance education* (pp. 3-21). New York: Pergamon Press.

Kegan, R. (1994). *In over our heads: The mental demands of modern life.* Cambridge: MA. Harvard University Press.

Kelton, A. J. Second Life: Reaching into the Virtual World for Real World Learning." *EDUCAUSE Center for Applied Research: Research Bulletin, 27*(17), 1-13.

Kember, D. (1997). A reconceptualisation of the research into university academics' conceptions of teaching. *Learning and Instruction, 7*(3), 255-275.

Kember, D., & Kwan, K. P. (2000). Lecturers' approaches to teaching and their relationship to conceptions of good teaching. *Instructional Science, 28*(5), 469-490.

Kett, J. F. (1994). *The pursuit of knowledge under difficulties.* Stanford CA: Stanford University Press.

Khine, M. S., & Lourdusamy, A. (2003). Blended learning approach in teacher education: combining face-to-face instruction, multimedia viewing and online discussion. *British Journal of Educational Technology, 34*(5), 671-675.

Kirkup, G., & Kirkwood, A. (2005). Information and communications technologies (ICT) in higher education teaching: A tale of gradualism rather than revolution. *Learning, Media and Technology, 30*(2), 185-199.

Kirscher, P.A., Sweller, J., & Clark, R. E. (2006). Why minimal guidance during instruction does not work: An analysis of the failure of constructivist, discovery, problem-based, experiential, and inquiry-based teaching. *Educational Psychologist, 41*(20), 75-86.

Kirschner, F., Paas, F., & Kirschner, P. A. (Submitted). Effects of individual and group-based learning.

Kirschner, P. A., Sweller, J., & Clark, R. E. (2006). Why minimal guidance during instruction does not work: An analysis of the failure of constructivist, discovery, problem-based, experiential, and inquiry-based teaching. *Educational Psychologist, 41*(2), 75–86.

Kiser, K. (2002). *Is blended best?* E-Learning Magazine. Retrieved July 31, 2006, from http://www.elearningmag.com/elearning/article/articleDetail.jsp?id=21259

Klemm, W. R. (1998). Eight ways to get students more engaged in online conferences. *T.H.E. Journal*, August. Retrieved, March 27, 2007, from http://www.thejournal.com/articles/14054.

Klett, F. (2002). Visual Communication in Web-based environment. *Educational Technology & Society, 5*(4). Retrieved March 8, 2008 from http://www.ifets.info/journals/5_4/klett.pdf

Knights, M. (2007, Feb. – Mar.) Web 2.0. *IET Communications Engineer* (pp. 30-35).

Knowles, M. S. (1970). *The modern practice of adult education: Andragogy versus pedagogy.* Follett: Association Press.

Knowles, M. S. (1975). *Self-directed learning: A guide for learners and teachers.* Englewood Cliffs: Prentice Hall Cambridge.

Knowles, M. S. (1976). *The modern practice of adult education: Andragogy versus pedagogy.* New York: Associated Press.

Knowles, M. S., & Others. (1984). *Andragogy in action: Applying modern principles of adult education.* San Francisco: Jossey Bass.

Knowlton, D. S., & Sharp, D. C. (Eds.) (2003). *Problem-based learning in the information age.* New Directions for Teaching and Learning, *95.* San Francisco: Jossey-Bass.

Kohn, A. (2000). *The schools our children deserve.* NY: Houghton-Mifflin.

Kolikant, Y. B. (2005). Students' alternative standards for correctness. *In Proceedings of the International Workshop on Computing Education Research* (pp. 37-43). New York: Association of Computing Machinery, 37-43. Retrieved October 3, 2008, from http://doi.acm.org/10.1145/1089786.1089790

Kolling, M., Quig, B., Patterson, A., & Rosenberg, J. (2003). The BlueJ system and its pedagogy. *Journal of Computer Science Education, special issue on learning and teaching object technology, 13*(4), 249-268.

Koory, M. A. (2003). Differences in learning outcomes for the online and F2F versions of "An Introduction to Shakespeare. *Journal of Asynchronous Learning Networks*, *7*(2). Retrieved August 28, 2008 from http://www.sloan-c.org/publications/jaln/v7n2/v7n2_koory.asp.

Koschmann, T. (1996) Paradigm shifts and instructional technology: An introduction. In T. Koschmann (Ed.) *CSCL: Theory and practice of an emerging paradigm* (pp. 1-22). Mahwah; NJ: Lawrence Erlbaum Associates, Publishers.

Koschmann, T. (1999). Toward a dialogic theory of learning: Bakhtin's contribution to learning in settings of collaboration. In *Proceedings of Computer Supported Collaborative Learning (CSCL '99)*, Palo Alto, CA (pp.308-313). Available at: http://kn.cilt.org/cscl99/A38/A38.HTM

Koschmann, T. (2001). Revisiting the paradigms of instructional technology. In G. M. Kennedy, C. McNaught & T. Petrovic (Eds.), *Meeting at the Crossroads. Proceedings of the 18th Annual Conference of the Australian Society for Computers in Learning in Tertiary Education* (pp. 15-22). Melbourne: Biomedical Unit, The University of Melbourne. Retrieved October 15, 2008, from http://www.ascilite.org.au/images/papers/conferences/melbourne01/pdf/papers/REVISITING%20THE%20PARADIGMS%20OF%20INSTRUCTIONAL%20TECHNOLOGY.pdf.

Koschmann, T. D., Kelson, A. C., Feltovich, P. J., & Barrows, H. S. (1996). Computer-supported Problem-based Learning: A Principled Approach to the Use of Computers in Collaborative Learning. In T. D. Koschmann (Ed.), *CSCL: Theory and Practice of an Emerging Paradigm* (pp. 83-124). Hillsdale, NJ: Lawrence Erlbaum Associates.

Koschmann, T., & Le Baron, C. (2003). Reconsidering common ground: Examining Clark's contribution theory in the OR. In K. Kuutti, E. Karsten, G. Fitzpatrick, P. Dourish, & K. Schmidt (Eds.), *ECSCW 2003: Proc. Eighth European Conference on Computer-Supported Cooperative Work.* Amsterdam: Kluwer.

Krauss, R. M., & Fussell, S. R. (1990). Mutual knowledge and communicative effectiveness. In J. Galegher, R. E. Kraut, C. Egido (Eds.), *Intellectual teamwork: Social and technological foundations of cooperative work.* Hillsdale, NJ: Erlbaum.

Kraut, R. (n.d.). Analyzing & improving online communities. *Educause*, (pp. 1-45).

Kreijns, K., & Kirschner, P. A. (2002). The social affordances of computer-supported collaborative learning environments. In *Proceedings of the 31th ASEE/IEEE Frontiers in Education Conference, 1*(T1F, pp. 12-17) October 10-13, 2002, Reno, NV, USA.

Krum, D. M., Omoteso, O., Ribarsky, W., Starner, T., & Hodges, L. F. (2002, Oct. – Nov.) *Evaluation of a multimodal interface for 3D terrain visualization* (p. 411). IEEE..

Kuhn, D., & Goh, W. (2005). Arguing on the computer. In: T. Koschmann, D. Suthers & T. W. Chan (Eds.), *Proceedings of the 2005 conference on Computer support for collaborative learning: the next 10 years!* (pp. 346 - 352). Mahwah, NJ: Lawrence Erlbaum Associates.

Kuhn, T. C. (1970). *The structure of scientific revolutions*. Chicago: University of Chicago Press.

Kuhn, T. S. (1962). *The Structure of Scientific Revolutions*. Chicago and London: The University of Chicago Press.

Kuzu, A. (2007). Views of pre-service teachers on blog use for instruction and social interaction *Turkish Online Journal of Education 8*(3), 34-51.

Lai, A. (2008, March 25). Adult students' perceptions of learning through online discussion. Paper present at the American Educational Research Association annual conference.

Lampert, M., & Ball, D. L. (1998). *Teaching, multimedia, and mathematics*. New York: Teachers College Press.

Lapadat, J. C. (2002). Written Interaction: A Key Component in Online Learning. *Journal of Computer Mediated Communication (JCMC), 7*(4). Retrieved April 6, 2007, from

Lather, P. (2001). Validity as an incitement to discourse: Qualitative research and the crisis of legitimation. In V. Richardson (Ed.), *Handbook of research on teaching* (4th ed., pp. 241-250). Washington DC: American Educational Research Association.

Laurillard, D. (1993). *Rethinking university teaching*. London: Routeledge.

Laurillard, D. (2002). *Rethinking university teaching: A conversational framework for the effective use of learning technologies* (2nd ed.). London & New York: Routledge.

Lave, J., & Wenger, E. (1991). *Situated learning: Legitimate peripheral participation*. Cambridge, MA: Cambridge University Press.

Lawrence, K. F., & Schraefel, M. C. (2006). *Freedom and restraint: Tags, vocabularies, and ontologies* (pp. 1745-1750). IEEE.

Lee, M. J., Mcloughlin, C., & Chan, A. (2008). Talk the talk: Learner-generated podcasts as catalysts for knowledge creation. *British Journal of Educational Technology, 39*(3), 501-521.

Legon, R. (2007). Online courses lead the way in quality assurance. *Sloan-C View: Perspectives in Quality Online Education 6*(1). Retrieved July 20, 2007 from http://www.sloan-c.org/publications/view/v6n1/viewv6n1.htm

Lei, T., Yang, Y., & Zhang, Y. (2006). The usability of multimedia interface based on user's mental models. *Proceedings of the 16th International Conference on Artificial Reality and Telexistence—Workshops (ICAT '06).*

Leu, D. J., Leu, D. D., & Len, K. (1997). *Teaching with the internet: Lessons from the classroom*. Noorwood, MA: Christopher-Gordon Publishers, Inc.

Lever-Duffy, J., McDonald, J. B., & Mizell, A. P. (2003). *Teaching and Learning with Technology*. Boston, MA: Allyn and Bacon.

Levin, B. L., He, Ye, & Robbins, H. H. (2006). Comparative analysis of preservice teachers' reflective thinking in synchronous versus asynchronous online case discussions. *Journal of Technology and Teacher Education, 14*(3), 439-60.

Lewandowski, G., Bouvier, D. J., McCartney, R., Sanders, K., & Simon, B. (2007). Commonsense computing (episode 3): concurrency and concert tickets. *In Proceedings of the Third international Workshop on Computing Education Research* (pp. 133-144). New York: Association of Computing Machinery. Retrieved October 3, 2008, from http://doi.acm.org/10.1145/1288580.1288598

Li, Y., Musilek, P., & Wyard-Scott, L. (2004). Fuzzy logic in agent-based game design (pp. 734-739). IEEE.

Liang, Y. D. (2006). *Fundamentals first introduction to Java programming* (6 ed.). Upper Saddle River, NJ: Prentice Hall.

Lieberman, A., & Grolnick, M. (1996). Networks and reform in American education. *Teachers College Record, 98*(1), 7-45.

Lieberman, A., & Miller, L. (1999). *Teachers – Transforming their world and their work*. New York: Teachers College Press.

Light, G., & Cox, R. (2006). *Learning & teaching in higher education: the reflective professional*. London: Sage Publications.

Ligorio, M. B., Annese, S., Spadaro, P. F., & Traetta, M. (2008). Building intersubjectivity and identity in on-line communities. In B. M. Varisco (Ed.), *Psychological, pedagogical and sociological models for learning and assessment in virtual communities of practice* (pp. 57-91). Milan: Polimetrica.

Lim, H. L. (2006). *Constructing learning conversations: A study of the discourse and learner experiences of online synchronous discussions.* Unpublished PhD Thesis, Murdoch University, Perth, Australia.

Lim, H. L. (2007). Community of inquiry in an online undergraduate information technology course. *Journal of Information Technology Education, 6*, 153-168.

Lim, H. L., & Sudweeks, F. (2008). Chatting to learn: A case study on student experiences of online moderated synchronous discussions in virtual tutorials. In S. Negash, M. Whitman, A. Woszcynski, K. Hoganson & H. Mattord (Eds.), *Handbook of distance learning for real-time and asynchronous information technology education* (pp. 170-191): Informing Science Reference, Idea Group.

Lin, L. (2008). An online learning model to facilitate learners' rights to education. *Journal of Asynchronous Learning Networks (JALN), 12*(1), 127-143.

Lin, L., Cranton, P., & Bridgall, B. (2005). Psychological type and asynchronous written dialogue in adult learning. *Teachers College Record. 107*(8), 1788-1813.

Linser, R., (2004). *Suppose you were someone else: The learning environment of a web-based role-play simulation.* Retrieved July 27, 2005, from http://www.simplay.net/papers/suppose.html

Lipman, M. (1991). *Thinking in education.* Cambridge: Cambridge University Press.

Lippincott, J. K. (2007, Nov. / Dec.). Student content creators: Convergence of literacies. E-Content. *Educause Review* (pp. 16-17).

Lister, R. & Leaney, J. (2003). Introductory programming, criterion-referencing, and bloom. In *Proceedings of the 34th SIGCSE Technical Symposium on Computer Science Education* (pp. 143-147). New York: Association of Computing Machinery. Retrieved October 3, 2008, from http://doi.acm.org/10.1145/611892.611954

Little, J. W. (2003). Inside teacher community: Representations of classroom practice. *Teachers College Record, 105*(6), 913-944.

Littlejohn, A., & Pegler, C. (2007). *Preparing for blended e-learning.* London: Routledge.

Liu, X., Bonk, C. J., Lee, S., & Su, B. (2005). Exploring four dimensions of online instructor roles: a program level case study. *Journal of Asynchronous Learning Networks, 9*(4). Retrieved July 27, 2008, from Sloan-C-Publications database http://www.sloan-c.org/publications/jaln/index.asp

Livingston, D. (2003, November). *Reclaiming Early Childhood Teacher Education: A Critical Constructivist Approach.* Paper presented at the Annual Meeting of the Southeastern Regional Association of Teacher Educators, Savannah, GA.

Lock, J. V. (2006). A new image: Online communities to facilitate teacher professional development. *Journal of Technology and Teacher Education, 14*(4), 663-678.

Lore, A., & Tait, A. (2004). Time: The new commodity in lifelong learning. *All About Mentoring*(Issue 27), 14-18.

Ludlow, P., & Wallace, M. (2007). *The Second Life Herald: The Virtual Tabloid that Witnessed the Dawn of the Metaverse.* Cambridge: The MIT Press.

Luebeck, J. L., & Bice, L. R. (2005). Online discussion as a mechanism of conceptual change among mathematics and science teachers. *Journal of Distance Education, 20*(2), 21–39.

Lund, A., & Smørdal, O. (2006, Aug.). Is there a space for the teacher in a WIKI (pp. 37-46). Association for Computing Machinery, Inc.

MacManus, R., & Porter, J. (2005, May). Web 2.0 for designers. *Web 2.0 Design: Bootstrapping the Social Web.*

Magolda, M. B. B. (2004). *Making their own way: Narratives for transforming higher education to promote self-development.* Sterling, VA: Stylus Publishing.

Maia, L. P., Machado, F. B., & Pacheco, A. C. (2005, June). A constructivist framework for operating systems education: A pedagogic proposal using the SOsim. IT-iCSE '05 (pp. 218-222).

Mäkitalo, K., Salo, P., Häkkinen, P., & Järvelä, S. (2001, August). *Analysing the mechanisms of a common ground in Web-based interaction.* Paper presented at the JURE pre-conference of the 9th European Conference for Research on Learning and Instruction, Fribourg, Switzerland.

Makkonen, P. (2000). Do WWW-based presentations support better (constructivistic) learning in the basics of informatics? *Proceedings of the 33rd Hawaii International Conference on System Sciences.*

Makri, K., & Kynigos, C. (2007). The role of blogs in studying the discourse and social practices of mathematics teachers. *Educational Technology and Society, 10*(1), 73-84.

Mandell, A., & Herman, L. (2007). The study and transformation of experience. *Journal of Transformational Education*, 5 (4), 339-353.

Mangu-Ward, K. (July, 2008). Education for profit: Why is everyone flaming the University of Phoenix? *Reason.* Retrieved July 10, 2008, from http://www.reason.com/news/show/126856.html.

Manninen T., & Kujanpää T. (2002) Non-Verbal Communication Forms in Multi-player Game Session. In X. Faulkner, J. Finlay, & F. Détienne, (Eds.), *Proceedings of HCI 2002 Conference* (pp. 383-401). Springer-Verlag,. Retrieved June 9, 2008 from http://www.tol.oulu.fi/~tmannine/publications/HCI2002_Tuppi3D_analysis_final.pdf.

Manning, M., Manning, G., & Long, R. (1994). *Theme immersion: Inquiry-based curriculum in elementary and middle schools.* Portsmouth, NH: Heineman.

Masie, E. (2006). The Blended Learning Imperative. In C. J. Bonk & C. R. Graham (Eds.), *The Handbook of Blended Learning. Global Perspectives, Local Designs* (pp. 22-26). San Francisco: Pfeiffer.

Matusov, E., Hayes, R., & Pluta, M. J. (2005). Using discussion webs to develop an academic community of learners. *Educational Technology & Society, 8*(2), 16-39. Retrieved May 29, 2008 from http://www.ifets.info/journals/8_2/ets_8_2.pdf#page=21

Mayes, J. T., & Fowler, C. J. H. (1999). Learning technology and usability: a framework for understanding courseware. *Interacting with Computers, 11*, 485-497.

McClintock, R. (1992). *Power and pedagogy: Transforming education through information technology.* New York: Teacher College Columbia University Institute for Learning Technologies.

McDowell, C., Hanks, B., & Werner, L. (2003). Experimenting with pair programming in the classroom. *ACM SIGCSE Bulletin Proceedings of the 8th annual conference on Innovation and technology in computer science education, 35*(3), (pp 60-64). ACM New York: Association for Computing Machinery. Retrieved October 3, 2008, from http://portal.acm.org/citation.cfm?id=961290.961531&coll=GUIDE&dl=GUIDE&CFID=5033070&CFTOKEN=27681204

McFedries, P. (2006). The web, take two. *IEEE Spectrum, 43*(6), 68-68.

McFerrin, K. M. (1999, February). Incidental Learning in a Higher Education Asynchronous Online Distance

Course. In J. D.Price, J. Willis, M. Jost, & S. Boger-Mehall (Eds.), SITE 99: *Proceedings of the Society for Information Technology & Teacher education International Conference*, San Antonio TX (Charlotteville VA, Association for the Advancement of Computers in Education (AACE)), (pp. 1418-1423).

McGookin, M. (2002). I can't let them down! Affiliative motivation and co-operative learning in higher education. In A. Goody, J. Herrington & M. Northcote (Eds.), *Proceedings of the 2002 Annual International Conference of the Higher Education Research and Development Society of Australia* (pp. 417-425). Perth, WA: HERDSA.

McKenna, P., & Laycock, B. (2004). Constructivist or instructivist: Pedagogical concepts practically applied to a computer learning environment (pp. 166-170). Association for Computing Machinery, Inc.

McKenzie, J. (2001). How teachers learn technology best. *The Educational Technology Journal, 10***(6)** [http://www.fno.org/mar01/howlearn.html]

McLuhan, M. (1964). *Understanding media: The extensions of man*. Cambridge, MA: MIT Press.

Merriam, S. B., & Brockett, R. G. (2007). *The profession and practice of adult education: an introduction*. San Francisco, CA: Jossey-Bass.

Merriam, S. B., Caffarella, R. S., & Baumgartner, L. M. (2006). *Learning in adulthood* (3rd ed). San Francisco: Jossey-Bass.

Merrill, M. D. (2000). Knowledge objects and mental models (pp. 244-246). IEEE.

Merrill, M. D. (2002). First principles of instructions. *Educational Technology Research and Development, 50*(3), 43-59.

Merrill, M.D. (1999). Instructional Transaction Theory (ITT): Instructional Design Based on Knowledge Objects. In Charles M. Reigeluth (Ed.), *Instructional Design Theories and Models: A New Paradigm of Instructional Technology.* Lawrence Erlbaum Associates.

Meyer, A. K. (2007). Student perceptions of face-to-face and online discussions: the advantage goes to… *Journal of Asynchronous Learning Networks, 11*(4). Retrieved July 27, 2008, from Sloan-C-Publications database http://www.sloan-c.org/publications/jaln/index.asp

Meyer, A. K. (2008). If higher education is a right, and distance education is the answer, then who will pay? *Journal of Asynchronous Learning Networks, 12*(1). Retrieved July 27, 2008, from Sloan-C-Publications database http://www.sloan-c.org/publications/jaln/index.asp

Meyer, E. & Land, R. (Ed) (2006). *Overcoming barriers to student understanding: Threshold concepts and troublesome knowledge*. London: Routledge.

Meyer, E., & Land, R. (2003). Threshold concepts and troublesome knowledge. ETL Project Occasional Paper 4 Retrieved March, 17, 2008 from http://www.ed.ac.uk/etl/docs/ETLreport4.pdf

Meyer, K. A. (2003). Face-to-face versus threaded discussions: The role of time and higher-order thinking. *Journal of Asynchronous Learning Networks*, 7(3), 55-65.

Meyer, K. A. (2004). Evaluating online discussions: Four different frames of analysis. *Journal of Asynchronous Learning Networks*, 8(2), 101-114.

Mezirow, J. (1991). *Transformative dimensions of adult learning.* San Francisco: Jossey-Bass.

Michelson, E., & Mandell, A., & contributors (2004). *Portfolio development and the assessment of prior learning.* Sterling VA: Stylus.

Miller, J. (1999). Critical thinking and asynchronous discussion. *Inquiry: Critical Thinking Across the Disciplines, 19*(1), 18-27.

Mingus, T. T. Y., & Grassi, R. M. (1998). Algorithmic and recursive thinking current beliefs and their implications for the future. In L. J. Morrow & M. J. Kenney (Eds.), *The teaching and learning of algorithms in school mathematics* (pp. 32-42). Reston, VA: National council of Teachers of Mathematics.

Miyoshi, M. (1998). "Globalization," culture, and the university. In F. Jameson & M. Miyoshi (Eds.), *The cultures of globalization* (pp. 247-270). Durham, NH: Duke University Press.

Moallem, M. (2001). Applying Constructivist and Objectivist Learning Theories in the Design of A Web-Based Course: Implications for Practice. *Educational Technology & Society,* 4(3), 113-125.

Molenda, M. (1997). Historical and Philosophical Foundations of Instructional Design. In Tennyson, R., Schott, F., Seel, N., & Dijkstra, S. (Eds), *Instructional Design Volume 1*, 41-54. New Jersey: Lawrence Erlbaum Associates.

Molenda, M. (2008). Historical Foundations. In J. Spector, D. Merrill, J. Van Merrienboer, & M. Driscoll (Eds.), *Handbook of Research on Educational Communications and Technology, 3rd ed.* (pp. 525-526). New York: Lawrence Erlbaum Accociates.

Moloney, J. F., Hickey, C. P., Bergin, A. L., Boccia, J., Polley, K., & Riley, J. E. (2007). Characteristics of successful local blended programs in the context of the Sloan-C pillars. *Journal of Asynchronous Learning Networks, 11*(1). Retrieved July 27, 2008, from Sloan-C-Publications database http://www.sloan-c.org/publications/jaln/index.asp

Moore, J. C., Sener, J, & Fetzner, M. (2006). Getting better: ALN and student success. *Journal of Asynchronous Learning Networks, 10*(3). Retrieved July 27, 2008, from Sloan-C-Publications database http://www.sloan-c.org/publications/jaln/index.asp

Moore, M. G., & Anderson, W. G. (2003). *Handbook of Distance Education.* NJ : L. Erlbaum Associates.

Moore, M., & Kearsley, G. (1996). *Distance education: A systems view.* California: Wadsworth Publishing Company.

Moreno, L., Gonzalez, C., Castilla, I., Gonzalez, E. J., & Sigut, J. (2007, May). Use of constructivism and collaborative teaching in an ILP processors course. IEEE Transactions on Education: 50 (2), 101-111.

Morrison, G., Ross, S., & Kemp, J. (2004). *Designing effective instruction.* New York: John Wiley & Sons.

Moss, J., Fearnley-Sander, M., & Hiller, C. (2001). Techno hero fiasco. In D. Murphy, R. Walker & G. Webb (Eds.). *Online learning and teaching with technology: Case studies, experience and practice* (pp. 155-161). London: Kogan Page.

Mossburger, K., Tolbert, C. J., & Stansbury, M. (2003). *Virtual inequality: Beyond the digital divide.* Washington, DC: Georgetown University Press.

Mueller, J. (2006). *What Is Authentic Assessment?* Retrieved June 9, 2008 from http://jonathan.mueller.faculty.noctrl.edu/toolbox/whatisit.htm#definitions.

Mueller, J. (2006). "Authentic Assessment Toolbox." Retrieved June 9, 2008 from http://jonathan.mueller.faculty.noctrl.edu/toolbox/.

Muilenburg, L., & Berge, Z. L. (2006). A framework for designing questions for online learning. *Welcome to emoderators.* Retrieved May 28, 2008, from http://www.emoderators.com/moderators/muilenburg.html

Murphy, D., Walker, R., & Webb, G. (Eds.). (2001). *Online learning and teaching with technology: Case studies, experience and practice.* London: Kogan Page.

Murphy, E. (2004). Identifying and measuring ill-structured problem formulation and resolution in online asynchronous discussions. *Canadian Journal of Learning and Technology*, 30(1), 5-20.

Myers, S. A. (2006). Can I use transformative pedagogy when teaching online? *College Teaching, 2*(1), 82-94.

Nachmias, R., Mioduser, D., Oren, A., & Lahav, A. (1999). Taxonomy of educational websites - A tool for supporting research, development and implementation of web-based learning. *International Journal of Educational Telecommunications, 5*(3), 193-210. Retrieved on 29.3.08 from http://muse.tau.ac.il/publications/taxonomy.html

Nardi, B. A., & Zarmer, C. L. (1991). *Beyond models and metaphors: Visual formalisms in user interface design* (p. 487). IEEE.

Natriello, G. (2005). Modest changes, revolutionary possibilities; distance learning and the future of education. *Teachers College Record, 107*(8), 1885-1904.

Negroponte, N. (1996). *Being digital.* NY: Vintage Press.

Nelson, M. R. (2008, Jan.). E-books in higher education: Nearing the end of the era of hype? *Educause Center for Applied Research: Research Bulletin.*

Newby, T. J. N., Stepich, D. A., Lehman, J. D., & Russell, J. D. (2000). *Instructional technology for teaching and learning: Designing instruction, integrating computers, and using media* (2nd ed.). Upper Saddle River, NJ: Merrill Prentice Hall.

Newfield, C. (2008). *Unmaking the public university: The forty year assault on the middle class.* Cambridge, MA: Harvard University Press.

Newman, D. R., Johnson, C., Cochrane, C., & Webb, B. (1996). *An experiment in group learning technology: Evaluating critical thinking in face-to-face and computer-supported seminars.* Retrieved May 28, 2008, from http://www.helsinki.fi/science/optek/1996/n1/newman.htm/contents.html

Newman, I., Daniels, M., & Faulkner, X. (2002). Open ended group projects a 'tool' for more effective teaching. Australian Computer Society, Inc.

Nielson, J. (2000). *Designing web usability: The practice of simplicity.* Indianapolis: New Riders Publishing.

Novak, J. D. (1977). *A theory of education.* Ithaca, NY: Cornell University Press.

Novak, J. D., & Gowin, D. B. (1984). *Learning how to learn.* New York: Cambridge University Press.

NQAI (National Qualifications Authority of Ireland) (2003). *The National Framework of Qualifications – An Overview.* Dublin: NQAI. Retrieved October 15, 2008, from http://www.nqai.ie/docs/publications/13.pdf

Ó Fathaigh, M. (2002, March). E-learning and access: Some issues and implications. Paper presented to UACE conference, University of Bath, UK. Retrieved May, 18, 2008 from http://www.ucc.ie/en/ace/Publications/DocumentFile,19890,en.pdf

Ó Fathaigh, M. (2003, November). E-learning and access: Some European/Irish issues and implications. *Sloan-C conference 9th International Conference on Asynchronous Learning Networks (ALN)*, Orlando, USA. Retrieved May, 18, 2008 from http://www.sloan-c.org/conference/proceedings/2003/pdf/1610.pdf

O' Brien, S., & Ó Fathaigh, M. (2007). Learning partnerships for social inclusion: Exploring lifelong learning contexts, issues and approaches. Cork: Oak Tree Press.

O'Reilly, T. (2005). What Is Web 2.0: Design Patterns and Business Models for the Next Generation of Software. Downloaded June 8, 2008 from http://www.oreillynet.com/pub/a/oreilly/tim/news/2005/09/30/what-is-web-20.html.

Oblinger, D., Barone, C. A., & Hawkins, B. L. (2001). *Distributed education and its challenges: An overview.* American Council on Education (ACE). Retrieved June, 2003: http://www.acenet.edu/bookstore/pdf/distributed-learning/distributed-learning-01.pdf

OECD (Organization for Economic Co-Operation and Development) (2001). *Starting Strong: Early childhood education and care.* Paris, OECD.

OECD (Organization for Economic Co-Operation and Development) (2006). *Starting Strong II: Early childhood education and care.* Paris, OECD.

Offerman, M., & Tassava, C. (2006). A Different Perspective on Blended Learning: Asserting the efficacy of Online Learning at Capella University. In C. J. Bonk & C. R. Graham (Eds.), *The Handbook of Blended Learning. Global Perspectives, Local Designs* (pp. 516-528). San Francisco: Pfeiffer.

Oliver, D., Dobele, T., Greber, M., & Roberts, T. (2004). This course has a Bloom Rating of 3.9. In R. Lister & A. Young (Eds.), *Proceedings of the Sixth Conference on Australasian Computing Education – Vol. 3.* (pp. 227-231). Darlinghurst, Australia: Australian Computer Society, Inc. Retrieved October 3, 2008, from http://portal.acm.org/citation.cfm?id=979968.979998&coll=GUIDE&dl=ACM&CFID=5033070&CFTOKEN=27681204

Oliver, R, Herrington, J., & Omari, A. (1997). Creating effective instructional materials for the World Wide Web. In *Proceedings of AusWeb 97 Conference*.

Oliver, R. (1999). Exploring strategies for online teaching and learning. *Distance Education, 20*(2), 240-254.

Oliver, R., & Herrington, J. (2003). Exploring the technology-mediated learning from a pedagogical perspective. *Journal of Interactive Learning Environments, 11*(2), 111-126.

Olssen, M. (1996, Sept.) Radical constructivism and its failings: Anti-realism and individualism. *British Journal of Educational Studies, 44*(3), 275-295.

Ong, W. J. (1982). *Orality and literacy: The technologizing of the word.* London: Routledge.

Otte, G. (2007). Online instruction as *local* education: CUNY's online baccalaureate. *Journal of Asynchronous Learning Networks, 11*(1). Retrieved July 27, 2008, from Sloan-C-Publications database http://www.sloan-c.org/publications/jaln/index.asp

Otting, H., & Zwaal, R. (2006). Critical task characteristics in problem-based learning. *Industry and Higher Education, 20*, 347.

Oulasvirta, A., Tamminen, S., & Höök, K. (2005). Comparing two approaches to context: Realism and constructivism (pp. 195-198). Association for Computing Machinery, Inc.

Owston, R., Garrison, D., & Cook, K. (2006). Blended learning at Canadian universities: issues and practices. In C. J. Bonk & C. R. Graham (Eds.), *The Handbook of Blended Learning. Global Perspectives, Local Designs* (pp. 338-350). San Francisco: Pfeiffer.

Paas, F., & Van Merriënboer, J. J. G. (1994). Instructional control of cognitive load in the training of complex cognitive tasks. *Educational Psychology Review, 6*, 51-71.

Paas, F., Renkl, A., & Sweller, J. (2003). Cognitive load theory and instructional design: Recent developments. *Educational Psychologist, 38*(1), 1-4.

Palincsar, A. S., & Brown, A. L. (1984). Reciprocal teaching of comprehension-fostering and comprehension-monitoring activities. *Cognition and Instruction, 2*, 117-175.

Palloff, R. M., & Pratt, K. (1999). *Building learning communities in cyberspace: Strategies for the online classroom.* San Francisco, CA: Jossey-Bass.

Palloff, R. M., & Pratt, K. (2001). *Lessons from the cyberspace classroom: The realities of online teaching.* San Francisco: Jossey-Bass.

Palloff, R. M., & Pratt, K. (2003). *The virtual student: A profile and guide to working with online learners.* San Francisco: Jossey-Bass.

Palloff, R. M., & Pratt, K. (2007). *Building online learning communities: Effective strategies for the virtual classroom.* San Francisco: Wiley.

Palloff, R., & Pratt, K. (2005). *Collaborating online: Learning together in community.* San Francisco: Jossey-Bass.

Palmer, P. (1998). *The Courage to Teach: Exploring the Inner Landscape of a Teacher's Life*. San Francisco, CA: Jossey-Bass.

Panda, S., & Juwah, C. (2006). Professional Development of Online Facilitators in Enhancing Interactions and Engagement: A Framework. In C. Juwah (Ed.), *Interactions in Online Education. Implications for Theory and Practice*. London: Routledge.

Papert, S. (1987). *Constructionism: A New Opportunity for Elementary Science Education.* Retrieved May 12, 2008 from http://nsf.gov/awardsearch/ showAward.do?AwardNumber=8751190

Papert, S. (1991). Situating Constructionism. In I. Harel & S. Papert (Ed.), *Constructionism* (pp.1-12). Norwood, NJ: Ablex Publisheng Corporation.

Pargas, R. P. (2006). Reducing lecture and increasing student activity in large computer science courses. *Proceedings of the 11th annual SIGCSE conference on In-*

novation and technology in computer science education, 38(3), (pp. 3-7). New York: Association for Computing Machinery. Retrieved October 3, 2008, from http://portal.acm.org/citation.cfm?id=1140124.1140129&coll=GUIDE&dl=ACM&CFID=5033070&CFTOKEN=27681204

Pellegrino, J. W., Chudowsky, N., & Glaser, R. (Eds.) (2001). *Knowing what students know: The science and design of educational assessment.* Washington, D.C.: National Academies Press.

Petraglia, J. (1998). *Reality by design: The rhetoric and technology of authenticity in education.* Mahwah, NJ: Lawrence Erlbaum Associates.

Philip, D. (2008). *Economics, knowledge building, and the Net Generation: A model of learning for Net Generation students.* Retrieved April 30, 2008 from: http://education-2008.org/speakers/don_philip/don_philip_abstract.cfm.

Phillips, D. (1995). The good, the bad and the ugly: The many faces of constructivism. *Educational Researcher, 24*(7), 5-12.

Phillips, D. C. (1995). The Good, the Bad and the Ugly: The Many Faces of Constructivism. *Educational Researcher, 24*(7), October, 5-12.

Phillips, D. C. (1995). The good, the bad, and the ugly: The many faces of constructivism. *Educational Researcher, 24*(7), 5-12.

Phillips, D. C. (2003). Subjectivity and Objectivity: An objective Inquiry. In M. Hammersley (Ed.), *Educational Research. Current issues* (pp. 57-72) London: Paul Chapman Publishing.

Phillips, D.C. (Ed). (2000). *Constructivism in Education: Opinions and Second Opinions on Controversial Issues.* Ninety-ninth Yearbook of the National Society for the Study of Education. Part I. Chicago, Illinois, The University of Chicago Press.

Picciano, A. G. (2002). Beyond student perceptions: Issues of interaction, presence and performance in an online course. *Journal of Asynchronous Learning Networks, 6*(1), 21-40.

Picciano, A. G., & Dziuban, C.D. (Eds.). (2007). *Blended learning: Research projects.* New York: Sloan-C.

Picciano, A. G., & Steiner, R. V. (2008). Bringing the real world of science to children: a partnership of the American Museum of Natural History and the City University of New York. *Journal of Asynchronous Learning Networks, 12*(1). Retrieved July 27, 2008, from Sloan-C-Publications database http://www.sloan-c.org/publications/jaln/index.asp

Pilkington, R. (2004). Developing discussion for learning. *Journal of Computer Assisted Learning, 20*(3), 161-164.

Poesio, M., & Traum, D. (1997, November). *Representing conversation acts in a unified semantic/pragmatic framework.* Paper presented at Working Notes: AAAI Fall Symposium on Communicative Action in Humans and Machines, Cambridge, MA.

Popper, K. (1968). *Conjectures and refutations.* New York: Harper

Portimojärvi, T. (2006) Synchronous and Asynchronous Communication in Online Problem-based Learning. In Verkkovirityksiä – ongelmaperustainen oppiminen ja projektioppiminen verkossa.

Postman, N. (1985). *Amusing ourselves to death.* New York: Penguin.

Postman, N. (1993). *Technopoly: The surrender of culture to technology.* New York: Vintage Books.

Preece, J. (2000). *Online communities: Designing usability and supporting sociability.* Hoboken, NJ: Wiley.

Prensky, M. (2002). The motivation of gameplay, or the REAL 21st century learning revolution. < http://www.marcprensky.com/writing/Prensky%20-%20The%20Motivation%20of%20Gameplay-OTH%2010-1.pdf >.

Prensky, M. (2006). Engage me or enrage me: What today's learner's demand. *Educause* October, 60-64. Retrieved September 17, 2008 from http://net.educause.edu/ir/library/pdf/erm0553.pdf

Preziosi, D. (1992). The question of art history. *Critical Inquiry*, 18(2). 363-386.

Prince, M. (2004). Does active learning work? A review of the research. *Journal of Engineering Education, 93*(3), 223-231.

Prosser, M., Ramsden, P., Trigwell, K., & Martin, E. (2003). Dissonance in Experience of Teaching and its Relation to the Quality of Student Learning. *Studies in Higher Education, 28*(1), 37-48.

Pulkkinen, J. (2004). The paradigms of e-Education: An analysis of the communication structures in the research on information and communication technology integration in education in the years 2000–2001. Unpublished doctoral dissertation, Faculty of Education, University of Oulu, Finland. Retrieved May 29, 2008 from http://herkules.oulu.fi/isbn9514272463/

Pusser, B., Breneman, D. W., Gansneder, B. M., Kohl, K. J., Levin, J. S., Milam, J. H., & Turner, S. E. (2007). Returning to learning: Adults' success in college is key to America's future. Retrieved May 18, 2008, from http://curry.edschool.virginia.edu/files/pathways/ReturntolearningApril2007.pdf

Puzziferro-Schnitzer, M. & Kissinger, J. (2005). Supporting online adjunct faculty: a virtual mentoring program. *Journal of Asynchronous Learning Networks, 9*(2). Retrieved July 27, 2008, from Sloan-C-Publications database http://www.sloan-c.org/publications/jaln/index.asp

Ramsden, P., Prosser, M., Trigwell, K., & Martin, E. (2007). University teachers' experiences of academic leadership and their approaches to teaching. *Learning and Instruction, 17*(2), 140-155.

Rankin, W. (2007). Transformations in 21st-Century education. *ITHINKED*. Retrieved May 30, 2008, from http://ithinked.com/archives/2007/10/transformations-in-21st-century-education-william-rankin/

Reeves, T. (1999). *"Keynote Address" Presented at ED-MEDIA 1999 World Conference on Educational Multimedia, Hypermedia & Telecommunications*. Seattle, Washington.

Reeves, T., Herrington, J., & Oliver, R. (2002). Authentic Activities and Online Learning. *Australian Journal of Educational Technology, 19*(1), 59-71

Reeves, T., Herrington, J., & Omari, A. (1996). *Creating effective instructional materials for the World Wide Web*. Retrieved May 31, 2008 from Reeves, T., http://elrond.scam.ecu.edu.au/oliver/docs/96/AUSWEB1d.pdf

Reigeluth, C. (2005). New Instructional Theories and Strategies for a Knowledge-Based Learning. In Spector, M., Ohrazda, C., Van Schaack, A., & Wiley, D (Eds.), *Innovations in Instructional Technology*, (pp. 207-218). New Jersey: Lawrence Erlbaum Associates.

Reiser, R., & Dempsey, J. (2002). *Trends and Issues in Instructional Design and Technology*. New Jersey: Pearson.

Resnick, M. (1991). Xylophones, Hamsters, and Fireworks: The Role of Diversity in Constructionist Activities. In I. Harel & S. Papert (Ed.), *Constructionism* (pp. 1-12). Norwood, NJ: Ablex Publisheng Corporation.

Resnick, M., & Ocko, S. (1991). LEGO/Logo: Learning Through and About Design. In I. Harel & S. Papert (Ed.), *Constructionism* (pp.1-12). Norwood, NJ: Ablex Publisheng Corporation.

Reynolds, C. L. (1999). Computer technologies and teaching and learning: What do the students and teachers say? In T. H. Murphy (Ed.), *Proceedings of the western region agricultural education research conference*. Texas A&M University Retrieved December, 10, 2007 from http://pubs.aged.tamu.edu/conferences/WRAERC1999/pdf/wr-1999-178.pdf

Richardson, J. C. & Swan, K. (2003). Examining social presence in online courses in relation to students' perceived learning and satisfaction. *Journal of Asynchronous Learning Networks, 7*(1), 68-88.

Richert, A. E. (1991). Case methods and teacher education: Using cases to teach teacher reflection. In B. R. Tabachnich & K. Zeichner (Eds.), *Issues and practices in inquiry-oriented teacher education* (pp. 130-150). London: Falmer Press.

Ridgway, J., McCusker, S., & Pead, D. (2004). *Literature Review of E-assessment.* Futurelab Retrieved June 8, 2008 from http://dro.dur.ac.uk/1929/.

Riegler, A. (2005). Editorial. The Constructivist Challenge. *Constructivist Foundations, 1*(1), 1-8.

Riggins, F. J., & Slaughter, K. T. (2006). The role of collective mental models in IOS adoption: Opening the black box of rationality in RFID deployment.

Riva, G. (2001). Communicating in CMC: making order out of miscommunication. In L. Anolli et al. (Eds.), *Say not to say: new perspectives on miscommunication* (pp. 203–233). Amsterdam: IOS Press.

Robinson, B., & Latchem, C. (2003). Teacher education: challenges and change. In B. Robinson, & C. Latchem (Eds.), *Teacher education through open and distance learning* (pp. 1-27). London: RoutledgeFalmer.

Roblyer, M. D. (2003). *Integrating educational technology into teaching* (3rd ed.). Upper Saddle River, NJ: Merrill Prentice Hall.

Rogers, D., & Swan, K. (2004). Self-regulated learning and internet searching. *Teachers College Record, 106*(9), 1804-1824.

Rogoff, B. (1990). *Apprenticeship in thinking.* New York: Oxford University Press.

Romiszowski, A. (1995). Use of hypermedia and telecommunications for case-study discussions in distance education. In: Lockwood, F. (Ed.) *Open and distance learning today.* New York: Routledge.

Romiszowski, A., & Mason, R. (1996). Computer-mediated communication. In D. Jonassen (Ed.), *Handbook of research for educational communications and technology* (pp. 438-456). New York: Simon & Schuster Macmillan.

Ronteltap, C. F. M., & Eurelings, A. M. C. (1997). POLARIS: The Functional Design of an Electronic Learning Environment to Support Problem-based Learning. In T. Müldner & T. C. Reeves (Eds.), *Educational Multimedia and Hypermedia.* Charlottesville: Association for the Advancement of Computing in Education (pp. 1802-1807).

Ronteltap, C. F. M., Goodyear, P., & Bartoluzzi, S. (2004). A pattern language as an instrument in designing for productive learning conversations. *World Conference on Educational Multimedia, Hypermedia and Telecommunications, 1*, 4271-4276.

Roschelle, J., Rosas, R. & Nussbaum, M. (n.d.). *Towards a design framework for mobile computer-supported collaborative learning* (pp. 520-524).

Ross, L., Greene, D., & House, P. (1977). The false consensus phenomenon: An attributional bias in self-perception and social perception processes. *Journal of Experimental Social Psychology, 13*, 279-301.

Rossett, A., Douglis, F., & Frazee, R. V. (2003). Strategies for building blended learning. *Learning Circuits.* Retrieved May 14, 2008, from http://www.learningcircuits.org/2003/jul2003/rossett.htm

Roszak, T. (1986). *The cult of information: The folklore of computers and the true art of thinking.* New York: Pantheon.

Rourke, L., & Anderson, T. (2002). Exploring social presence in computer conferencing. *Journal of Interactive Learning Research, 13*(3), 259-275.

Rourke, L., Anderson, T. Garrison, D. R., & Archer, W. (1999). Assessing Social Presence in Asynchronous, Text-Based Computer Conferencing. *Journal of Distance Education, 14*(3), 51-70.

Rourke, L., Anderson, T., Archer, W., & Garrison, D. R. (1999). Assessing social presence in asynchronous, text-based computer conferences. *Journal of Distance Education, 14*(3), 51-70.

Rourke, L., Anderson, T., Garrison, D. R., & Archer, W. (2001). Assessing social presence in asynchronous text-based computer conferencing. *Journal of Distance Education, 14*(2), 50-71.

Rovai, A. P., & Ponton, M. K. (2005). An examination of sense of classroom community and learning among

African American and Caucasian graduate students. *Journal of Asynchronous Learning Networks, 9*(3). Retrieved July 27, 2008, from Sloan-C-Publications database http://www.sloan-c.org/publications/jaln/index.asp

Roxanne-Hiltz, S., & Goldman, R. (Eds.) (2005). *Learning Together Online: Research on Asynchronous Learning Networks*. Mahwah, NJ: Lawrence Erlbaum Associates.

Rubin, D. (1995). Constructivism, Sexual Harassment, and Presupposition: A (Very) Loose Response to Duit, Saxe and Spivey. In P. Steffe & J. Gale (Eds.), *Constructivism in Education* (pp. 355-366). Hillsdale, NJ: Lawrence Erlbaum Associates.

Rumelhart, D. E. (1989). The architecture of mind: A connectionist approach. In M. I. Posner (Ed.), *Foundations of cognitive science* (pp. 133-159). Cambridge, MA: MIT Press.

Säljö, R. (2003). Representational tools and the transformation of learning. In B. Wasson, S. Ludvigsen & U. Hoppe (Eds.), *Designing for change in networked learning environments: Proceedings of the international conference on CSCL*. Dordrecht: Kluwer Academic Publishers.

Salmon, G. (2000 & 2004). *E-moderating: The key to teaching and learning online*. London: Taylor and Francis.

Salmon, G. (2000). *E-Moderating. The Key to Teaching and Learning Online.* London and Sternling (USA): Kogan

Salmon, G. (2001). E-*moderating: The key to teaching and learning online*. London: Kogan Page.

Salmon, G. (2002). *E-tivities. The Key to Active Online Learning.* London and Sternling (USA): Kogan

Salmon, G. (2002). Mirror, mirror, on my screen..Exploring online reflections. *British Journal of Educational Technology, 33*(4), 379-391.

Salomon, G. (Ed.) (1993). *Distributed cognitions*. Cambridge, UK: Cambridge University Press.

Sandars, J., & Schroter, S. (2007). Web 2.0 technologies for undergraduate and postgraduate medical education: an online survey. *Postgraduate Medical Journal, 83*(986), 759-762.

Sanderson, P. M., & Murtagh, J. M. (1989). Troubleshooting with an inaccurate mental model (pp. 1238-1243). IEEE.

Saraiya, P., North, C., & Duca, K. (2005, July – Aug.) An insight-based methodology for evaluating bioinformatics visualizations. IEEE Transactions on Visualization and Computer Graphics, *11*(4), 443.

Savery, J. R. & Duffy, T. M. (2001). *Problem based learning: An instructional model and its constructivist framework*. Indiana University: CRLT Technical Report No. 16-01. Retrived May 29, 2008 from http://cee.indiana.edu/publications/journals/TR16-01.pdf

Savery, J., & Duffy, T.(1995). Problem based learning: An instructional model and its constructivist framework. *Educational Technology, 35*(5), 31-38.

Savin-Baden, M. (2006). The challenge of using problem-based learning online. In M. Savin-Baden & K. Wilkie (Eds.) *Problem-based Learning Online* (pp. 3-13). Maidenhead: Open University Press.

Scardamalia, M., & Bereiter, C. (1994). Computer Support for Knowledge-Building Communities. *The Journal of the Learning Sciences, 3*(3), 256-283.

Schank, P., Ranney, M., & Hoadley, C. (1995). Convince Me [Computer Programe and Manual]. In J.R. Jungck, N. Peterson & J.N. Calley (Eds.), *The BioQuest Library.* College Park, MD: Academic Software Development Group, University of Maryland.

Schegloff, E. A. (1987). Recycled Turn Beginnings: A Precise Repair Mechanism in Conversation's Turn-Taking Organization. In G. Button, L. Lee (Eds.), *Talk and Social Organization*. Clevelon, England: Multilingual Matters.

Schön, D. (1983). *The reflective practitioner*. New York: Basic Books.

Schön, D. (1987). *Educating the reflective practitioner: Toward a new design for teaching and learning in the professions*. San Francisco: Jossey-Bass.

Schribner, S. (1984). Studying working intelligence. In B. Rogoff & J. Lave (Eds.), *Everyday cognition: Its development in social context* (pp. 9-40). Cambridge, MA: Harvard University Press.

Schroth, C. (2007). Web 2.0 versus SOA: Converging concepts enabling seamless cross-organizational collaboration. *The 9th IEEE International Conference on E-Commerce Technology and The 4th IEEE International Conference on Enterprise Computing, E-Commerce and E-Services (CEC-EEE 2007).*

Schrum, L., & Hong, S. (2002). Dimensions and strategies for online success: voices from experienced educators. *JALN, 6 (1).* Retrieved August 11, 2002 from http://www.aln.org/alnWeb/journal/jaln-vol6issue1.htm#schrum

Schuh, K., & Barab, S. (2008). Handbook of Research on Educational Communications and Technology. In J. M. Spector, M. D. Merrill, J. van Merrienboer, & M. Driscoll (Eds.), *Philosophical Perspective* (pp. 67-81). New York, NY: Taylor & Francis.

Schwandt, T. (2001). *Dictionary of qualitative inquiry* (2nd ed.). Thousand Oaks, California: Sage.

Schwartz, D. L., & Lin, X. D. (2000). Computers, productive agency, and the effort after shared meaning. *Journal of Computing in Higher Education, 12*(2), 3-33.

Schwarz, B. B., & Glassner, A. (2003). The blind and the paralytic: Fostering argumentation in everyday and scientific issues. In J. Andriessen, M. Baker, & D. Suthers (Eds.), *Arguing to learn: Confronting cognitions in computer-supported collaborative learning environments* (pp. 227-260). Dordrecht: Kluwer Academic Publishers.

Schweber, C. (2008). Determined to learn: accessing education despite life-threatening disasters. *Journal of Asynchronous Learning Networks, 12*(1). Retrieved July 27, 2008, from Sloan-C-Publications database http://www.sloan-c.org/publications/jaln/index.asp

Scott, T. (2003). Bloom's taxonomy applied to testing in computer science classes. *Journal for Computing in Small Colleges, 19*(1), 267-274.

Second Life. http://www.secondlife.com.

Second Life--North American User Statistics. SL Wikispaces. https://secondter.wikispaces.com/Second+Life--North+American+User+Statistics?responseToken=206937f26492c9a5fbe6b2046cdc01f6

Seel, N. (2005). Cognition Designing a Model-Centered Learning Environments: Hocus Pocus or the Focus Must Strictly be on Locus. In M. Spector, C. Ohrazda, A. Van Schaack, & D. Wiley, (Eds.), *Innovations in Instructional Technology*, (pp. 95-90). New Jersey: Lawrence Erlbaum Associates.

Sennett, R. (2006). *The culture of the new capitalism.* New Haven: Yale University Press.

Sennett, R. (2008). *The craftsman.* New Haven: Yale University Press.

Shafer-Landau, R. (2003). *Moral Realism.* Oxford: Clarendon Press.

Shang, Y., Shi, H-C., & Chen, S-S. (2001, Summer). An intelligent distributed environment for active learning. *ACM Journal of Educational Resources in Computing, 1*(2), 3, 5-6.

Shaochun X., & Rajlich, V. (2004). Cognitive process during program debugging. *Proceedings of the Third IEEE International Conference on Cognitive Informatics, (pp.* 176-182). Retrieved October 3, 2008, from http://www2.computer.org/portal/web/csdl/doi/10.1109/COGINF.2004.1327473

Sharpe, R., Benfield, G., Roberts, G., & Francis, R. (2006). *The undergraduate experience of blended E-Learning: a review of UK literature and practice*, Retrieved December 12, 2006, from http://www.heacademy.ac.uk/4884.htm

Shea, P. (2006). A study of students' sense of learning community in online environments. *Journal of Asynchronous Learning Networks, 10*(1), 35-44.

Shea, P. J., Pickett, A. M., & Pelz, W. E. (2003). A follow-up investigation of "teaching presence" in the SUNY Learning Network. *Journal of Asynchronous Learning Networks, 7*(2), 61-80.

Shea, P., & Bidjeramo, T. (2008). *Community of inquiry as a theoretical framework to foster "epistemic engagement" and "cognitive presence" in online education.* Paper presented at the Annual Meeting of the American Educational Research Association, New York, March.

Shea, P., Li, C. S., & Pickett, A. (2006). A study of teaching presence and student sense of learning community in fully online and web-enhanced college courses. The *Internet and Higher Education*, 9(3), 175-190.

Shea, P., Li, C., Swan K., Pickett, A. (2005) Developing Learning Community in Online Asynchronous College Courses: The Role of Teaching Presence. *Journal of Asynchronous Learning Networks, 9(*4). Retrieved May 1, 2008 from: http://www.sloan-c.org/publications/jaln/v9n4/v9n4_shea.asp

Shen, J., Hiltz, S. R., & Bieber, M. (2006, Nov.). Collaborative online examinations: Impacts on interaction, learning, and student satisfaction. *IEEE Transactions on Systems, Man, and Cybernetics—Part A: Systems and Humans, 36*(6), 1045-1053.

Sherry, L. (2000). The nature and purpose of online conversations: A brief synthesis of current research. *International Journal of Educational Telecommunications, 6*(1), 19-52.

Short, J., Williams, E., & Christie, B. (1976). *The Social Psychology of Telecommunications.* Toronto: Wiley.

Show, Don't Tell: Video and Accountability (2002). Coalition of Essential Schools National Web. Retrieved June 9, 2008 from http://www.essentialschools.org/cs/resources/view/ces_res/229.

Shulman, L. S. (2005). Signature pedagogies in the professions. *Daedalus*, 134(3), 52–59.

Siebert, H. (2004). Sozialkonstruktivismus: Gesellschaft als Konstruktion. *Sowi-online.e.v., 2.* Retrieved May 5, 2008, from http://www.sowi-online.de/journal/2004-2/sozialkonstruktivismus_siebert.htm

Siebert, H. (2005). *Pädagogischer Konstruktivismus. Lernzentrierte Pädagogik in Schule und Erwachsenenbildung.* Weinheim und Basel: Beltz Verlag

Simon, H. (1982). *Sciences of the Artificial.* Cambridge, MA: The MIT Press.

Simon, H. (2001). Learning to research about learning. In S. Carver & D. Klahr (Eds.), *Cognition and Instruction: Twenty-fice years of Progress* (pp. 205-26). Mahwah, NJ: Lawernce Erlbaum Associates.

Smith, B. (2003). The use of communication strategies in computer-mediated communication. *System, 31*, 29-53.

Smith, M. K. (2002). Malcolm Knowles, informal adult education, self-direction and anadragogy, the encyclopedia of informal education. Retrieved December, 10, 2007 from http://www.infed.org/thinkers/et-knowl.htm

Smith, P. L., & Ragan, T. J. (2005*). Instructional design.* Upper Saddle River, NJ: Merrill/Prentice Hall.

Smith, P., & Ragan, T. (1999). *Instructional design* (2nd ed.). New York: John Wiley & Sons, Inc.

Smyth, J. (1989). Developing and sustaining critical reflection in teacher education. *Journal of Teacher Education*, (pp. 2-9).

Soloway, E., Jackson, S. L., Klein, J., Quintana, C., Reed, J., Spitulnik, J., Stratford, S. J., Studer, S., Eng, J., & Scala, N. (1996). *Learning theory in practice: Case studies of learner-centered design* (pp. 189-196). Association for Computing Machinery, Inc.

Somekh, B. (2001). The great courseware gamble. In D. Murphy, R. Walker & G. Webb (Eds.). *Online learning and teaching with technology: Case studies, experience and practice* (pp. 121-128). London: Kogan Page.

Somerville, M. M., & Nino, M. (2007). Collaborative co-design: A user-centric approach for advancement of organizational *learning. Performance Measurement & Metrics, 8*(3), 180-188.

Sosnoski, J. J. (2006). Configuring African-American Culture as Virtual Experiences of History. In J. J. Sosnoski, P. Harkin, & B. Carter (Eds.), *Configuring History: Teaching the Harlem Renaissance through Virtual Reality Cityscapes* (pp. 31-41). New York: Peter Lang.

Sosnoski, J. J., & Portlock, T. (2006). Design for Narrating History in Virtual Reality Scenarios. In J. J. Sosnoski, P.Harkin, & B. Carter (Eds.), *Configuring History: Teaching the Harlem Renaissance through Virtual Reality Cityscapes* (pp. 61-68). New York: Peter Lang.

Spector, M. (2000). *A Philosophy of Instructional Design for the 21st Century.* Invited paper presentation at the American Educational Research Association Annual Meeting, 27 April 2000, New Orleans, Louisiana.

Stager, G. S. (2002, July). Computationally-rich constructivism and at-risk learners. Australian Computer Society, Inc. *CRPIT '02: Proceedings of the Seventh World Conference on Computers in Education conference on Computers in Education,* 8

Stake, R. (1988). Case study methods in educational research: Seeking Sweet Water. In R. Jaeger (Ed.), *Contemporary methods for research in education* (pp. 253-265). Washington, D.C.: American Educational Research Association.

Stake, R. (2000). Case studies. In N. Denzin & Y. Lincoln (Eds.), *Handbook of qualitative research* (2nd ed., pp. 435-454). London: Sage Publications.

Stake, R. E. (1995). *The art of case study research.* Thousand Oaks, CA: Sage Publications.

Stary, C., & Totter, A. (n.d.). On learner control in e-learning. *Trust and Control in Complex Socio-Technical Systems* (pp. 41-48).

Staver, J. R. (1998). Constructivism: Sound theory for explicating the practice of science and science teaching. *Journal of Research in Science Teaching, 35*(5), 501-520.

Steffe, L. P., & Gale, J. E. (Eds.) (1995). *Constructivism in Education.* Hillsdale, NJ: Lawrence Erlbaum.

Steiger, N. M., & Steiger, D. M. (2007). Knowledge management in decision making: Instance-based cognitive mapping. *Proceedings of the 40th Hawaii International Conference on System Sciences* (pp. 1-2).

Steimberg, Y., Ram, J., Nachmia, R., & Eshel, A. (2006). An online discussion for supporting students in preparation for a test. *Journal of Asynchronous Learning Networks, 10*(4). Retrieved July 27, 2008, from Sloan-C-Publications database http://www.sloan-c.org/publications/jaln/index.asp

Stein, D. G. (Ed.). (2004). *Buying in or selling out: The commercialization of the American research university.* Rutgers, NJ: Rutgers University Press.

Stiler, G., & Phileo, T. (2003). Blogging and blogspots: An alternative format for encouraging reflective practices among preservice teachers. *Education, 123*(4), 789-797.

Stone, A. R. (1995). Sex and death among the disembodied: VR, cyberspace, and the nature of academic discourse. In S. L. Star (Ed.), *The cultures of computing* (pp. 243-255). Oxford, UK; Blackwell Publishers.

Stotskopf, L. D., & Moorash, A. (2005). EArmyU: expanding education access and excellence to highly mobile online learners. *Journal of Asynchronous Learning Networks, 9*(2). Retrieved July 27, 2008, from Sloan-C-Publications database http://www.sloan-c.org/publications/jaln/index.asp

Stross, C. (2005). *Accelerando.* New York: Ace Books.

Sudweeks, F. (2003). Connecting students with group work. In C. Constantinou & Z. Zacharia (Eds.), *Computer-based learning in science* (Vol. 1, pp. 173-183). Nicosia, Cyprus: University of Cyprus.

Sudweeks, F. (2004). *Development and leadership in computer-mediated collaborative groups.* Unpublished PhD Thesis, Murdoch University, Perth, Australia.

Sudweeks, F., & Simoff, S. (2000). Participation and reflection in virtual workshops. In J. Burn (Ed.), *Proceed-*

ings of 3rd Western Australian Workshop on Information Systems Research. Perth, Australia (CD-ROM).

Sun Microsystems. (2008). Current Reality and Future Vision: Open Virtual Worlds. A Sun Services White Paper. http://www.sun.com/service/projectdarkstar/index.jsp.

Susstein, C. R. (2006). *Infotopia: How many minds produce knowledge.* Oxford, UK: Oxford University Press.

Swan, K. (2002). Building communities in online courses: the importance of interaction. *Education, Communication and Information, 2*(1), 23-49.

Swan, K. (2003). Developing social presence in online discussions. In S. Naidu (Ed.), *Learning and Teaching with Technology: Principles and* Practices (pp. 147-164). London: Kogan Page.

Swan, K. (2003). Learning effectiveness: What the research tells us. In J. Bourne & J. C. Moore (Eds.), *Elements of quality online education: Practice and direction* (pp. 13–45). Needham, MA: Sloan Center for Online Education.

Swan, K. (2005). A constructivist model for thinking about leaning online. In J. Bourne & J. C. Moore (Eds.), *Elements of Quality Education Online: Engaging Communities.* Needham: Sloan C.

Swan, K. (2007). Research on online learning. *Journal of Asynchronous Learning Networks, 11*(1). Retrieved July 27, 2008, from Sloan-C-Publications database http://www.sloan-c.org/ publications /jaln/index.asp

Swan, K., & Shih, L-F. (2005). On the nature and development of social presence in online course discussions. Journal of Asynchronous Learning Networks, 9(3), 115-136.

Swan, K., Richardson, J.C, Ice, P., Garrison, D. R., Cleveland-Innes, M., & Arbaugh, J. B. (2008). Validating a measurement tool of presence in online communities of inquiry. *E-Mentor, 2*(24), 1-12. http://www.e-mentor.edu.pl/e_index.php?numer=24&all=1

Swan, K., Shea, P., Fredericksen, E., Pickett, A., Pelz, W., & G. Maher (2000). Building knowledge building communities: consistency, contact and communication in the virtual classroom. *Journal of Educational Computing Research, 23*(4), 389-413.

Sweller, J., Kirschner, P. A., & Clark, R. E. (2007). Why minimally guided teaching techniques do not work: A reply to commentaries. *Educational Psychologist, 42*(2), 115-121.

Tagg, J. (2003). *The learning paradigm college.* Bolton, MA: Anker Publishing Company Inc.

Tapscott, D., & Williams, A.D. (2008). *Wikinomics: How mass collaboration changes everything.* NY: Portfolio/Penguin.

Tennyson, R., & Schott, F. (1997). Instructional Design Theory, Research, and Models. In R. Tennyson, F. Schott, N., Seel, & S. Dijkstra, (Eds.), *Instructional Design Volume 1,* (pp. 1-18). New Jersey: Lawrence Erlbaum Associates.

The Chronicle of Higher Education (2008, September 11). *Professor uses web 'Widgets' to share course content.* Retrieved October 10, 2008, from http://chronicle.com/wiredcampus/article/3307/professor-uses-web-widgets-to-share-course-content.

Thomas, Angela. (2007). *Youth Online: Identity and Literacy in the Digital Age.* New York: Peter Lang.

Thomas, D. M., & Bostrom, R. P. (2007). The role of a shared mental model of collaboration technology in facilitating knowledge work in virtual teams. *Proceedings of the 40th Hawaii International Conference on System Sciences* (pp. 1-8).

Thompson, (2007). Web 2.0 *takes on colleges and universities: The dawn of education 2.0.* Retrieved April 29, 2008 from http://www.masternewmedia.org/news/2007/04/20/Web_20_takes_on_colleges.htm

Thorne, S. L. (2003). Artifacts and cultures-of-use in intercultural communication. *Language Learning and Technology, 7*(2), 38-67.

Thorpe, M. (2002). From independant learning to collaborative learning: new communities of practice in open, distance and distributed learning. In M. Lea and K. Nicoll (Eds.), *Distributed Learning: Social and Cultural Approached to Practice* (pp. 131-52). London: RoutledgeFalmer.

Tight, M. (2004). Research into higher education: an a-theoretical community of practice? *Higher Education Research & Development, 23*(4), 395-411.

Tishman, S., & Palmer, P. (2005). Visible thinking. *Leadership Compass*, 2-3. Retrieved November, 30, 2006 from http://www.pz.harvard.edu/Research/ArtThink.htm

Torres, J. T. (2008, May 3). Stop dreaming and think: A 24-year-old's advice for high-school seniors. *Orlando Sentinel* (p. A19).

Tough, A. (1968). *Why adults learn: A study of the major reasons for beginning and continuing a learning project.* Toronto: Ontario Institutions for Studies in Education.

Tu, C. H. (2000). On-line learning migration: From social learning theory to social presence theory in CMC environment. *Journal of Network and Computer* Applications, *23*(1), 27–37.

Tu, C-H. & McIsaac, M. (2002). The relationship of social presence and interaction in online classes. *The American Journal of Distance Education, 16*(3), 131–150.

Tufekci, Z. (2008). Grooming, gossip, Facebook and Myspace. *Information, Communication & Society, 11*(4), 544-564.

Twomey-Fosnot, C. (1996). Constructivism: A psychological theory of learning. (pp. 8-33). In C. Twomey-Fosnot (Ed.), *Constructivism: Theory, perspectives, and practice.* New York: Teachers College Press.

U.S. Census Bureau (2008). *Question & answer center.* Retrieved May 18, 2008, from https://ask.census.gov/cgi-bin/askcensus.cfg/php/enduser/std_adp.php?p_faqid=284 &p_created=1077815981&p_sid=nl3o5k3j&p_accessibility=0&p_redirect=&p_lva=&p_sp=cF9zcmNoPTEmcF9zb3J0X2J5PSZwX2dyaWRzb3J0PSZwX3Jvd19jbnQ9MjYsMjYmcF9wcm9kcz0mcF9jYXRzPSZwX3B2PSZwX2N2PSZwX3BhZ2U9MSZwX3NlYXJjaF90ZXh0PWNvbGxlZ2UgZ3JhZHVhdGVz&p_li=&p_topview=1

U.S. Department of Education (2006). *A test of leadership: Charting the future of U.S. education.* Retrieved May 14, 2008, from http://www.ed.gov/about/bdscomm/list/ hiedfuture/reports/final-report.pdf

Uden, L., & Beaumont, C. (2006). *Technology and Problem-Based Learning.* Hershey, PA, USA: Information Science Publishing.

University Continuing Education Association (2006). *Lifelong learning trends: A profile of continuing education.* Washington, DC: National Center for Educational Statistics.

University of Washington (n.d.). *WOT history project.* Retrieved May 18, 2008, from http://depts.washington.edu/wtohist/about_project.htm

Valaitis, R. K., Sword, W. A., Jones, B., & Hodges, A. (2005). Problem-based learning online: perceptions of health science students. *Advances in Health Sciences Education, 10*, 231-252.

Valasidou, A., Sidiropoulos, D., & Makridou-Bousiou, D. (2005). The constructivist perspective in distance learning environments. In P. Kommers & G. Richards (Eds.), *Proceedings of World Conference on Educational Multimedia, Hypermedia and Telecommunications 2005* (pp. 1932-1935). Chesapeake, VA.

Valentine, T., & Darkenwald, G. G. (1990). Deterrents to participation in adult education: Profiles of potential learners. *Adult Education Quarterly, 41*(1), 29-41.

Van der Pol, J. (2002). *Identifying and modelling variables in complex CSCL-situations. Case study: The use of asynchronous electronic discussions.* Paper presented at the Conference on computer support for collaborative learning (CSCL2002), Boulder, CO.

Van der Pol, J. Admiraal, W. & Simons, P. R. J. (2006). The affordance of anchored discussion for the collaborative processing of academic texts. *International journal of Computer Supported Collaborative Learning, 1*(3), 339-357.

Van der Pol, J. Admiraal, W. & Simons, P. R. J. (2006). Context Enhancement for co-intentionality and co-reference. *Journal of artificial intelligence & society, 20*(3), 301-313.

Van der Pol, J., & Admiraal, W. F. (2003). Het succesvol inzetten van asynchrone elektronsiche discussies [Succesfully implementing asynchronous online discussions]. *Onderzoek van Onderwijs, 32*(2), 26–31.

Van Eck, Richard. (2007). Digital Game-Based Learning: It's Not Just the Digital Natives Who are Restless. In J. J. Hirschbuhl & J. Kelley, (Eds.), *Annual Editions: Computers in Education. Twelfth Edition* (pp. 55-63).

Van Ments, M. (1983). *The effective use of Role-Play*. London, UK: Kondon Page.

Vander Valk, F. (2008). Identity, power, and representation in virtual environments. *Online Journal of Learning and Teaching, 4*(2). Retrieved Sept. 20, 2008, from http://jolt.merlot.org/vol4no2/abstracts.htm

Vanderstraeten, R., & Biesta, G. (1998). Constructivism, Educational Research and John Dewey. *In Paideia Project on-line, 20th World Congress of Philosophy*. Retrieved May 5, 2008, from http://www.bu.edu/wcp/Papers/Amer/AmerVand.htm

Vaughan, N., & Garrison, D. R. (2006). How blended learning can support a faculty development community of inquiry. *Journal of Asynchronous Learning Networks, 10*(4), 139-152.

Veerman, A. L. (2000). *Computer-supported collaborative learning through argumentation*. Unpublished doctoral dissertation, Utrecht University, Utrecht.

Vignare, K., Geith, C., & Schiffman, S. (2006). *Business models for online learning: an exploratory* survey. *Journal of Asynchronous Learning Networks, 10*(2). Retrieved July 27, 2008, from Sloan-C-Publications database http://www.sloan-c.org/publications/jaln/index.asp

Vinge, V. (2006). *Rainbows end.* New York: Tor Books.

Virilio, P. (2000). The information bomb. (C. Turner, Trans.). Verso: London and New York.

von Glasersfeld, E. (1989). Cognition, construction of knowledge, and teaching. *SYNTHESE, 80*(1), 121-140.

Von Glasersfeld, E. (1995). A Constructivist Approach to Teaching. In L. P. Steffe, & J. E. Gale, (Eds.), *Constructivism in Education*. Hillsdale, NJ: Lawrence Erlbaum.

von Glasersfeld, E. (1995). A constructivist approach to teaching. In L. Steffe & J. Gale (Eds.), *Constructivism in education* (pp. 3-15). Hillsdale, New Jersey: Lawrence Erlbaum Associates, Inc., Publishers.

Von Glasersfeld, E. (1996). Introduction: Aspects of Constructivism. In C. Twomey-Fosnot (Ed.). *Constructivism: Theory, perspectives and practice* (pp. 3-7). Teachers College, Columbia University.

von Glasersfeld, E. (1997, May 1997). *Piaget's legacy: Cognition as adaptive activity.* Paper presented at the Presented at International Congress "Does Representation need Reality?" Vienna.

Von Glasersfeld, E. (2005). Introduction: Aspects of Constructivism. In C. Fostenot, (Ed.), *Constructivism: Theory, Perspective, and Practice.* New York: Columbia University.

Vosniadou, S., & Ortony, A. (1989). *Similarity and Analogical Reasoning.* Cambridge University Press.

Vrasidas, C., Zembylas, M. and Legaspi, J. M. B. (2004). EmpowerICT: eLearning for in-service teacher education and support, *International Journal of the Computer, the Internet and Management, 12,* 2, 81-86. Retrieved May 26.2008 from http://www.ijcim.th.org/v12n2/pdf/p81-86-vrasidas-EmpowerICT.pdf

Vygotsky, L. (1962). *Thought and language*. Cambridge, MA: MIT Press. Revised and edited by A. Kozulin, 1986.

Vygotsky, L. S. (1978). *Mind in Society: The Development of Higher Psychological Process.* Cambridge, MA: Harvard University Press.

Vygotsky, L.S. (1934/1986) *Thought and language.* Cambridge, MA: MIT Press.

Vygotsky, L.S. (1978). *Mind and society: The development of higher mental processes*. Cambridge, MA: Harvard University Press.

Wagner, C., & Prasarnphanich, P. (2007). Innovating collaborative content creation: The role of altruism and wiki technology. Proceedings of the 40th Hawaii International Conference on System Sciences 2007 (pp. 1-10). IEEE.

Wagner, E. D. (2006). On Designing Interaction Experiences for the Next Generation of Blended Learning. In C. J. Bonk & C. R. Graham (Eds.) *The Handbook of Blended Learning. Global Perspectives, Local Designs* (pp. 41-55). San Francisco: Pfeiffer.

Walters, S. (2000). Globalization, adult education, and development. In Stromquist & Monkman (Eds.), *Globalization and education: Integration and contestation across cultures* (pp. 197-215). Lanham, MD: Rowman & Littlefield.

Walther, J. (1994). Interpersonal effects in computer mediated interaction. *Communication Research, 21*(4), 460-487.

Wang, F-Y., Zeng, D., Carley, K. M., & Mao, W. (2007, Mar. – Apr.). *Trends and controversies: Social computing: From social informatics to social intelligence* (pp. 79-83). IEEE.

Wang, M. (2007). Designing online courses that effectively engage learners from diverse cultural backgrounds *British Journal of Educational Technology,* 38(2), 294-311. Retrieved August, 10, 2007 from http://www.blackwellsynergy.com/doi/pdf/10.1111/j.1467-8535.2006.00626.x

Wang, M., & Kang, M. (2006). Cybergogy for engaged learning: A framework for creating learner engagement through information and communication technology . In M. S. Khint & D. Hung (Eds.), *Engaged learning with emerging technologies.* (pp. 225-253). Springer: New York.

Wartfoskij, M. (1973). *Models*. Dordrecht, NL: D. Reidel.

Watson, K., & Harper, C. (2008, Feb.). Supporting knowledge creation: Using wikis for group collaboration. *Educause Center for Applied Research: Research Bulletin: 2008*(3), 2-13.

WCET, (2008). *WCET Annual Conference*. Retrieved January 15, 2008, from:http://www.wcet.info/2.0/.

Webb, N. M., & Mastergeorge, A. M. (2003). Promoting effective helping behavior in peer-directed groups. *International Journal of Educational Research, 39*, 73-97.

Webster, L., & Murphy, D. (2001). The reluctant software developers. In D. Murphy, R. Walker & G. Webb (Eds.), *Online learning and teaching with technology: Case studies, experience and practice* (pp. 114-120). London: Kogan Page.

Wegerif, R. (1998). The social dimension of asynchronous learning networks. *Journal of Asynchronous Learning Networks, 2*(1), 34-49.

Wegerif, R. (2007). From Dialectic to Dialogic: A response to Wertsch and Kazak. In T. Koschmann (Ed.), *Theorizing Learning Practice*. Mahwah, NJ: Lawrence Erlbaum Associates.

Weiland, S. (2008). Teacher education toward liberal education. In M. Cochran-Smith, S. Feiman-Nemser & D. McIntyre (Eds.), *Handbook of research on teacher education: Enduring questions and changing contexts* (3rd ed.) (pp. 1204-1227). New York: Routledge.

Weimer, M. (2002). *Learner-centered teaching: Five key changes to practice*. San Francisco CA: Jossey-Bass.

Weinberger, D. (2008). *Everything is miscellaneous: The power of the new digital disorder.* New York: Holt Paperbacks.

Weiner, M. & Mehrabian, A. (1968). *Language Within Language: Immediacy, a Channel in Verbal Communication.* New York: Appleton-Century-Crofts.

Wells, G., Chang, G. L. M., & Maher, A. (1990). Creating classroom communities of literate thinkers. In S. Sharan (Ed.), *Cooperative learning: Theory and research.* (pp. 95-121), New York, NY: Praeger Publishers.

Welton, M. (2005). *Designing the just learning society: A critical inquiry.* Leicester, UK: NAICE.

Wenger, E. (1998). *Communities of practice, learning, meaning, and identity.* Cambridge, MA: Cambridge University Press.

Wenger, E. (1998). *Communities of Practice. Learning, Meaning and Identity.* Cambridge, UK: Cambridge University Press.

Wenger, E. (2006). Communities of practice, a brief introduction. Retrieved October 2, 2008 from http://www.ewenger.com/theory/communities_of_practice_intro_WRD.doc

Werry, C. (1996). Linguistic and interactional features of Internet Relay Chat. In S. Herring (Ed.), *Computer-mediated communication* (pp. 47-64). Philadelphia, USA: John Benjamins Publishing Company.

West, R. E., Wright, G., Gabbitas, B., & Graham, C. R. (2006). Reflections from the introduction of blogs and RSS feeds into a preservice instructional technology course. *TechTrends, 50*(4), 54-60.

Westbrook, R. B. (1991). *John Dewey and American Democracy.* Ithaca, New York: Cornell University Press.

Westwood, P. (2008). *What teachers need to know about teaching methods.* Melbourne: ACER Press.

Wiggins, G. P (1990). The case for authentic assessment. *Practical Assessment, Research & Evaluation, 2*(2). Retrieved June 8, 2008 from http://PAREonline.net/getvn.asp?v=2&n=2.

Wiggins, G. P. (1993). *Assessing student performance.* San Francisco: Jossey-Bass.

Wilkerson, M., Griswold, W.G., & Simon, B. (2005). *Ubiquitous Presenter: Increasing student access and control in a digital lecturing environment* (pp 116-120). St. Louis, Missouri: Association for Computing Machinery. Retrieved October 3, 2008, from http://portal.acm.org/citation.cfm?id=1047344.1047394&coll=GUIDE&dl=ACM&CFID=5033070&CFTOKEN=27681204

Wilson, B. G. (1996). What is a constructivist learning environment? In B. G. Wilson (Ed.), *Constructivist learning environments: Case studies in instructional design* (pp. 3–8). Englewood Cliffs NJ: Educational Technology Publications

Wilson, B., & Lowry, M. (2000). *Constructivist Learning on the Web. New Directions for Adult and Continuing Education.* Retrieved February 19, 2007 from http://www3.interscience.wiley.com/cgi-bin/jhome/86011352

Wilson, D., & Sperber, D. (1995). *Relevance: communication & cognition.* Blackwell publishing, Oxford.

Wilson, S. M., & Peterson, P. L. (2006). *Theories of learning and teaching: What do they mean for educators?* Best practices working paper, Washington: National Education Association.

Winterbottom, C., & Blake, E. (2004). Designing a VR interaction authoring tool using constructivist practices (pp. 67-71). Association for Computing Machinery, Inc.

Winters, C. (2007). Second Life User Statistics. *Social Signal*, March 13, 2007. Retrieved June 9, 2008 from http://www.socialsignal.com/blog/catherine-winters/second-life-user-statistics.

Witte, J. (2008). Aspired convergence, cherished diversity: Dealing with the contradictions of Bologna. *Tertiary Education and Management, 14*(2), 81–93.

Wlodkowski, R. J., & Kasworm, C. C. (2003). *Accelerated learning for adults: The promise and practice of intensive educational formats.* San Francisco: Jossey-Bass

Woo, Y., & Reeves, T. C. (2007). Meaningful interaction in web-based learning: A social constructivist interpretation. *The Internet and Higher Education, 10*(1), 15-25.

Woo, Y., & Reeves, T. C. (2007). Meaningful interaction in web-based learning: A social *constructivist* interpretation. *Internet & Higher Education, 10*(1), 15-25.

Woodhead, M. (2006). Changing perspectives on early childhood: theory, research and policy, background paper for UNESCO EFA Global Monitoring Report 2007. *International Journal of Equity and Innovation in Early Childhood, 4*(2), 5–48.

Woolfolk, A. E. (1995). *Educational psychology* (6th ed.). Boston, MA: Allyn & Bacon.

Woolfolk, A. E. (2008). *Educational psychology: Active learning edition* (10th ed.). Boston, MA: Allyn & Bacon.

Wu, D., & Hiltz, S. R. (2004). Predicting learning from asynchronous online discussions. *Journal of Asynchronous Learning Networks, 8*(2), 139–152.

Wulf, T. (2005). Constructivist approaches for teaching computer programming. In *Proceedings of the 6th Conference on information Technology Education,* (pp. 245-248). New York, Association for Computing Machinery. Retrieved October 3, 2008, from http://www2.computer.org/portal/web/csdl/abs/proceedings/icci/2004/2190/00/21900176abs.htm

Yang, S. J. H., Zhang, J., & Chen, I. Y. L. (2007). Web 2.0 services for identifying communities of practice through social networks. *2007 IEEE International Conference on Services Computing (SCC 2007).*

Yates, S. (1997). Gender, identity and CMC. *Journal of Computer Assisted Learning, 13*(4), 281-90.

Yee, N. (2006). The Psychology of Massively Multi-User Online Role-Playing Games: Motivations, Emotional Investment, Relationships and Problematic Usage. In R. Schroeder & A. Axelsson (Eds.), *Avatars at Work and Play: Collaboration and Interaction in Shared Virtual Environments* (pp. 187-207). London: Springer-Verlag.

Yin, R. (1994). *Case study research: Design and methods* (Vol. 5). Thousand Oaks, CA: Sage Publications.

Young, G., & Cronin, J. (2005). Using digital images in an art history department: the experience of University College. Cork' TASI (Technical Advisory Service for Images) Institute for Learning and Research Technology, University of Bristol Retrieved August, 20, 2008 from http://www.tasi.ac.uk/resources/cork.html

Young, M. (1993). Instructional design for situated learning. *Educational Technology Research & Development, 41*(1), 43-58.

Zarnic, B. (2001). Learning to learn: An Epistemological Paradox in Education. *Synthesis Philosophica, 32,* 355-362.

Zauder, K., Lazić, J. L., & Zorica, M. B. (2007, June). Collaborative tagging supported knowledge discovery. *Proceedings of the ITI 2007 29th International Conference on Information Technology Interfaces* (pp. 437-442). Cavtat, Croatia.

Zellermayer, M., Salomon, G., Globerson, T. & Givon, H. (1991). Enhancing writing-related metacognitions through a computerized writing partner. *American Educational Research Journal, 28*(2), 372-391.

Zemsky, R., & Massey, W. F. (2004). *Thwarted Innovation: What Happened to eLearning and Why.* Philadelphia, PA: The Learning Alliance, University of Pennsylvania.

Zhou, G., Varnhagen, S., Sears, M. R. Kasprzak, S., Shervey, G. (2007). Online professional development for inservice teachers in Information and Communication Technology: Potentials and challenges, *Canadian Journal of Learning and Technology, 33*(2). Retrieved August 15, 2008 from http://www.cjlt.ca/content/vol33.2/zhou.html

Zualkernan, I. A. (2006). A framework and a methodology for developing environments. *Educational Technology & Society, 9*(2), 198-212.

About the Contributors

Carla R. Payne is Professor Emerita of Graduate Studies at Union Institute & University, where she supervised low residency MA students in individualized interdisciplinary studies, and designed and implemented an online Master's program. She earned the BA in Philosophy from Barnard College, and the MA and PhD, also in Philosophy, from SUNY/Buffalo. Publications include "Good Practice and Motivation in Online Learning," in *Motivating & Retaining Adult Learners Online* (2002); "Teaching and Technology for Human Development," co-authored with A.W. Chickering and G. Poitras, in *Educational Technology* (2001); **"**Design for Success: Applying Progressive Educational Principles Online,**"** in *Current Perspectives on Applied Information Technologies: Preparing Teachers to Teach with Technology* (2004); "What Do They Learn?" in *Flexible Learning in an Information Society* (2006); "Can We Talk? Course Management Software and the Construction of Knowledge," co-authored with Cornel J. Reinhart, in *On the Horizon (2008)* She continues to teach ethics and philosophy online for the Community College of Vermont.

* * *

Radwan Ali earned his doctorate in Instructional Technology from the University of Georgia in 2008. He is an assistant professor of Management Information Systems at Kennesaw State University. He is a member of the National Business Education Association (NBEA) and Beta Gamma Sigma Honor Society.

Naomi R. Boyd holds a PhD in interdisciplinary education and a MA in special education. As the Director, Teaching and Learning Technologies, University of South Florida Polytechnic she assists instructors with the process of integrating technology into the teaching and learning process. She facilitates the training and development of both faculty and students in the appropriate platforms necessary for course participation as well as supports learning tools such as Blackboard, Elluminate, and Chalk & Wire. Dr. Boyer teaches in the Adult Education programs. Her research interests on the use of self-directed frameworks in online settings, the impact of technology on educational settings, and the faculty development programs.

Xenia Coulter is currently an Emeritus Professor from Empire State College, where she now is a part-time mentor in the Center for International Programs and a part-time instructor in the Center for Distance Learning. With a PhD in Experimental Psychology, she also taught for ten years at the State University at Stony Brook. She has written articles, many for *All About Mentoring*, plus chapters such as in Rosenzweig (Ed.) *Neural Mechanisms and Learning and Memory* (MIT Press, 1976, with Byron Campbell), Spear & Campbell (Eds.), *Ontogeny of Learning and Memory* (Erlbaum, 1980), and Distefano, Rudestam, & Silverman (Eds.), *Encyclopedia of Distributed Learning* (Sage, 2004, with Alan Mandell) and made numerous presentations over the years in the areas of learning & memory, adult development, experiential learning, and other topics related to higher education.

James G. R. Cronin is a founding member of History of Art, University College Cork where he is currently employed as Visual Resources Officer. He is also currently employed as Course Co-ordinator in the Centre for Adult and Continuing Education, University College Cork. Cronin graduated in History and English from University College Cork (1991) and broadcast media from the University of Central Lancashire (2000). He has taught in higher education since 1992. He is interested in assessing the ways technologies can enhance critical learning and reflective practice. He is a member of AONTAS (Irish National Adult Learning Organisation) and DRHA (**Digital Resources for the Humanities and Arts)** based in the United Kingdom. He is involved in a NAIRTL (National Academy for the Integration of Research and Teaching and Learning) research project focused on exploring connections in teaching and learning across disciplines in Irish higher education.

Barbara de la Harpe is associate professor and Dean Academic Development in the College of Design and Social Context at RMIT University, Melbourne, Australia. In this role, she provides academic leadership to over 1,000 faculty and 20,000 students. Her background is in Science Education and Educational Psychology and her PhD study was on student learning. Her fields of expertise include learning; generic skill development; quality and change management; and staff professional development. She has been involved for a number of years in teaching and academic professional development in higher education and has overseen many large organizational change and funded research projects. She is the recipient of numerous awards (including an Australian Learning and Teaching Council citation for sustained and outstanding contribution to student learning) and is widely published in learning and teaching.

Roisin Donnelly has been lecturing and researching in higher education since 1992. She has taught in universities in Northern Ireland and was a lecturer and visiting research fellow in the University of New South Wales, Sydney. She is currently Programme Co-ordinator for DIT's MSc Applied eLearning, and tutors and supervises on the PG Certificate, Diploma and MA in Third Level Learning and Teaching. She also is an online tutor on courses with Oxford Brookes University and The University of Queensland. She has a range of conference proceedings, chapter and journal publications to reflect her teaching and research interests, including online academic development, designing eLearning, supporting virtual communities, transformative pedagogies and blended problem-based learning. Her first co-edited book is published in 2008: *Applied eLearning and eTeaching in Higher Education*. Hershey, PA: Information Science Reference. She is currently writing and editing a second textbook in the field of eLearning within academic development.

Nicholas Eastham, BA, MA, is currently working toward completion of a PhD in Educational Technology from the University of Northern Colorado. He received his BA in Literature (minor Journalism) from Texas Christian University, and an M.A. in General Education from Eastern New Mexico University. His career includes serving as a school technology coordinator, where he taught public school staff and students a wide variety of computer applications on multiple platforms, and designed and maintained websites for public schools and non-profit organizations. He is currently the webmaster for the Center for the Enhancement of Teaching and Learning (CETL) website at the University of Northern Colorado, where he also conducts individual and group technology workshops for the CETL. Nicholas' research has focused on on-line learning and visual literacy. He has authored articles on assistive technology and on-line collaborative work.

Noel Fitzpatrick is a Learning Development Officer at the Dublin Institute of Technology Ireland. After graduating from University College Dublin, Ireland, with a BA in English and Philosophy, and an MA in Philosophy, he obtained an MPhil from the University of Paris VII, and his PhD with first class honors. His research interests include eLearning, integration of new technologies in learning and teaching, philosophy of education, computer mediated communication, discourse analysis, philosophy of language and pragmatic linguistics. Noel taught at the University of Paris VII for twelve years in the applied languages department. He has worked in the development of eLearning projects in both the public and private sectors since the mid 1990s and has specialized in the use of eLearning in language teaching and training.

M. Evan Gaffney is a high school physics teacher in Bergen County, NJ with a degree in physics and minors in mathematics and education. His past research includes areas of mathematics, physics, and education. Recent publications include "Investigating Cutsets in the Boolean Lattice, Order B_7" and "Tsar Bomba: Power Unimaginable." He is currently spearheading a project to develop a unified physical science curriculum for his school district.

D. Randy Garrison is the Director of the Teaching & Learning Centre and a Professor in the Faculty of Education at the University of Calgary. Dr. Garrison has published extensively on teaching and learning in higher, adult and distance education contexts. His most recent books are: *E-Learning in the 21st Century* (2003) and *Blended Learning in Higher Education* (2008).

Alessio Gaspar is associate professor with the University of South Florida's Information Technology department and holds a PhD in Computer Science from the University of Nice Sophia Antipolis (France, 2000). He is an ACM SIG member for the Special Interest Groups in Computer Science (SIGCSE) and Information Technology (SIGITE) Education. His research interests are in evolutionary computation and computing education (e.g. developing technologies to enable new pedagogies, investigating constructivist frameworks applicability to computing). His research has been so far federally funded from the National Science Foundation's CCLI division for which he also serves regularly as panelist.

Nóirín Hayes, PhD, is Head of the School of Social Science and Law at the Dublin Institute of Technology, Ireland, and a former Head of Learning Development within the Faculty of Applied Arts. She is a developmental psychologist and lectures on children and childhoods, early education and research methods. She is Director of the Centre for Social and Educational Research (CSER) at the DIT and leads a number of research projects in the field of early childhood. She is the author of a number of publications on practice and policy in early childhood education and an active member of the Irish Children's Rights Alliance.

Ida Heilweil, received her PhD in Computers in Education from University of Leicester, England. She is a lecturer in the Computers in Education Department at Levinsky College of Education, where she also works as an educational designer in the Digital Learning Center. In addition, she teaches in the professional development program for teacher educators on ICT in teacher education at the Mofet Institute where she is also a member of the ICT research team Her research focuses on e-learning and ICT in teacher education.

David Kendrick, BA, MA, MS, EdD (Director, Center for the Enhancement of Teaching and Learning, University of Northern Colorado) David Kendrick received his BA in English and MA in English/Communications from Colorado State University, an MS in Educational Technology, and EdD in Higher Education Administration & Research from Texas A&M University-Corpus Christi. His career has included teaching and administration in k-12 and higher education. After teaching in the public schools in Colorado, he spent five years teaching post-secondary students of the United Arab Emirates. Upon his return to the U.S., he served as the technology director of a large, South Texas school district, shortly thereafter returning to higher education at Texas A&M University-Corpus Christi where he taught courses in educational technology and served as Assistant Director of the Center for Teaching Excellence. His research has focused on e-learning environments and the cost benefits of the transformation to web-based courses, programs, and systems.

Sarah Langevin is a student at the University of South Florida. Her interests include psychology, biology, computing education research and open source technologies. She is particularly interested in the application of educational and cognitive psychology theories to introductory and intermediate programming courses for regular students and adults education. Sarah is also interested in open source technologies (e.g. virtualization) which can support and enable pedagogical innovations in computing by allowing students to learn by doing and engage in authentic social apprenticeship learning experiences. The confluence of technology and pedagogy is what makes computing education an interesting research endeavor.

Jennifer Lee is an Assistant Director of Academic Services for the College of Education at the University of North Texas. She is a doctoral student in Learning Technologies. Her research interest is in distributed learning, academic advising, and new media and technologies.

M. Beatrice Ligorio is associate professor at the University of Bari (IT). She graduated in Psychology at the University of Rome and, in the 1999, she received her PhD in Psychology of Communication at the University of Bari. She has been a NATO fellow for collaboration in the Community of Learners model, designed by Ann Brown (University of Berkeley, CA). She also received a Marie Curie grant to develop virtual educational environments at the University of Nijmegen (NL). She has collaborated on European, international, and national projects and has published many articles in international journals. Recently she has edited two books in Italian about educational technology. Professor Ligorio has also edited the Italian translation of *Cultural Psychology. A once and future discipline* (Cole, 1996). Her main research interests are in educational technology, models of community, cultural psychology, socio-constructivism, virtual environments, dialogical approach, digital identity, blended learning, m-learning.

Lin Lin is an Assistant Professor in Learning Technologies at the University of North Texas. Lin's research interests include new media and technology, computer-mediated communication, cognition, online learning, professional development and adult learning. Lin holds an EdD degree in Communication, Computing and Technology in Education from Teachers College, Columbia University. Please visit for more information.

Linda Lohr received a BS in Vocational Education from Colorado State University, an MS in Consumer Science (minor Computer Science) from Texas Women's University, and a EdD from the University of Memphis in Instructional Design and Technology. Her career has spanned many areas in instruction and technology, including professional work as a junior high school teacher, college-level software instructor, school district technology supervisor, assessment provider, instructional interface design analyst and lead instructional designer for commercial instructional software, instructional software consultant to numerous companies (including Sun Microsystems, Boston Market, MCI, HP, Shell Oil and Safeway), and most recently as Associate Professor of Educational Technology at the University of Northern Colorado (UNC). Among her professional activities, she is a book review editor and consulting editor (and past editorial board member) for Educational Technology Research and Development, and is the portfolio reviews editor for TechTrends. Prof. Lohr's research emphasis has been on instructional software usability and on the use of visuals in instruction. She has authored 2 textbooks, 7 book chapters, and 37 juried articles.

Hwee Ling Lim is an Assistant Professor at The Petroleum Institute (Abu Dhabi, UAE) which specializes in engineering education and research. She has Bachelor (English Language, Literature) and Master (English Language) degrees, a post-graduate Diploma in Education from The National University of Singapore, and a PhD (Information Technology) from Murdoch University (Perth, Western Australia). Her areas of research interest are educational technology, computer-mediated communication, and distance learning. She can be contacted at hlim@pi.ac.ae, 971-2-6075324, The Petroleum Institute, P.O. Box 2533, Abu Dhabi, United Arab Emirates.

John Paul McMahon is currently reading for a doctorate degree in art history at University College Cork, Ireland. His thesis examines the work and writings of Vito Acconci and Michael Fried. McMahon has published articles on Acconci and Fried, William Blake, and Chinese art since the death of Mao Zedong. Currently, McMahon is completing an article on the theme of madness and art history. This article looks specifically at the early writing of Michel Foucault. McMahon teaches on the Diploma in European Art History, University College Cork and has tutored on the university's art history degree programme. He has given papers internationally on a variety of subjects from semiotics to the work of Doris Salcedo.

Alan Mandell is College Professor of Adult Learning and Mentoring and currently Susan Turben Chair in Mentoring at SUNY Empire State College. For more than thirty years, Mandell has served as administrator, mentor in the social sciences, and director of the college's Mentoring Institute. With colleague Elana Michelson, he is the author of *Portfolio Development and the Assessment of Prior Learning* (2nd edition, Stylus, 2004). With colleague Lee Herman, he has written many essays and book chapters on mentoring, including *From Teaching to Mentoring: Principle and Practice, Dialogue and Life in Adult Education (*Routledge, 2004). Mandell and Herman's essay, "Mentoring: When Learners

Make the Learning" will be included in the forthcoming *Transformative Learning in Action* edited by Mezirow and Taylor (Jossey-Bass, 2009).

John Miller teaches literature and writing at National University in Costa Mesa, California. He has been designing and teaching online courses in both subjects since 1999. His published scholarly work has included studies of English Renaissance prose, fantasy and science fiction, hypertext fiction, and online pedagogy.

Nili Mor has a Ph.D in Computers in Education from Haifa University. She teaches in the graduates programs of two higher education institutions in Israel: Levinsky College of Education and The Open University. In addition, she coordinates the professional development program for teacher educators on ICT in teacher education at the Mofet Institute. Her research focuses on E-learning, ICT in teacher education and the professional developent of teachers educators.

Laura Nicosia, PhD is an Assistant Professor of English and the Director of English Education at Montclair State University in New Jersey. She teaches American literature and explores educational uses of technology to improve pedagogy. Her most recent literary publications examine the narrative constructs of Philip Roth, John Dufresne and Gloria Naylor, including the book chapter, "Gloria Naylor's *The Women of Brewster Place*: Evolution of a Short Story Cycle," in *Narratives of Community: Women's Short Story Sequences*. Her recent techno-pedagogical publications focus on constructivist teaching and learning using the virtual reality platform of Second Life. These include: "Literature Alive! Using Second Life to Teach American Literature," in *The EDUCAUSE Review* 43.5 and "Adolescent Literature and Second Life: Teaching Young Adult Texts in the Digital World," in *New Literacies: A Professional Development Wiki for Educators*.

Kevin C. O'Rourke is Head of eLearning Support at the Dublin Institute of Technology, Ireland. He holds a BA and MA in philosophy from University College Dublin, Ireland, and completed a PhD in nineteenth-century intellectual history at University College London. His research interests include the philosophy and history of education, and the socio-cultural impact of information and communications technologies. Prior to joining the DIT in 2002, he worked for Fathom.com, the eLearning consortium founded by Columbia University.

G. Andrew Page, earned a doctorate of philosophy at the University of Georgia (2004) with an emphasis on adult education, instructional technology, and research methods. Currently, he is an assistant professor in the College of Education at the University of Alaska Anchorage. Andy's research interests include how to effectively use emerging and assistive technologies to promote positive learning outcomes in traditional and hybrid courses. He has designed, developed, and manages a new e-Learning Graduate Certificate program to help train facilitators of education with the tools of technology. In 2008 he was the co-principal investigator for a National Science Foundation grant that focuses on developing a model for cross-hemispheric collaboration and between paired classrooms in distant communities (Tasmania and Alaska, USA) using information and communication technologies. He has professional memberships in the American Educational Research Association and the Association for the Advancement of Computing in Education.

Maria Luisa Pérez Cavana Born in Madrid in 1961 she studied Philosophy (BA Hons) and continued her studies in Philosophy, History and Education in Germany where she received her doctorate (PhD Ed) for her thesis on Georg Simmel and Ortega y Gasset. She has worked as a research associate at the Madrid Complutense University in different research projects and she has numerous publications on German Enlightenment and feminist epistemology. Recently she has focused on the practice of teaching German as a foreign language and she is particularly interested in how to improve language learning from the point of view of learner autonomy. She has been developing an electronic European Language Passport for the Open University (UK) where she works as lecturer in German and Spanish. Other research interests are the use of new technologies (ejournals, Web 2.0) in language learning.

J. Fiona Peterson is Director of Learning and Teaching in the School of Creative Media at RMIT University, Melbourne, Australia. She has 29 years' experience as a teacher and educational facilitator spanning high school, vocational education and training, as well as undergraduate and postgraduate higher education programs. Her background is in Communication Studies and she has a PhD in collaborative learning networks. Her research interests include strategic knowledge networks, Mode 2 knowledge, virtual communities, global education and learning and teaching in higher education contexts. She has been principal investigator in a number of SoTL projects, both nationally and internationally. She is a recipient of an Australian Learning and Teaching Council citation for sustained and outstanding contribution to student learning – for development of innovative curriculum incorporating virtual teamwork and collaboration with industry worldwide that prepared graduates for new leadership roles in the global networked environment.

Jennifer Richardson is an Associate Professor in the College of Education at Purdue University. Jennifer's research focuses on distance education, in particular online learning environments. Specific areas of research include measuring learning in online environments and the impacts of social presence and interactions on students' perceptions and learning.

Carol R. Rinke is an assistant professor at Gettysburg College, with a specialization in science and math education. A former middle school and high school teacher, she has a deep interest in preparing new teachers for the classroom. Her areas of research include science and math teacher recruitment, retention, education, and professional development and she has a special interest in urban education. Recent publications include "Making adequate yearly progress: Teacher learning in school-based accountability contexts" appearing in *Teachers College Record* with Linda Valli. She is currently working on a project investigating the life histories of teacher education candidates.

Beth Rubin is an assistant professor and the director of online programs at the School for New Learning at DePaul University, where she conducts research in effective online learning, teaches and advises students. She has worked as a curriculum director, faculty manager, and Director of Assessment at Cardean University, as well as Program Dean and Dean of Academic Administration at DeVry University Online. She was awarded the Associate's Award for Excellence in Teaching and received tenure in the Asper School of Business at the University of Manitoba, and has taught at LeMoyne College, Syracuse University and Michigan State University. She earned an M.A. and PhD in Industrial/Organizational Psychology from Michigan State University, and a B.A. in Psychology from Cornell University.

Nadia Sansone is a training research assistant in Psychology at the University of Bari (IT). She had a training internship with the Collaborative Knowledge Building Group (CKGB) (www.ckbg.org). Her research work is focused on planning blended university course, with a main interest on virtual role-play. Her expertise is in tutoring online, and she has been acting as e-tutor for several blended university courses. She has participated in a several Italian projects about dialogical approach and educational technology and has collaborated in several local projects about educational uses of the Internet for children and teenagers. Ms Sansone has also published articles about Computer Supported Collaborative Learning and online tutorship in national journals. Other research interests are in virtual environments and digital identity, blended learning, and socio-constructivism.

Mark Schulman has been President of Goddard College in Plainfield, Vermont, since January 1, 2003. Dr. Schulman received his Doctor of Philosophy in Communications from The Union Institute and University and a Specialist and Master's in Educational Media from Indiana University. He has extensive background in communications and education consulting and higher education administration. Prior to joining Goddard College, he was President and Professor of Humanities at Antioch University Southern California, Academic Vice President at Pacific Oaks College in Pasadena, CA, and Chair of the Media Studies Department at The New School for Social Research. He has been a faculty member and department administrator at Mount Vernon College, Antioch College, Saint Mary's College of California, and The City College of New York.

Lauren E. Schaeffer is a science teacher at North Carroll High School in Carroll County, Maryland. She is a 2008 graduate of Gettysburg College where she received her Bachelor of Science in Biology. Her research on blogging and the importance of cooperation and collaboration has helped develop her as a teacher and aided in her contributions to her peer learning community.

Shalin Hai-Jew works as an instructional designer at Kansas State University, and she teaches through WashingtonOnline (WAOL). She has BAs in English and psychology, and an MA in English, from the University of Washington. She has an Ed.D. (2005) from Seattle University, where she was a Morford Scholar. She was a tenured professor of Communications and English at Shoreline Community College. She taught at Jiangxi Normal University (1988- 1990 in Nanchang, Jiangxi Province) and Northeast Agriculture University (1992 – 1994, in Harbin, Heilongjiang Province), with the latter two years under the auspices of the UNDP United Nations Volunteer Programme. She originated and maintains the Instructional Design Open Studio (IDOS) blog. She is currently writing a book titled *Digital Imagery and Informational Graphics in E-Learning: Maximizing Visual Technologies* for IGI-Global. She is interested in the design of immersive spaces, simulations, and trainings for high reliability organizations.

Divonna M. Stebick is an assistant professor at Gettysburg College, with a specialization in literacy and special education. A former classroom teacher, literacy coach, and staff developer, she has a deep interest in preparing new teachers to meet the needs of all students. Her areas of research include comprehension and vocabulary instruction, the culture of change, and reading and writing across the content areas. Recent publications include "Comprehension Strategies for Your K-6 Literacy Classroom" published by Corwin Press and co-authored with Joy Dain. She is currently working on a project investigating how school districts manage transitions during the period of change when implementing the Response to Intervention regulations.

Karen Swan is the Stukel Distinguished Professorship of Educational Leadership in the College of Education and Human Services at the University of Illinois Springfield. Dr. Swan has published and presented widely in the general area of technology and learning, most currently on online learning and on student learning in ubiquitous computing environments. Her most recent book is *Ubiquitous Computing in Education* (2007).

Fay Sudweeks is an Emeritus Associate Professor at Murdoch University and Adjunct Associate Professor at the Australian National University, Australia. She has Bachelor of Arts (Psychology, Sociology), Master of Cognitive Science, PhD (Communication Studies) degrees. Her research interests are social, cultural, and economic aspects of CMC and CSCW, group dynamics, and e-learning. She has published authored book, 5 edited books, 15 edited proceedings, and more than 80 papers in journals, books, and conference proceedings. She is an *Australian Computer Society* member, on editorial boards of *Journal of Computer-Mediated Communication, New Media and Society, International Journal of e-Learning, Human Communication Research, Open Communication Journal, International Journal of Education and Development using Information and Communication Technologies, Journal of Electronic Commerce Research* and *Journal of Electronic Commerce in Organizations*. With Charles Ess, she has co-chaired five international conferences on *Cultural Attitudes towards Technology and Communication* (CATaC).

Michael Waldron graduated with a joint-honours degree in English and History of Art from University College Cork in 2005. His specific interest in English was in the Anglo-Irish literary tradition, especially, lesbian and gay fiction. In History of Art, his interests were in the fountain sculptures of Gianlorenzo Bernini and the Orientalist works of David Roberts. He subsequently completed a taught Master of Arts degree in English in 2006, focusing in particular on nineteenth and twentieth century Anglo-Irish fiction and theory, culminating in a thesis on Kate O'Brien. He has taught on the Diploma in European Art History programme. He is currently embarking on doctoral research focusing on the Anglo-Irish author Elizabeth Bowen, with particular interest in her literary output during World War II.

Michal Zellermayer is a professor of teacher education at Levinsky College of Education in Tel Aviv. She currently heads the MEd program in Teaching and Learning and serves a coordinator of academic relations at the College. Until recently, Professor Zellermayer served as a co-coordinator of the Teaching and Teacher Education Special Interest Group at the European Association for Research on Learning and Instruction. She is a member of the editorial board of the Educational Research Review as well as Teachers and Teaching, Theory and Practice. She has published numerous papers on writing instruction, on developing rich communicative environments, on teacher learning and on action research.

Index

A

B

C